Cereal Grain Crops

Editor: Corey Aiken

R CALLISTO REFERENCE

www.callistoreference.com

Callisto Reference,
118-35 Queens Blvd., Suite 400,
Forest Hills, NY 11375, USA

Visit us on the World Wide Web at:
www.callistoreference.com

ISBN: 978-1-63239-784-3 (Hardback)

Cataloging-in-publication Data

Cereal grain crops / edited by Corey Aiken.
    p. cm.
Includes bibliographical references and index.
ISBN 978-1-63239-784-3
1. Grain. 2. Food crops. 3. Agricultural innovations. 4. Grain--Biotechnology. I. Aiken, Corey.
SB189 .C47 2017
633.1--dc23

Table of Contents

Permissions

List of Contributors

Index

Preface

Cereal grain crops are members of the grass family that are grown for their edible and scratchy seeds. Cereals include the most consumed food articles such as rice, maize, wheat, etc. and are hence part of staple crops. Topics included in the book deal with production of cereals as well as the study of different crop patterns and farming methods. Cereal farming is universally practiced and hence, there is an abundance of techniques of harvest, irrigation and storage. Scientists and students actively engaged in this field will find this book full of crucial and unexplored concepts. This book is an essential guide for both academicians and those who wish to pursue this discipline further. It includes contributions of experts and scientists which will provide innovative insights into this field.

This book has been a concerted effort by a group of academicians, researchers and scientists, who have contributed their research works for the realization of the book. This book has materialized in the wake of emerging advancements and innovations in this field. Therefore, the need of the hour was to compile all the required researches and disseminate the knowledge to a broad spectrum of people comprising of students, researchers and specialists of the field.

At the end of the preface, I would like to thank the authors for their brilliant chapters and the publisher for guiding us all-through the making of the book till its final stage. Also, I would like to thank my family for providing the support and encouragement throughout my academic career and research projects.

Editor

Transcriptome Assembly and Analysis of Tibetan Hulless Barley (*Hordeum vulgare* L. var. *nudum*) Developing Grains, with Emphasis on Quality Properties

Xin Chen[1,2,3], **Hai Long**[1], **Ping Gao**[2], **Guangbing Deng**[1], **Zhifen Pan**[1], **Junjun Liang**[1], **Yawei Tang**[4], **Nyima Tashi**[4], **Maoqun Yu**[1]*

1 Chengdu Institute of Biology, Chinese Academy of Sciences, Chengdu, Sichuan, China, 2 College of Life Sciences, Sichuan University, Chengdu, Sichuan, China, 3 University of Chinese Academy of Sciences, Beijing, China, 4 Tibet Academy of Agricultural and Animal Husbandry Sciences, Lhasa, Tibet, China

Abstract

Background: Hulless barley is attracting increasing attention due to its unique nutritional value and potential health benefits. However, the molecular biology of the barley grain development and nutrient storage are not well understood. Furthermore, the genetic potential of hulless barley has not been fully tapped for breeding.

Methodology/Principal Findings: In the present study, we investigated the transcriptome features during hulless barley grain development. Using Illumina paired-end RNA-Sequencing, we generated two data sets of the developing grain transcriptomes from two hulless barley landraces. A total of 13.1 and 12.9 million paired-end reads with lengths of 90 bp were generated from the two varieties and were assembled to 48,863 and 45,788 unigenes, respectively. A combined dataset of 46,485 All-Unigenes were generated from two transcriptomes with an average length of 542 bp, and 36,278 among were annotated with gene descriptions, conserved protein domains or gene ontology terms. Furthermore, sequences and expression levels of genes related to the biosynthesis of storage reserve compounds (starch, protein, and β-glucan) were analyzed, and their temporal and spatial patterns were deduced from the transcriptome data of cultivated barley Morex.

Conclusions/Significance: We established a sequences and functional annotation integrated database and examined the expression profiles of the developing grains of Tibetan hulless barley. The characterization of genes encoding storage proteins and enzymes of starch synthesis and (1–3;1–4)-β-D-glucan synthesis provided an overview of changes in gene expression associated with grain nutrition and health properties. Furthermore, the characterization of these genes provides a gene reservoir, which helps in quality improvement of hulless barley.

Editor: Mark Gijzen, Agriculture and Agri-Food Canada, Canada

Funding: This study is supported by National Natural Foundation of China (31101150), the West Light Foundation of the Chinese Academy of Sciences and the National Science and Technology Supporting Programs (2012BAD03B01). The funders had no role in study design, data collection and analysis, decision to publish, or preparation of the manuscript. National Natural Foundation of China: www.nsfc.gov.cn; Chinese Academy of Sciences: english.cas.cn/; Ministry of National Science and Technology of China: www.most.gov.cn/eng/.

Competing Interests: The authors have declared that no competing interests exist.

* E-mail: yumaoqun@cib.ac.cn

Introduction

Barley (*Hordeum vulgare* L.) is among the most ancient cereal crops [1] and currently ranks fourth in terms of harvested area and tonnage of the world cereal production (http://faostat.fao.org). However, barley is the least utilized cereal for human food consumption and is usually cultivated either in regions unsuitable for wheat growing, or where barley is preferred for cultural reasons [2]. It was also neglected by plant breeders in Europe during the period of intensive crop improvement in the 20th Century. However, it is currently gaining attention as a health food in Europe, North America and other non-traditional barley growing areas [3,4]. Barley grains are rich in minerals; proteins and lysine and have a high β-glucan content, which inhibits cholesterol synthesis [5–7]. Hulless (naked) barley with caryopses that thresh free from the pales is preferred for human consumption [8–10].

Hulless barley also allows to omit a processing step, thus, providing an additional advantage for the food industry [11,12]. Therefore, hulless barley is a potential resource for breeding new healthy food worldwide. The grain of barley is the major storage tissue. Different end uses require alternative quality characteristics of barley grain in terms of molecular composition of starch and proteins. So far, there has been limited research regarding metabolic profiling and gene expression patterns related to the metabolism of storage compounds during barley grain development.

The Qinghai-Tibet Plateau in western China has abundant hulless barley resources [13] and is considered as one of the main regions of domestication and diversity of cultivated barley [14,15]. In the past millennia, people continuously modified local hulless barley populations to develop cultivars with increased grain yield. However, more efficient methods of barley production are needed

Table 1. Summary of *de novo* assemblies for two accessions.

Samples	Total Reads	Total Nucleotides (nt)	Unigenes	All-Unigenes
XQ754	13,069,860	1,176,287,400	48,863	46,485
Nimubai	12,918,520	1,162,666,800	45,788	

to meet the increasing food demand imposed by climate change, potential food shortage, and demand for the use of grains as a renewable energy resource. The study of the genetic basis of agronomically important genes in hulless barley would certainly aid in developing better cultivation methods.

Genome sequencing is considered pivotal for solving key questions in crops and investigating the molecular mechanisms related to yield and quality. The International Barley Sequencing Consortium (IBSC) has made great achievements in the genomic sequencing of barley [16]. Meanwhile, numerous molecular technologies have also been applied to generate a greater functional understanding of barley, including microarrays [17–19], Affymetrix arrays [20,21], cDNA-AFLP [22], SAGE [23,24] and molecular markers [25]. These technologies have helped in generating data from more than 15 tissues or organs at various developmental stages and under diverse environmental conditions [17,18]. However, the primary focus of these studies is usually on malting and feed characteristics. In this study, we conducted *de novo* transcriptome sequencing and analyses of the developing grains from two Tibetan hulless barley landraces, which have long been used as human food. A large number of unigenes were assembled, functionally annotated, and their expression accumulation was also calculated. We further analyzed the transcripts related to seed storage protein, starch, and β-glucan synthesis along with those identified in the Morex transcriptome data set [16]. This study provides abundant resources for identification of genes required for quality improvement in barley.

Materials and Methods

Ethics Statement

No specific permits were required for the described field studies as well as for the location where the experimental materials were planted. No endangered or protected species were involved in our field studies. The GPS coordinates of the three planting fields were 30°34′N, 103°53′E.

Plant materials and RNA isolation

Two local varieties of Tibetan hulless barley, XQ754 and Nimubai (used and known as tribute barley), were conserved by the Tibet Academy of Agricultural and Animal Husbandry Sciences. Nimubai has a higher amylose content (33.9%) and β-glucan content (7.5%) as compared to XQ754, which had 27.2% amylose and 6.0% β-glucan (data collected from 2009–2010 in Chengdu). The hulless barley plants were cultivated in October, 2010 and grown under normal conditions in the three fields in Chengdu, Sichuan Province of China.

Grains of Nimubai and XQ754 plants were sampled at 5, 10, 15, 20, and 25 days after pollination (dap) for RNA extraction. Each sample consisted of grains from nine individuals. Total RNA was extracted from the grains using Trizol Reagent (Takara) and Fruit-mate for RNA purification (Takara), according to the manufacturer's instructions. The concentration and quality of RNA samples were determined using a Nano Drop 2000

micro-volume spectrophotometer (Thermo Scientific, Waltham, MA, USA). Equal amounts of RNA from each sample of the identical accessions were pooled to construct two cDNA libraries [26,27].

De novo transcriptome sequencing, assembly and evaluation

The library construction and sequencing were performed by the Beijing Genomics Institute (BGI)-Shenzhen, Shenzhen, China (http://www.genomics.cn). Briefly, beads with Oligo (dT) were used to isolate poly(A) mRNA from total RNA. Fragmentation buffer was added to breakdown mRNA into short fragments. Random hexamer-primers were added to the shortened fragments (~200 bp), and first-strand cDNA was synthesized. The second-strand cDNA was synthesized using buffer, dNTPs, RNaseH and DNA polymerase I. Short fragments were purified with QiaQuick PCR extraction kit after resolution with agarose gel electrophoresis. Sequencing adapters were ligated to the cDNA strands and suitable fragments were selected for the PCR amplification as templates. After PCR amplification, the pair-end sequencing (90 bp in length) was carried out using Illumina HiSeq 2000.

Raw sequence data was generated by the Illumina pipeline and clean reads were generated by filtering out adaptor-only reads, reads containing more than 5% unknown nucleotides, and low-quality reads (reads containing more than 50% bases with Q-value ≤20). Only clean reads were used in the following analysis. The sequences from the Illumina sequencing were deposited in the NCBI Sequence Read Archive (Accession numbers: SRR1032035, SRR1032036, SRX375649 and SRX378862).

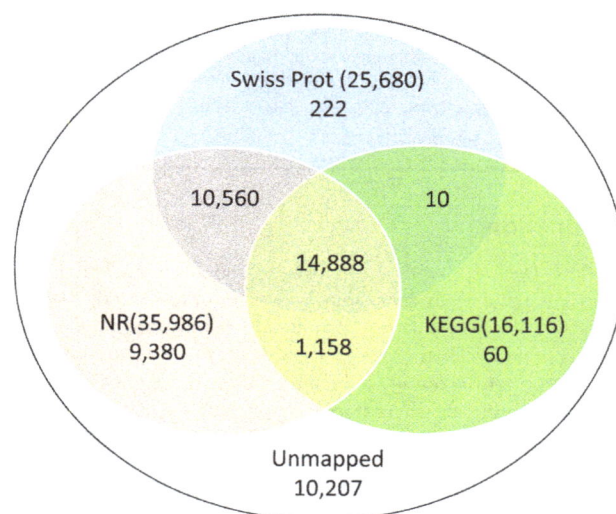

Figure 1. Distribution of Homologous genes in three public databases. The numbers of annotated and unmapped unigenes are indicated in the ellipses, respectively.

To reduce the data complexity, each library was assembled to unigenes separately with the program Trinity [28] using the follow parameters: group_pairs_distance = 250, path_reinforcement_distance = 70, min_glue = 2, min_kmer_cov = 2 and other default parameters. After assembly by Trinity, all contigs from two samples were combined, and the redundancy of contigs was removed by the TGICL [29] and Phrap assemblers (http://www.phrap.org/) for obtaining distinct sequences (All-Unigenes). The following parameters were used to ensure quality of assembly: a minimum of 95% identity between contigs, a minimum of 35 overlapping bases, a minimum of 35 scores and a maximum of 20 unmatched overhanging bases at sequence ends.

In addition to the evaluation of the quality of the assemblies, the known 26,159 high-confidence genes [16] combined of RNA-seq-derived and barley flcDNAs-derived sequences were considered as references in this study, and were used to Blast against each assembly with Blastn (E-value <1e-10) [30]. Based on the Blast results, the averages of sensitivity and accuracy of each assembly were considered. Sensitivity or transcriptome coverage was determined as the ratio of the sum of all uniquely aligned segment lengths to the reference length. Accuracy was determined as the ratio of the sum of all unique aligned segment lengths to the assembled transcript lengths.

Functional annotation and classification

Blastx alignment (E-value <1e-5) between unigenes and protein databases such as nr, Swiss-Prot, KEGG, COG and GO was performed, and the best-aligning results were used to determine the sequence direction and coding regions (CDS) and its amino acid sequence of unigenes. When different databases conflicted, the results were prioritized in the order: nr, Swiss-Prot, KEGG, GO and COG. When a unigene did not align to any of the databases, ESTScan [31] was used to decide its sequence direction and CDS.

A non-redundant unigene set "All-Unigenes" assembled from the two unigene sets were aligned by Blastx to protein databases (nr, Swiss-Prot, KEGG and COG) with E-value<1e-5, and

proteins (including their protein functional annotations) having the highest sequence similarity with the given unigenes were retrieved. With nr annotation, the Blast2GO program [32] was used to get GO annotation of the All-Unigenes. WEGO software [33] using the GO functional classification for all All-Unigenes was used to understand the distribution of gene functions. The KEGG database (V56.0, Oct. 1, 2010) [34,35] was employed to annotate the pathway of these unigenes.

SNPs Identification

To detect the single nucleotide polymorphisms (SNPs) of XQ754 and Nimubai compared to the ESTs of barley (NCBI), 525,781 ESTs were downloaded from NCBI website (http://www.ncbi.nlm.nih.gov/). For the ESTs have high redundancy, clustering and assembly were performed by TGICL [29] and Phrap assemblers with the same parameters as mentioned previously, and a reference data set of 61,902 unigenes was generated. Thereafter, we realigned all the clean reads from each library onto the reference sequence separately using SOAP aligner with default parameters. SNPs were detected using SOAPsnp [36] with default parameters. To ensure the quality of SNP, we used the follow cutoff to filters: MinQual (minimal Quality form SOAPsnp) ≥ 20; Max_soap_rep <1.5; MinDist ≥ 5; MinDepth ≥ 5; MaxDepth <10000 [36,37]. 29 SNPs in the CDS of eight genes encoding enzymes for starch and β-glucan synthesis were validated using Sanger sequencing.

Differential Gene Expression Analysis

For gene expression analysis, the number of reads that uniquely aligned to a unigene was calculated and then normalized to RPKM (reads per kb per million reads) [38]. The RPKM method eliminates the influence of different gene lengths and sequencing levels on the calculation of gene expression. Therefore the calculated gene expression can be directly used for comparing the difference of in gene expression among samples. To identify differentially expressed genes between two samples, a statistical analysis of the frequency of each unique-match read in each

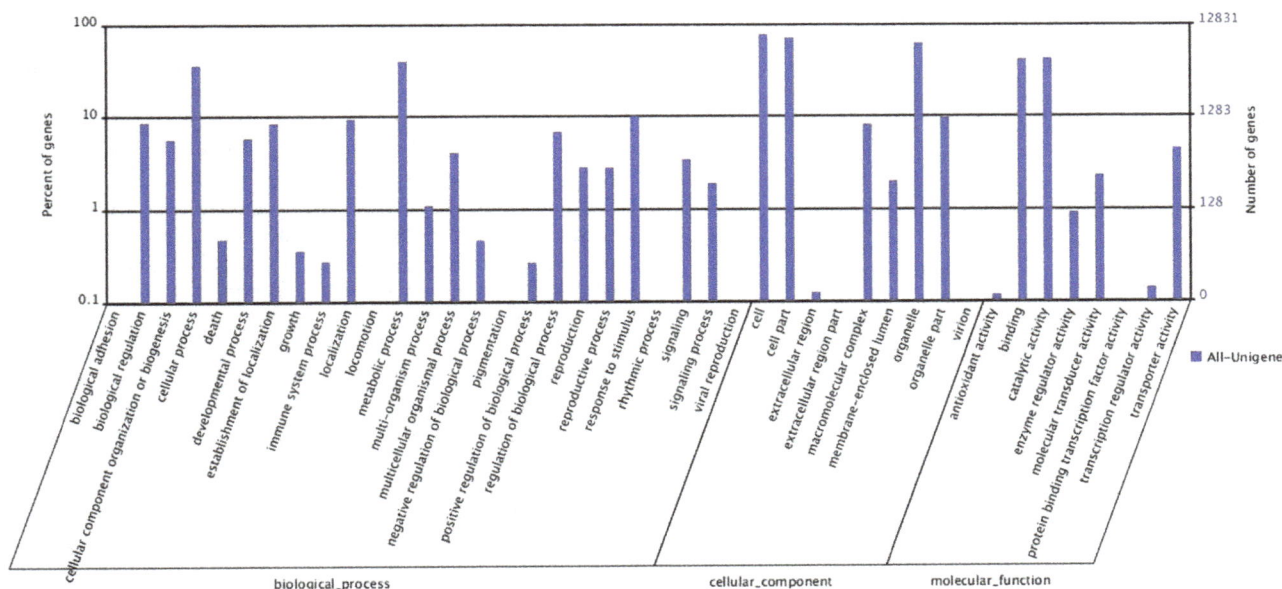

Figure 2. Go annotation of transcriptome. The x-axis indicates the categories and the y-axis indicates the number and proportion of All-Unigenes.

library was performed by referring to "the significance of digital gene expression profiles" [39]. The P value was used to identify differentially expressed genes following the described formula [39], wherein N1 and N2 represent the total clean read numbers of unique-match reads in Samples 1 and 2, respectively, and gene A holds x and y unique-match reads in Samples 1 and 2, respectively.

$$2\sum_{i=0}^{y} p(i|x) \left(while \sum_{i=0}^{y} p(i|x) \le 0.5 \right)$$

or

$$2 \times \left(1 - \sum_{i=0}^{y} p(i|x)\right) \left(while \sum_{i=0}^{y} p(i|x) > 0.5 \right)$$

$$p(i|x) = \left(\frac{N_2}{N_1}\right)^{i} \frac{(x+i)!}{x!i!\left(1+\frac{N_2}{N_1}\right)^{(x+i+1)}}$$

FDR (False Discovery Rate) was used in multiple hypothesis testing to correct for P value [40]. Following the formula below, assuming R differentially expressed genes had been selected, S genes of those were really differential expressed, whereas V genes indicated no difference which were false positive. The FDR value should not exceed 0.01, if the error ratio (Q = V/R) was required to be below a specified cutoff (0.01). FDR-values were calculated according to the previous algorithm [40].

$$FDR = E(Q) = E\{V/(V+S)\} = E(V/R)$$

To judge the significance of gene expression differences, we used FDR ≤ 0.001, the Ratio ≥ 2 (the ratio of RPKM values). The genes with significant differential expression levels were subjected to GO function and KEGG pathway analyses.

Q-PCR validation

The expression levels of ADP-glucose pyrophosphorylase small subunit gene, starch synthase IIa gene, 13s globulin gene and seven randomly selected genes were comfirmed using quantitative real-time PCR (Q-PCR). Q-PCR was performed using the same samples used for RNA-seq analysis. First-strand cDNA was synthesized using M-MLV reverse transcriptase (TaKaRa) according to the manufacturer's instructions. The cDNA was used as a template for Q-PCR. Unigenes and primers (designed using Primer Premier 5.0, Premier Biosoft International, Palo Alto, CA, U.S.) are listed in Table S1. The cDNA reaction mixture was diluted to five folds. The Q-PCR mixture (20 µl total volume) contained 10 µl of iQ SYBR green supermix (Bio-Rad), 0.5 µl of each primer (10 µM), 2 µl of cDNA, and 7 µl of RNase-free water. The reactions were performed on Chromo4 real-time PCR detector system (Bio-Rad, United States) according to the manufacturer's instructions. The Q-PCR program was performed after pre-incubation at 95°C for 5 min, followed by 40 cycles of denaturation at 95°C for 15 s, annealing at 60°C for

15 s, and extension at 72°C for 15 s. Template free controls for each primer pair were included in each run. The specificity of Q-PCR primers was confirmed by melting curve. The data were managed with the Gene Expression Analysis for iCycler iQ Real-Time PCR Detection System (Bio-Rad, Hercules, CA, USA) and normalized to that of the housekeeping gene EF (elongation factor 1α). The correlation coefficient (Pearson) of differential expression ratios between RNA-Seq and qRT-PCR was analyzed by using SPSS software 18.0 (http://www-01.ibm.com/software/analytics/spss/).

Differentially expressed genes (DEGs) related to grain quality and expression pattern

Sequence similarity searches were performed using publicly available sequences from monocot species and *Arabidopsis* by Blastn (E-value <1e-10) to identify unigenes related to seed storage proteins and enzymes of starch and cellulose synthesis.

Patterns of gene expression in the germinating grain (4 day) embryos (EMB Embryo), roots (ROO) shoots from seedlings (LEA) (10 cm stage), early developing inflorescences (5 mm (INF 1) & 15 mm (INF 2)), developing tiller internodes (NOD) (six- leaf stage; sectioned between arrows), immature grains [5day post anthesis (dpa) (CAR5) & 15 dpa (CAR15)] were determined by RNA-seq in barley cv. Morex [16]. Representative transcript for one gene was chosen as those that had the maximum ORF extension. A transcript with the RPKM level above 0.4 was viewed as an expressed transcript.

Results

Transcriptome sequencing, *de novo* assembly, and quality evaluation

Sequencing of the XQ754 and Nimubai transcriptomes resulted in 13,069,860 and 12,918,520 clean reads, both with Q20 scores of 92.2% (Table 1). The GC contents of the two varieties were 56.5% and 56.2%, respectively. *De novo* assembly of XQ754 and Nimubai transcriptomes resulted in 48,863 and 45,788 unigenes with the average transcript length of 444 bp and 413 bp, respectively.

For the annotation, the two datasets were combined to form a non-redundant collection (All-Unigenes) containing 46,485 unigenes with an average length of 542 bp. About 62.0% (28,631) of the All-Unigenes were in the range of 300–500 bp; 11.8% (5,487) were longer than 1,000 bp, and no All-Unigene was shorter than 200 bp (Figure S1). Sequence similarity analysis was performed using the barley high-confidence gene set [16] to assess the assembly quality as queries for local Blast against the assembled unigenes. The average values of sensitivity and accuracy of the final assembly were 0.73 and 0.88, respectively, suggesting that the assembly was satisfactory.

Characterization of the unigenes and CDS (coding sequences) prediction

The All-Unigenes were aligned to three public protein databases (nr, Swiss-Prot and KEGG), and a total 36,278 unigenes were annotated, in which 35,986 (77.41%), 25,680 (55.24%) and 16,116 (34.67%) unigenes were annotated by nr, Swiss-Prot, and KEGG databases, respectively (Figure 1). The sequences direction of CDS (coding region sequences) and their amino acid sequences were acquired for among 38,229 unigenes, among which 36,307 (78.1%) unigenes were determined by Blastx (E-value <1e-5) against the public protein databases of nr, Swiss-Prot, KEGG and COG, and 1,922 (4.1%) were predicted by ESTScan [31].

Functional classification

Among the 35,986 nr annotated All-Unigenes, only 12,831 could be further annotated with at least one GO term using Blast2GO [32], indicating that a large part of the nr annotation from hulless barley was not available for GO classifications. These 12,831 All-Unigenes were sorted in 42 GO terms (Figure 2), which were functionally assigned with the three GO terms as of Biological Process (19,010), Cellular Components (29,344) and Molecular Function (11,667). Within the biological process category, All-Unigenes were primarily assigned to GO terms of metabolic process (5,084 unigenes), cellular process (4,588 unigenes), response to stimulus (1,289 unigenes), biological regulation (1,105 unigenes) and establishment of localization (1,083 unigenes). With regard to the cellular component category, most All-Unigenes were assigned to cell (9,836 unigenes), cell part (9,066 unigenes), and organelle (7,879 unigenes). In the molecular function category, the major GO terms were catalytic activity (5,384 unigenes) and binding (5,263 unigenes). A similar profile was found in seeds of oat [41].

Clusters of Orthologous Groups of proteins (COGs) were delineated by comparing protein sequences encoded in complete genomes, representing major phylogenetic lineages. Each COG consisted of individual proteins or orthologous groups from at least three lineages and thus corresponded to an ancient conserved domain. The All-Unigenes were compared to the COG database using the Blastx algorithm specifying E-values of less than 10^{-5}. A total of 13,579 All-Unigenes were annotated with 1,398 functional annotations in the COG database, which could be grouped into 25 functional categories belonging to cellular structure, molecular processing, biochemistry metabolism, signal transduction, etc. (Figure S2). Most All-Unigenes were assigned to general function prediction (4,256), followed by transcription (3,209), function unknown (3668), translation, ribosomal structure and biogenesis (3,207), posttranslational modification, protein turnover and chaperones (2,530), signal transduction mechanisms (1,096, 10.8%), cell wall/membrane/envelope biogenesis (2,381), replication, recombination and repair (2,340), cell cycle control, cell division and chromosome partitioning (2,248). Furthermore, 6,612 unigenes which might affect the quality of the grains were also identified. These unigenes were assigned to carbohydrate transport and metabolism; amino acid transport and metabolism; lipid transport and metabolism; energy production and conversion; and secondary metabolites biosynthesis, transport and catabolism.

We further analyzed biochemical pathways represented by the collection of unigenes. Using the KEGG database, which categorizes gene functions with emphasis on biochemical pathways, a total of 120 pathways represented by 16,116 All-Unigenes were predicted. These pathways in the developing grain of hulless barley have significant roles in biochemical for compound biosynthesis, assimilation, degradation, and utilization and pathways involved in generation of precursor metabolites and energy. Plant metabolites are crucial for both plant life and human nutrition. Furthermore, these metabolites are important for enzymes involved in all steps in the major plant metabolic pathways including the Calvin cycle, TCA cycle, glycolysis, gluconeogenesis and the pentose phosphate pathway represented by unigenes derived from the hulless barley grain dataset. The functional significance of secondary metabolites in reproductive plant parts, particularly seeds of plants in natural ecosystems, is not well known. However, our study highlighted the unigenes associated with these parts, which can enhance our understanding of these metabolites. Furthermore, several unigenes involved in other important secondary metabolite biosynthesis pathways were found. These included the flavonoid biosynthesis pathway, which plays important roles in a number of biological processes and confers health-promoting effects against chronic diseases, such as cardiovascular diseases. Unigenes associated with carotenoid biosynthesis, which is indispensable to plants and plays a critical role in human nutrition and health were also found. Moreover, unigenes involved in several signaling pathways including ethylene pathway, programmed cell death (PCD), and abscisic acid (ABA)-mediated maturation were also found.

Gene expression patterns

On the basis of RPKM, five expression patterns on relative expression levels were classified for 46,485 All-Unigenes. Pattern 1 contains eight unigenes in XQ754 and 14 unigenes in Nimubai with dramatically high RPKM values of 10,000 and 27,000, respectively. Pattern 2 consists of seven unigenes in XQ754 and five unigenes in Nimubai with very high RPKM value from 5,000 to 10,000. The two patterns include the barley stripe mosaic virus genes, resistance genes, hordein genes and a probable cytochrome P450 monooxygenase gene. There are 115 unigenes in XQ754 and 135 unigenes in Nimubai with high RPKM values (pattern 3) from 1,000 to 5,000. Some of these 115 unigenes are involved in grain development, response to stimulus, ribosome biogenesis, metabolic process, cation binding and gene expression (data not shown). There are 1,618 unigenes in XQ754 and 1,514 unigenes in Nimubai with RPKM value from 100 to 1,000 (pattern 4) and more than 80% unigenes of the two accessions have the RPKM value below 100 (pattern 5), and genes of these two patterns mainly function in grain development and nutrition biosynthesis. Over all, the pathways with most abundant transcripts according to the RPKM value are metabolic pathways, spliceosome, ribosome, plant-pathogen interaction, endocytosis, starch and sucrose metabolism and protein processing in the endoplasmic reticulum.

We also compared the expression patterns of the two accessions and found 4,532 (9.7%) differently expressed unigenes. Of this, 1,381 unigenes were expressed at higher levels and 3,151 unigenes were expressed at comparatively lower levels in Nimubai as compared to those in XQ754 (Figure 3). The GO analysis of the differentially expressed unigenes revealed that within the biological process category (Figure S3), differential expressed unigenes were primarily assigned to GO terms of metabolic process (574 unigenes), cellular process (480 unigenes), response to stimulus (153 unigenes), biological regulation (118 unigenes) and localization (117 unigenes). In the cellular component category, most differentially expressed unigenes were assigned to cell (1,031 unigenes), cell part (937unigenes) and organelle (802 unigenes). In the molecular function category, the major GO terms were binding (553 unigenes) and catalytic activity (553 unigenes).

According to the annotations of nr, Swiss-Prot, KEGG, COG and GO, data mining of genes related to barley grain quality was performed. Altogether, 373 quality related transcripts belonging to starch metabolism (starch biosynthesis or degradation), grain storage protein synthesis (hordeins, globulins and glutelin), essential amino acids biosynthesis and degradation (asparagine, aspartate, lysine, methionine, and threonine), seed maturation, and seed development were identified (Table S2). We analyzed the expression levels of these unigenes in the developing grains of the two landraces and found that most of the unigenes showed little or no change in expression. Only 44 (11.8%) unigenes showed differences in expression, wherein 11 unigenes were expressed at higher levels, and 33 unigenes were expressed at comparatively lower levels in Nimubai than those in XQ754. In the two accessions, differentially expressed genes were mainly involved in biosynthesis and degradation of the aspartate family amino acids

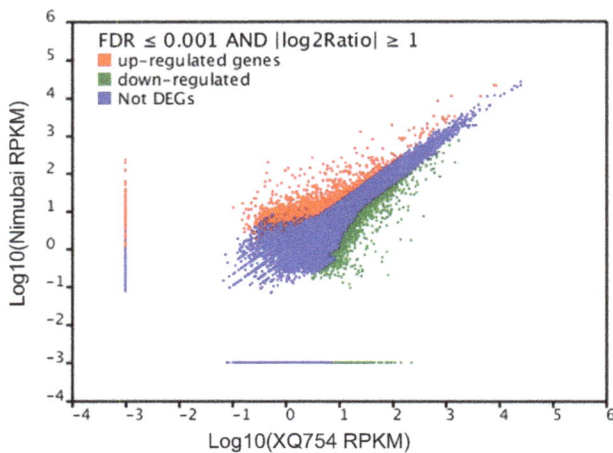

Figure 3. Gene expression levels of XQ754 and Nimubai. The differentially expressed genes are shown in red and green. Genes without expression changes are shown in blue. FDR ≤0.001 and ratio larger than 2.

and starch metabolism. Furthermore, a remarkable expression of enzymes involved in methionine metabolism revealed the availability of sulfur-containing amino acids for protein synthesis during grain development. This is significant in designing strategies for modifying the nutritional value of barley seeds. Further research is needed to explain the specific functions of these genes on barley grain quality.

Genes involved in starch biosynthesis

We further studied the transcripts involved in the synthesis of main storage nutrient in hulless barley grain. Starch comprises 70% of the dry weight of cereal seeds and provides up to 80% of the calories consumed by humans. Starch biosynthesis in the barley grains requires the coordinated activities of several core enzymes [42–47]. The All-Unigenes dataset and the transcriptome dataset of barley cultivar Morex [16] were searched by Blastn (E-value <1e-10) using the known enzyme sequences of Arabidopsis, maize, and rice as query. A total of 19 All-Unigenes relevant to starch biosynthesis enzymes were detected, including ADP-glucose pyrophosphorylase (AGPase), granule-bound starch synthase (GBSS), soluble starch synthase (SS), starch branching enzyme (SBE), starch debranching enzyme (DBE), isoamylase (ISA) and the pullanase (or beta-limit dextrinase; PUL) (Figure 4).

The AGPase, a heterotetrameric enzyme composed of two small (AGP-S) and two large (AGP-L) subunits, catalyzes the first key regulatory step in the starch biosynthetic pathways in all higher plants. Transcripts of *AGP-S1*, *AGP-S2*, *AGP-L1* and *AGP-L2* were detected in the two accessions and in all tested tissues of Morex (Figure 4). The *AGPS1* apparently encodes the transcripts for AGPS1a and AGPS1b, which differ only in their first exons. *AGP-S1a* and *AGP-S2* were abundantly expressed in the starchy grains of the two accessions, whereas *AGP-S1b* was found to be present only at a moderate level in the grain. *AGP-L1* had expression above 80 RPKM, while *AGP-L2* had expression below 10 RPKM in the developing grains of both XQ754 and Nimubai. Peak expression of *AGP-S1*, *AGP-S2* and *AGP-L1* was attained in 15 dpa grain (CAR15) and all AGPase transcripts except *AGP-L2* were strongly up-regulated at the grain filling stage (Figure 4).

Of the two currently known *GBSS* isoforms in barley, *GBSSIa* had a much higher expression level (>30 times) than *GBSSIb* in

Nimubai and XQ754 grains. However, there were no significant differences between the two accessions. Furthermore, Morex data revealed that *GBSSIa* was mainly expressed in storage tissues and strongly up-regulated in 15 dap grain, whereas *GBSSIb* were not detected in grain but were found in transitory starch accumulated tissues, especially in INF1 and INF2 (Figure 4).

The transcriptome database screen also identified the unigenes of *SSI*, *SSIIa*, *SSIIb*, *SSIIIa*, *SSIIIb*, and a fraction of *SSIV* (Figure 4). In Morex, the gene expression of *SSIIIa* and *SSIIa* was restricted to grains compared with *SSI*, *SSIIb*, *SSIIIb* and *SSIV* which were also expressed in other tissues. In addition, the transcripts of *SSIIb*, *SSIIIb* and *SSIV* had an accumulation peak in the node but were expressed at relatively low levels during grain developing. *SSI* and *SSIIa* had the highest RPKM values as compared to the others in the two accessions accounting for more than 70% of the total SS expression. However, *SSI*, *SSIIa* in 5 dpa grain and *SSI*, *SSIIa and SSIIIa* in 15 dpa grain of Morex had the highest RPKM than other SSs (Figure 4). Nevertheless, the differentially expressed transcripts were not found among these SS enzymes between the two accessions.

Sequences of the corresponding transcripts of three *SBE*, three different *ISA* and the *PUL* were recovered. *SBE1* was expressed at remarkably high levels in 15 dap grains but was expressed at low levels in other tissues. A moderate level of *SBE2a* expression was found in all tissues but this expression peaked at 15 dap in the grain. *SBE2b* transcripts were only detected in the developing barley grains with the highest expressed level in 15 dap grain (Figure 4). *ISA1* transcripts were abundant in 15 dap grain and had low expression level in other tissues while *ISA3* transcripts were abundant in node and early grain. *ISA2* was barely expressed in all tissues involved; the *PUL* gene was highly expressed in 15 dap grain but had low expression levels in other tissues (Figure 4). Moreover, the expression levels of these unigenes did not show a notable difference between the two accessions.

Genes related to β-glucan synthesis

The β-glucans can significantly reduce the risk of serious human diseases such as type II diabetes, cardiovascular disease and colorectal cancer. Barley grain is particularly high in β-glucans and has a claimed usage in health products in more developed countries [16]. Two members of cellulose synthase-like (CSL) super family, *CslF* and *CslH*, have proved implication in β-glucan biosynthesis [48,49]. In Morex, eight transcripts with close sequence similarity to known genes of *CslF* and *CslH* family [48,49] were found, while a new transcript showed 64% identity to *CslF4* and another new transcript showed 70% identity to *CslF9* were also found. The two new transcripts were designated as *CslF4-like* and *CslF9-like* respectively (Figure 5). *CslF6* showed highest expression levels in all tissues tested, while *CslF9* showed second highest expression levels in grains. The expression of *CslF8* and *CslH1* were barely detected in immature grains but were high in roots and nodes, which is consistent to previous results obtained by quantitative PCR [50]. Meanwhile, *CslF3*, *CslF4*, *CslF7* and *CslF10* were not expressed in developing grains. In our investigation, four *Csl* genes, *CslF6*, *CslF8*, *CslF9*, and *CslH1* were detected in the two hulless accessions (Figure 5). *CslF6* showed highest expression levels followed by *CslF9*. *CslF8* and *CslH1* showed very low expression levels. The expression levels of *CslF9* in XQ were higher than those in NM while vice versa in the expression levels of *CslF8* and *CslH1*.

Genes encoding grain storage proteins

Globulins are found in the embryo and outer aleurone layer of the endosperm. The structure and properties of the globulins are

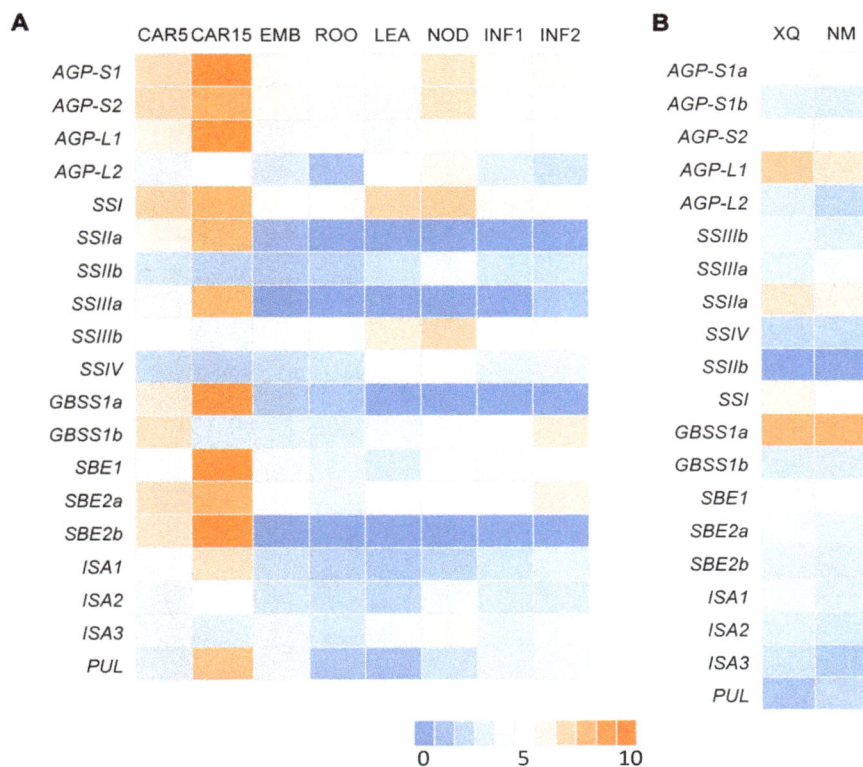

Figure 4. Heat map showing expression profiles of genes involved in starch biosynthesis. A) Gene expression profiles in eight tissues of Morex. B) Gene expression profiles of XQ (XQ754) and NM (Nimubai). Red color shows high expression level, while blue marks low expression level.

similar to the 7S vicilins of legumes [51]. Transcripts for eight *globulin* genes were found in XQ754, Nimubai, and Morex, including one *BEG1*, one *BEG2*, two *11S-like globulins*, one *12S-like globulin*, one *19kDa-like globulin* highly homologous to *19kDa globulin* gene of rice, and two transcripts with high homology to the *Setariaitalica 13S globulin* (Figure 6). The *BEG1* transcript shares 99% identity with previously reported barley embryo globulin gene which exhibits sequence similarity to 7S seed globulins of both monocots and dicots [52]. Distinct from *BEG1* (only 38% identity), a novel globulin transcript, temporarily designated as *BEG2*, was identified. *BEG2* was found to be homologous to the maize *GLB2*. Among the globulin genes, *BEG1* and *BEG2* were the most abundant transcripts followed by transcripts of a 13S-like and a 12S-like globulin in Nimubai and XQ754. *BEG1*, *BEG2* and the *12S-likeglobulin* transcript showed remarkably high accumulatio-nin15 dap grain but were rarely expressed in 5 dap grain and other tested tissues of Morex. The *19kDa-like globulin* was expressed at comparatively lower levels in Nimubai, XQ754 and Morex but showed similar expression pattern as *BEG1*, *BEG2* and the *12S-like globulin* in Morex. One *11S globulin-like* transcript which was rarely expressed in the two accessions was not expressed in the grains of Morex, but showed high expression levels in embryo and leaf, while the other one lowly expressed in the two accessions showed low expression levels in all tested tissues of Morex. Furthermore, the expression of one *13S globulins-like* was ubiquitous in all tested tissues at a low level in Morex but at a comparatively high level in the grains of the two accessions. However, the transcript that was undetected in Morex showed a lower expressed in Nimubai and XQ754. With the exception of one *11S globulin-like*, there was no significant difference in the globulin transcript between the two accessions.

Hordein accounts for ~50% of the total protein in the mature grains, and could be classified into four groups named B, C, D and γ-hordeins based on their electrophoretic mobilities [53]. In Nimubai and XQ754, four *B-hordeins*, seven *C-hordeins*, five *D-hordeins*, and two *γ-hordeins* transcripts were found and most of them were highly expressed. Morex shows different transcript numbers of B, C, and D types. Only one transcript of D-hordein was detected and its expression level is unavailable. The five D-hordein transcripts of the two accessions shared over 92% identity with the transcript of D-hordein of Morex and 86% identity with the wheat γ-type high molecular weight glutenin subunit gene.

Validation of RNA-Seq data

Ten differentially expressed genes were selected to demonstrate the RNA-seq results using QPCR (Table S1). The Q-PCR data showed the similar trends with RNA-Seq samples. Linear regression [y = αx+ b, (y = Q-PCR value; x = RNA-seq value)] analysis showed a high correlation (R =0.8391), indicating that the gene expression differences observed in transcript abundance between the two samples were highly credible (Figure S4).

SNPs identification

By comparing our data with the public expressed sequence data of barley, we roughly found 17,608 and 14,121 SNPs in 7,335 and 6,285 unigenes of Nimubai and XQ754, respectively. Among them, a total of 8,893 SNPs were shared by both accessions and 13,943 SNPs were found between two hulless barley landraces. Within the detected SNPs, the transitions were much more common than transversions (about 2:1). Meanwhile, a similar number of A/G and C/T transitions and four transversion types

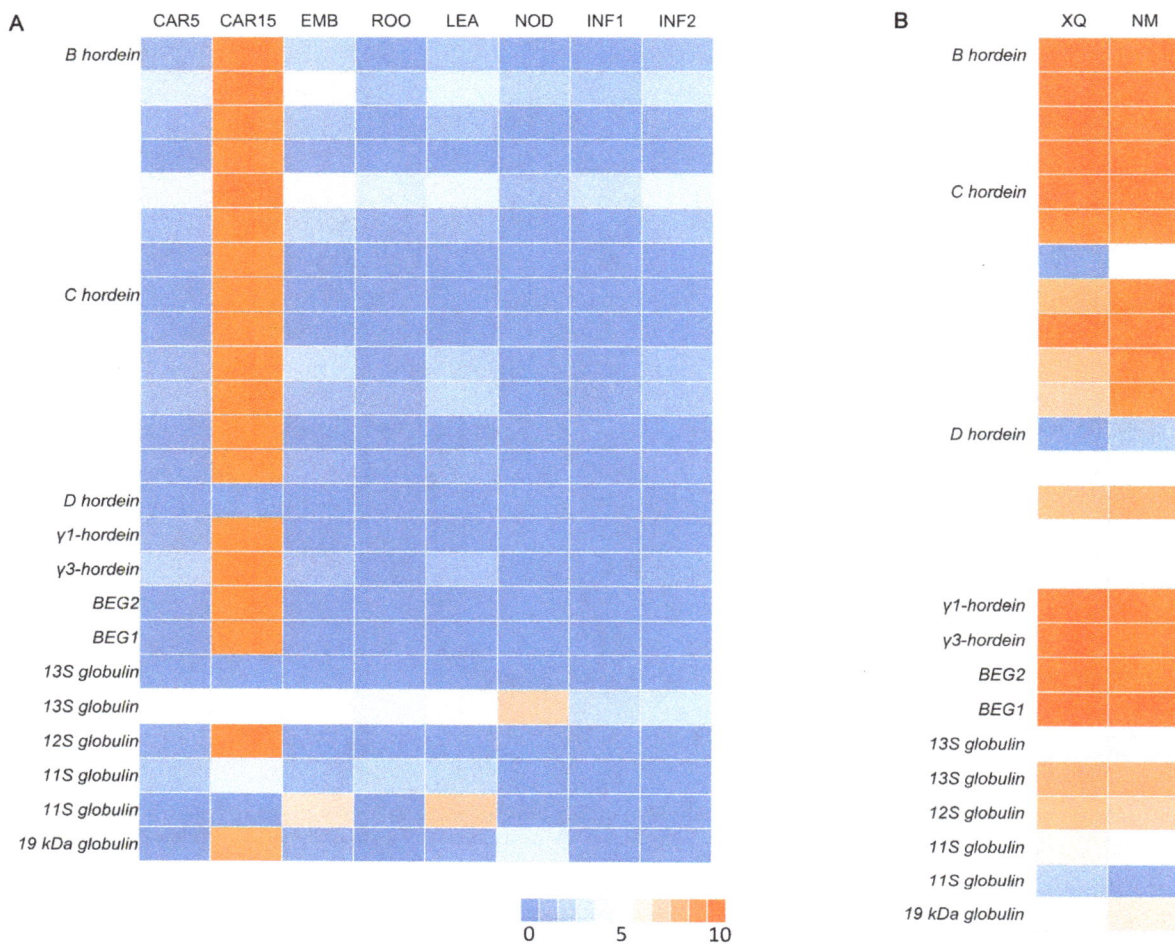

Figure 5. Heat map showing expression profiles of genes encoding cereal grain storage proteins. A) Gene expression profiles in eight tissues of Morex. B) Gene expression profiles of XQ (XQ754) and NM (Nimubai). Red color shows high expression level, while blue marks low expression level.

(A/T, A/C, G/T, and C/G) were detected. We identified 29 SNPs in the CDS of eight genes encoding enzymes for starch and β-glucan synthesis. Fourteen SNPs were found between the two accessions, in which 3 and 11 occurred in Nimubai and XQ754, respectively, and 15 SNP were shared by both accessions (Table 2). Nine SNPs (~31% of total) were nonsynonymous and resulted in

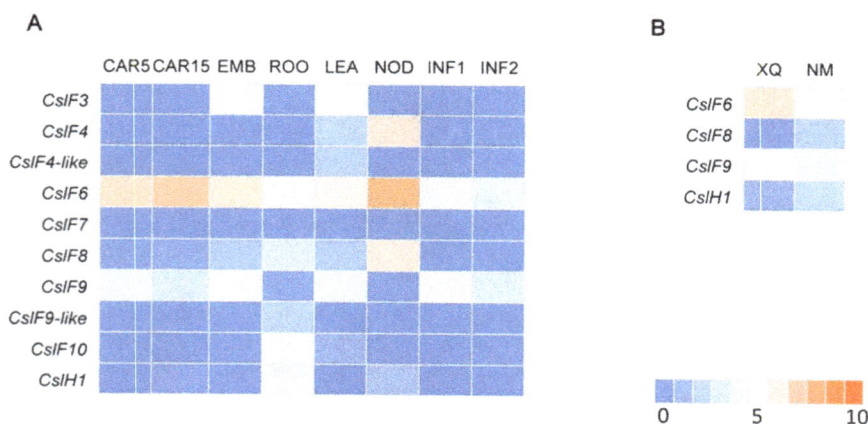

Figure 6. Heat map showing expression profiles of *HvCslF* and *HvCslH* gene families. A) Gene expression profiles in eight tissues of Morex. B) Gene expression profiles of XQ (XQ754) and NM (Nimubai). Red color shows high expression level, while blue marks low expression level.

Table 2. SNPs of genes involved in starch and β-glucan biosynthesis.

	Transcripts length	Coordinate	Ref. Nuc	SNP xQ Nuc	SNP xQ depth(hit/total)	NM Nuc	NM depth(hit/total)	AA alteration
AGP-S1a	1545	501	T	C*	24/24	C*	20/20	
AGP-L1	1572	360	T	C	99/101	C	47/64	
SSIIb	448	271	C	T	21/23	T	14/15	
		303	T	C	17/19	C	19/19	
		336	T	C	16/16	C	24/24	
GBSS1a	1829	529	T	C*	9/9	C*	14/14	
		789	A	G*	255/474	G*	255/593	
		795	T	C*	255/459	C*	255/540	
		837	A	G*	255/451	G*	255/538	
		1077	A	G*	104/104	G*	107/107	
		1272	C	A	239/239	A	255/267	
		1383	G	A*	128/138	A*	184/189	
SBE1	1811	711	A	G*	8/8	G*	17/17	
CslF6	2451	512	A	C	69/69	C	49/49	M-L
CslF9	1829	1311	A	C	69/69	C	37/37	N-H
AGP-L1	1572	519	A	G*	93/97	-		
CslF6	2451	439	T	C	14/14	-		
CslF9	1829	1682	G	A	8/8	-		
SSIIIa	1112	429	T	-		C	15/15	F-L
		437	G	-		C	15/15	
		460	G	-		T	8/8	C-F
		806	T	-		C	22/22	
		943	T	-		C	14/14	F-S
SSIIb	448	421	C	-		G	10/10	
GBSS1a	1829	374	A	-		G*	11/11	I-V
		423	G	-		A*	110/110	S-N
		451	C	-		T*	138/140	
SBE1	1811	629	C	-		T*	11/11	S-L
CslF9	1829	1815	T	-		G	14/14	W-G

Note: Nuc, nucleotide; AA, amino acid; *, SNPs confirmed using Sanger sequencing. – indicates that the nucleotide is identical with the reference.

nine amino acid changes. All these 29 SNPs were validated in Nimubai, XQ754, and other 10 hulless barley landraces by Sanger sequencing (data not shown). Among these, 13 SNPs were also variable (Table 2) and the others are identical among all accessions of hulless barley tested.

Discussion

Hulled cultivated barley has been used in the brewing industry worldwide, however, lesser attention was paid on the grain quality of the hulless barley, which is the staple food at some barren regions or highland. Hulless barley has gained significant attention in recent years because of its potential health benefits such as higher β-glucan content than the hulled barley. Comparing to a long growing history and rich diversity in the Qinghai-Tibet Plateau, very few hulless barley cultivars have been developed for the modern UK or European agricultural systems. Thus, exploitation of germplasm resources and revealing the formation mechanism of grain quality in hulless barley will aid in the development of better hulless cultivars with desirable dietary characteristics. Here, we used high-throughput deep sequencing technology to profile the grain transcriptome of two Tibetan hulless barley landraces Nimubai and XQ754. We assembled 48,863 and 45,788 unigenes in two samples and constructed a combined non-redundant data set of 46,485 All-Unigenes. A total of 36,278 All-Unigenes could be functionally annotated, and the CDS and directions of 38,229 All-Unigenes were predicted.

Using Blast search and functional annotation, new transcripts with homology to the genes previously reported in other species could be identified. For instance, six new globulin transcripts (*BEG2*, two *11S-like globulins*, two *13S-like globulins* and one *19kDa-like globulin*) were predicted in the All-unigene dataset and Morex, respectively. Furthermore, two new transcripts *CslF9-like* and *CslF4-like* were detected in Morex. The deduced amino acid sequences of these new transcripts were compared with other known sequences and domains from NCBI (Figure S5–S11). Most of these new transcripts were validated by highly homogenous ESTs (Table S3) from full-length cDNAs in barley [54,55]. Although their functional roles need further verification, all novel transcripts will help us to study the storage proteins and β-glucans synthesis. They will also provide valuable insights for identifying new genes that influence the grain quality and seed development.

We attempted to characterize the sequences and transcript accumulation of grain quality related genes encoding the seed storage proteins and the enzymes involved in starch and β-glucan biosynthesis in grains. Nineteen unigenes relevant to starch biosynthetic enzymes were detected. Among them, *AGP-S1* and *AGP-L1* were mainly expressed in the developing grain at high levels, suggesting their importance at the first step of starch biosynthesis. Moreover, they possibly associate to form a heterotetrameric cytosolic AGPase, similar to AGP-S2b and AGP-L2 of rice [56]. The chain elongation of amylose and amylopectin are distinctively catalyzed by the starch granule-bound form of starch synthase (GBSS) and soluble form of starch synthase (SS), respectively. Of the two GBSS isoforms, *GBSSIb* functions in non-storage plant tissues in which transitory starch accumulates, while *GBSSIa* is confined to storage tissues and has a much higher expression level than *GBSSIb* in grains of Nimubai and XQ754. *GBSSIa* then acts as the main limiting enzyme in the endosperm amylose production. This result is consistent with previous research in barley, rice and wheat [42,43,57]. However, the expression levels of *GBSSIa* in Nimubai and XQ754 were not significantly different in our study.

Among the SSs, *SSIV* gene was expressed in diverse tissues and at relatively low levels during grain filling and similar expression profiles were found in a Morex and rice [57]. The *SSIV* mutants of Arabidopsis show a striking reduction in the number of starch granules but an increase in starch granule size, indicating that SSIV could be selectively involved in the priming of starch granule formation [58]. Furthermore, the *SSIV* gene may not play typical roles as other SSs in the elongation of amylopectin chains during starch biosynthesis in barley. *SSI* and *SSIIa* of the two accessions and *SSI*, *SSIIa*, *SSIIIa* of Morex had the highest expression level among SSs.

In rice endosperm, *SSI* and *SSIIIa* are the major SS enzymes and *SSI* activity is higher than that of *SSIIIa*, constituting about 70% of the SS activity [59], which is consistent with other data of wheat [60] and maize [61]. Contrastingly, *SSII* and *SSIII* account for the major SS activities in potato tubers [62] and pea embryos [63]. In barley, we found that SSI and SSIIIb act extensively in diverse tissues, whereas SSIIa and SSIIIa mainly function during seed development. This suggests that the expression level of *SSI*, *SSIIa* and *SSIIIa* may be divergent among species, and their coordinated action might play a critical role in the grain amylopectin chain biosynthesis.

Comprehensively, *AGP-S1*, *AGP-L1*, *GBSSIa*, *SSI*, *SSIIa*, *SSIIIa*, *SBE1*, *SBE2b*, *ISA1* and *PUL*, which are mainly expressed in barley grain may significantly affect the starch biosynthesis in barley endosperm. There were no differentially expressed transcripts relevant to starch biosynthesis enzymes (except *AGP-S2*) between XQ754 and Nimubai. In starch biosynthetic pathway, each enzyme plays a distinct role, but presumably functions as part of a complex network. In this synthesis network, genes controlling amylopectin and amylose synthesis possibly interact [64,65]. Thus, even though there is no divergence among the expression levels of the associated unigenes, the two accessions might have a different percentage of amylose mediated by multiple genes. In rice, the association analysis with individual starch synthesis-related genes revealed that *Wx* (*GBSS*) and *SSII-3* mainly control amylose content. *Wx* is likely the major gene and *SSII-3* acts as a minor effector. Under the same Wx background, varieties with different allelic *SSII-3* states show diverse amylose content [66]. SSIIa of barley accounts for the majority of amylopectin polymer elongation activity [67] and is highly homologous to SSII-3 of rice. In our results, Nimubai, which contains higher amylose content, also showed a higher RPKM ratio of *GBSSIa* to *SSIIa* as compared to XQ754. The elongation reactions for the chains of amylose and amylopectin are distinctively catalyzed by GBSS and SSs, respectively, thus the ratio of expression levels of *GBSSIa* to *SSIIa* might influence the ratio of amylose to amylopectin in barley.

β-glucan is a major constituent of the endosperm cell wall in barley grains [68,69]. High content of β-glucan in barley grains has a negative effect on malting and pearling processes but is desirable for barley used as human food. Our analysis indicated that transcripts for the *CslF6* were the most abundant in developing barley grains, indicting its key role in controlling β-glucan synthesis in endosperm, which was also supported by analysis in barley β-glucanless mutants [70] and RNAi inhibition of *CslF6* in wheat grains [71]. Transcripts of the *CslF9* peaked earlier than *CslF6* and the previous study also described that the *CslF9* gene was transcribed at a stage when cellularization of the endosperm was completed and starch deposition had commenced, but disappeared somewhere between 12 and 15 days post-pollination [72]. In this study, we found that the *CslF9* transcript was expressed at a higher level in XQ754 than that in Nimubai (higher β-glucan content). This result is consistent with the previous study that *CslF9* appeared to be much more abundant in the elite malting variety 'Sloop' (lower) than the hulless barley 'Himalaya' (higher) [72]. This result suggests that CslF9 might not be a determinant of the β-glucan content and its role in β-glucan synthesis needs further study. Consequently, *CslF6* gene appears to encode the major β-glucan synthase, because of being constitutively expressed at much higher levels than all the other *CslF* genes in all tested tissues of barley. Other *CslF* genes may function as modifier in different stages of development or different tissues and organs. The *CslH1* has a proven function in β-glucan synthesis in barley. In this study, *CslH1* exhibited low expression levels in both hulless landraces, as well as in Morex, which is consistent with previous report. However, we noted that it is expressed at significantly higher level (~2.7-fold) in Nimubai than that in XQ754. These results imply that *CslH1* may affect the total accumulation of β-glucan in barley grains independent of *CslF6*.

Cereal seed proteins are a source of primary nutrition for humans and livestock and have a great influence on the utilization of the grains in food processing. They usually account for about 10–15% of the dry weight of the seed and are mainly composed of globulins and prolamins [73,74]. Eight globulins related transcripts were identified that showed similar expression patterns in hulled and hulless barley with the exception of one *13S globulin*. The *BEG1* and *BEG2* and *12S-like globulin* transcripts were highly expressed in hulled and hulless barley grains specifically. They encode globulins containing two 'Cupin' domains as those in13S-like globulins. This is consistent with prior research that the accumulation of *Beg1* mRNA was noted beginning 15–20 dpa of the developing barley grain [75]. Thus BEG1, and BEG2 and 12S-like globulins appear to function solely as main storage globulins.

Prolamins are the major endosperm storage proteins in most cereal grains. The allelic variation observed in hordeins and its influence on the food making, and malting quality is noteworthy. The B-hordeins and C-hordeins, encoded by the *Hor2* loci and *Hor1* loci, consist of 20–30 genes per haploid of barley genome [76,77]. However, the D-and γ-hordeins, encoded by the *Hor3* and *Hor5* loci [78,79], have minor members and the extent of polymorphism is unclear. The transcript numbers of B-hordein, C-hordein, and D-hordein between hulless and hulled genotypes were diverse and showed high variability. One D-hordein transcript was found in Morex; the sequence analysis of a 120-kb D-hordein region reported one D-hordein in that region [80], whereas five expressed D-hordein transcripts were found in the two hulless barleys. It is not known whether the increased number of D-hordein transcripts is caused by diverse members in the two accessions or improper sequence assembling.

In this study, we roughly identified more than ten thousand SNPs in the two hulless barley landraces. Twenty-nine SNPs identified in eight starch and β-glucan synthesis related genes were confirmed to be valid, indicating the high accuracy of SNP identification by transcriptome data. Thus, compared to the large-scale genomic sequencing, the transcriptome sequencing serves as an economic way for diversity detection. Furthermore, originating from expressed genes, all these transcriptome derived SNP might have great potential in the function associated analysis in the future.

Supporting Information

Figure S1 Length distribution of All-Unigenes. The x-axis indicates the sequence length of unigenes and the y-axis indicates the number of unigenes, and the numbers of unigenes with a certain length are indicated on the top of the rectangle bars. (PDF)

Figure S2 COG function classification. The capital letters in x-axis indicate the COG categories as listed on the right of the histogram and the y-axis indicates the number of unigenes. (PDF)

Figure S3 Go annotation of differential expression unigenes. The x-axis indicates the categories and the y-axis indicates the number and proportion of differentially expressed unigenes. (PDF)

Figure S4 Coefficient analysis between expression ratios obtained from RNA-seq and Q-PCR data of two landraces. ** indicates a significant difference at $p \leq 0.01$. (PDF)

Figure S5 Alignment of amino acid sequences of putative7S globulin from barley cultivar Morex and the two accessions. Domains are indicated by bars and labels below the Alignment. (PDF)

Figure S6 Alignment of amino acid sequences of putative11S-1 globulin from barley cultivar Morex and the two accessions. Domains are indicated by bars and labels below the Alignment. (PDF)

Figure S7 Alignment of amino acid sequences of putative11S-2 globulin from barley cultivar Morex and the two accessions. Domains are indicated by bars and labels below the Alignment. (PDF)

Figure S8 Alignment of amino acid sequences of putative13S globulin from barley cultivar Morex and the two accessions. Domains are indicated by bars and labels below the Alignment. (PDF)

Figure S9 Alignment of amino acid sequences of putative19KD globulin from barley cultivar Morex and the two accessions. Domains are indicated by bars and labels below the Alignment. AAI_SS: Alpha-Amylase Inhibitors (AAIs) and Seed Storage (SS)protein subfamily; composed of cereal-type AAIs and SS proteins. (PDF)

Figure S10 Alignment of amino acid sequences of putative CslF4 and CslF4-like proteins of barley cultivar Morex and the two accessions. Domains are indicated by bars and labels below the Alignment. Glycosyltransferase family A (GT-A) includes diverse families of glycosyltransferaseswith a common GT-A type structural fold. (PDF)

Figure S11 Alignment of amino acid sequences of putativeCslF9 and CslF9-like proteins from barley cultivar Morex and the two accessions. Domains are indicated by bars and labels below the alignment. Glycosyltransferase family A (GT-A) includes diverse families of glycosyltransferases with a common GT-A type structural fold. (PDF)

Table S1 Validation of ten differentially expressed genes using Q-PCR validation. Note: NM, Nimubai; XQ, XQ754. (DOCX)

Table S2 List of genes related to seed quality. (XLSX)

Table S3 New transcripts validated by highly homogenous ESTs of nr database. (DOCX)

Acknowledgments

We thank Dr. Garry Rosewarne of International Maize and Wheat Improvement Centre (CIMMYT) for his help to revision the manuscript. We wish to thank the two anonymous reviewers for helpful comments and constructive suggestions that improved the manuscript.

Author Contributions

Conceived and designed the experiments: MQY HL PG. Performed the experiments: XC GBD ZFP JJL. Analyzed the data: XC HL. Contributed reagents/materials/analysis tools: XC GBD ZFP JJL YWT NT. Wrote the paper: XC.

References

1. Zohary D, Hopf M (2000) Domestication of Plants in the Old World: The Origin and Spread of Cultivated Plants in West Asia, Europe, and the Nile Valley: Oxford University Press.

2. Fischbeck G (2003) Chapter 3 Diversification through breeding. In: Roland von Bothmer TvHHK, Kazuhiro S, editors. Developments in Plant Genetics and Breeding: Elsevier. 29–52.

3. Liu CT, Station IAE (1996) Hulless Barley: A New Look for Barley in Idaho: University of Idaho, College of Agriculture, Cooperative Extension System, Agricultural Experiment Station.

4. Dickin E, Steele K, Edwards-Jones G, Wright D (2012) Agronomic diversity of naked barley (Hordeum vulgare L.): a potential resource for breeding new food barley for Europe. Euphytica 184: 85–99.

5. Jadhav SJ, Lutz SE, Ghorpade V, Salunkhe DK (1998) Barley: Chemistry and Value-Added Processing. Critical Reviews in Food Science and Nutrition 38: 123–171.

6. Hecker KD, Meier ML, Newman RK, Newman CW (1998) Barley β-glucan is effective as a hypocholesterolaemic ingredient in foods. Journal of the Science of Food and Agriculture 77: 179–183.

7. Edney MJ, Tkachuk R, Macgregor AW (1992) Nutrient composition of the hull-less barley cultivar, condor. Journal of the Science of Food and Agriculture 60: 451–456.

8. Baik B-K, Ullrich SE (2008) Barley for food: Characteristics, improvement, and renewed interest. Journal of Cereal Science 48: 233–242.

9. Abdel-Aal E, Wood PJ (2005) Specialty grains for food and feed. St. Paul, MN: American Association of Cereal Chemists.

10. Newman RK, Newman CW (2008) Barley: Taxonomy, Morphology, and Anatomy. Barley for Food and Health: John Wiley & Sons, Inc. 18–31.

11. Sharma P, Gujral HS (2010) Milling behavior of hulled barley and its thermal and pasting properties. Journal of Food Engineering 97: 329–334.

12. RS B (1993) Physicochemical properties of roller-milled barley bran and flour. Cereal Chemistry Journal: 397–402.

13. Sun L, Lu W, Zhang J, Zhang W (1999) Investigation of barley germplasm in China. Genetic Resources and Crop Evolution 46: 361–369.

14. Badr A, M K, Sch R, Rabey HE, Effgen S, et al. (2000) On the Origin and Domestication History of Barley (Hordeum vulgare). Molecular Biology and Evolution 17: 499–510.

15. Yin YQ, Ma DQ, Ding Y (2003) Analysis of genetic diversity of hordein in wild close relatives of barley from Tibet. Theoretical and Applied Genetics 107: 837–842.

16. Mayer KF, Waugh R, Brown JW, Schulman A, Langridge P, et al. (2012) A physical, genetic and functional sequence assembly of the barley genome. Nature 491: 711–716.

17. Sreenivasulu N, Radchuk V, Strickert M, Miersch O, Weschke W, et al. (2006) Gene expression patterns reveal tissue-specific signaling networks controlling programmed cell death and ABA- regulated maturation in developing barley seeds. Plant J 47: 310–327.

18. Sreenivasulu N, Altschmied L, Radchuk V, Gubatz S, Wobus U, et al. (2004) Transcript profiles and deduced changes of metabolic pathways in maternal and filial tissues of developing barley grains. Plant J 37: 539–553.

19. Oztur ZN, Talame V, Deyholos M, Michalowski CB, Galbraith DW, et al. (2002) Monitoring large-scale changes in transcript abundance in drought- and salt-stressed barley. Plant Mol Biol 48: 551–573.

20. Druka A, Muehlbauer G, Druka I, Caldo R, Baumann U, et al. (2006) An atlas of gene expression from seed to seed through barley development. Funct Integr Genomics 6: 202–211.

21. Close TJ, Wanamaker SI, Caldo RA, Turner SM, Ashlock DA, et al. (2004) A new resource for cereal genomics: 22K barley GeneChip comes of age. Plant Physiol 134: 960–968.

22. Leymarie J, Bruneaux E, Gibot-Leclerc S, Corbineau F (2007) Identification of transcripts potentially involved in barley seed germination and dormancy using cDNA-AFLP. J Exp Bot 58: 425–437.

23. Ibrahim AF, Hedley PE, Cardle L, Kruger W, Marshall DF, et al. (2005) A comparative analysis of transcript abundance using SAGE and Affymetrix arrays. Funct Integr Genomics 5: 163–174.

24. White J, Pacey-Miller T, Crawford A, Cordeiro G, Barbary D, et al. (2006) Abundant transcripts of malting barley identified by serial analysis of gene expression (SAGE). Plant Biotechnol J 4: 289–301.

25. Zhang X-Q, Li C, Tay A, Lance R, Mares D, et al. (2008) A new PCR-based marker on chromosome 4AL for resistance to pre-harvest sprouting in wheat (Triticum aestivum L.). Molecular Breeding 22: 227–236.

26. Peng X, Wood CL, Blalock EM, Chen KC, Landfield PW, et al. (2003) Statistical implications of pooling RNA samples for microarray experiments. BMC Bioinformatics 4: 26.

27. Liu S, Lin L, Jiang P, Wang D, Xing Y (2011) A comparison of RNA-Seq and high-density exon array for detecting differential gene expression between closely related species. Nucleic Acids Res 39: 578–588.

28. Grabherr MG, Haas BJ, Yassour M, Levin JZ, Thompson DA, et al. (2011) Full-length transcriptome assembly from RNA-Seq data without a reference genome. Nat Biotech 29: 644–652.

29. Pertea G, Huang X, Liang F, Antonescu V, Sultana R, et al. (2003) TIGR Gene Indices clustering tools (TGICL): a software system for fast clustering of large EST datasets. Bioinformatics 19: 651–652.

30. Altschul SF, Madden TL, Schäffer AA, Zhang J, Zhang Z, et al. (1997) Gapped BLAST and PSI-BLAST: a new generation of protein database search programs. Nucleic Acids Res 25: 3389–3402.

31. Iseli C, Jongeneel CV, Bucher P. ESTScan: a program for detecting, evaluating, and reconstructing potential coding regions in EST sequences; 1999. 138–147.

32. Conesa A, Gotz S, Garcia-Gomez JM, Terol J, Talon M, et al. (2005) Blast2GO: a universal tool for annotation, visualization and analysis in functional genomics research. Bioinformatics 21: 3674–3676.

33. Ye J, Fang L, Zheng H, Zhang Y, Chen J, et al. (2006) WEGO: a web tool for plotting GO annotations. Nucleic Acids Res 34: W293–297.

34. Rismani-Yazdi H, Haznedaroglu BZ, Bibby K, Peccia J (2011) Transcriptome sequencing and annotation of the microalgae Dunaliella tertiolecta: pathway description and gene discovery for production of next-generation biofuels. BMC Genomics 12: 148.

35. Kanehisa M, Goto S (2000) KEGG: kyoto encyclopedia of genes and genomes. Nucleic Acids Res 28: 27–30.

36. Li R, Li Y, Fang X, Yang H, Wang J, et al. (2009) SNP detection for massively parallel whole-genome resequencing. Genome research 19: 1124–1132.

37. Rasmussen M, Li Y, Lindgreen S, Pedersen JS, Albrechtsen A, et al. (2010) Ancient human genome sequence of an extinct Palaeo-Eskimo. Nature 463: 757–762.

38. Mortazavi A, Williams BA, McCue K, Schaeffer L, Wold B (2008) Mapping and quantifying mammalian transcriptomes by RNA-Seq. Nature methods 5: 621–628.

39. Audic S, Claverie J-M (1997) The significance of digital gene expression profiles. Genome research 7: 986–995.

40. Benjamini Y, Hochberg Y (1995) Controlling the false discovery rate: a practical and powerful approach to multiple testing. Journal of the Royal Statistical Society Series B (Methodological): 289–300.

41. Gutierrez-Gonzalez JJ, Tu ZJ, Garvin DF (2013) Analysis and annotation of the hexaploid oat seed transcriptome. BMC Genomics 14: 471.

42. Vrinten PL, Nakamura T (2000) Wheat granule-bound starch synthase I and II are encoded by separate genes that are expressed in different tissues. Plant physiology 122: 255–264.

43. Radchuk VV, Borisjuk L, Sreenivasulu N, Merx K, Mock H-P, et al. (2009) Spatiotemporal profiling of starch biosynthesis and degradation in the developing barley grain. Plant physiology 150: 190–204.

44. Wei K-S, Zhang Q-F, Cheng F-M, Chen N, Xie L-H (2009) Expression Profiles of Rice Soluble Starch Synthase(SSS) Genes in Response to High Temperature Stress at Filling Stage. Acta Agronomica Sinica 35: 18–24.

45. Zhao N-C (2009) Characteristics of Starch Synthesis in Grains and Translocation of Car-bohydrate in Leaves and Sheaths at Filling Stage for Low Phytic Acid Mutant Rice. Acta Agronomica Sinica 34: 1977–1984.

46. Asare EK, Jaiswal S, Maley J, Baga M, Sammynaiken R, et al. (2011) Barley grain constituents, starch composition, and structure affect starch in vitro enzymatic hydrolysis. J Agric Food Chem 59: 4743–4754.

47. Eggert K, Pawelzik E (2011) Proteome analysis of Fusarium head blight in grains of naked barley (Hordeum vulgare subsp. nudum). Proteomics 11: 972–985.

48. Doblin MS, Pettolino FA, Wilson SM, Campbell R, Burton RA, et al. (2009) A barley cellulose synthase-like CSLH gene mediates (1, 3; 1, 4)-β-D-glucan synthesis in transgenic Arabidopsis. Proceedings of the National Academy of Sciences 106: 5996–6001.

49. Burton RA, Wilson SM, Hrmova M, Harvey AJ, Shirley NJ, et al. (2006) Cellulose synthase-like CslF genes mediate the synthesis of cell wall (1, 3; 1, 4)-β-D-glucans. Science 311: 1940–1942.

50. Doblin MS, Pettolino FA, Wilson SM, Campbell R, Burton RA, et al. (2009) A barley cellulose synthase-like CSLH gene mediates (1, 3; 1, 4)-β-D-glucan synthesis in transgenic Arabidopsis. Proceedings of the National Academy of Sciences 106: 5996–6001.

51. Kriz AL (1999) 7S globulins of cereals. Seed Proteins: Springer. 477–498.

52. Heck GR, Chamberlain AK, Ho T-HD (1993) Barley embryo globulin 1 gene, Beg1: characterization of cDNA, chromosome mapping and regulation of expression. Molecular and General Genetics MGG 239: 209–218.

53. Shewry P, Kreis M, Parmar S, Lew E-L, Kasarda D (1985) Identification of γ-type hordeins in barley. FEBS letters 190: 61–64.

Transcriptome Assembly and Analysis of Tibetan Hulless Barley (Hordeum vulgare L. var. nudum)...

13

54. Sato K, Shin T, Seki M, Shinozaki K, Yoshida H, et al. (2009) Development of 5006 full-length cDNAs in barley: a tool for accessing cereal genomics resources. DNA research 16: 81–89.

55. Matsumoto T, Tanaka T, Sakai H, Amano N, Kanamori H, et al. (2011) Comprehensive sequence analysis of 24,783 barley full-length cDNAs derived from 12 clone libraries. Plant Physiol 156: 20–28.

56. Ohdan T, Francisco PB, Sawada T, Hirose T, Terao T, et al. (2005) Expression profiling of genes involved in starch synthesis in sink and source organs of rice. Journal of experimental botany 56: 3229–3244.

57. Hirose T, Terao T (2004) A comprehensive expression analysis of the starch synthase gene family in rice (Oryza sativa L.). Planta 220: 9–16.

58. Roldán I, Wattebled F, Mercedes Lucas M, Delvallé D, Planchot V, et al. (2007) The phenotype of soluble starch synthase IV defective mutants of Arabidopsis thaliana suggests a novel function of elongation enzymes in the control of starch granule formation. The Plant Journal 49: 492–504.

59. Fujita N, Yoshida M, Asakura N, Ohdan T, Miyao A, et al. (2006) Function and characterization of starch synthase I using mutants in rice. Plant physiology 140: 1070–1084.

60. Li Z, Mouille G, Kosar-Hashemi B, Rahman S, Clarke B, et al. (2000) The structure and expression of the wheat starch synthase III gene. Motifs in the expressed gene define the lineage of the starch synthase III gene family. Plant physiology 123: 613–624.

61. Cao H, Imparl-Radosevich J, Guan H, Keeling PL, James MG, et al. (1999) Identification of the soluble starch synthase activities of maize endosperm. Plant physiology 120: 205–216.

62. Marshall J, Sidebottom C, Debet M, Martin C, Smith AM, et al. (1996) Identification of the major starch synthase in the soluble fraction of potato tubers. The Plant Cell Online 8: 1121–1135.

63. Tomlinson K, Craig J, Smith AM (1997) Major differences in isoform composition of starch synthase between leaves and embryos of pea (Pisum sativum,L.). Planta 204: 86–92.

64. Fulton DC, Edwards A, Pilling E, Robinson HL, Fahy B, et al. (2002) Role of granule-bound starch synthase in determination of amylopectin structure and starch granule morphology in potato. Journal of Biological Chemistry 277: 10834–10841.

65. van de Wal M, D'Hulst C, Vincken J-P, Buléon A, Visser R, et al. (1998) Amylose is synthesized in vitro by extension of and cleavage from amylopectin. Journal of Biological Chemistry 273: 22232–22240.

66. Tian Z, Qian Q, Liu Q, Yan M, Liu X, et al. (2009) Allelic diversities in rice starch biosynthesis lead to a diverse array of rice eating and cooking qualities. Proceedings of the National Academy of Sciences 106: 21760–21765.

67. Morell MK, Kosar-Hashemi B, Cmiel M, Samuel MS, Chandler P, et al. (2003) Barley sex6 mutants lack starch synthase IIa activity and contain a starch with novel properties. The Plant Journal 34: 173–185.

68. Fincher G (1975) Morphology and chemical composition of barley endosperm cell walls. Journal of the Institute of Brewing 81: 116–122.

69. Fincher G (1976) Ferulic acid in barley cell walls: a fluorescence study. Journal of the Institute of Brewing 82: 347–349.

70. Taketa S, Yuo T, Tonooka T, Tsumuraya Y, Inagaki Y, et al. (2012) Functional characterization of barley betaglucanless mutants demonstrates a unique role for CslF6 in (1, 3; 1, 4)-β-D-glucan biosynthesis. Journal of experimental botany 63: 381–392.

71. Nemeth C, Freeman J, Jones HD, Sparks C, Pellny TK, et al. (2010) Down-regulation of the CSLF6 gene results in decreased (1, 3; 1, 4)-β-D-glucan in endosperm of wheat. Plant physiology 152: 1209–1218.

72. Burton RA, Jobling SA, Harvey AJ, Shirley NJ, Mather DE, et al. (2008) The genetics and transcriptional profiles of the cellulose synthase-like HvCslF gene family in barley. Plant physiology 146: 1821–1833.

73. Konzak CF (1977) Genetic control of the content, amino acid composition, and processing properties of proteins in wheat. Adv genet 19: 407–582.

74. Saastamoinen M, Plaami S, Kumpulainen J (1989) Pentosan and β-glucan content of Finnish winter rye varieties as compared with rye of six other countries. Journal of Cereal Science 10: 199–207.

75. Heck GR, Chamberlain AK, Ho TH (1993) Barley embryo globulin 1 gene, Beg1: characterization of cDNA, chromosome mapping and regulation of expression. Mol Gen Genet 239: 209–218.

76. Bunce N, Forde B, Kreis M, Shewry P (1986) DNA restriction fragment length polymorphism at hordein loci: application to identifying and fingerprinting barley cultivars. Seed science and technology 14: 419–429.

77. Shewry P, Bunce N, Kreis M, Forde B (1985) Polymorphism at the Hor 1 locus of barley (Hordeum vulgare L.). Biochemical genetics 23: 391–404.

78. Shewry P, Finch R, Parmar S, Franklin J, Miflin B (1983) Chromosomal location of Hor 3, a new locus governing storage proteins in barley. Heredity 50.

79. Piston F, Dorado G, Martin A, Barro F (2004) Cloning and characterization of a gamma-3 hordein mRNA (cDNA) from Hordeum chilense (Roem. et Schult.). Theoretical and Applied Genetics 108: 1359–1365.

80. Gu YQ, Anderson OD, Londeorë CF, Kong X, Chibbar RN, et al. (2003) Structural organization of the barley D-hordein locus in comparison with its orthologous regions of wheat genomes. Genome 46: 1084–1097.

2

Assessment of Genetic Diversity among Barley Cultivars and Breeding Lines Adapted to the US Pacific Northwest, and Its Implications in Breeding Barley for Imidazolinone-Resistance

Sachin Rustgi[1*◗], Janet Matanguihan[1◗], Jaime H. Mejías[1,2], Richa Gemini[1], Rhoda A. T. Brew-Appiah[1], Nuan Wen[1], Claudia Osorio[1], Nii Ankrah[1], Kevin M. Murphy[1], Diter von Wettstein[1,3,4*]

1 Department of Crop & Soil Sciences, Washington State University, Pullman, Washington, United States of America, 2 Instituto de Investigaciones Agropecuarias INIA, Vilcún, Chile, 3 School of Molecular Biosciences, Washington State University, Pullman, Washington, United States of America, 4 Centre for Reproductive Biology, Washington State University, Pullman, Washington, United States of America

Abstract

Extensive application of imidazolinone (IMI) herbicides had a significant impact on barley productivity contributing to a continuous decline in its acreage over the last two decades. A possible solution to this problem is to transfer IMI-resistance from a recently characterized mutation in the 'Bob' barley *AHAS* (*acetohydroxy acid synthase*) gene to other food, feed and malting barley cultivars. We focused our efforts on transferring IMI-resistance to barley varieties adapted to the US Pacific Northwest (PNW), since it comprises ~23% (335,000 ha) of the US agricultural land under barley production. To effectively breed for IMI-resistance, we studied the genetic diversity among 13 two-rowed spring barley cultivars/breeding-lines from the PNW using 61 microsatellite markers, and selected six barley genotypes that showed medium to high genetic dissimilarity with the 'Bob' *AHAS* mutant. The six selected genotypes were used to make 29–53 crosses with the *AHAS* mutant and a range of 358–471 F_1 seeds were obtained. To make informed selection for the recovery of the recipient parent genome, the genetic location of the *AHAS* gene was determined and its genetic nature assessed. Large F_2 populations ranging in size from 2158–2846 individuals were evaluated for herbicide resistance and seedling vigor. Based on the results, F_3 lines from the six most vigorous F_2 genotypes per cross combination were evaluated for their genetic background. A range of 20%–90% recovery of the recipient parent genome for the carrier chromosome was observed. An effort was made to determine the critical dose of herbicide to distinguish between heterozygotes and homozygotes for the mutant allele. Results suggested that the mutant can survive up to the 10× field recommended dose of herbicide, and the 8× and 10× herbicide doses can distinguish between the two *AHAS* mutant genotypes. Finally, implications of this research in sustaining barley productivity in the PNW are discussed.

Editor: Tianzhen Zhang, Nanjing Agricultural University, China

Funding: This work was supported by the Washington Grain Commission Grant #13C-3019-3590. The funders had no role in study design, data collection and analysis, decision to publish, or preparation of the manuscript.

Competing Interests: The authors have declared that no competing interests exist.

* Email: rustgi@wsu.edu (SR); diter@wsu.edu (DvW)

◗ These authors contributed equally to this work.

Introduction

Barley is a short-season, early maturing annual grain crop with some degree of tolerance to drought and salinity, which allows its production in a wide range of climatic zones including both irrigated and dryland production areas [1]. Barley is the third major feed grain crop produced in the United States, after corn and sorghum [2]. Spring barley is a preferred rotational crop in the US Pacific Northwest (PNW) for two- or three-year rotations with winter wheat (*Triticum aestivum* L.), pea (*Pisum sativum* L.), lentil (*Lens culinaris* L.), or fallow [1,3]. A cropping system like spring wheat-fallow or winter wheat-fallow is generally practiced in the PNW, which encourages populations of summer and winter

annual-grassy weeds, respectively [4]. These weed cycles can be broken with a winter wheat-barley-fallow rotation [6]. Depending upon the management practices followed in an area, this cropping system results in a buildup of crown and root rot pathogens including *Fusarium*, *Rhizoctonia* and *Phythium* species, which frequently result in significant yield losses [5]. Similarly, in an eight-year dryland no-till cropping systems experiment conducted near Ritzville, Washington, a significant drop in the incidence of bare patches caused by *Rhizoctonia* was observed by adaptation of a two-year spring wheat rotation with spring barley. A significant gain in average yield of spring wheat was also documented with this change [5]. Likewise, in continuous cropping systems, spring barley fits well after winter wheat because the time interval between harvesting the barley crop and planting winter wheat is usually sufficient to allow soil moisture recharge to support an optimum winter wheat stand [6,7]. In addition to its agronomical relevance and commercial value as a feed or malt grain crop, barley is regaining popularity as human food due to the antioxidant and β-glucan (dietary fiber) rich grains [8,9]. Despite its agronomical importance and rising market value, barley acreage in the US has declined from 8.94 million acres in 1991 to 3.48 million acres in 2013 [10]. In Washington State alone the acreage has dropped significantly from 500,000 acres planted in 1999 to 180,000 acres in 2013 [10].

The significant drop in barley acreage during the last two decades can be partly attributed to the wide scale application of imidazolinone herbicides in combination with the introduction of imidazolinone (IMI)-resistant crops, and the residual activity of the herbicides of this family [1]. The decline in acreage can also be explained by the overlapping distribution of regions under barley cultivation in the PNW and the regions under extensive application of Imazamox (Beyond) and/or Imazethapyr (Pursuit) [11]. Collectively, the major reason for the decline in barley acreage is its sensitivity to commonly used herbicides. Many of the widely used herbicides, which impose barley plant-back restrictions, belong to the group B herbicides [12]. Thus, identification of IMI-resistant mutant(s) in barley and its transfer to relevant feed, food and/or malting barley cultivars adapted to the PNW is of extreme importance to sustain barley productivity in this region and elsewhere.

The group of herbicides belonging to the imidazolinone family targets acetohydroxyacid synthase (AHAS) or acetolactate synthase (ALS), an octameric enzyme with four catalytic and four regulatory subunits [13]. The enzyme AHAS catalyses two parallel reactions in the synthesis of branched chain amino acids. The first reaction is condensation of two pyruvate molecules to yield acetolactate leading to the production of valine and leucine, and the other reaction is the condensation of pyruvate and α-ketobutyrate that give rise to acetohydroxybutyrate, which subsequently results in the synthesis of isoleucine [14]. The AHAS-inhibiting herbicides are known to bind at the substrate access channel, blocking the path of substrate to the active site. When AHAS is inhibited, deficiency of the amino acids (valine, leucine and isoleucine) causes a decrease in protein synthesis, which in turn slows down the rate of cell division. This process eventually kills the plant, with symptoms observed in meristematic tissues where biosynthesis of amino acids primarily takes place [12]. In most cases, resistant plants have a reduced sensitivity to these herbicides due to amino acid substitution(s) in AHAS that give rise to catalytically active isoforms of the enzyme. Most AHAS isoenzymes resistant to the herbicides carry substitutions for the amino acid residues Ala122, Pro197, Ala205, Asp376, Trp574 or Ser653 (amino acid numbering refers to the sequence in *Arabidopsis thaliana*) [13]. Amino acid substitutions at Ala122 and Ser653 confer high levels of resistance to imidazolinone herbicides, whereas substitutions at Pro197 endow high level of resistance against sulfonylureas and provide low-level resistance against imidazolinone and triazolopyrimidine herbicides. Likewise, substitutions at Trp574 provide high levels of resistance to imidazolinones, sulfonylureas and triazolopyrimidines, while substitutions at Ala205 confer resistance against all AHAS-inhibiting herbicides [15].

In the case of barley, there is no IMI-resistance reported for any of the varieties cultivated in the PNW. Thus, introduction of a barley variety with IMI-resistance will provide greater flexibility to barley as a rotational crop after winter wheat [11]. An IMI-resistant mutant was earlier isolated by our group from an extensive screening of two million seeds of 'Bob' treated with sodium azide. Molecular characterization of the mutant revealed an amino acid substitution in the substrate access channel of the catalytic subunit of the AHAS enzyme, changing a serine to asparagine at amino acid location 653 [16]. This mutation in the substrate access channel does not allow imazamox to block the path of the substrate to the active site, thus allowing the plant to survive with no obvious effects on plant fitness even when exposed to field recommended dose of herbicide used on the IMI-tolerant winter wheat (i.e., 0.118 L/Acre Beyond with 1% non-ionic surfactant).

In view of the agronomical importance of this trait and the great demand for IMI-resistant barley cultivars in the PNW, this study was undertaken with the following objectives: i) estimation of genetic diversity among the 13 two-rowed spring barley cultivars/breeding-lines adapted to the US PNW using 61 microsatellite markers to select for lines showing sufficient genotypic differences with the 'Bob' AHAS mutant, to be used in the crossing program; and (ii) transfer the IMI-resistance to selected food, feed and malting barley cultivars using marker-assisted foreground and background selections.

Materials and Methods

Plant material

Seeds of the 13 two-rowed spring barley cultivars or breeding lines were procured from the variety testing program at the Washington State University (WSU), Pullman. Of the 13 genotypes selected for genetic analysis, eight are feed barleys, three are food barleys and the remaining two are malting barleys (Table 1).

Crossing scheme

To transfer IMI-resistance from the 'Bob' AHAS mutant, crosses were made between the mutant and each of the six barley genotypes, selected on the basis of genetic diversity analysis performed using microsatellite markers specific to chromosome 6H (see later for details). Twenty nine to fifty three crosses were made per genotype combination during the summers of 2012 at the Spillman Agronomy Farm (WSU, Pullman) and a range of 358 to 471 F_1 grains were harvested. The F_1 plants were propagated in 48-well flats in the glasshouse to obtain F_2 seeds. Subsequently, a range 2158 to 2846 F_2 plants per cross combination were evaluated for herbicide resistance by spraying two-week-old seedlings with 0.236 L/Acer Beyond (twice the field recommended dose applied to the IMI-tolerant winter wheat) with 1% methylated seed oil (MSO). A month after herbicide spray, the survivors (i.e., resistant plants) were evaluated for plant height as an indicator of early vigor and the 250 top ranking lines per cross combination were raised to maturity for seeds. Later, one to three F_3 plants each from the six most vigorous F_2 lines per cross combination were evaluated for the genotype at the AHAS locus by DNA sequencing, and the percent recovery of the recipient parent genome using chromosome 6H-specific SSR markers.

DNA extraction and PCR amplification

DNA was extracted from the one-month-old seedlings of each of the 13 barley genotypes, and the two-week-old seedlings of the F_3 progeny of selected F_2 lines, using the modified CTAB (Cetyl Trimethyl Ammonium Bromide) method [17]. DNA was treated by RNAse and purified by phenol extraction (25 phenol: 24 chloroform: 1 isoamyl alcohol, v/v/v) followed by ethanol precipitation [18]. Concentration of DNA samples was adjusted

Table 1. List of two-rowed spring barley varieties/breeding lines used in the study.

Genotype	Pedigree	Class
Baronesse	([(Mentor×Minerva)×mutant of Vada] ×[(Carlsberg×Union)× (Opavsky×Salle)×Ricardo]) ×(Oriol×6153 P40)	hulled, feed barley (originally released as malting barley)
Bob	(Lewis somaclonal line)/Baronesse	hulled, feed barley
Champion	Baronesse/Camas	hulled, feed barley
Clearwater	Baronesse*2/pmut882//HB317 (CDC Dawn sib)	hulless, low phytate, food barley
Lenetah	94Ab12981/Criton	hulled, feed barley
Conrad	B1215/B88–5336	hulled, malting barley
Radiant	*ant29–667* (an induced mutant in Harrington)/Baronesse	hulled, malting barley, pro-anthocyanidine-free
Spaulding	Vanguard/Imber//Zephyr/3/ Heavyweight/4/VD403582	hulled, feed barley
WAS4	01WA-13862.3/Radiant	hulless, food barley
05WA-316.99	Baronesse/Spaulding	hulled, feed barley
Lyon	Baronesse/Spaulding	hulled, feed barley
07WA-682.1	WA 10701–99/AC Metcalfe	hulled, feed barley
Meresse	Merlin/Baronesse	hulless, food barley

to 50 ng μl^{-1} using *Hind* III digested λ DNA as a marker. DNA amplification was carried out on a C1000 thermal cycler (Bio-Rad Laboratories). The PCR reactions were performed in 20 µl reaction mixtures, each containing 50 ng template DNA, 0.25 µM primers, 200 µM dNTPs, 1.5 mM $MgCl_2$, 1×PCR buffer and 0.5 U Ex *Taq* DNA polymerase (TAKARA, Bio Inc.) using the following PCR profile: initial denaturation at 95°C for 3 min followed by 40 cycles at 95°C for 30 sec, 53–61°C (depending upon the primer pair used) for 30 sec (for primer details, cf. [19]), 72°C for 45 sec, and a final extension at 72°C for 5 min. The amplification products were resolved on 10% polyacrylamide denaturing gels followed by silver staining [20]. A hundred base pair ladder was used as a size marker (New England BioLabs, Inc., Beverly, USA). The amplified product/ allele sizes were determined using Fragment Size Calculator available at http://www.basic.northwestern.edu/biotools/ SizeCalc.html.

DNA sequencing and sequence analysis

To determine the genotype at the *AHAS* locus, genomic DNA extracted from the F_3 progeny of selected F_2 lines was amplified using the *AHAS* gene-specific sequence tagged site (STS) primers that flank the point mutation responsible for the IMI-resistance (for primer details, cf. [16]). The amplification product was resolved on 1% agarose gel. A 100-bp ladder was used as a size marker (New England BioLabs). The band of expected size was excised from the gel, and DNA was eluted from the band using the Geneclean kit following the manufacturer's instructions (MP Biomedicals). The eluted DNA was used as a template for the sequencing reaction using either forward or reverse primers in separate reactions. The sequencing reactions were carried out at the DNA Sequence Core, WSU, Pullman. Alignment of the DNA sequences was performed using the Vector NTI AdvanceTM 9.1 (Invitrogen).

Determination of the polymorphic information content (PIC) and genetic diversity

For each microsatellite or simple sequence repeat (SSR) locus, PIC was calculated using the following equation: $PIC = 1 - \Sigma(Pi)^2$, where Pi is the proportion of genotypes carrying the ith allele [21]. For dissimilarity analysis, null alleles were scored as zero (0) and other microsatellite alleles (length variants) were each scored in the form of single bands of expected sizes, which were later converted into the number of repeat units as allele codes (all modalities were given equal weight during the analysis). The numerical data thus obtained was used to calculate Sokal and Michener dissimilarity indices ($di-j$) [22]. The dissimilarity indices between pairs of accessions using genotypic data were calculated on the basis of the following equation: $di-j = (n11+n00)/(n11+n01+n10+n00)$, where n11 is the number of fragments present in both *i* and *j*, n01 and n10 is the number of fragments present in one accession but absent in the other, and n00 is the number of fragments absent in both *i* and *j*. From the obtained distance matrix, an un-weighted Neighbor-Joining tree [23] was computed using the Darwin 5.0 software [24] and branch robustness was tested using 1000 bootstraps.

Enzyme extraction

Soluble proteins from 'Bob' and 'Bob' *AHAS* mutant were extracted following Singh et al. [25], with minor modifications. Briefly, two batches of 500 mg of the fresh leaf tissue were pulverized each with 5 mL of the protein extraction buffer [consisting of 100 mM potassium phosphate buffer (pH 7.5), 10 mM sodium pyruvate, 5 mM $MgCl_2$, 5 mM EDTA, 100 µM flavin adenine dinucleotide (FAD) and 10% Glycerol], using a polypropylene mesh bag (supplied with the P-PER Plant Protein Extraction Kit, Thermo Scientific). After adding the extraction buffer to the leaf tissue, the bag was rubbed from the outside with a ceramic pestle until a homogeneous mixture of the tissue was obtained. Later, the lysate was suctioned from the bag using a pipette and placed into a 15 mL conical tube and centrifuged at 22,000×g for 20 min at 4°C. The supernatant was transferred to a new tube and mixed with an equal volume of saturated $(NH_4)_2SO_4$. The mixture was incubated on ice for 30 min, and then centrifuged at 4°C for 20 min at 22,000×g. The supernatant was discarded and the pellet containing protein was re-suspended in 700 μL of the buffer solution containing 50 mM potassium phosphate (pH 7.5), 100 mM sodium pyruvate, 10 mM $MgCl_2$, 1 mM EDTA, 10 μM FAD, 100 mM NaCl and 1 mM thiamine pyrophosphate (TPP).

After extraction, protein concentration was determined using Bradford colorimetric micro-assay by mixing 80 μL of protein extract with 20 μL of the Bradford reagent (containing 1 mL of concentrated Bradford solution in 4 mL of deionized water), and measuring absorbance at 590 nm wavelength. The presence of the enzyme in the extract was also confirmed by loading protein extracts on 10% sodium dodecyl sulfate (SDS) polyacrylamide gel. For this purpose 15 μL of protein extract was mixed with 3 μL of the loading buffer, and electrophoresed on polyacrylamide gel for 2 h at 120 volts. After electrophoresis, the gel was stained with Coomassie brilliant blue reagent (80% Coomassie and 20% methanol, v/v) for 24 h. A protein band of ~65 kDa was observed, which corresponds with the size of AHAS enzyme monomers, confirming its presence in the extract.

Colorimetric enzyme activity assay

Enzyme activity was tested by using five different doses of Beyond (i.e., 1×, 4×, 6×, 8× and 10× the field recommended dose applied on IMI-tolerant winter wheat) with 0.25% (v/v) nonionic. surfactant (NIS). Initial reaction was performed in 1.5 mL microfuge tube by adding 52 μL of enzyme (in extraction buffer containing the substrate and co-factors, see above for the buffer composition) to equal volume of herbicide and incubating the mixture at 37°C for 1 h to facilitate acetolactate production. Later, the reaction was stopped by adding 21 μL of 5% H_2SO_4, and incubating at 60°C for 15 min. After incubation, tubes were spiked with 175 μL of color change solution containing 0.32 g of NaOH, 0.12 g of 1-naphtol and 0.01 g of creatine in 4 mL of deionized water, and the mixture was re-incubated at 60°C for 15 min. After incubation, 200 μL sub-samples of the reaction mixture were added to a 96-well microtiter plate (Falcon cat#353077) to determine the enzyme activity by studying color change using a microplate reader spectrophotometer (Spectra Max, M2, Molecular Devices) at 520 nm wavelength.

Results and Discussion

Chromosomal assignment of the gene encoding catalytic subunit of barley AHAS enzyme

The AHAS holoenzyme (~548 kDa) consists of two halves where one half, known as the large or catalytic subunit, is comprised of a homotetramer of ~65 kDa polypeptides, and the second half, known as the small or regulatory subunit, consists of homo-tetramer/-pentamer of polypeptides of ~52 kDa each [16,26,27]. The regulatory subunit stimulates enzyme activity and is required for the feedback regulation of the branched-chain amino acid biosynthesis, whereas the catalytic subunit is solely responsible for the enzyme activity and is also the site of point mutation(s) that confers resistance against IMI-herbicides [16]. Due to the importance of the catalytic subunit in providing IMI-resistance, the genes encoding it have been studied in common wheat and assigned to group 6 chromosomes [6A (*imi3*), 6B (*imi2*) and 6D (*imi1*)], using nulli-tetrasomic lines [28]. Later, the genetic location of *imi1* gene on the long arm of chromosome 6D was determined using three mapping populations, namely Cashup/cv.

9804, Madsen/cv. 9804 and Opata 85/W7984 [28]. However, the genetic location of the *AHAS* gene in barley remains unknown. Therefore, we used the map location of the *AHAS* gene in wheat to decipher its location in barley, which is possible in this particular case due to the shared ancestry of the two genera, and high levels of synteny as well as colinearity between them [29]. The availability of common markers between wheat and barley maps allowed an approximation of the barley *AHAS* gene location on chromosome 6H (Fig. S1). Moreover, we used the complete *AHAS* gene sequence we had previously obtained to blast against the barley genomic DNA sequences available in the public domain (http://webblast.ipk-gatersleben.de/barley/). The BLASTn search (score = 2834 and E-value = 0.0) allowed unambiguous assignment of the gene to genetically anchored 'Morex' BAC contig numbered 40275 on chromosome 6H at 67.917 cM (Fig. S1). In addition, the initial genotyping of the F_3 progeny of selected F_2 lines (carrying the *AHAS* mutant allele in hetero-/homozygous state) from all six cross combinations with chromosome 6H specific microsatellite markers showed higher recovery rate (50–72%) of recipient parent alleles for markers mapping to the non-proximal long arm in comparison with the short arm and the centromeric region (37–58%) (see next section for details). This is an indication of suppressed recombination, likely due to selection for the trait of interest. Collectively, the *in silico* and experimental data strongly indicate that the gene encoding the catalytic subunit of the AHAS enzyme maps to the sub-centromeric region of the barley chromosome arm 6HL.

Polymorphism survey using chromosome 6H-specific microsatellite markers

The level of genetic diversity among 13 two-rowed spring barley cultivars/breeding lines adapted to the PNW was assessed using microsatellite or simple-sequence repeat (SSR) markers specific to the barley chromosome 6H. Out of the 13 genotypes selected for the analysis, eight are feed barleys, three are food barleys and two are malting barleys (Table 1). The 61 SSR markers selected for the analysis are evenly distributed along the entire length of chromosome 6H (Table 2) [19]. The major reason behind selecting markers from chromosome 6H lies in the fact that this chromosome carries the gene encoding for the catalytic subunit of acetohydroxyacid synthase (AHAS) enzyme and the mutation providing IMI-resistance (see above). It is known through trait-introgression studies that due to linkage-drag, it always takes longer (several backcrossing and selfing generations) to recover the recipient parent genotype for the carrier chromosome in comparison with non-carrier chromosomes, which assort independently [30]. Thus, to identify the rare recombinant(s) carrying the precise gene introgression in the early generation, it is important to screen large segregating populations with the markers derived from the carrier chromosome.

Of the 61 markers used for analysis, two markers (*HvWaxy4* and *GBM1319*) were non-functional (no amplification observed in any of the genotypes), three markers (*HVM22*, *GBM1215* and *GMS6*) were monomorphic, and 56 markers were polymorphic. These polymorphic markers allowed us to detect 62 loci. Of the 56 polymorphic markers, one marker detected three loci, another marker detected two loci, while the remaining 52 markers each detected a single locus. (Fig. S2). These 56 markers amplified 1 to 12 alleles from the 13 barley genotypes (Table 2). The number of alleles detected by each marker and their frequencies were used to calculate the polymorphic information content (PIC) of the marker. The PIC value, which depends on the number of detectable alleles and the distribution of their frequency, indicates the marker's utility in detecting polymorphism within a population

Table 2. List of chromosome 6H specific microsatellite markers used for the genetic diversity analysis and marker-assisted background selection, their repeat elements, respective locations in the genetic-linkage map [17], number of alleles detected and their polymorphic information content (PIC).

Marker/loci	Repeat element	Position (cM)	PIC	Allele#
Af166121	$(A)_{10}$	0.00	0.38	3
84c21j33	$(T)_{10}$	0.00	0.14	2
Bmac0316	$(AC)_{19}$	7.16	0.80	7
scssr09398	$(CTT)_9$	7.16	0.43	2
Bmag0500	$(AG)_6CG(AG)_{29}(AGAGGG)_3(AG)_6$	31.65	0.72	6
GBM1270	$(GCC)_8$	36.52	0.56	4
GBM1355	$(GCA)_7$	40.43	0.14	2
GBM1212	$(AGG)_5$	55.10	0.14	2
Bmag0807	$(TC)_{18}$	56.11	0.39	4
Bmag0173	$(CT)_{29}$	57.79	0.86	9
GBM1423	$(CGGCTC)_5$	58.46	0.36	2
HVM31	$(AC)_9$	60.90	0.57	3
Bmac0040	$(AC)_{20}$	61.07	0.77	7
Bmag0174	$(AG)_9$	61.40	0.72	6
EBmac0560	$(AC)_7$	61.70	0.77	5
GBM1267	$(TTG)_9$	61.70	0.69	4
Bmac0018	$(AC)_{11}$	61.79	0.77	7
Bmac0144	$(AT)_4(AC)_{20}$	61.79	0.91	12
Bmac0175	$(AC)_{12}$	61.79	0.57	3
GBM5012	$(ACG)_7$	61.95	0.49	2
Ebmac0674	$(TG)_{18}(AG)_9$	61.96	0.46	3
EBmac0874.1	$(CA)_8AA(CA)_4CG(CA)_8AA(CA)_7AA(CA)_9(TA)_8$	61.96	0.67	4
EBmac0874.2	$(CA)_8AA(CA)_4CG(CA)_8AA(CA)_7AA(CA)_9(TA)_8$	61.96	0.80	7
HVM65	$(GA)_{10}$	62.11	0.71	6
Bmag0009	$(AG)_{13}$	62.21	0.57	3
Ebmac0639	$(TG)_5(TG)_8$	62.21	0.57	4
EBmatc0028.1	$(ATC)_3N3(ATC)_6$	62.21	0.50	2
EBmatc0028.2	$(ATC)_3N3(ATC)_6$	62.21	0.58	4
EBmatc0028.3	$(ATC)_3N3(ATC)_6$	62.21	0.67	5
Bmac0297.1	$(AC)_9(AC)_{10}$	62.23	0.58	4
Bmac0297.2	$(AC)_9(AC)_{10}$	62.23	0.77	5
Bmac0297.3	$(AC)_9(AC)_{10}$	62.23	0.46	3
Bmac0047	$(AC)_{16}$	62.27	0.47	2
Bmac0127	$(AC)_{26}$	62.27	0.47	2
GBM1389	$(GCCT)_5$	62.27	0.26	2
HVM14	$(CA)_{11}$	62.28	0.50	2
Bmag210	$(AG)_7T(AG)_{13}$	62.28	0.57	3
HVM34	$(GA)_{10}$	62.43	0.36	2
HVM74	$(GA)_{13}$	62.66	0.56	3
Bmag0003.1	$(AG)_{28}$	63.49	0.78	6
Bmag0003.2	$(AG)_{28}$	63.49	0.59	6
Bmag0004	$(AG)_{14}$	64.71	0.91	12
BMG001	$(G)_{10}$	64.71	0.52	3
Bmgt0001	$(GTTTTT)_5$	64.71	0.27	3
scssr02093	$(GA)_{18}$	67.20	0.49	2
Bmag0344	$(CT)_{10}GT(CT)_{16}$	67.80	0.66	6
GBM1400	$(CACG)_5$	67.80	0.27	3
Bmac0251	$(AC)_{12}A(AC)_{13}$	69.25	0.38	3

Table 2. Cont.

Marker/loci	Repeat element	Position (cM)	PIC	Allele#
Bmag0613	(GA)$_{17}$	69.82	0.66	6
Bmac0218	(AC)$_{14}$	71.99	0.79	6
Bmac602	(AC)$_9$AT(AC)$_7$(AG)$_9$	75.42	0.49	4
GBM1256	(GA)$_8$	75.46	0.52	3
HVM11	(GGA)$_3$(GGA)(GAA)$_2$	88.47	0.56	3
scssr05599	(AAG)$_4$	96.34	0.63	4
GBM1140	(ATC)$_5$	97.31	0.52	3
GBM1356	(GTG)$_7$	98.38	0.57	3
scssr00103	(GT)$_{10}$	105.26	0.59	3
GBM1274	(TCG)$_7$	123.45	0.27	3
GBM1275	(TGC)$_7$	124.29	0.15	2
GBM1276	(TGC)$_7$	124.29	0.46	3
GBM1087	(AGG)$_5$	127.70	0.54	3
GBM1404	(TATG)$_5$	129.76	0.46	3

[21]. The PIC values ranged from 0.14 (*84c21j33, GBM1355, GBM1212*) to 0.91 (*Bmac0144, Bmag0004*) (Table 2). When the PIC value for each marker was plotted against its location on the genetic-linkage map, it showed a multimodal distribution, with low levels of PIC values observed at the sub-telomeric and centromeric regions of the chromosome (Fig. 1). This distribution shows the level of nucleotide diversity along the entire length of the chromosome and suggests the possibility of identifying a polymorphic marker from a specific region of the chromosome. The type of repeat element, chromosomal location, number of repeat units, and sequence of repeat element can influence the level of nucleotide diversity. Thus, we classified the SSR markers according to the type of repeat element into simple and compound repeats. Whenever two or more repeat runs were present adjacent to each other or microsatellite array of same repeat was interrupted by non-repeat base(s) the repeat was classified as compound repeat. We further classified simple repeats into mono-, di-, tri-, tetra-, penta- and hexa-nucleotide repeats and reported their mean PIC values. Compound repeats in general showed higher PIC values in comparison with simple repeats, whereas, among simple repeats the di-nucleotide repeats showed highest PIC values (Table 3). To distinguish the effect of chromosomal location from the microsatellite element type, the PIC values obtained for different microsatellite types (i.e, mono-, di-, tri, tetra-, hexa-nucleotide repeats and compound repeats) were individually plotted against their respective location on the genetic-linkage map. The analysis revealed reduced levels of nucleotide diversity in the peri-centromeric region for di-nucleotide repeats and in sub-telomeric regions for the tri-nucleotide repeats (Fig. S3). However, it was apparent from the analysis that the number of repeat units does not have any influence on the number of alleles detected per locus.

Preferential association of different SSR elements of variable sequences and lengths (i.e., total number of repeat units) with physical chromosome landmarks like the centromere, telomere, heterochromatin and euchromatin, and their relevance in determining chromosome function, has been extensively documented in literature [31–33]. Thus, the influence of the genomic locations of these markers on their evolvability and/or divergence is plausible. For instance, a low level of nucleotide diversity was observed in the proximal chromosomal regions of both *Triticum aestivum* and wild emmer (*Triticum turgidum* ssp. *dicoccoides*) [34]. Moreover, the effect of direct or indirect selection on genomic diversity is also a likely cause of observed fluctuations in genetic diversity along the chromosome length. Similar regions of low diversity associated with sites of domestication loci and genomic regions under selection in later breeding efforts were reported in maize [35]. Since barley genotypes selected in this study were bred in the PNW, they share some common ancestry. Thus, the regions of low diversity observed in the present study are likely to represent the genomic regions providing adaptive advantage to these genotypes. However, this aspect needs further investigation.

Assessment of genetic diversity among barley genotypes

The genetic relationships among the barley genotypes were evaluated based on the combined profiles of 62 SSR loci. The genetic dissimilarity coefficient (GD) values were calculated for all possible 78 pairs of genotypes, and ranged from 0.339 (between Bob and Baronesse) to 0.806 (between WAS4 and Conrad) with a mean of 0.601 (Fig. S4). All 13 genotypes were grouped into three clusters (Fig. 2). Two clusters were further subdivided into two sub-clusters each. As expected on the basis of pedigree information (Table 1), Bob, Baronesse, Meresse, 05WA-316.99 and Clearwater formed a single cluster (middle), where the first three genotypes grouped into one sub-cluster and the latter two genotypes grouped into the other sub-cluster. Clustering of these genotypes in a single group can be explained by the presence of Baronesse in their lineages. The cultivars Radiant, Champion, Lenetah, 07WA-682.1, WAS4 and Conrad formed another cluster (top), where the first five genotypes formed a sub-cluster and Conrad alone formed a sub-cluster. The first sub-cluster was further divided into two sub-sub clusters, the former containing Radiant, Champion and Lenetah, and the later containing 07WA-682.1 and WAS4. The remaining two genotypes Spaulding and Lyon formed a separate cluster (bottom), which is well justified due to the Spaulding lineage of Lyon. The above diversity analysis proved useful in selecting lines to cross with the Bob *AHAS* mutant to transfer IMI-resistance, and will also prove useful in future breeding efforts where these lines will be used. Nevertheless, Baronesse has been extensively used in barley breeding programs in the PNW; the

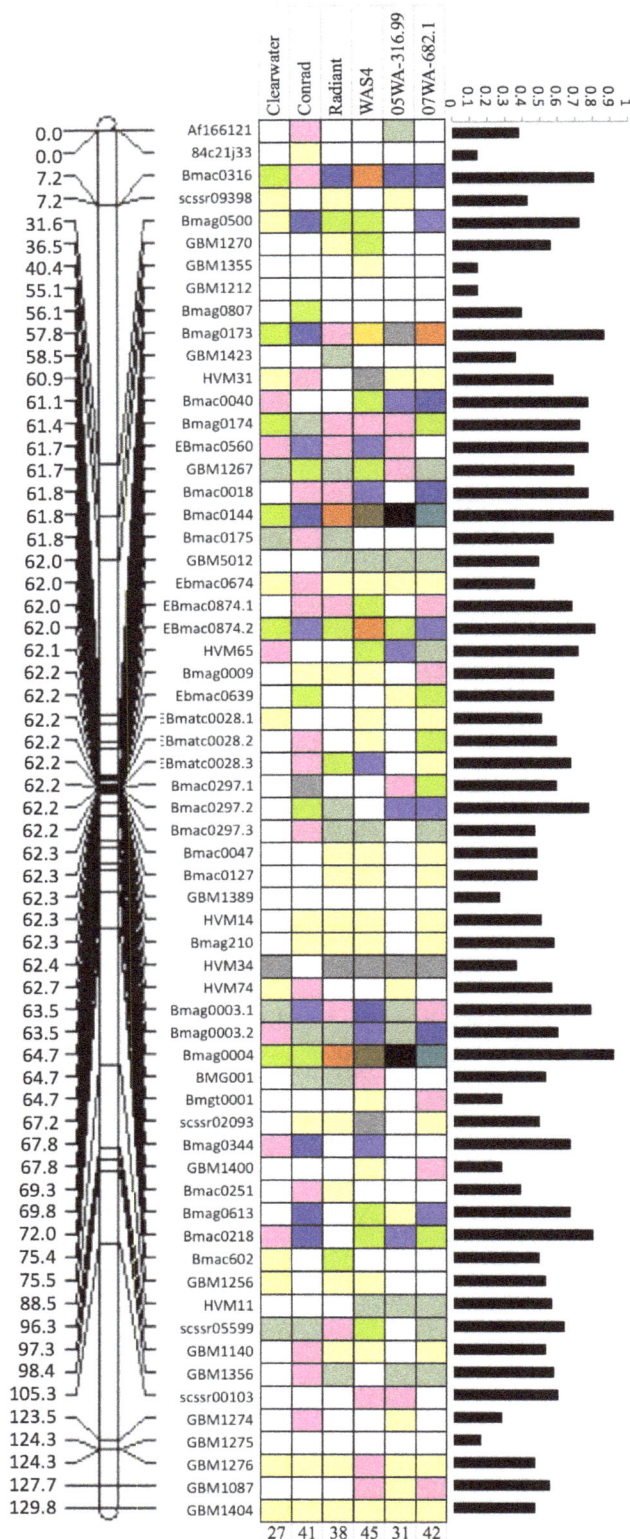

Figure 1. Genetic linkage map of chromosome 6H showing the respective locations of 61 microsatellite markers used in the present study (left). Various alleles detected from six barley genotypes used for crossing with the Bob *AHAS* mutant are indicated by colored boxes (middle), where each color represents a unique allele and the white color represents the 'Bob'-type allele. The total number of polymorphic markers identified per genotype pair with the Bob mutant

is shown below. The PIC value calculated for each marker was plotted against its location on the genetic linkage map (right) to indicate the level of nucleotide diversity observed using 13 barley genotypes, and its distribution along the entire length of chromosome 6H.

results clearly demonstrated high level of genetic diversity among studied genotypes, which is very important for the success of any breeding program. Thus, this study uniquely provides information about the genetic makeup of cultivars/breeding lines developed in the US PNW.

In summary, the polymorphism survey and diversity analysis i) allowed determination of genetic relationships of barley genotypes adapted to the US PNW; ii) provided data to make informed selection of barley genotypes used for crossing with the Bob *AHAS* mutant; iii) allowed identification of the most divergent pair of genotypes with the Bob mutant to be used for the genetic mapping of the *AHAS* gene; iv) allowed identification of the polymorphic markers for each pair of genotypes with Bob mutant to uniquely track and reconstitute the genetic-background of the recipient genotype; and v) allowed determination of the level of nucleotide diversity along the entire length of the barley chromosome 6H. This information not only proved useful during the present study but will also prove useful in later studies.

Determination of the critical dose of herbicide

From previous experience we know that the 0.118 L/acre dose of Beyond is sufficient to distinguish the susceptible barley genotypes from the resistant ones [16]. However, a critical herbicide dose, which could discriminate between the heterozygous and homozygous states of the *AHAS* mutation, remains unknown. Thus, in the present study, an attempt was made to determine the critical herbicide dose by spraying 0.118, 0.236 and 0.295 L/acre doses of Beyond on the segregating F_2 population derived from WAS4×Bob mutant cross. A non-significant deviation from the 2:1 segregation ratio (at $p<0.05$) of surviving vs dead plants was observed at each herbicide dose, which indicates the semi-dominant nature or dominant transmission of this mutation with incomplete penetrance (see next section for details). Subsequently, an effort was made to determine the maximum dose of herbicide, which can be tolerated by the IMI-resistant AHAS isoform. In order to achieve this objective, crude enzyme extracted from the leaf tissues of the Bob *AHAS* mutant was fed with the substrate (pyruvate) in presence of the increasing concentrations of the herbicide (see Materials and Methods). The assay suggested that the mutant enzyme can survive up to 1.18 L/acre Beyond that is 10 times field recommended dose applied on the IMI-tolerant winter wheat (Fig. 3). The assay also allowed discrimination of homozygotes from heterozygotes at 8× and 10× field recommended doses of the herbicide, displayed in the test by the intensity of red color as determined by the spectrophotometer. The heterozygotes took longer to produce same intensity of color that homozygotes produced in shorter duration of time (data not shown). However, these high doses of herbicide are impractical for use in glasshouse and field trials. In actual field conditions, the plant only receives a maximum of 0.236 L/acre dose, especially in the overlapping areas. Thus, for rest of the analyses, we used 0.236 L/acre herbicide dose.

Collectively these results suggested that the mutant AHAS enzyme can survive up to 10× field recommended dose of herbicide, which makes it unlikely to find a critical herbicide dose that can discriminate homozygotes from heterozygotes at the *AHAS* locus.

Table 3. Microsatellite markers classified according to repeat element type.

Repeat type	SSR markers used	Mean PIC	Number of repeats
Simple	45	0.41	-
Mononucleotide	3	0.35	10
Dinucleotide	24	0.63	7 to 29
Trinucleotide	13	0.43	4 to 9
Trtranucleotide	3	0.33	5
Hexanucleotide	2	0.31	5
Compound	17	0.61	-

Number of simple sequence repeats (SSRs) or microsatellites falling in each category is listed and the range of alleles detected by SSRs in these categories and their average PIC (polymorphic information content) values are shown.

Transfer of the IMI-resistance to other barley cultivars

A large collection of recombinants was screened in order to transfer IMI-resistance to selected genotypes in a single generation, and to identify rare recombinants carrying a small chromosomal segment with the gene of interest introgressed in the desired genetic background (Table 4). This will alleviate the need of backcrossing and avoid overriding the 'Breeder's Code of Ethics'. As mentioned in the Materials and Methods, the F_1s were grown to obtain F_2 seeds and a range of 2158 to 2846 F_2 lines per cross combination were evaluated for the presence of the mutant allele. This has been achieved by spraying the F_2 populations with $2\times$ equivalent to the field recommended dose of Beyond used on the IMI-tolerant winter wheat (i.e., 0.236 L/Acre Beyond with 1% methylated seed oil), and by phenotyping the resistant plants for early vigor a month after spraying with the herbicide. The expected 3:1 ratio of resistant vs susceptible plants, an indicative of the dominant nature of the mutation was not observed with any of the six segregating populations. Instead, the crosses between WAS4, Radiant, and Clearwater with the Bob mutant showed a

2:1 segregation ratio of resistant vs susceptible plants (>0.05 probability). Collectively, the observed segregation ratios obtained from the greenhouse herbicide tests of the six segregating populations, at the best suggested a semi-dominant nature of the mutation or incomplete penetrance of the trait (Table 4). This low trait penetrance could be explained due to the cumulative effect of a number of factors like genetic differences for leaf and/or culm wax coating in the parental genotypes of a population, though this possibility needs further investigation.

The semidominant nature of the mutant prompted us to determine the genotype at the *AHAS* locus (the foreground selection) by DNA sequencing of the *AHAS* gene fragment from 1 to 3 F_3 lines each from the six most vigorous F_2 plants selected per cross combination (Figs. 4 and 5). Although, an allele-specific agarose based assay exists for genotyping of segregating populations for the *AHAS* mutant allele, it is unsuitable for use in this situation due to its dominant nature (i.e., incapability of distinguishing between a heterozygote and a mutant type homozygote) [16]. Later, the six F_3 plants showing the *AHAS* mutant allele in homo- or heterozygous state were selected to

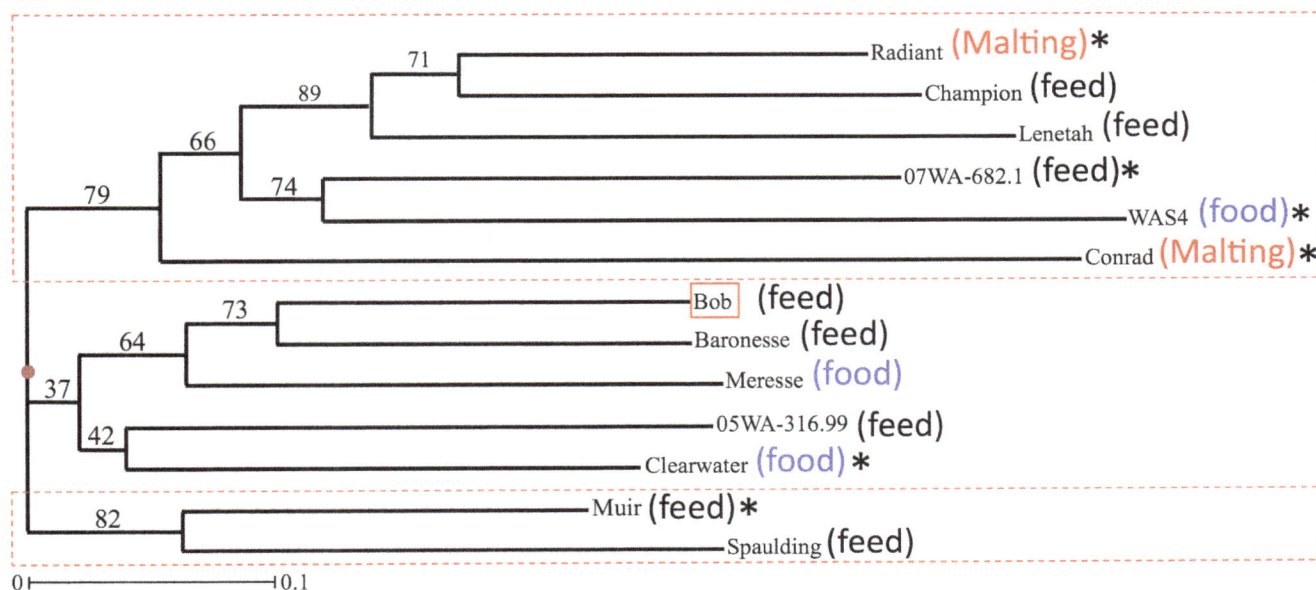

Figure 2. A dendrogram showing the clustering pattern of the 13 barley genotypes (see Table 1 for genotype details). Genetic distances for the dendrogram were estimated using SSR (simple sequence repeat) polymorphism data. Bootstrap values are indicated at each node. The genotypes selected for crossing with the Bob *AHAS* mutant are marked with asterisk.

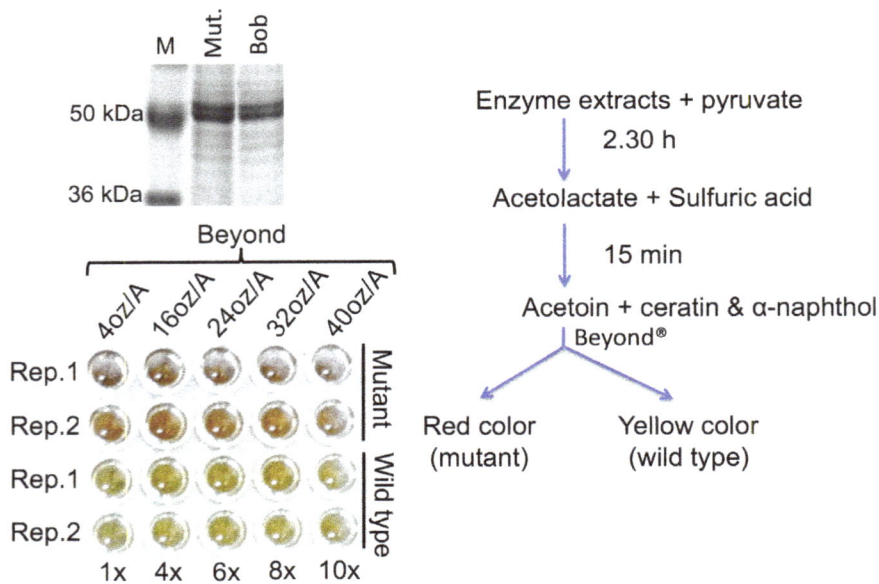

Figure 3. Results of the *in vitro* colorimetric enzyme activity assay performed for AHAS enzyme in the presence of inhibiting concentrations of an imidazolinone herbicide Beyond. Upper left, the crude enzyme was extracted from the leaves of 'Bob' and 'Bob' *AHAS* mutant and loaded on 10% SDS-polyacrylamide gel to check for the presence of the AHAS enzyme in the extracts. The presence of the enzyme in the extract was confirmed by a ~65 kDa protein band on the gel. Lower left, the enzyme activity assay further confirms the presence of enzyme in the extracts from wild type and mutant. In the assay, the enzyme was fed with pyruvate (substrate) in presence of increasing concentrations of Beyond. The assay clearly showed that the mutant AHAS enzyme can survive up to 1.18 L/acre dose of herbicide and show equal activity if measured 15 min after addition of the color change solution (for more details, see Materials and Methods). The assay was performed as summarized in the line-diagram on the right.

check for carrier chromosome recovery using 10–12 SSR markers specific to barley chromosome 6H. A range of 20 to 90% recovery of the recipient parent genome for the carrier chromosome was observed in the different cross combinations (Fig. 5). Collectively, this pilot study clearly demonstrates the feasibility of transferring IMI-resistance to desired barley genotypes in a single generation with the possibility of finding lines showing good recovery of the recipient parent genome.

Conclusion

Results of the study are of high significance not only to growers in the Pacific Northwest but also to growers in other parts of the US and the world, wherever IMI-herbicides are applied and IMI-resistant crops are cultivated. In this study we determined the genetic diversity among 13 barley cultivars/breeding lines, which benefitted the present study and is expected to prove useful in future breeding efforts. Chromosomal localization of the gene encoding the catalytic subunit of the barley AHAS enzyme will

Table 4. List of the number of crosses made, the F_1 seeds obtained per cross combination and the F_2 lines screened for imidazolinone (IMI)-resistance.

Female	Male	Crosses	F_1 harvested	F_2 sampled	F_2 screened against herbicide	
					Susceptible	Resistant
Feed						
05WA-316.99	Bob mutant	53	471	2815	856	1959
07WA-682.1	Bob mutant	29	394	2251	813	1438
Malting						
Radiant*	Bob mutant	37	358	2130	669	1461
Conrad	Bob mutant	35	445	2671	806	1865
Food						
Clearwater*	Bob mutant	38	394	2336	790	1546
WAS4*	Bob mutant	38	403	2342	736	1606
Total		230	2465	14545	4670	9875

*Fitted in 2:1 (resistant vs susceptible) segregation ratio at 0.05 significance level.

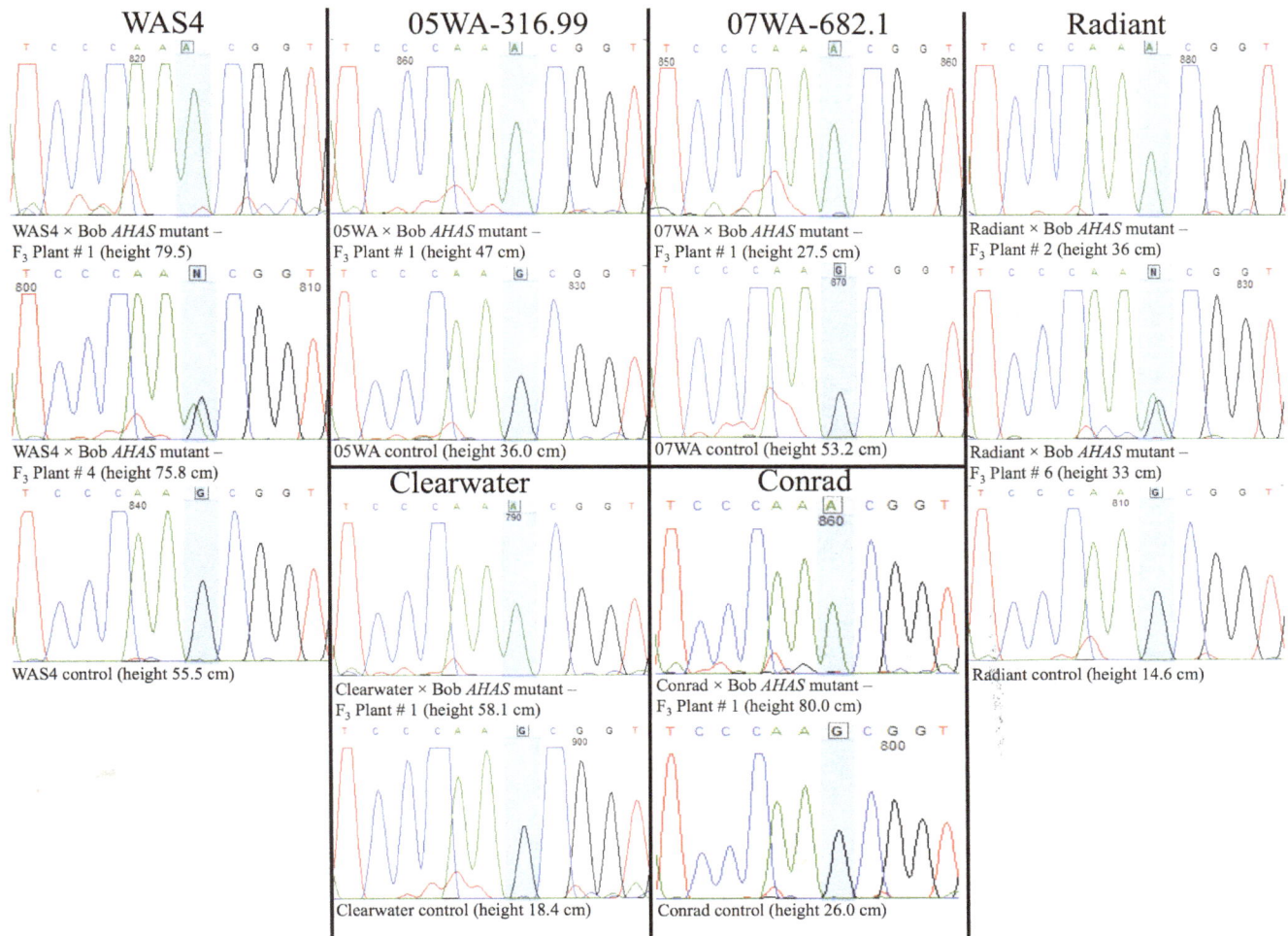

Figure 4. A part of the DNA sequence of the *AHAS* gene showing the point mutation responsible for IMI-resistance (highlighted in blue). The DNA sequencing results clearly demonstrated the transfer of IMI-resistance to two feed barleys, 05WA-316.99 and 07WA-682.1, two food barleys, Clearwater and WAS4, and two malting barleys, Radiant and Conrad.

also prove useful in future gene-transfer studies leading to the development of herbicide-resistant cultivars with other agronomically important traits. Determination of the working dose of herbicide used for phenotypic screening of this trait will be used in future breeding efforts to transfer IMI-resistance. This pilot study with a limited number of selected F_2 lines shows that it is possible to identify genotypes showing good recovery of the recipient parent genome by screening large $F_{2:3}$ populations and following a strategic selection scheme (Fig. 5).

Our future objective is to take the recently developed IMI-resistant food, feed and malting barley genotypes from the glasshouse to the field by i) screening large numbers of F_3 families, representing the 250 top ranking F_2 lines selected per cross combination, based on their vigor a month after herbicide spray, for their genetic backgrounds using DNA markers; ii) fixing heterozygosity (which confounds phenotypic evaluations) in selected lines by doubled haploid (DH) production; iii) field evaluation of the DH lines for their performance on herbicide residue and under spray trials. This will allow identification of barley lines showing more genetic proximity to their respective recipient parents.

For the first objective, F_3 seeds belonging to the 250 F_2 lines which survived the herbicide spray (at the rate of 0.236 L/acre Beyond with methylated seed oil) and showed early vigor a month after spray are currently being propagated in herbicide treated soil in the glasshouse. Cultivating plants on herbicide treated soil will allow elimination of susceptible individuals, which are expected in a segregating population at a proportion of one in four individuals. Genotype of the survivors will be determined at the AHAS locus by DNA sequencing following the procedure described above. It is of considerable importance to differentiate homozygotes from heterozygotes at the AHAS locus, as the two genotypic states at this locus are undistinguishable from each other using herbicide treatment alone. This is due to the semi-dominant nature of the AHAS mutation. The lines possessing the mutant allele(s) at the AHAS locus either in homo- or heterozygous state will be evaluated for their genetic background in a stepwise fashion first using 10 carrier chromosome (6H) specific microsatellite markers followed by 4 DNA markers per non-carrier chromosomes (2 markers per arm). The second step of background selection will be performed on the F_3 plants showing good carrier chromosome recovery in the first step. The lines showing good recovery of recipient parent genome will be converted to doubled haploids via

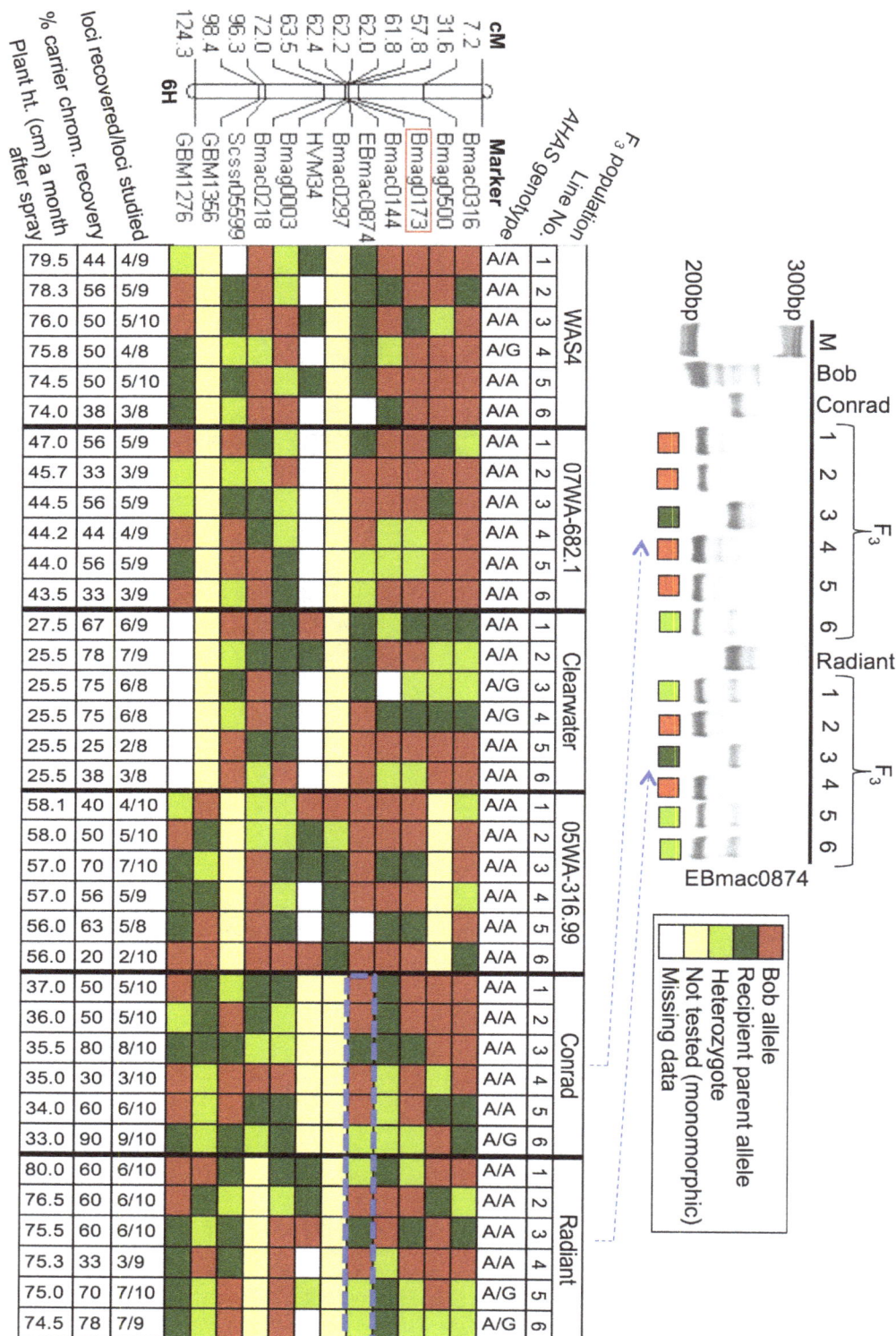

Figure 5. Diagrammatic representation of the results of the marker assisted background selection on the F₃ progeny of the selected F₂ lines. After foreground selection, the six F₃ lines per cross combination were evaluated for the recovery of recipient parent background by genotyping each line with 10 to 12 chromosome 6H-specific microsatellite markers. The markers were selected on the basis of polymorphism data obtained earlier during diversity analysis and their respective location on chromosome 6H. Map locations of selected markers are shown on the left. Each column in the picture represents a F₃ line and each cell represents the marker genotype in an individual. The marker genotype is represented by a color code: a) light green color denotes heterozygotes carrying marker alleles from both parents; b) dark green color denotes a marker allele similar to the recipient parent; and c) red color denotes the marker allele of the donor parent. Thus, a column with more dark and light green cells represent a genotype showing high percentage of the recipient parent genome, as observed for the 6th F₃ individual in the Conrad×Bob mutant cross, which had 90% recovery of the Conrad alleles. In contrast, a column showing more red cells represents

a donor-like carrier chromosome as examplified by the 6th F_3 individual in the 05WA-316.99×Bob mutant cross, which had only 20% recovery of the 05WA-316.99 parent genome.

a microspore culture based method following Kasha et al. [36]. The resultant doubled haploids will be evaluated for their performance in the field on herbicide residue and herbicide spray trials.

The major outcome of this project will be the development of IMI-resistant barley varieties and germplasm with a combination of beneficial traits including resistance for various biotic and abiotic stresses, higher grain yield and better quality. Moreover, adding imidazolinone resistance to barley cultivars adapted to the PNW will certainly improve the sustainability of barley, which is one of the best rotational crops for this region.

Supporting Information

Figure S1   Comparative mapping of wheat chromosome 6D and barley chromosome 6H to determine approximate location of the *AHAS* gene on chromosome 6H. (a) Genetic linkage map of wheat chromosome 6D. (b) Physical map of wheat chromosome 6D. Short arm is at the top, and the black circle indicates the centromere. Deletion-line breakpoints and fraction lengths (FLs) are indicated by the horizontal line to the left. Breakpoint positions are drawn approximately to scale. Darkened areas within chromosome arms are C-bands (cf. Endo and Gill. 1996. Journal of Heredity 87:295). (c) Microsatellite consensus map of barley chromosome 6H (modified from Varshney et al. 2007. Theoretical and Applied Genetics 114:1091). (d) Genetic location of the *AHAS* gene determined on the basis of *in silico* analysis. The gene was assigned to the 'Morex' BAC-contig #40275 anchored to the consensus genetic linkage map at 67.917 cM (cf. Close et al. 2009. BMC Genomics 10:582). (TIFF)

Figure S2   Consensus map of barley chromosome 6H (left; Varshney et al. 2007 Theoretical and Applied Genetics 114:1091) used to select simple sequence repeat (SSR)-markers for diversity analysis of two-rowed spring barley genotypes. Amplification profile of a few SSR markers used for analysis of barley genotypes are shown on left, and their locations on the genetic-linkage map are highlighted by red rectangles. Different SSR alleles are coded by different numbers and shown on the bottom of each SSR profile. (TIFF)

Figure S3   Polymorphic information content (PIC) values for different SSRs (classified according to repeat element type) are plotted against their respective location (in cM) on the genetic linkage map, showing variation in nucleotide diversity observed along the entire length of chromosome 6H. (TIFF)

Figure S4   The dissimilarity coefficient (GD) values calculated for 78 pairs of genotypes. High to low dissimilarity coefficient values with 'Bob' are shown on a red to white scale, with the highest value (0.726) shaded with the darkest red color, and the lowest value (0.435) in white. (TIFF)

Acknowledgments

The authors would like to thank Patrick Reisenauer, Dennis Pittmann, Shanshan Wen, Mingming Yang and Jayanthi Mohan for assistance in the glasshouse and laboratory experiments, and Dr. Ian Burke for useful discussions.

Author Contributions

Conceived and designed the experiments: SR KMM DvW. Performed the experiments: SR JM JHM RATB-A NW CO NA. Analyzed the data: SR RG JM JHM. Contributed to the writing of the manuscript: SR JM KMM DvW.

References

1. Garstang JR, Spink JH, Suleimenov M, Schillinger WF, McKenzie RH, et al. (2011) Cultural practices: Focus on major barley-producing regions. In: Ullrich SE, editor. Barley: Production, improvement, and uses, Wiley-Blackwell, Oxford, UK. 221–281.

2. Brester G (2012) Barley profile. Available: http://www.agmrc.org/commodities__products/grains __oilseeds/barley-profile/#. Accessed 2014 Mar 1.

3. Smiley RW, Ogg AG, Cook RJ (1992) Influence of glyphosate on Rhizoctonia root rot, growth, and yield of barley. Plant Disease 76: 937–942.

4. Janosky J, Young D, Schillinger W (2002) Economics of conservation tillage in a wheat-fallow rotation. In: Pacific Northwest Conservation Tillage Handbook Series No. 16, Chapter 10 - New Technology Access, Adaptation and Economics, February 2002. Available: http://pnwsteep.wsu.edu/tillagehandbook/chapter10/101602.htm. Accessed 2014 Mar 1.

5. Schillinger WF, Paulitz TC (2006) Reduction of Rhizoctonia bare patch in wheat with barley rotations. Plant Disease 90: 302–306.

6. Black AL, Bauer A (1986) Optimizing winter wheat management in the Northern Great Plains region. In: Proc. A Hands-On Workshop for Implementing Maximum Economic Yield (MEY) Systems, North Dakota Coop. Ext. Serv., Kirkwood Inn, Bismarck, ND, July 8–10, 1986.

7. Turner WJ, Yenish J, Line RF, Chen X (2001) Crop profile for barley in Washington. Available: http://www.ipmcenters.org/cropprofiles/docs/wabarley.html. Accessed 2014 Mar 1.

8. Baik BK, Newman CW, Newman RK (2011) Food uses of barley. In: Ullrich SE, editor. Barley: Production, improvement, and uses, Wiley-Blackwell, Oxford, UK. 532–562.

9. Newman RK, Newman CW (2008) Barley for food and health: Science, technology and products. Wiley Publishers, New York.

10. United States Department of Agriculture, Economic Research Service (2013) Available: http://www.ers.usda.gov/data-products/feed-grains-database/feed-grains-yearbook-tables.aspx.Accessed 2014 Mar 1.

11. Rustgi S (2013) Bringing barley back in crop rotation by breeding for imidazolinone resistance. Medicinal & Aromatic Plants 2: e148.

12. Shaner DL, Singh BK (1997) Acetohydroxyacid synthase inhibitors. In: Roe RM, Burton JD, Kuhr RJ, editors. Herbicide activity: Toxicity, biochemistry and molecular biology, IOS Press, Washington DC. 69–110.

13. Duggleby RG, McCourt JA, Guddat LW (2008) Structure and mechanism of inhibition of plant acetohydroxyacid synthase. Plant Physiology and Biochemistry 46: 309–324.

14. Duggleby RG, Pang SS (2000) Acetohydroxyacid Synthase. Journal of Biochemistry and Molecular Biology 33: 1–36.

15. Tan S, Evans RR, Dahmer ML, Singh BK, Shaner DL (2005) Imidazolinone-tolerant crops: history, current status, and future. Pest Management Science 61: 246–257.

16. Lee H, Rustgi S, Kumar N, Burke IC, Yenish JP, et al. (2011) Single nucleotide mutation in the barley *acetohydroxy acid synthase* (*AHAS*) gene confers resistance to imidazolinone herbicides. Proceedings of the National Academy of Sciences of the United States of America 108: 8909–8913.

17. Saghai-Maroof MA, Soliman KM, Jorgensen RA, Allard RW (1984) Ribosomal DNA spacer length polymorphisms in barley: Mendelian inheritance, chromosomal location, and population dynamics. Proceedings of the National Academy of Sciences of the United States of America 81: 8014–8018.

18. Sambrook J, Fritsch EF, Maniatis T (1989) Molecular Cloning: a laboratory manual, 2nd edition. Cold Spring Harbor Laboratory Press, Plainview, New York, USA.

19. Varshney RK, Marcel TC, Ramsay L, Russell J, Röder MS, et al. (2007) A high density barley microsatellite consensus map with 775 SSR loci. Theoretical and Applied Genetics 114: 1091–1103.

20. Tegelstrom H (1992) Detection of mitochondrial DNA fragments. In: Hoelzel AR, editor. Molecular genetic analysis of populations: a practical approach. Oxford: IRL Press. 89–114.

21. Botstein D, White RL, Skolnick M, Davis RW (1980) Construction of a genetic linkage map in man using restriction fragment length polymorphisms. American Journal of Human Genetics 32: 314–331.

22. Sokal RR, Michener CD (1958) A statistical method for evaluating systematic relationships. University of Kansas Science Bulletin 28: 1409–1438.

23. Saitou N, Nei M (1987) The neighbor-joining method: a new method for reconstructing phylogenetic trees. Molecular Biology and Evolution 4: 406–425.

24. Perrier X, Flori A, Bonnot F (2003) Data analysis methods. In: Hamon P, Seguin M, Perrier X, Glaszmann JC, editors. Genetic diversity of cultivated tropical plants. Montpellier (France): Enfield, Science Publishers. 43–76.

25. Singh BK, Stidham MA, Shaner DL (1988) Assay of acetohydroxyacid synthase. Analytical Biochemistry 171: 173–179.

26. McCourt JA, Pang SS, King-Scott J, Guddat LW, Duggleby RG (2006) Herbicide-binding sites revealed in the structure of plant acetohydroxyacid synthase. Proceedings of the National Academy of Sciences of the United States of America 103: 569–573.

27. Lee YT, Duggleby RG (2001) Identification of the regulatory subunit of *Arabidopsis thaliana* acetohydroxyacid synthase and reconstitution with its catalytic subunit, Biochemistry 40: 6836–6844.

28. Anderson JA, Matthiesen L, Hegstad J (2004) Resistance to an imidazolinone herbicide is conferred by a gene on chromosome 6DL in the wheat line cv. 9804. Weed Science 52: 83–90.

29. Mayer KF, Martis M, Hedley PE, Simková H, Liu H, et al. (2011) Unlocking the barley genome by chromosomal and comparative genomics. Plant Cell 23: 1249–1263.

30. Randhawa HS, Mutti JS, Kidwell K, Morris CF, Chen X, et al. (2009) Rapid and targeted introgression of genes into popular wheat cultivars using marker-assisted background selection. PLoS ONE 4(6): e5752.

31. Cuadrado A, Cardoso M, Jouve N (2008) Physical organization of simple sequence repeats (SSRs) in Triticeae: structural, functional and evolutionary implications. Cytogenetics and Genome Research 120: 210–219.

32. Molnár I, Cifuentes M, Schneider A, Benavente E, Molnár-Láng M (2011) Association between simple sequence repeat-rich chromosome regions and intergenomic translocation breakpoints in natural populations of allopolyploid wild wheats. Annals of Botany 107: 65–76.

33. Bandopadhyay R, Rustgi S, Chaudhuri RK, Khurana P, Khurana JP, et al. (2011) Use of methylation filtration and C_0t fractionation for analysis of genome composition and comparative genomics in bread wheat. Journal of Genetics and Genomics 38: 315–325.

34. Akhunov ED, Akhunova AR, Anderson OD, Anderson JA, Blake N, et al. (2010) Nucleotide diversity maps reveal variation in diversity among wheat genomes and chromosomes. BMC Genomics. 11: 702.

35. Hufford MB, Xu X, van Heerwaarden J, Pyhäjärvi T, Chia JM, et al. (2012) Comparative population genomics of maize domestication and improvement. Nature Genetics 44: 808–811.

36. Kasha KJ, Simion E, Oro R, Yao QA, Hu TC, et al. (2001) An improved in vitro technique for isolated microspore culture of barley. Euphytica 120: 379–385.

3

Crosstalk between SNF1 Pathway and the Peroxisome-Mediated Lipid Metabolism in *Magnaporthe oryzae*

Xiao-Qing Zeng[1][9], Guo-Qing Chen[2][9], Xiao-Hong Liu[1], Bo Dong[3], Huan-Bin Shi[1], Jian-Ping Lu[4], Fucheng Lin[1,5]*

1 State Key Laboratory for Rice Biology, Biotechnology Institute, Zhejiang University, Hangzhou, China, 2 State Key Laboratory of Rice Biology, China National Rice Research Institute, Hangzhou, China, 3 Institute of Virology and Biotechnology, Zhejiang Academy of Agricultural Science, Hangzhou, China, 4 College of Life Sciences, Zhejiang University, Hangzhou, China, 5 China Tobacco Gene Research Center, Zhengzhou Tobacco Institute of CNTC, Zhengzhou, China

Abstract

The SNF1/AMPK pathway has a central role in response to nutrient stress in yeast and mammals. Previous studies on SNF1 function in phytopathogenic fungi mostly focused on the catalytic subunit Snf1 and its contribution to the derepression of cell wall degrading enzymes (CWDEs). However, the MoSnf1 in *Magnaporthe oryzae* was reported not to be involved in CWDEs regulation. The mechanism how MoSnf1 functions as a virulence determinant remains unclear. In this report, we demonstrate that MoSnf1 retains the ability to respond to nutrient-free environment via its participation in peroxisomal maintenance and lipid metabolism. Observation of GFP-tagged peroxisomal targeting signal-1 (PTS1) revealed that the peroxisomes of *ΔMosnf1* were enlarged in mycelia and tended to be degraded before conidial germination, leading to the sharp decline of peroxisomal amount during appressorial development, which might impart the mutant great retard in lipid droplets mobilization and degradation. Consequently, *ΔMosnf1* exhibited inability to maintain normal appressorial cell wall porosity and turgor pressure, which are key players in epidermal infection process. Exogenous glucose could partially restore the appressorial function and virulence of *ΔMosnf1*. Toward a further understanding of SNF1 pathway, the β-subunit MoSip2, γ-subunit MoSnf4, and two putative Snf1-activating kinases, MoSak1 and MoTos3, were additionally identified and characterized. Here we show the null mutants *ΔMosip2* and *ΔMosnf4* performed multiple disorders as *ΔMosnf1* did, suggesting the complex integrity is essential for *M. oryzae* SNF1 kinase function. And the upstream kinases, MoSak1 and MoTos3, play unequal roles in SNF1 activation with a clear preference to MoSak1 over MoTos3. Meanwhile, the mutant lacking both of them exhibited a severe phenotype comparable to *ΔMosnf1*, uncovering a cooperative relationship between MoSak1 and MoTos3. Taken together, our data indicate that the SNF1 pathway is required for fungal development and facilitates pathogenicity by its contribution to peroxisomal maintenance and lipid metabolism in *M. oryzae*.

Editor: Richard A. Wilson, University of Nebraska-Lincoln, United States of America

Funding: This work was supported by grants (No. 31371890 and 31370171) funded by the National Natural Science Foundation of China. The funders had no role in study design, data collection and analysis, decision to publish, or preparation of the manuscript.

Competing Interests: The authors have declared that no competing interests exist.

* Email: fuchenglin@zju.edu.cn

[9] These authors contributed equally to this work.

Introduction

The conserved SNF1/AMP-activated protein kinase (AMPK) family is well known to serve as the cellular energy sensor and regulator of carbon metabolism in eukaryotes [1,2,3]. The yeast SNF1 kinase is a heterotrimer, composed of a catalytic α-subunit Snf1, a regulatory γ-subunit Snf4, and one of the three β-subunit isoforms, Sip1, Sip2, or Gal83, which tethers Snf1 and Snf4 together to form the functional kinase complex [4]. The best documented function of SNF1 kinase is to respond to glucose limitation and enable yeast cells to utilize non-preferred carbon sources when glucose is deprived [3,5]. The kinase activity of SNF1 is activated by its upstream kinases, Sak1, Tos3, and Elm1 in yeast, which phosphorylate the activation-loop residue Thr210 of the Snf1/α subunit [6,7]. Although Sak1 is the major kinase in this activation, only simultaneous absence of the three Snf1-activating kinases confers completely abolished growth on non-

preferred carbon sources, indicating a partially redundant function among them [7,8,9]. One of the best-studied targets of yeast SNF1 is the transcriptional repressor Mig1, which represses the expression of pivotal enzymes involved in the utilization of alternative sugars [10]. Upon glucose depletion, SNF1 is activated by its upstream kinases and thereafter phosphorylates the repressor Mig1, resulting in the translocation of Mig1 from the nucleus to the cytoplasm and the relief of transcriptional repression imposed by Mig1 [3,10]. Besides Mig1, SNF1 also regulates transcriptional activators, such as Cat8, Adr1, and Sip4, which activate the expression of many genes involved in peroxisome biosynthesis, gluconeogenesis, the glyoxylate cycle, as well as β-oxidation [11,12,13]. In addition to nutrient stress, the yeast SNF1 pathway also participates in environmental stress resistance, aging, invasive and pseudohyphal growth [3,14,15].

To date, most studies of Snf1 function in phytopathogenic fungi focused on its contribution to the derepression of cell wall degrading enzymes (CWDEs) [16,17,18,19]. For plant pathogens, the major barrier of penetration to the host is plant cell wall. Many pathogenic fungi successfully overcome the obstacle by employing mechanical forces or enzymatic methods or a combination of both [19,20]. Cell wall degrading enzymes (CWDEs), which can depolymerize the different constituents of plant cuticle, occupy an important position in pathogenesis [20]. Production of these enzymes is subject to carbon catabolite repression [21], and in some phytopathogenic fungi Snf1 is required for relieving such repression and upregulating CWDEs expression when invasion occurs [16,17,18,19]. Removal of *SNF1* was reported to cause loss or significant reduction of pathogenicity in some plant pathogens, such as *Cochliobolus carbonum* [16], *Ustilago maydis* [19], and *Verticillium dahliae* [18]. The essential role of Snf1 imposed on virulence is partially attributed to its derepression of CWDEs, and another potential contributing factor is the Snf1-dependent utilization of alterative sugars, which can be acquired from the host to drive infection [16,17,18,22]. However, the MoSnf1 in *Magnaporthe oryzae* is not involved in derepression of CWDEs or the metabolism of alterative sugars [23]. Therefore the key player employed by MoSnf1 in pathogenesis remains mysterious. Furthermore, except the catalytic subunit Snf1, the function of other components incorporated in the critical pathway has been rarely reported in filamentous fungi hitherto.

M. oryzae, a heterothallic ascomycete fungus, is the causal agent of the rice blast disease [24]. Upon recognition of environmental cues, the fungus differentiates a well-specialized cell structure, appressorium, from the end of a short germ tube after the contact and germination of a conidium on the host leaf [25]. During maturation, the appressorium becomes melanin-pigmented and accumulates substantial glycerol to generate hydrostatic turgor of up to 8 MPa through vacuolar degradation of lipid reserves [26]. Relying on the turgor pressure, the fungus elaborates a penetration peg to mechanically punch the plant cuticle, and subsequently colonizes host tissues [27]. Since the differentiation of infection structures is in an environment without exogenous nutrients, it underlines the fact that the early stages of plant infection are fuelled by compounds reserved in the conidium. At the onset of appressorial development, translocation of mass lipid bodies from conidium to appressorium occurs, accompanied by rapid lipolysis in appressorial vacuole. Triglycerides, the most abundant form of lipids, are degraded to fatty acids and glycerol under the catalysis of triacylglycerol lipases [26,28]. Consequently, a requirement for fatty acid β-oxidation and subsequent activation of the glyoxylate cycle and gluconeogenesis has been proposed [26,28,29]. The fatty acid β-oxidation, occurred predominantly in peroxisome, leads to the generation of acetyl-CoA pool, which is available for melanin biosynthesis pathway and also fuels fungal cell wall biosynthesis via the glyoxylate bypass and gluconeogenesis during plant infection [30,31,32]. The importance of such process is highlighted by the fact that mutants impaired in peroxisome biosynthesis [30,33,34], β-oxidation [35], carnitine acetyl transfer system [31,36], or glyoxylate cycle [29] are non-pathogenic. Meanwhile, recent transcriptomic analysis independently confirms the pivotal role of peroxisomal acetyl-CoA production during appressorial development [37]. In yeast, both lipid metabolism and peroxisomal proliferation are under the regulation of SNF1 pathway [5,38]. As opposed to the wealthy information in yeast, the function of SNF1 in these critical processes has not yet been studied in filamentous fungi.

In this study, the *M. oryzae* SNF1 pathway was systematically characterized by targeted deletions of the three SNF1 complex subunits and two putative upstream Snf1-activating kinases. Through investigation of GFP-PTS1 signals, we found SNF1 pathway is indispensable for peroxisomal maintenance, which might account for its essential role in lipid metabolism. Furthermore, the interruption of SNF1 pathway resulted in enlarged size of appressorial wall pore and decreased turgor pressure, ultimately the loss of pathogenicity. Our results highlight the importance of SNF1 complex integrity, the upstream kinases, and their contributions to energy homeostasis.

Results

Identification of the SNF1 complex components and two putative Snf1-activating kinases in M. oryzae

In *Saccharomyces cerevisiae*, the SNF1 kinase complex consists of three subunits, the catalytic subunit Snf1, the γ-subunit Snf4, and one of the three β-subunit isoforms, Gal83, Sip1, or Sip2 [4]. There are three upstream Snf1-activating kinases (Sak1, Elm1, and Tos3), each of which is sufficient to activate the SNF1 complex. MoSnf1 (MGG_00803), the catalytic subunit of SNF1 complex in *M. oryzae*, had been identified and proved functionally homologous to *S. cerevisiae* Snf1 [23]. In this study, we additionally sought for other *M. oryzae* orthologs involved in SNF1 pathway to obtain a further understanding of its function. Using protein sequences of *S. cerevisiae* counterparts for BLASTP searches, we identified only one β subunit MoSip2 (MGG_06930), the γ subunit MoSnf4 (MGG_04005), and two upstream kinases, MoSak1 (MGG_07003) and MoTos3 (MGG_06421) in the *M. oryzae* genome (http://www.broadinstitute.org/annotation/genome/magnaporthe_comparative/MultiHome.html), named after the best match (Figure 1A and Table S1). Domains identified by the InterPro database (http://www.ebi.ac.uk/interpro/) of these *M. oryzae* proteins exhibited high conservation (Table S1), including GBD (glycogen-binding domain) and ID (kinase interaction domain) in the C-terminal region of MoSip2, two pairs of cystathionine-beta-synthase (CBS) repeats integrated in MoSnf4, and kinase domains in MoSak1 and MoTos3.

Yeast two hybrid (Y2H) assays were carried out to clarify the interacting network of SNF1 pathway in *M. oryzae*. The results provided evidence that MoSnf1, MoSip2, and MoSnf4 physically interacted with each other, hinting the importance of the complex integrity (Figure 1B). Whereas no interaction was observed between MoSnf1 and MoSak1 or MoTos3 in Y2H (Figure 1B), unlike the stable association between Snf1 and Sak1 in yeast [39]. The result may be caused by the reason that the interaction between MoSnf1 and its upstream kinases was too transient to detect or only occurred under certain conditions.

The expression profiles of SNF1 pathway components in M. oryzae

To obtain some insights into the potential function of SNF1 pathway components, gene expression patterns of the upstream Snf1-activating kinases (*MoSAK1* and *MoTOS3*) and SNF1 complex subunits (*MoSNF1*(α), *MoSIP2*(β), and *MoSNF4*(γ)), were examined by quantitative real-time RT-PCR (qRT-PCR) in vegetative hyphae, conidia, appressoria (8 hpi), and infected barley leaves (72 hpi). When normalized by the vegetative growth stage, the transcription of all the tested genes was up-regulated, although with varied up-regulation folds (Figure 1C). In comparison with the moderate increases of the two regulatory subunits, *MoSIP2* and *MoSNF4*, the expression levels of *MoSNF1* were greatly induced in conidia (4.34-fold) and appressoria (12.97-fold), suggesting a key role of MoSnf1 in the kinase complex. While during invasive growth, the transcription levels of *MoSNF1* and

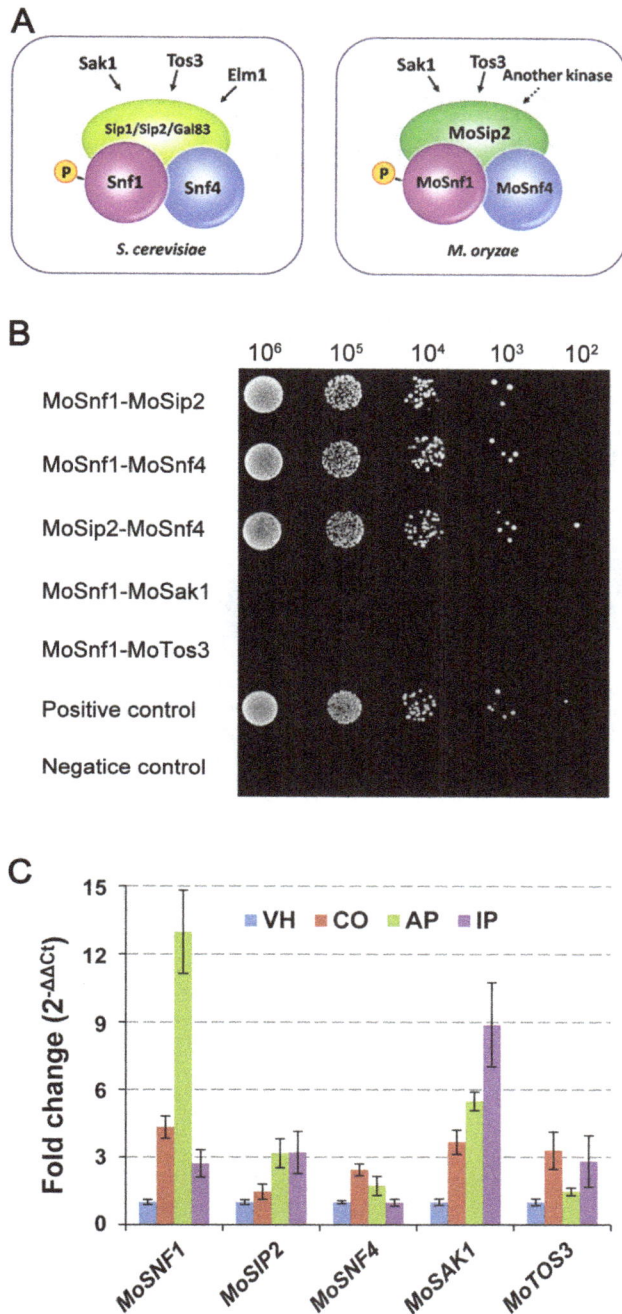

Figure 1. Protein interaction and gene expression analyses of SNF1 kinase complex components and its activating kinases in M. oryzae. (**A**) Different composition of the heterotrimeric SNF1 kinase complex and upstream kinases between *S. cerevisiae* and *M. oryzae*. (**B**) MoSnf1, MoSip2, and MoSnf4 interacted with each other, while no interaction was observed between MoSnf1 and its activating kinases in yeast two-hybrid assay. Yeast transformants expressing MoSnf1 plus MoSip2, MoSnf1 plus MoSnf4, MoSip2 plus MoSnf4, MoSnf1 plus MoSak1, or MoSnf1 plus MoTos3 were 10-fold serially diluted with a starter culture of 10^6 cells/ml and then spotted (5 µl) onto SD-Trp-Leu-His-Ade medium. (**C**) Gene expression profiles of *MoSNF1*, *MoSIP2*, *MoSNF4*, *MoSAK1*, and *MoTOS3* among different developmental stages. Tested fungal tissues included vegetative hyphae (VH), conidia (CO), appressoria 8 hpi (AP), and invasive hyphae (72 hpi), which were within infected plant leaves (IP). Gene expression data, obtained from quantitative RT-PCR analysis, were normalized by using β-tubulin as an internal control and calibrated against the transcript abundances of VH stage.

MoSIP2 were elevated similarly, which was consistent with the observation that both were essential for pathogenicity (see below). Except similar transcript abundance in conidia, *MoSAK1* had much higher expression level increase than *MoTOS3* in appressoria and infected barley leaves, indicating the predominant MoSnf1-regulating position of MoSak1 over MoTos3. The SNF1 pathway up-regulation profile was indicative of its wide influence upon pathogenesis-related processes.

The critical role of SNF1 complex integrity and its upstream kinases in conidiogenesis, aerial hyphae development, appressorial formation and morphology of *M. oryzae*

To further study the biological function of SNF1 pathway in *M. oryzae*, five null mutant strains specific to *MoSNF1*, *MoSIP2*, *MoSNF4*, *MoSAK1*, and *MoTOS3* were generated and verified by Southern blot analysis (Figure S1). Double deletion mutant *ΔMosak1ΔMotos3* was also constructed to determine whether the two upstream kinases have functional overlap. Finally, we acquired at least two independently targeted gene deletions for each mutant (*ΔMosnf1*, *ΔMosip2*, *ΔMosnf4*, *ΔMosak1*, *ΔMotos3*, and *ΔMosak1ΔMotos3*), and selected one strain for detailed phenotypic analysis.

In brief, the morphological phenotypes of SNF1 complex mutants (including *ΔMosnf1*, *ΔMosip2*, and *ΔMosnf4*), were similar to each other, including abnormal spore appearance (Figure 2C), poor sporulation (Figure 2B and Figure 3A), reduced appressorial size (Figure 3B), shortened aerial hyphae (Figure 2A and 2B), but normal growth rate (Figure 2A) in complete media as compared to WT (wild-type) and complemented strains. Furthermore, all the three mutants exhibited delayed conidial germination and appressorial formation, the rates of which were significantly lower than WT strain even the incubation time was prolonged to 24 h (Figure 3C). Nevertheless, the defects in *ΔMosip2* and *ΔMosnf4* were more alleviated than that in *ΔMosnf1*.

The two upstream kinase mutants, *ΔMosak1* and *ΔMotos3*, exhibited different phenotypes during fungal development. *ΔMosak1* possessed sparse aerial hyphae with extremely poor conidiation, which was decreased by 5-fold compared to WT and the complemented strain (Figure 2B and Figure 3A). Besides, the process of spore germination and appressorial formation was greatly affected (Figure 3C). In contrast, the other upstream kinase mutant, *ΔMotos3*, formed much denser aerial hyphae with short branch, producing large scales of normal spores as many as 3-fold of WT (Figure 2 and Figure 3A). Re-introduction of *MoTOS3* into the mutant recovered these phenotypic changes. Whereas other phenotypes of *ΔMotos3* were indistinguishable from those of the wild-type strain (Figure 3B and 3C). In *ΔMosak1ΔMotos3* double mutant, the colony had an almost flat appearance, due to the attenuated aerial hyphae, and a slower growth rate (Figure 2A and 2B). The conidiogenesis and appressorial formation defects of the double mutant were more severe than *ΔMosak1* and comparable to the null SNF1 complex mutants (Figure 3), indicative of a collaborative relationship between MoSak1 and MoTos3 on the regulation of the SNF1 complex.

These results suggest that the SNF1 complex integrity is essential for aerial hyphae development, conidiogenesis, appressorial formation and morphology in *M. oryzae*. Meanwhile, the two upstream Snf1-activating kinases, MoSak1 and MoTos3, play distinct roles during above processes.

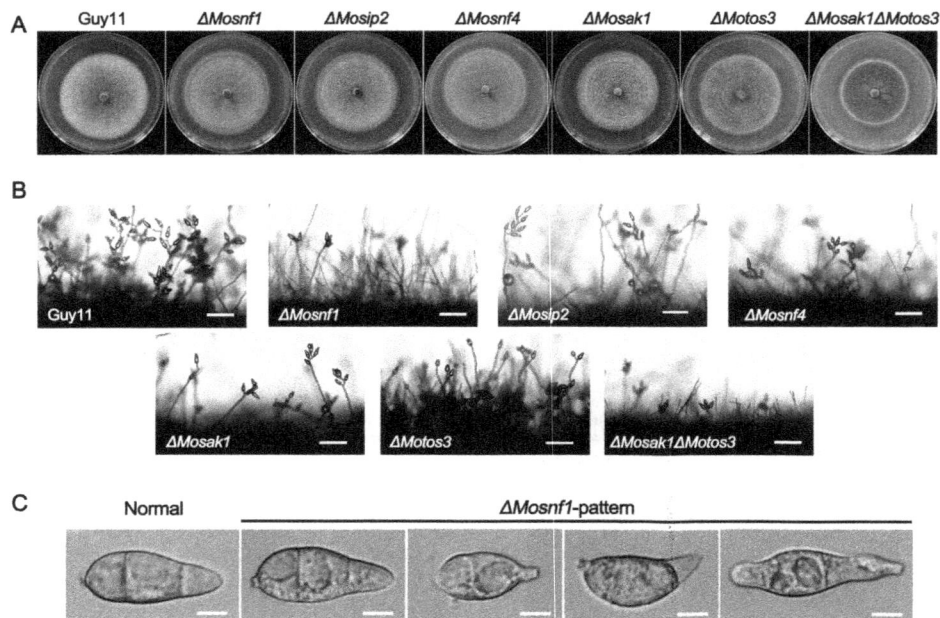

Figure 2. Comparison of the SNF1 pathway mutants with regard to colony morphology and conidial development. (A) Strains were cultured on CM plates at 25°C for 10 days. *ΔMosak1ΔMotos3* exhibited a decreased mycelial growth rate, while no significant difference in the colony size was observed between other mutants and Guy11. (B) Microscopic observation of conidial development. Significant reduction in conidial production was observed in *ΔMosnf1*, *ΔMosip2*, *ΔMosnf4*, *ΔMosak1*, and *ΔMosak1ΔMoto3* at 24 hpi. However, *ΔMotos3* developed short, yet dense conidiophores with plenty of spores arrayed thereon. Bars = 50 μm. (C) Conidia of WT and the mutants were harvested and observed under the light microscope. Conidial shape of *ΔMosip2*, *ΔMosnf4*, *ΔMosak1*, and *ΔMosak1ΔMoto3* was identical to that of *ΔMosnf1* (*ΔMosnf1*-pattern), whereas there was no measurable difference between *ΔMotos3* and Guy11 (Normal). Bars = 5 μm.

Disruption of SNF1 pathway suppressed the mycelial growth on media with non-fermentable carbon sources

In yeast, *Δsnf1* is unable to survive on media with non-fermentable carbon as sole carbon source [5]. To determine whether *M. oryzae* SNF1 pathway contributes to non-fermentable carbon metabolism, mycelial agar plugs of the SNF1 pathway mutants were incubated on minimal media supplemented with various non-fermentable carbons as sole carbon source.

Growth tests revealed that all the tested mutants, except *ΔMotos3*, exhibited severe defects in utilization of acetate, Tween 80 (the principal component is oleate), triolein (one of the typical triglycerides), and olive oil (long chain fatty acids) (Figure 4 and Table 1). The growth rate of *ΔMotos3* was slightly affected except on Tween 80-contained medium (Figure 4 and Table 1). Among other mutants, the deficiency degree varied upon carbon type. Overall, the defects of *ΔMosip2* and *ΔMosnf4* were slightly relieved as compared to *ΔMosnf1* (Figure 4 and Table 1). The mycelial growth of *ΔMosak1ΔMotos3* was comparable to *ΔMosak1* except on Tween 80 medium, with 88% and 75% growth reduction, respectively (Figure 4 and Table 1). These data indicate that the SNF1 pathway is of great importance to the efficient utilization of non-fermentable carbon sources in *M. oryzae*.

The maintenance of peroxisomes is dependent on the SNF1 pathway

The single-membrane organelle peroxisome is well known for its function in fatty acid β-oxidation, the glyoxylate cycle, and removal of reactive oxygen species (ROS) [40]. In *M. oryzae*, mutants with abnormal peroxisomes such as *ΔMopex6* [30], *ΔMopex5* [34], *ΔMopex7* [33], and *ΔMopex19* [41] showed

inability to metabolize fatty acids and NaAC. In addition, yeast *Δsnf1* and *Δsnf4* were reported to be devoid of peroxisomal structures [38]. Thus, we carried out to assess the association between SNF1 pathway and peroxisomal biogenesis in *M. oryzae*.

Peroxisomal matrix proteins usually include specific motifs known as peroxisomal targeting signals (PTSs), which could be recognized by the import machinery and targeted to peroxisome. PTS1 is a conserved tripeptide sequence (S/A/C) (H/R/K) (I/L/M) at the C terminus of most known peroxisomal matrix proteins [34,40]. In this study, PTS1 (SKL) signal was employed to visualize peroxisome by introducing GFP-PTS1 vector (kindly provided by Dr. Jiaoyu Wang) into *ΔMosnf1*, *ΔMosip2*, *ΔMosnf4*, *ΔMosak1*, *ΔMotos3*, *ΔMosak1ΔMotos3*, and the wild type, respectively. Subcellular localization of GFP-PTS1 was then investigated in the transformed strains. During vegetative growth phase, both Guy11 and the SNF1 pathway mutants performed punctate GFP fluorescence, indicative of peroxisomal structures, however, the puncta size seemed to be different among them. In *ΔMosnf1*, *ΔMosip2*, *ΔMosnf4*, *ΔMosak1*, and *ΔMosak1ΔMotos3* background, enlarged peroxisomes were more frequently observed than in WT and *ΔMotos3*, suggesting some aberrant changes therein (Figure 5A). Likewise, numerous GFP-PTS1 labeled peroxisomes were observed as punctate spots in the conidia of WT and *ΔMotos3* (Figure 5B), and began to enter the vacuolar lumens during appressorial differentiation (Figure S2); in the meantime, their incipient appressoria were peroxisome-rich (Figure S2). While in other mutants, GFP-PTS1 had already been mis-localized to the CMAC-stained vacuoles even before conidial germination, leaving sharp decline of fluorescent spots in the conidial cytoplasm (77.6±5.1%, 64.3±4.6%, 62.2±5.7%, 35.5±5.3%, 76.9±6.6% spores with fluorescent vacuoles in *ΔMosnf1*, *ΔMosip2*, *ΔMosnf4*, *ΔMosak1*, and *ΔMosak1ΔMotos3*,

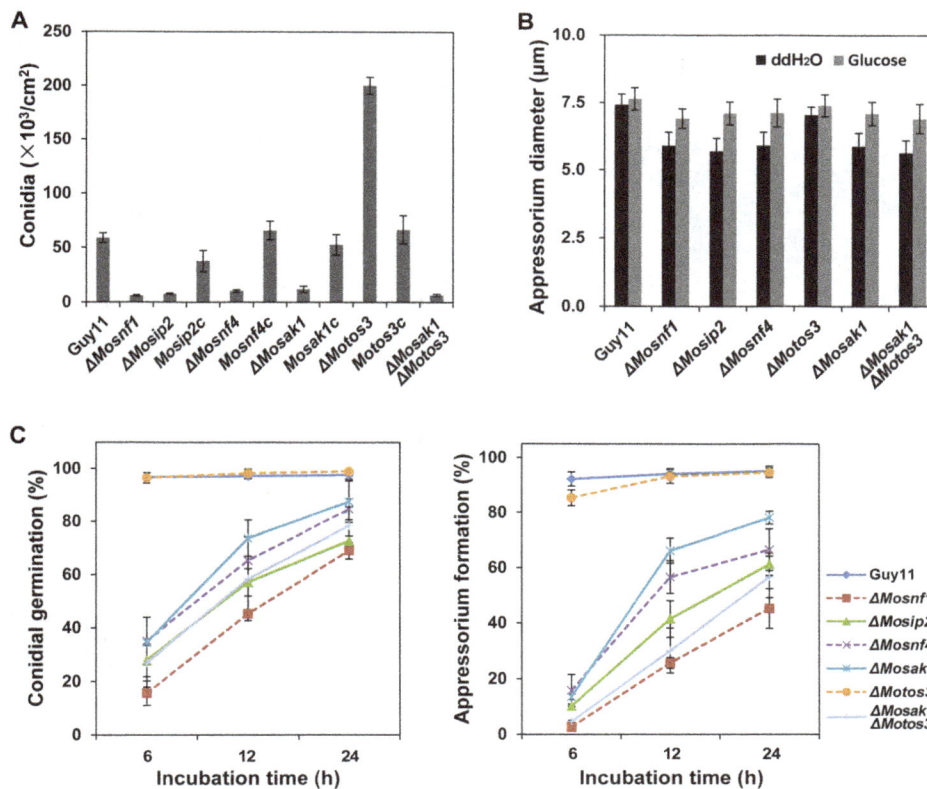

Figure 3. SNF1 pathway mutants exhibited dramatic defects in sporulation, appressorial size, conidial germination, and appressorium formation. (**A**) Conidial production of *ΔMosnf1*, *ΔMosip2*, *ΔMosnf4*, *ΔMosak1*, and *ΔMosak1ΔMoto3* was severely impaired, while *ΔMotos3* showed an elevated conidiation in comparison to Guy11. The complemented strains restored the sporulation defects of the corresponding mutants. Conidia were collected and counted from 10-day-old cultures grown on CM plates. The means and standard deviations of three independent experiments are presented as columns with error bars. (**B**) Appressorial diameters were measured and statistically analyzed after 48 h incubation on the artificial surfaces. Addition of 2.5% glucose to conidial suspensions partially restored the mutant defect in appressorial size. (**C**) Disruption of SNF1 pathway genes impaired conidial germination and appressorium formation in *M. oryzae*. Conidial suspensions harvested from 8-day-old CM cultures were incubated on hydrophobic surfaces and observed under a light microscope at indicated time points.

respectively, vs. 1.6±0.7%, 1.4±0.6% in WT and *ΔMotos3*, respectively) (Figure 5B). Further investigation of the GFP-PTS1 sequential localization revealed that the fluorescent signals were almost invisible in *ΔMosnf1* appressoria but arrested in conidial spherical vacuoles (Figure S2).

These findings indicate that the *M. oryzae* SNF1 pathway plays an important role in peroxisomal dynamics during fungal development.

Lipid mobilization was dramatically retarded in *ΔMosnf1*, *ΔMosip2*, *ΔMosnf4*, *ΔMosak1*, and *ΔMosak1ΔMotos3*

During appressorial development, lipid reserves in conidia are rapidly transferred to appressoria where they are degraded and supplied as the source of glycerol and other intermediates, a process demanding peroxisomal function [28,35]. To determine whether the impairment of peroxisomal maintenance affected the lipid droplets mobilization in the SNF1 pathway mutants, lipid bodies were monitored during appressorial morphogenesis via Nile red staining.

It was observed that lipid droplets had been entirely translocated from conidia to appressoria within 24 h and the majority was degraded at 48 hpi in Guy11 (Figure 6). In *ΔMotos3*, no distinguishable difference was detected in the transfer efficiency of lipid bodies as compared to WT, but the degradation rate was

slightly decreased, with about 20% appressoria stained by Nile red compared to only 1.67% in Guy11 at 96 hpi (Figure 6B). On the contrary, the mobilization of lipid reserves was significantly retarded in *ΔMosnf1*, *ΔMosip2*, *ΔMosnf4*, *ΔMosak1*, and *ΔMosak1ΔMotos3*, ranging from 64.0% to 85.2% conidia still containing large lipid deposits after 48 h of incubation (Figure 6A, 6B left). The degradation rates were also severely influenced, with large merged lipid droplets still observed in more than 74% appressoria of the mutants at 96 hpi (Figure 6A, 6B right). We therefore conclude that the SNF1 pathway is indispensible for lipid droplets translocation and degradation.

SNF1 pathway is responsible for turgor genesis and normal porosity of the appressorial wall

Appressorium-mediated penetration in *M. oryzae* is a turgor-driven process required substantial glycerol accumulation. The predominant cellular source of glycerol is from lipid droplets degradation during appressorium maturation [31,42]. Since SNF1 pathway participates in the degradation of lipid bodies, cytorrhysis assay [43] was performed to test whether the SNF1 pathway mutants had the ability to generate turgor pressure comparable to WT. The appressorial collapse rates of Guy11 and *ΔMotos3* were similar under a serial concentrations of glycerol (from 1 M to 4 M) (Figure 7A), while the appressoria of the other mutants were much

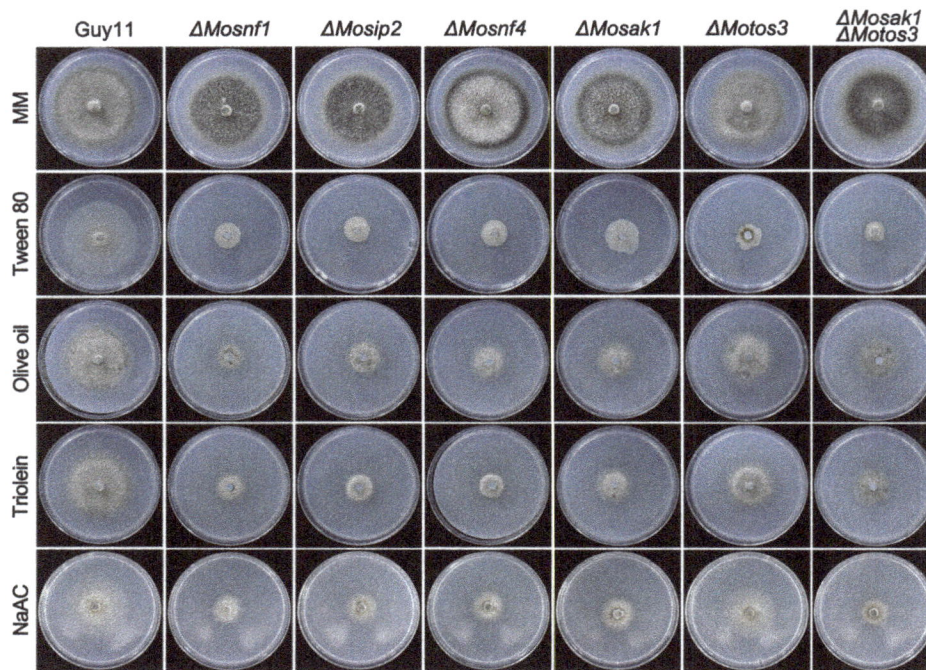

Figure 4. Mutations in SNF1 pathway affected the utilization of non-fermentable carbons. Strains were cultured on MM plates supplemented with 1% Glucose, 1% Tween 80, 1% Olive oil, 1% Triolein, or 50 mM Sodium acetate as sole carbon source for 10 d at 25°C.

more fragile and vulnerable to collapse, with the severity from high to low as follows: *ΔMosnf1/ΔMosak1ΔMotos3*, *ΔMosip2*, *ΔMosnf4*, and *ΔMosak1* (Figure 7A). So the impairment in SNF1 function dose affect the accumulation of appressorial turgor pressure.

When dipped in glycerol solution, plasmolysis was frequently observed in the appressoria of the mutants. Thus we speculated the SNF1 pathway might also play a key role in appressorial cell wall integrity. To explore the possibility, cytorrhysis/plasmolysis test was carried out by applying solutions of polyethylene glycols (PEGs) with different molecular weights but the same osmotic pressure (4 MPa) to appressoria at 48 hpi. Through calculating the ratio of plasmolysis to cytorrhysis, we found for the wild type the proportion of appressoria showing cytorrhysis was dominant over all the tested PEG types (Figure 7B). However, in *ΔMosnf1* and *ΔMosak1ΔMotos3*, the ratio of plasmolysis/cytorrhysis declined

along with the diameter of external solute species increased, and was eventually similar to that of WT in PEG3350 or larger (Figure 7B). Furthermore, the plasmolysis was more severe in mutants than WT as shown in Figure 7C. These data suggest the SNF1 function is essential for maintaining proper porosity of the appressorial wall. Transmission electron microscopy (TEM) was also performed to observe the appressorial structures in detail. As a result, electron-dense melanin layer, with identical thickness, was distinctly observed in both wild type and the mutants (Figure S3), suggesting the appressorial wall defects of the SNF1 pathway mutants are not associated with melanin layer biosynthesis.

Lipid degradation not only liberates glycerol but also feeds the acetyl-CoA pool produced by beta-oxidation of fatty acids. The end product acetyl-CoA can be shuttled to the glyoxylate cycle and gluconeogenesis which enable it to synthesize components of cell wall such as glucans and chitin [31,32]. Hence, we inferred why

Table 1. Growth rate of the SNF1 pathway mutants on non-fermentable carbon media.

Strain	Olive oil	Tween-80	Triolein	NaAC
Guy11	76.5±2.4[a]	70.0±2.2[a]	78.4±0.7[a]	69.5±3.1[a]
ΔMosnf1	21.5±3.1[d]	18.2±0.6[c]	23.3±0.6[e]	24.6±2.1[d]
ΔMosip2	31.3±0.6[c]	17.2±0.6[cd]	25.4±1.1[de]	27.7±2.2[cd]
ΔMosnf4	34.0±2.3[c]	18.9±1.3[c]	29.8±6.2[cd]	26.9±1.2[cd]
ΔMosak1	34.7±3.3[c]	25.3±1.6[b]	30.6±0.6[cd]	31.0±0.8[c]
ΔMotos3	53.5±1.8[b]	13.8±3.4[de]	50.8±3.1[b]	56.3±4.7[b]
ΔMosak1ΔMotos3	35.6±2.4[c]	11.9±4.0[e]	32.2±2.4[c]	26.5±0.8[cd]

Vegetative growth rate (%) = (the diameter of strains on MM with various carbon sources/the diameter of cultures on regular MM plates) ×100; colony diameters on regular MM plates are set as 100% control. Data were collected from 10 days postincubation and presented as means±SD from three independent experiments. Duncan's multiple range tests were used to determine significance at the 0.05 level of probability. The same letters in a column mean no significant difference.

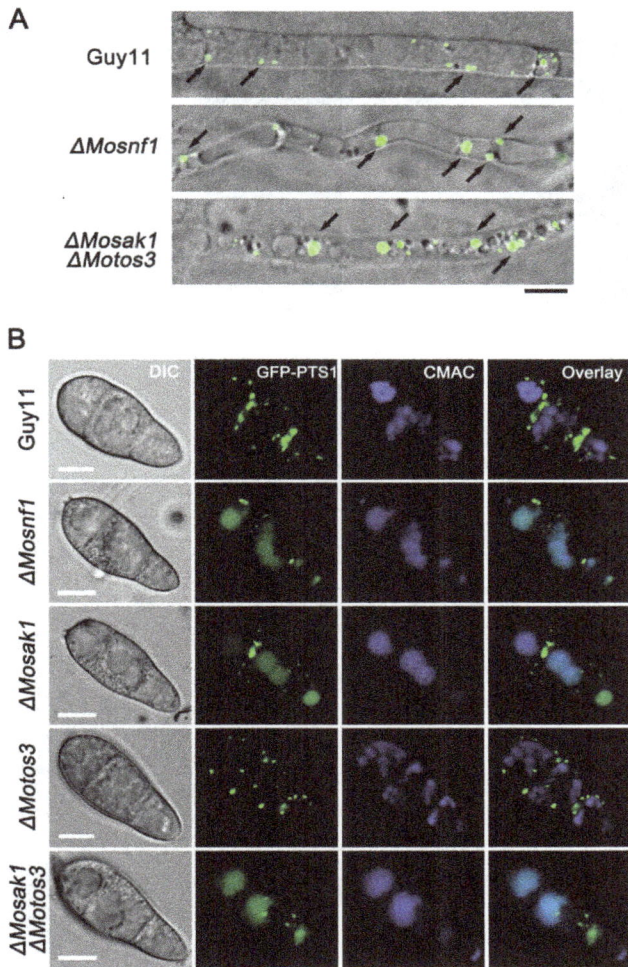

Figure 5. Effects of SNF1 pathway mutations on GFP-PTS1 distribution. (A) Confocal microscopic observation of mutant strains expressing GFP-PTS1. Images shown were representative of the majority of vegetative hyphae. Enlarged peroxisomes were more frequently observed in *ΔMosnf1* and *ΔMosak1ΔMotos3* than WT. Arrows point to peroxisomes. Bar = 5 μm. (B) Colocalization of GFP-PST1-positive peroxisomes and CMAC-stained vacuoles. The amount of cytoplasmic peroxisomes was decreased dramatically in the conidia of *ΔMosnf1*, *ΔMosak1*, and *ΔMosak1ΔMotos3*, while in WT and *ΔMotos3*, numerous peroxisomal puncta were observed with the absence of vacuolar GFP fluorescence. The localization patterns of GFP-PTS1 in *ΔMosip2* and *ΔMosnf4* conidia were indistinguishable from that in *ΔMosnf1* conidia. Bars = 5 μm.

the SNF1 pathway mutants failed to maintain sufficient turgor pressure was caused by the reduced accumulation of intracellular osmolites and increased appressorial cell wall porosity, both of which were the results of lipid metabolism inability.

Different components of SNF1 pathway make unequal contributions to the pathogenicity

Due to the significant influence on turgor genesis, the pathogenicity of *ΔMosnf1* was severely impaired (Figure 8 and Figure 9), consistent with findings from previous study [23]. However, the exact reason why *ΔMosnf1* lost virulence has not been determined. Besides, other proteins of SNF1 pathway do not perform equal functions as MoSnf1 does in various processes, it is

therefore necessary to investigate whether they contribute differently to plant infection by *M. oryzae*.

Conidial suspensions, freshly collected from Guy11 and the mutants, were appropriately diluted and set to infect host leaves. Similar to *ΔMosnf1*, both *ΔMosip2* and *ΔMosak1ΔMotos3* failed to elicit any visible disease lesions on barley and rice leaves (Figure 8 and Figure 9), while the virulence of *ΔMosnf4* and *ΔMosak1* was sharply attenuated, causing only tiny and restricted lesions on rice leaves (7.0% and 8.9% diseased leaf area, respectively, vs. 45.4% in Guy11) (Figure 8 and Figure 9). In contrast, inoculation with spores from Guy11, *ΔMotos3*, and the complemented strains resulted in the development of typical rice blast symptoms (Figure 8, Figure 9, and Figure S4). Subsequently, barley leaves were abraded to examine the invasive growth of the mutants *in planta*. Interestingly, *ΔMosip2*, *ΔMosnf4*, and *ΔMosak1* all developed extensive lesions although with compromised spread rate compared to Guy11 and *ΔMotos3* (Figure 8). However, the wounded leaf areas underneath conidial droplets of *ΔMosnf1* and *ΔMosak1ΔMotos3* remained healthy, suggesting the two mutants also lost the ability to proliferate within host tissues (Figure 8).

To further test the appressorial function, penetration and invasive growth were microscopically observed after inoculating barley leaves with conidial suspensions from WT and the mutants. Consistent with the pathogenicity assay, the penetration rates of *ΔMosnf4* and *ΔMosak1* (11.4% and 39.8%, respectively) were strongly compromised when compared to Guy11 (77.0%) at 48 hpi (Figure 10C), although some mutant appressoria developed aggressive invasive hyphae (Figure 10A). The rate of appressoria forming infectious hyphae in *ΔMotos3* (70.8%) was comparable to WT (Figure 10C). Whereas *ΔMosnf1*, *ΔMosak1ΔMotos3*, and the majority of *ΔMosip2* (98.3%) appressoria were incapable of penetrating barley epidermal cells at 48 hpi (Figure 10A and 10C), revealing that their reduced virulence was the result of perturbed appressorial function.

The above observations indicate the different components of SNF1 pathway play unequal roles in pathogenicity-related processes. Although the phenotypic defects of *ΔMosak1* were much more severe than *ΔMotos3*, additional deletion of *MoTOS3* led to a similar phenotype as what occurred in *ΔMosnf1*, suggesting that MoSak1 together with MoTos3 make contributions to MoSnf1 activation.

External carbon supplement relieved the defects of mutants with impaired SNF1 function

Since the infection of *M. oryzae* occurs in a nutrient-free environment, the development of appressoria and subsequent infectious hyphae must be nutritionally supported by the degradation of conidial reserves [31]. Previous studies have proved the generation of acetyl-CoA via fatty acid β-oxidation is a prerequisite for appressorium-mediated plant infection, as acetyl-CoA is the substrate to synthesize glucans and chitin, which are required for cell wall biosynthesis [31,35]. Given the decreased amount of peroxisomes during appressorial development, we investigated whether the mutant defects in appressorial function and invasive growth could be rescued by external carbon source, which can supply acetyl-CoA and other intermediates via the glycolytic pathway and citric acid cycle independent of peroxisomal fatty acid β-oxidation.

As expected, pathogenic defects of all the mutants were partially complemented in the presence of exogenous glucose (Figure 8 and Figure 9). In order to explore the exact reason, we monitored the changes in appressorial structure and function in detail. When supplemented with 2.5% glucose, the SNF1 pathway mutants

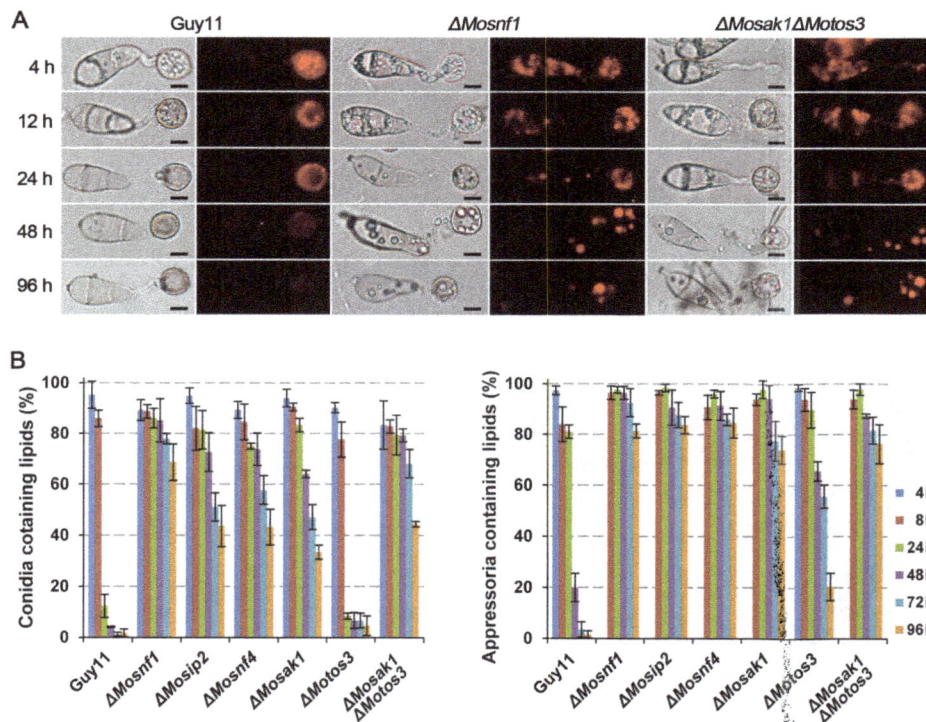

Figure 6. Intracellular mobilization of lipid droplets in WT and SNF1 pathway mutants during appressorium morphogenesis. Conidial suspensions were incubated on the surfaces of hydrophobic films and stained with Nile red to observe the status of lipid droplets movement and distribution at the indicated time points under epifluorescence microscope. (**A**) *ΔMosnf1* and *ΔMosak1ΔMotos3* showed significant delays in lipid mobilization and degradation with the presence of Nile red-stained lipid bodies even at 96 hpi, while fluorescent signals were almost invisible in WT at 48 hpi. Bars =5 μm. (**B**) Percentages of conidia (left) or appressoria (right) that contained lipid droplets. Varied degrees of defect in lipid mobilization were observed among the mutants.

were restored to a large extent in terms of appressorial size (Figure 3B) and cell wall porosity (the ratio of plasmolysis to cytorrhysis in 1.7 M glycerol =0.24±0.05 and 0.26±0.07 in *ΔMosnf1* and *ΔMosak1ΔMotos3*, respectively, dramatically lower than the data listed in Figure 7B). The appressorial penetration ability was also raised significantly in *ΔMosnf1* (0 to 26.7%), *ΔMosip2* (1.69% to 33.1%), *ΔMosnf4* (11.4% to 35.3%),

ΔMosak1 (39.8% to 50.2%), and *ΔMosak1ΔMotos3* (0 to 22.1%), compared to the slight improvement in Guy11 (77.0% to 80.7%) and *ΔMotos3* (70.8% to 72.5%) at 48 hpi (Figure 10C). Furthermore, exogenous glucose enabled *ΔMosnf1*, *ΔMosip2*, and *ΔMosak1ΔMotos3* to develop branched secondary hyphae, in striking contrast to the almost inhibited penetration observed in the absence of nutrient (Figure 10A and 10B).

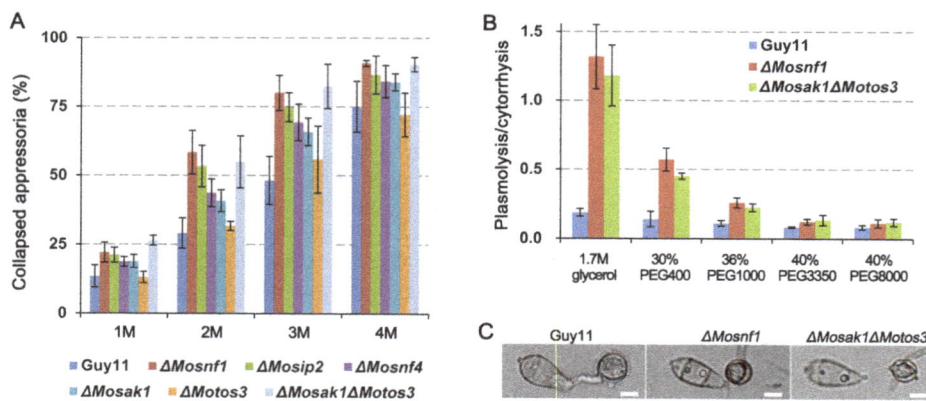

Figure 7. Indirect assessment of turgor pressure and appressorial porosity. (**A**) To measure appressorial turgor pressure, incipient cytorrhysis assay was performed on induced appressoria at 48 hpi with glycerol solutions of varying concentrations (1–4 M). (**B**) Plasmolysis/cytorrhysis assay with osmotic solutions of different average molecular weights. The solutions were adjusted to the denoted concentrations to exert 4 MPa osmotic pressure on appressoria at 48 hpi. (**C**) Plasmolyzed appressoria at 48 hpi were photographed after soaked in 1.7 M glycerol solution for 10 min. Bars =5 μm.

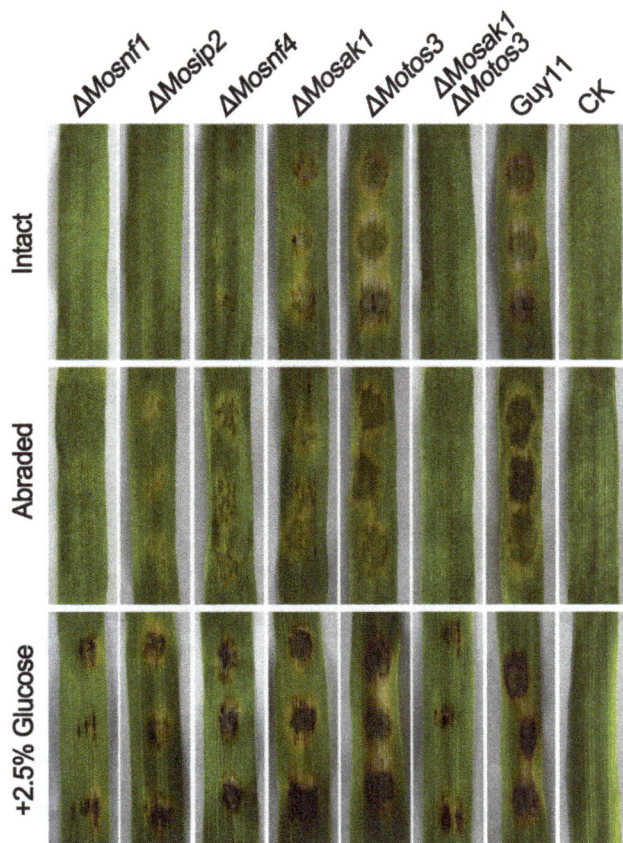

Figure 8. Pathogenicity assay on detached barley leaves. Intact and abraded barley leaves were inoculated with 20 μl conidial suspensions (1×10^5 conidia/ml) of the tested strains for 4 days before photography. ΔMosnf1, ΔMosip2, ΔMosnf4, ΔMosak1, and ΔMosak1Δ-Motos3 were deficient in appressorium-mediated infection of intact barley leaves. Elevated virulence was observed in ΔMosip2, ΔMosnf4, and ΔMosak1 when tested on abraded barley leaves, while ΔMosnf1 and ΔMosak1ΔMoto3 were still non-pathogenic. Addition of 2.5% glucose to conidial suspensions could evidently, yet partially, restored the virulence of the defective mutants.

These results indicate that external glucose metabolism could partially compensate the insufficient peroxisomal function and rescue the phenotypic defects of SNF1 pathway mutants, which is

consistent with the effect of extra glucose on Δpex6 mutants [35,44].

Discussion

As a hemibiotrophic pathogen, M. oryzae has to regulate its cellular metabolic activities to adapt to nutrient unavailability during the early infection stage [24]. In yeast, the SNF1 signaling pathway plays a central role in regulating energy status by its involvement in carbon catabolite derepression, a mechanism to ensure the utilization of unfavorable carbon sources when glucose is deprived [3]. Previous studies on SNF1 function in phytopathogenic fungi have uncovered its great influences on the expression of cell wall degrading enzymes (CWDEs) and the utilization of alternative sugars, both of which are subject to carbon catabolite repression (CCR) [16,17,18,19]. However, the M. oryzae MoSnf1 protein was reported not to preserve such regulatory role [23]. Recently, trehalose-6-phosphate synthase 1 (Tps1) was recognized as a glucose-6-phosphate sensor, which cooperates with its downstream inhibitors Nmr1-3 to mediate CCR regulatory system in M. oryzae [45]. Besides, the expression pattern of CWDEs was found to be disturbed in MoTPS1 disruption mutant [45]. These reports suggest the components of metabolic regulatory systems have changed greatly in M. oryzae, likely not involving MoSnf1 [46]. This study set out to investigate how MoSnf1 acts as a virulence determinant and delve into the function of SNF1 pathway in M. oryzae.

In fungi, where non-fermentable compounds like fatty acids and acetate can serve as sole source of carbon and energy, the acetyl-CoA must be converted to C4 compounds via the glyoxylate cycle, allowing gluconeogenesis [5,47]. Peroxisome plays an essential role in this process, as it could serve as the location where fatty acid beta-oxidation occurs to generate acetyl-CoA [48], meanwhile many glyoxylate cycle enzymes are also peroxisomal [49]. In M. oryzae [30,33,34,41], Colletotrichum lagenarium [44], and Fusarium graminearum [50], mutants with aberrant peroxisome function showed severe defects in utilization of lipids, fatty acids, and acetate. Yeast Δsnf1 is devoid of peroxisomal structures and fails to survive on media with non-fermentable carbon sources [11,38]. However, we found ΔMosnf1 possessed abnormal other than abolished peroxisomes with enlarged size in the mycelia. The poor growth of ΔMosnf1 on fatty acids or NaAC-contained media suggested an aberrant function of the enlarged peroxisomes therein. In S. cerevisiae, PEX1/PEX11, FOX2, and ICL1, all of which are regulated by Snf1 [11], play an important role in

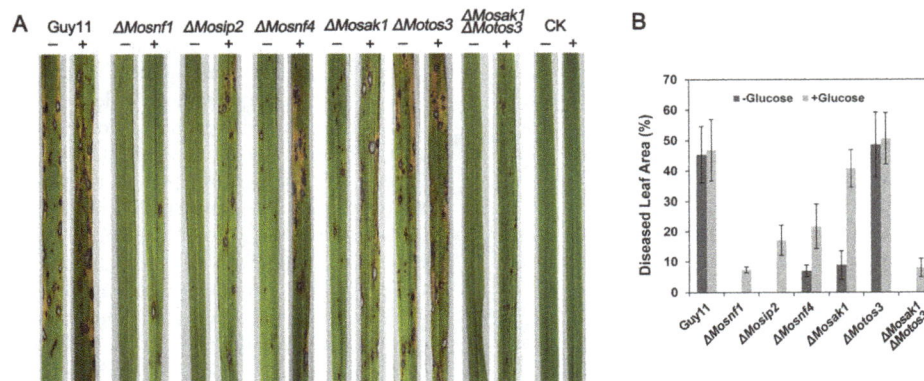

Figure 9. Spray inoculation assay with rice seedlings. Conidial suspensions (1×10^5 conidia/ml) with (+) or without (-) 2.5% glucose were evenly sprayed onto rice leaves for 7 days before photographing the typical infected leaves (**A**) and calculating the percentage of diseased leaf area (**B**).

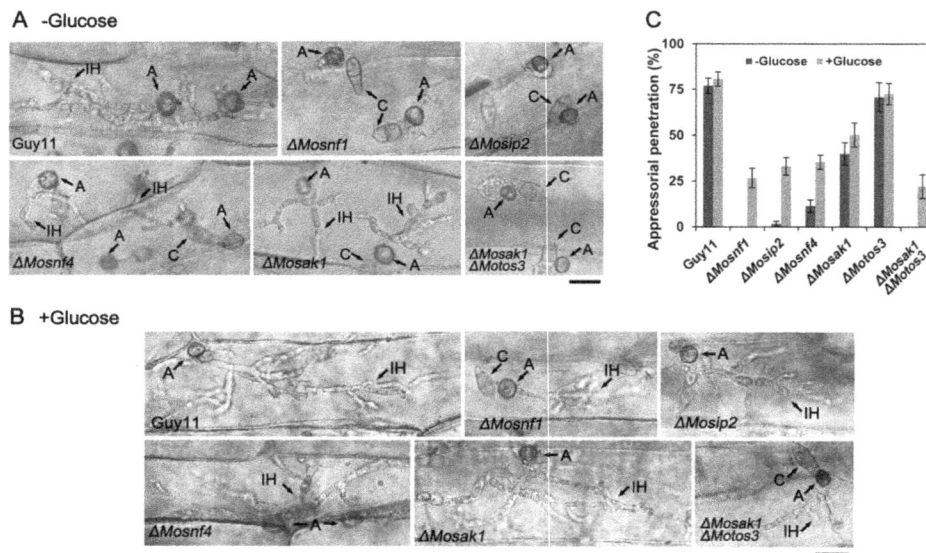

Figure 10. Penetration assay on barley epidermal cells. Barley leaves were inoculated with conidial suspensions supplemented without (**A**) or with (**B**) 2.5% glucose for 48 h, and then decolored by methanol before observation. A, appressorium; C, conidium; IH, invasive hyphae. Bars = 15 μm. (**C**) Percentage of appressoria capable of penetration was counted at 48 hpi, and the results were presented as means and standard deviations.

peroxisomal biosynthesis and proliferation [51], peroxisomal fatty acid beta-oxidation [48], and the glyoxylate cycle [52], respectively. However, gene expression profiling by qRT-PCR revealed that *M. oryzae* homologs, *MoPEX1* (MGG_09299), *MoPEX11* (MGG_08896), *MoFOX2/MoMFP1* (MGG_06148), and *MoICL1* (MGG_04895), were up-regulated when induced by olive oil or triolein in both WT and *ΔMosnf1*, while no significant differences in the up-regulation folds were observed (Figure S5). Thus, the regulatory mechanism controlling peroxisome function by MoSnf1 remains obscure, which may involve the derepression of other key factors not tested in this study.

Consistent with previous studies [53], peroxisomal targeting signals of WT and *ΔMotos3* were distributed as cytoplasmic punctate spots, but not visible in the vacuoles of germinating conidia until over 6 hpi. However, accelerated degradation of PTS1 signals was observed in the ungerminated conidia of SNF1 function-disturbed mutants. Unlike *ΔMopex* mutants with dispersed GFP-PTS1 localization in cytoplasm [30,34,35], punctate fluorescent spots were still present in *ΔMosnf1*, suggesting it retained the integral PTS1 import pathway and peroxisomal formation ability but failed to maintain adequate peroxisomes. One reason we deduced why *ΔMosnf1* performed the abnormal GFP-PTS1 sequential localization could be caused by the premature pexophagy. Another likelihood might be that *ΔMosnf1* was incapable to form adequate peroxisomes as Guy11 did, and the surplus GFP-PTS1 had to be delivered to vacuolar lumen for degradation directly.

Insufficient amount of peroxisomes could give rise to serious consequence in *M. oryzae*. It is known that enormous turgor pressure is required for the pathogen to physically penetrate the host surface, and its genesis relies on the substantial glycerol accumulation via rapid lipolysis during appressorial maturation [26]. In order to maintain the efficient lipid mobilization, fatty acids resulting from lipolysis demand to be transported to peroxisomes where they are metabolized via β-oxidation to form acetyl-CoA [29,31,35]. The resulting acetyl-CoA is not only the precursor in melanin synthesis but also supplies substrates to synthesize chitin and glucans via the glyoxylate shunt and

gluconeogenesis [30,32]. So peroxisome, the major β-oxidation location and acetyl-CoA sink, plays a central role in appressorial morphogenesis and function [37]. Mutants with dysfunctional peroxisomes display significant defects in lipid metabolism, melanin layer formation, appressorial wall porosity, turgor generation, and ultimately pathogenicity [30,33,34,44]. Possibly due to the sharp decline of peroxisomal number, *ΔMosnf1* performed a significant retard in lipid droplets mobilization, and formed larger appressorial wall pore, coupled with the failure to generate enormous turgor and the loss of virulence. However, a distinct melanin layer, without significant differences in thickness, was observed between cell wall and plasma membrane in both *ΔMosnf1* and Guy11 appressorium, indicating that the defect in porosity of appressorial wall seemed not to be related to melanin biosynthesis. Consistent with *M. oryzae* mutants disturbed in peroxisomal function [30,34], *ΔMosnf1* partially restored the appressorial morphology and pathogenicity when supplemented with external carbon source, which could compensate acetyl-CoA via the glycolytic pathway and citric acid cycle independent of peroxisomal function. Further quantification of diseased leaf area and penetration rate on hosts revealed that the restoration by external glucose was incomplete in *ΔMosnf1*. Thus, except for peroxisomal maintenance, MoSnf1 may be involved in other cellular processes required for virulence.

Apart from elucidating how MoSnf1 was involved in pathogenesis, we additionally elaborated the SNF1 pathway cascade in *M. oryzae* through targeted gene deletion strategy. The solo β and γ subunits of SNF1 complex, MoSip2 and MoSnf4 respectively, were identified and further confirmed based on the phenotypic evidences and their strong interaction with MoSnf1, as well as each other. The mutants without β or γ subunit exhibited Snf1‾-like phenotype, but with weaker phenotypic defects compared to *ΔMosnf1*, indicating that MoSnf1 plays a major role in the kinase complex, while the complex integrity promotes MoSnf1 function conversely. In accordance with our findings, truncated yeast Snf1 kinase domain (residues 1–309), *snf1* (1–309), does not interact with the β or γ subunit, but has partial Snf1 function [54]. These results suggest the Snf1 kinase activity might occur by a mechanism independent of the β and γ subunits, although in a limited level. Besides, we additionally characterized two putative upstream Snf1-activating kinases, MoSak1 and MoTos3. However, the two kinase mutants acquired unique properties. ΔMosak1 was defective in conidiogenesis, while the sporulation ability of ΔMotos3 was three times as high as WT, suggesting an antagonistic effect exists between them. On the other hand, only removal of them both resulted in completely eliminated virulence, indicative of a collaborative relationship between the two kinases. The phenotypic changes in ΔMosak1ΔMotos3 were comparable to those of ΔMosnf1 in most aspects, but differences still existed such as shorter aerial hyphae and conidiophores, and reduced mycelial growth rate, implying the two kinases may regulate more processes apart from the SNF1 pathway. In S. cerevisiae, for example, the three upstream kinases, Sak1, Elm1, and Tos3, are also found to participate in the regulation of G protein signaling [55]. Furthermore, the sensitivity to Ca^{2+} in ΔMosak1ΔMotos3 was comparable to that in ΔMosak1, but slighter than the SNF1 complex mutants (Figure S6), while the growth rate of ΔMotos3 on calcium supplemented medium was similar to Guy11 (Figure S6). Therefore, there might be another kinase together with MoSak1 to regulate SNF1 activity under the certain stress. In yeast, although partially redundant function exists within the three Snf1-activating kinases, there is a clear preference to Sak1 in Snf1 activation [6,7]. Likewise, the significant defects in ΔMosak1 suggested MoSak1 occupies a predominant position in MoSnf1 regulation in M. oryzae, while MoTos3 plays an auxiliary role to our knowledge. The conclusion was further supported by the transcription pattern of MoSAK1, with a relatively higher expression increase than MoTOS3 during infection-related processes.

Taken together, the integral SNF1 complex is crucial for various development patterns in M. oryzae, simultaneously the two upstream kinases are of great importance to SNF1 activity with the predominant status of MoSak1. We also demonstrate the SNF1 pathway retains the conserved role to enable the fungus to adapt to nutrient-free environment via its participation in peroxisomal maintenance and lipid metabolism, thus acting as an important pathogenicity-related module. However, the exact mechanism by which SNF1 pathway interplays with peroxisomal dynamics remains elusive in M. oryzae. This study provides some insights for further research on the conservation and divergence of SNF1 pathway among fungi.

Materials and Methods

Strains and culture conditions

M. oryzae wild-type strain Guy11 and all the derivative transformants were grown routinely on complete medium (CM) at 25°C with a 16 h fluorescent light photophase [56]. Growth phenotypic comparisons of Guy11 and the mutant strains were performed on MM supplemented with glucose-substituted non-fermentable carbon substrates (1% Tween 80, 1% olive oil, 1% triolein, or 50 mM sodium acetate) for 10 d, or on CM in addition with 0.3 M Calcium chloride for 7 d.

Quantitative RT-PCR analysis

Fungal tissues used for qRT-PCR assay included fresh mycelia cultured in liquid CM for 3 days, conidia harvested from 10-day-old CM cultures, appressoria incubated on hydrophobic surfaces 8 hours postincubation (hpi), and infected barley leaves collected at 72 hpi. For lipid induced transcription analysis, strains were first incubated in liquid CM for 48 h, and then transferred to liquid MM-C supplemented with 1% glucose, 1% triolein, or 1% olive oil to induce 6 h. The extraction of total RNAs from above samples followed a previously described protocol [56] with the Trizol reagent (Takara). First-strand cDNA was synthesized from 800 ng total RNA according to SYBR ExScriptTM RT-PCR kit (Takara). Quantitative real-time PCR was performed as previously described [56] with SYBR Premix ExTaq (Takara) on a Mastercycler ep realplex thermo cycler (Eppendorf). Primers used for qRT-PCR assays were listed in Table S2. The relative expression level of each gene was calculated by the $2^{-\Delta\Delta Ct}$ method [57] with β-tubulin (MGG_00604) as reference. Data were collected from three independent experiments with three replicates, and a representative set of results was displayed.

Generation of gene knockout mutants and complementation

The gene deletion vectors were constructed based on double-joint PCR strategy [58]. Take the construction of MoSIP2 knockout vector for example: approximate 1.1 kb up- and down-stream regions of MoSIP2 locus were amplified from M. oryzae genomic DNA using primers SIP2up-1/2 and SIP2dn-1/2. A 1.4 kb hph cassette was cloned from pCB1003 with primers HPH-1/2. The three fragments were joined together in the second round PCR, the product of which acted as the template for the final amplification with nest primers SIP2-N1/2. The 3.4 kb double-joint PCR product which contained the flanking sequences and hph cassette was inserted into the XhoI/XbaI sites of pCAMBIA1300 to obtain the targeted gene deletion vector. Using a similar construction strategy, targeted gene deletion vectors of MoSNF1, MoSNF4, MoSAK1, and MoTOS3 were generated. These vectors were introduced into M. oryzae WT strain via Agrobacterium tumefaciens-mediated transformation (ATMT) [59]. To generate the double mutant, the MoTOS3 deletion vector carrying sulfonylurea resistance allele of Magnaporthe ILV1 gene (amplified with primes SUR-1/2) was introduced into ΔMosak1 background. Gene deletion events were screened by PCR and further confirmed by Southern blot analysis. The complementary fragments, which contained the entire targeted genes and their native promoter and terminator regions, were amplified by PCR with primers C1/2 (Table S2) and inserted into a modified pCAMBIA1300 vector, which contained a geneticin resistance gene. The resultant constructs were randomly inserted into the genome of the corresponding mutants via ATMT. The complemented transformants were screened on the selective media containing 800 μg/ml geneticin. As the complementation of MoSNF1 had been carried out before [23], we didn't repeat the work here. Southern blot analysis was performed according to the digoxigenin (DIG) high prime DNA labeling and detection starter Kit I (Roche). Primer pairs used in DNA manipulation events were listed in Table S2.

Yeast two-hybrid assay

The Yeast two-hybrid assay was carried out according to the BD Matchmaker Library Construction & Screening Kits instruction (Clontech). The coding sequence of each candidate gene was amplified with primer pairs listed in Table S2, each of which was incorporated with 15 bases of homology with the ends of the linearized vector. The construct strategy was according to the In-Fusion HD Cloning Kit (Clontech). Consequently, the bait vector pGBKT7-MoSnf1 and each of the prey vector pGADT7-MoSip2, MoSnf4, MoSak1, MoTos3, were co-transformed into yeast strain AH109. The positive tranformants on SD-Leu-Trp medium were further tested on SD-Ade-His-Leu-Trp medium. In order to test the interaction between MoSip2 and MoSnf4, the bait vector pGBKT7-MoSip2 was constructed and co-transformed with pGADT7-MoSnf4 into AH109 as described above. The positive and negative control strains used in the assay were from above mentioned Kit.

Assays for sporulation, appressorium formation, and plant infection

Quantitative measurement of conidiation was assayed with 10-day-old cultures grown on CM plates [23], while the aerial hyphal and conidial development was monitored as described previously [56,60]. To allow appressorium formation, 40 μl of conidial suspensions adjusted to 5×10^4 conidia/ml were placed on plastic cover slips (Fisher) under humid conditions at room temperature. The conidial germination and appressorium formation were observed under a light microscope after 6 h, 12 h, and 24 h postincubation. To monitor appressorium-mediated penetration and invasive hyphae growth, leaf explants of barley (*Hordeum vulgare* cv. ZJ-8) were inoculated with 20 μl of the above conidial suspensions and treated according to previously described protocols [61] at 48 hpi before observation.

For rice infection assay, conidial suspensions were diluted to 1×10^5 conidia/ml in 0.2% gelatin, and 4 ml of each suspension was sprayed onto the 3–4 leaf stage rice seedlings (*Oryza sativa* cv.CO39). Diseased leaves were examined and imaged at 7 d after inoculation. For further disease severity assessment, the diseased leaf area (%) was quantified by calculating the pixels under lesion and healthy areas of diseased leaf blades using the Histogram command of Photoshop CS5 [62]. For barley infection assay, 20 μl 1×10^5 conidia/ml suspensions were deposited on leaf segments of 7-day-old seedlings of barley ZJ-8 incubated in a humid chamber at 25°C for 4 d. For wounded leaf inoculation, the barley leaf surfaces were slightly abraded to remove the cuticle before infection assay. Each test was repeated three times.

Staining methods and microscopy

Eight-day-old conidia were used in the cytological studies. Conidia suspensions diluted to 1×10^5/ml were incubated on hydrophobic films to form appressoria in a humid chamber at 25°C. To stain vacuolar lumen, CMAC (7-amino-4-chloromethyl-coumarin) was used as previously described [63]. Observation of lipid bodies during appressorial development was carried out at 4 h, 12 h, 24 h, 48 h, and 96 h postincubation with Nile Red staining as previously described [28]. Because the melanin layer of the appressorium might interfere with the lipid droplets visualization, samples were treated with 10 μg/ml tricyclazole, a melanin biosynthesis inhibitor, before induced. To investigate the localization of PTS1-containing proteins, the GFP-PTS1 fusion vector p1300NMGFPA (kindly provided by Dr. Jiaoyu Wang), which carried geneticin resistance and was under control of *MPG1* promoter, was introduced into Guy11, *ΔMosnf1*, *ΔMosip2*, *ΔMosnf4*, *ΔMosak1*, *ΔMotos3*, and *ΔMosak1ΔMotos3*, respectively via ATMT. For quantitative analysis of lipid droplets mobilization and PTS1 localization based on fluorescence, more than 100 conidia and appressoria were analyzed for each strain in triplicate. The light and epifluorescence microscopic examination was conducted under an Eclipse 80i microscope (Nikon) equipped with a Plan APO VC 100X/1.40 oil objective. To visualize the GFP-PTS1 signals and CMAC-stained vacuoles in detail, ZEISS LSM780 inverted confocal microscope (Carl Zeiss Inc.) equipped with a 30 mW Argon laser was used. To detect the melanin layer of the appressorium under a JEM-1230 electron microscope (JEOL, Tokyo, Japan) operating at 70 kV, conidia collected from WT and mutants were induced on barley leaves, and the 48 h postincubation segments were treated as described in [43].

Measurement of appressorial turgor and cell wall porosity

Forty microlitre of conidial suspensions with a density of 1×10^5 conidia/ml were dropped on plastic cover slips (Fisher) and incubated in a moist chamber for 48 h. Appressorial turgor was estimated by performing incipient cytorrhysis (cell collapse) assays as described previously [43]. The porosity of an appressorial wall was evaluated by plasmolysis/cytorrhysis assay as described previously [64], where polyethylene glycols (PEGs) of different average molecular weights were added to 48 h appressoria to generate an external osmotic pressure of 4 MPa. Each assay was repeated three times.

Supporting Information

Figure S1 Targeted gene replacements of *MoSNF1*, *MoSIP2*, *MoSNF4*, *MoSAK1*, and *MoTOS3*. (**A**) Targeted gene deletion of *MoSAK1*. The gene deletion vector was constructed based on double joint PCR. The orientations and positions of primers SAK1up-1/2, SAK1dn-1/2, HPH-1/2, and SAK1-N1/2 are indicated as 1–8, respectively, with small arrows. And the deletion event was verified by Southern blot analysis. *Xho*I-digested genomic DNAs were hybridized with the 1.1 kb probe amplified with primes SAK1pb-1/2. As expected, a single band was shifted from WT 2.6 kb to 4.6 kb in *ΔMosak1* and *ΔMosak1ΔMotos3*. The complemented transformant with a single-copy epic insertion was confirmed by two distinct bands observed. The targeted gene replacements of *MoTOS3* (**B**), *MoSNF1* (**C**), *MoSIP2* (**D**), and *MoSNF4* (**E**) were carried out by the similar strategy.
(TIF)

Figure S2 Subcellular localization of GFP-PTS1 in Guy11 and *ΔMosnf1* during appressorial development. Conidia of Guy11 and *ΔMosnf1* were incubated on the surfaces of hydrophobic films and observed at the indicated time points. Contrast to the punctate peroxisomes in the nascent appressoria of WT, GFP-PTS1 was almost absent in the *ΔMosnf1* appressoria. Arrows denote fluorescence-contained vacuoles of WT conidia. Bars = 5 μm.
(TIF)

Figure S3 Ultrastructural analysis of the appressorium cell wall. Appressoria were allowed to form on barley leaves for 48 h, and the ultrathin sections were processed for transmission electron microscopy. The melanin layers (indicated by the arrows) were detected in the wild type and mutant strains. Bar = 0.5 μm.
(TIF)

Figure S4 Recovery of pathogenicity in the complemented strains. (**A**) Pathogenicity assay on detached barley leaves. Diseased leaves were photographed at 4 dpi. (**B**) Spray inoculation assay on rice leaves. Diseased leaves were photographed at 7 dpi.
(TIF)

Figure S5 Expression profiles of *MoPEX1*, *MoPEX11*, *MoMFP1*, and *MoICL1* in the wild type (A) and *ΔMosnf1* (B) strains after induced in fatty acid media for 6 h.
(TIF)

Figure S6 Effect of calcium excess on the growth rate of SNF1 pathway mutants. CM agar plates added with 0.3 M $CaCl_2$ were used to culture strains for 7 days. Different degrees of sensitivity existed between the SNF1 complex mutants and the upstream kinase mutants.
(TIF)

Table S1 Characteristics of SNF1 complex subunits and the upstream Snf1-activating kinases in *M. oryzae*.
(DOC)

Table S2 List of primers used in this study.
(DOCX)

Author Contributions

Conceived and designed the experiments: XZ GC XL FL. Performed the experiments: XZ HS BD. Analyzed the data: XZ JL XL GC. Contributed reagents/materials/analysis tools: XZ JL XL GC. Contributed to the writing of the manuscript: XZ GC JL XL FL.

References

1. Hardie DG, Carling D, Carlson M (1998) The AMP-activated/SNF1 protein kinase subfamily: metabolic sensors of the eukaryotic cell? Annu Rev Biochem 67: 821–855.
2. Hardie DG (2007) AMP-activated/SNF1 protein kinases: conserved guardians of cellular energy. Nat Rev Mol Cell Biol 8: 774–785.
3. Hedbacker K, Carlson M (2008) SNF1/AMPK pathways in yeast. Front Biosci 13: 2408–2420.
4. Amodeo GA, Rudolph MJ, Tong L (2007) Crystal structure of the heterotrimer core of *Saccharomyces cerevisiae* AMPK homologue SNF1. Nature 449: 492–495.
5. Schuller HJ (2003) Transcriptional control of nonfermentative metabolism in the yeast *Saccharomyces cerevisiae*. Curr Genet 43: 139–160.
6. McCartney RR, Schmidt MC (2001) Regulation of Snf1 kinase. Activation requires phosphorylation of threonine 210 by an upstream kinase as well as a distinct step mediated by the Snf4 subunit. J Biol Chem 276: 36460–36466.
7. Hong SP, Leiper FC, Woods A, Carling D, Carlson M (2003) Activation of yeast Snf1 and mammalian AMP-activated protein kinase by upstream kinases. Proc Natl Acad Sci U S A 100: 8839–8843.
8. McCartney RR, Rubenstein EM, Schmidt MC (2005) Snf1 kinase complexes with different beta subunits display stress-dependent preferences for the three Snf1-activating kinases. Curr Genet 47: 335–344.
9. Perez-Sampietro M, Casas C, Herrero E (2013) The AMPK family member Snf1 protects *Saccharomyces cerevisiae* cells upon glutathione oxidation. PLoS One 8: e58283.
10. Zhang M, Galdieri L, Vancura A (2013) The yeast AMPK homolog SNF1 regulates acetyl coenzyme A homeostasis and histone acetylation. Mol Cell Biol 33: 4701–4717.
11. Young ET, Dombek KM, Tachibana C, Ideker T (2003) Multiple pathways are co-regulated by the protein kinase Snf1 and the transcription factors Adr1 and Cat8. J Biol Chem 278: 26146–26158.
12. Vincent O, Carlson M (1998) Sip4, a Snf1 kinase-dependent transcriptional activator, binds to the carbon source-responsive element of gluconeogenic genes. EMBO J 17: 7002–7008.
13. Ratnakumar S, Young ET (2010) Snf1 dependence of peroxisomal gene expression is mediated by Adr1. J Biol Chem 285: 10703–10714.
14. Hong SP, Carlson M (2007) Regulation of snf1 protein kinase in response to environmental stress. J Biol Chem 282: 16838–16845.
15. Palecek SP, Parikh AS, Huh JH, Kron SJ (2002) Depression of *Saccharomyces cerevisiae* invasive growth on non-glucose carbon sources requires the Snf1 kinase. Mol Microbiol 45: 453–469.
16. Tonukari NJ, Scott-Craig JS, Walton JD (2000) The *Cochliobolus carbonum* SNF1 gene is required for cell wall-degrading enzyme expression and virulence on maize. Plant Cell 12: 237–248.
17. Ospina-Giraldo MD, Mullins E, Kang S (2003) Loss of function of the *Fusarium oxysporum SNF1* gene reduces virulence on cabbage and Arabidopsis. Curr Genet 44: 49–57.
18. Tzima AK, Paplomatas EJ, Rauyaree P, Ospina-Giraldo MD, Kang S (2011) VdSNF1, the sucrose nonfermenting protein kinase gene of *Verticillium dahliae*, is required for virulence and expression of genes involved in cell-wall degradation. Mol Plant Microbe Interact 24: 129–142.
19. Nadal M, Garcia-Pedrajas MD, Gold SE (2010) The snf1 gene of *Ustilago maydis* acts as a dual regulator of cell wall degrading enzymes. Phytopathology 100: 1364–1372.
20. Walton JD (1994) Deconstructing the Cell Wall. Plant Physiol 104: 1113–1118.
21. Ruijter GJ, Visser J (1997) Carbon repression in *Aspergilli*. FEMS Microbiol Lett 151: 103–114.
22. Lee SH, Lee J, Lee S, Park EH, Kim KW, et al. (2009) GzSNF1 is required for normal sexual and asexual development in the ascomycete *Gibberella zeae*. Eukaryot Cell 8: 116–127.
23. Yi M, Park JH, Ahn JH, Lee YH (2008) MoSNF1 regulates sporulation and pathogenicity in the rice blast fungus *Magnaporthe oryzae*. Fungal Genet Biol 45: 1172–1181.
24. Wilson RA, Talbot NJ (2009) Under pressure: investigating the biology of plant infection by *Magnaporthe oryzae*. Nat Rev Microbiol 7: 185–195.
25. Tucker SL, Talbot NJ (2001) Surface attachment and pre-penetration stage development by plant pathogenic fungi. Annu Rev Phytopathol 39: 385–417.
26. Weber RW, Wakley GE, Thines E, Talbot NJ (2001) The vacuole as central element of the lytic system and sink for lipid droplets in maturing appressoria of *Magnaporthe grisea*. Protoplasma 216: 101–112.
27. Howard RJ, Ferrari MA, Roach DH, Money NP (1991) Penetration of hard substrates by a fungus employing enormous turgor pressures. Proc Natl Acad Sci U S A 88: 11281–11284.
28. Thines E, Weber RW, Talbot NJ (2000) MAP kinase and protein kinase A-dependent mobilization of triacylglycerol and glycogen during appressorium turgor generation by *Magnaporthe grisea*. Plant Cell 12: 1703–1718.
29. Bhadauria V, Banniza S, Vandenberg A, Selvaraj G, Wei Y (2012) Peroxisomal alanine: glyoxylate aminotransferase AGT1 is indispensable for appressorium function of the rice blast pathogen, *Magnaporthe oryzae*. PLoS One 7: e36266.
30. Ramos-Pamplona M, Naqvi NI (2006) Host invasion during rice-blast disease requires carnitine-dependent transport of peroxisomal acetyl-CoA. Mol Microbiol 61: 61–75.
31. Bhambra GK, Wang ZY, Soanes DM, Wakley GE, Talbot NJ (2006) Peroxisomal carnitine acetyl transferase is required for elaboration of penetration hyphae during plant infection by *Magnaporthe grisea*. Mol Microbiol 61: 46–60.
32. Wang ZY, Thornton CR, Kershaw MJ, Debao L, Talbot NJ (2003) The glyoxylate cycle is required for temporal regulation of virulence by the plant pathogenic fungus *Magnaporthe grisea*. Mol Microbiol 47: 1601–1612.
33. Goh J, Jeon J, Kim KS, Park J, Park SY, et al. (2011) The PEX7-mediated peroxisomal import system is required for fungal development and pathogenicity in *Magnaporthe oryzae*. PLoS One 6: e28220.
34. Wang J, Zhang Z, Wang Y, Li L, Chai R, et al. (2013) PTS1 peroxisomal import pathway plays shared and distinct roles to PTS2 pathway in development and pathogenicity of *Magnaporthe oryzae*. PLoS One 8: e55554.
35. Wang ZY, Soanes DM, Kershaw MJ, Talbot NJ (2007) Functional analysis of lipid metabolism in *Magnaporthe grisea* reveals a requirement for peroxisomal fatty acid beta-oxidation during appressorium-mediated plant infection. Mol Plant Microbe Interact 20: 475–491.
36. Yang J, Kong L, Chen X, Wang D, Qi L, et al. (2012) A carnitine-acylcarnitine carrier protein, MoCrc1, is essential for pathogenicity in *Magnaporthe oryzae*. Curr Genet 58: 139–148.
37. Soanes DM, Chakrabarti A, Paszkiewicz KH, Dawe AL, Talbot NJ (2012) Genome-wide transcriptional profiling of appressorium development by the rice blast fungus *Magnaporthe oryzae*. PLoS Pathog 8: e1002514.
38. Simon M, Binder M, Adam G, Hartig A, Ruis H (1992) Control of peroxisome proliferation in *Saccharomyces cerevisiae* by *ADR1*, *SNF1* (*CAT1*, *CCR1*) and *SNF4* (*CAT3*). Yeast 8: 303–309.
39. Liu Y, Xu X, Carlson M (2011) Interaction of SNF1 protein kinase with its activating kinase Sak1. Eukaryot Cell 10: 313–319.
40. Titorenko VI, Rachubinski RA (2001) The life cycle of the peroxisome. Nat Rev Mol Cell Biol 2: 357–368.
41. Li N, Wang J, Zhang Z, Wang Y, Liu M, et al. (2014) MoPex19, which Is Essential for Maintenance of Peroxisomal Structure and Woronin Bodies, Is Required for Metabolism and Development in the Rice Blast Fungus. PLoS One 9: e85252.
42. Wang ZY, Jenkinson JM, Holcombe LJ, Soanes DM, Veneault-Fourrey C, et al. (2005) The molecular biology of appressorium turgor generation by the rice blast fungus *Magnaporthe grisea*. Biochem Soc Trans 33: 384–388.
43. Liu XH, Lu JP, Zhang L, Dong B, Min H, et al. (2007) Involvement of a *Magnaporthe grisea* serine/threonine kinase gene, *MgATG1*, in appressorium turgor and pathogenesis. Eukaryot Cell 6: 997–1005.
44. Kimura A, Takano Y, Furusawa I, Okuno T (2001) Peroxisomal metabolic function is required for appressorium-mediated plant infection by *Colletotrichum lagenarium*. Plant Cell 13: 1945–1957.
45. Fernandez J, Wright JD, Hartline D, Quispe CF, Madayiputhiya N, et al. (2012) Principles of carbon catabolite repression in the rice blast fungus: Tps1, Nmr1-3, and a MATE-family pump regulate glucose metabolism during infection. PLoS Genet 8: e1002673.
46. Fernandez J, Marroquin-Guzman M, Wilson RA (2013) Mechanisms of Nutrient Acquisition and Utilization During Fungal Infections of Leaves. Annu Rev Phytopathol.
47. Hynes MJ, Murray SL, Duncan A, Khew GS, Davis MA (2006) Regulatory genes controlling fatty acid catabolism and peroxisomal functions in the filamentous fungus *Aspergillus nidulans*. Eukaryot Cell 5: 794–805.
48. Hiltunen JK, Mursula AM, Rottensteiner H, Wierenga RK, Kastaniotis AJ, et al. (2003) The biochemistry of peroxisomal beta-oxidation in the yeast *Saccharomyces cerevisiae*. FEMS Microbiol Rev 27: 35–64.
49. Kunze M, Pracharoenwattana I, Smith SM, Hartig A (2006) A central role for the peroxisomal membrane in glyoxylate cycle function. Biochim Biophys Acta 1763: 1441–1452.
50. Min K, Son H, Lee J, Choi GJ, Kim JC, et al. (2012) Peroxisome function is required for virulence and survival of *Fusarium graminearum*. Mol Plant Microbe Interact 25: 1617–1627.
51. Lazarow PB (2003) Peroxisome biogenesis: advances and conundrums. Curr Opin Cell Biol 15: 489–497.

52. Fernandez E, Moreno F, Rodicio R (1992) The *ICL1* gene from *Saccharomyces cerevisiae*. Eur J Biochem 204: 983–990.

53. Deng Y, Qu Z, Naqvi NI (2013) The role of snx41-based pexophagy in *magnaporthe* development. PLoS One 8: e79128.

54. Ruiz A, Liu Y, Xu X, Carlson M (2012) Heterotrimer-independent regulation of activation-loop phosphorylation of Snf1 protein kinase involves two protein phosphatases. Proc Natl Acad Sci U S A 109: 8652–8657.

55. Clement ST, Dixit G, Dohlman HG (2013) Regulation of yeast G protein signaling by the kinases that activate the AMPK homolog Snf1. Sci Signal 6: a78.

56. Chen G, Liu X, Zhang L, Cao H, Lu J, et al. (2013) Involvement of *MoVMA11*, a Putative Vacuolar ATPase c' Subunit, in Vacuolar Acidification and Infection-Related Morphogenesis of *Magnaporthe oryzae*. PLoS One 8: e67804.

57. Livak KJ, Schmittgen TD (2001) Analysis of relative gene expression data using real-time quantitative PCR and the 2(-Delta Delta C(T)) Method. Methods 25: 402–408.

58. Yu JH, Hamari Z, Han KH, Seo JA, Reyes-Dominguez Y, et al. (2004) Double-joint PCR: a PCR-based molecular tool for gene manipulations in filamentous fungi. Fungal Genet Biol 41: 973–981.

59. Rho HS, Kang S, Lee YH (2001) *Agrobacterium tumefaciens*-mediated transformation of the plant pathogenic fungus, *Magnaporthe grisea*. Mol Cells 12: 407–411.

60. Lau GW, Hamer JE (1998) Acropetal: a genetic locus required for conidiophore architecture and pathogenicity in the rice blast fungus. Fungal Genet Biol 24: 228–239.

61. Lu JP, Feng XX, Liu XH, Lu Q, Wang HK, et al. (2007) Mnh6, a nonhistone protein, is required for fungal development and pathogenicity of *Magnaporthe grisea*. Fungal Genet Biol 44: 819–829.

62. Wilson RA, Gibson RP, Quispe CF, Littlechild JA, Talbot NJ (2010) An NADPH-dependent genetic switch regulates plant infection by the rice blast fungus. Proc Natl Acad Sci U S A 107: 21902–21907.

63. Ohneda M, Arioka M, Nakajima H, Kitamoto K (2002) Visualization of vacuoles in *Aspergillus oryzae* by expression of CPY-EGFP. Fungal Genet Biol 37: 29–38.

64. Jeon J, Goh J, Yoo S, Chi MH, Choi J, et al. (2008) A putative MAP kinase kinase kinase, *MCK1*, is required for cell wall integrity and pathogenicity of the rice blast fungus, *Magnaporthe oryzae*. Mol Plant Microbe Interact 21: 525–534.

4

Comparative Transcriptome Profiling of Two Tibetan Wild Barley Genotypes in Responses to Low Potassium

Jianbin Zeng, Xiaoyan He, Dezhi Wu, Bo Zhu, Shengguan Cai, Umme Aktari Nadira, Zahra Jabeen, Guoping Zhang*

Department of Agronomy, Key Laboratory of Crop Germplasm Resource of Zhejiang Province, Zhejiang University, Hangzhou, China

Abstract

Potassium (K) deficiency is one of the major factors affecting crop growth and productivity. Development of low-K tolerant crops is an effective approach to solve the nutritional deficiency in agricultural production. Tibetan annual wild barley is rich in genetic diversity and can grow normally under poor soils, including low-K supply. However, the molecular mechanism about low K tolerance is still poorly understood. In this study, Illumina RNA-Sequencing was performed using two Tibetan wild barley genotypes differing in low K tolerance (XZ153, tolerant and XZ141, sensitive), to determine the genotypic difference in transcriptome profiling. We identified a total of 692 differentially expressed genes (DEGs) in two genotypes at 6 h and 48 h after low-K treatment, including transcription factors, transporters and kinases, oxidative stress and hormone signaling related genes. Meanwhile, 294 low-K tolerant associated DEGs were assigned to transporter and antioxidant activities, stimulus response, and other gene ontology (GO), which were mainly involved in starch and sucrose metabolism, lipid metabolism and ethylene biosynthesis. Finally, a hypothetical model of low-K tolerance mechanism in XZ153 was presented. It may be concluded that wild barley accession XZ153 has a higher capability of K absorption and use efficiency than XZ141 under low K stress. A rapid response to low K stress in XZ153 is attributed to its more K uptake and accumulation in plants, resulting in higher low K tolerance. the ethylene response pathway may account for the genotypic difference in low-K tolerance.

Editor: Keqiang Wu, National Taiwan University, Taiwan

Funding: The authors are grateful for financial support by Natural Science Foundation of China (31330055 and 31171544), "948" Project of Ministry of Agriculture, China (2012-Z25) and China Agriculture Research System (CARS-05). The funders had no role in study design, data collection and analysis, decision to publish, or preparation of the manuscript.

Competing Interests: The authors have declared that no competing interests exist.

* Email: zhanggp@zju.edu.cn

Introduction

Mineral nutrition is crucial for plant growth and development. However, many plants are often subjected to nutrition stress due to insufficient nutrient supply in soils. Like nitrogen (N), potassium (K) is one of the most abundant elements in plants and performs vital functions in growth, stress adaptation and metabolism, as it is involved in stoma movement, enzyme activation, maintenance of cytosolic pH homeostasis, and stabilization of protein synthesis, etc [1–6]. Although K is quite abundant in the lithosphere and soils, being nearly 10 times higher than N and phosphorus (P) in terms of absolute content, most of them (90–98%) exists in the form of unavailability for plants [6,7]. In other words, available potassium content in soils is commonly very low. In China, most soils show K deficiency for crops, and the case become more severe in recent decades, accompanied by a wide planting of hybrid rice, as it absorbs more K and is more sensitive to low K than inbred rice [8,9].

On the other hand, plants have developed the strategies of coping with low-K stress. It has been well documented that there is a dramatic difference among plant species and genotypes within a species in the response to low-K stress [6,7], indicating that K nutrition in plants is a genetically-controlled trait, and can be improved by genetic manipulation. Thus, it is imperative for us to reveal the mechanism or to explore the relevant genes of high K use efficiency. However, narrower genetic diversity in cultivated barley has become a bottleneck for genetic improvement [10]. The Tibetan Plateau is one of the centers of cultivated barley, and well known for its extreme environment [11]. The Tibetan annual wild barley (referred to wild barley thereafter) has been proved to be rich in genetic diversity and high tolerance to abiotic stresses, such as drought and salinity [12–14]. In the previous experiments, we found that wild barley grew well in the soils with poor fertility and less fertilizer application. Therefore it is possible that wild barley has the special mechanisms in tolerance to low-K stress.

Transcriptome analysis has been widely used in studies of functional genomics. There are two major approaches in the studies of transcriptomes, i.e. sequencing-based and hybridization-based. With the rapid advancement of High-throughput sequencing or so-called Next Generation Sequencing (NGS), RNA-Seq has recently become an attractive method. Compared with hybridization-based tool, such as microarray, RNA-Seq emerges as higher sensitivity, greater dynamic range of expression and base-pair resolution for transcription profiling [15–17]. Furthermore, it also shows clear advantages in revealing novel transcribed regions, single nucleotide polymorphisms (SNPs), the precise location of transcription boundaries and splice isoforms [18,19]. This technique has been used in many plants to reveal gene annotation and expression under biotic and abiotic stresses [20–24].

Previous studies suggested that there is a considerable genetic variation in low-K tolerance among the wild barley accessions [25]. However, a comprehensive transcriptomic analysis of wild barley in response to low-K stress is still not done up to date. Based on the evaluation of low-K tolerance of 99 wild barley genotypes (accessions), we selected 2 wild barley accessions (XZ153, low-K-tolerant and XZ141, low-K-sensitive) as materials in transcriptome analysis using the Illumina RNA-Seq method. The objectives of this study are to determine (1) the possible difference in transcriptome profiles of two wild barley accessions in response to low-K stress; and (2) the signaling pathways and regulatory networks related to low-K tolerance.

Materials and Methods

Plant materials and low-K stress

A hydroponic experiment was conducted in a greenhouse with natural light at Zijingang Campus, Zhejiang University, China. Seeds of two wild barley accessions (XZ153, low-K-tolerant and XZ141, low-K-sensitive) were sterilized with 2% H_2O_2 for 30 min and rinsed with distilled water for three times, then soaked for 6 h at room temperature. The seeds were germinated on moistened filter papers in the germination boxes, placed into a plant growth chamber (22/18°C, day/night). Ten-days-old seedlings were transplanted into plastic pots (5L) for hydroponic incubation. The full-strength nutrient solution contains: 1 mM Ca(-NO_3)$_2$.4H$_2$O, 1 mM KCl, 1 mM MgSO$_4$, 0.25 mM NH$_4$H$_2$PO$_4$, 50 μM CaCl$_2$, 20 μM Fe-citrate.nH$_2$O, 12.5 μM H$_3$BO$_3$, 0.5 μM H$_2$MoO$_4$, 0.5 μM CuSO$_4$.5H$_2$O, 2 μM MnCl$_2$.4H$_2$O, 2 μM ZnSO$_4$.7H$_2$O. The pH was adjusted to 6.0±0.1 as required. Plants were supplied with half-strength of the hydroponic solution in the first week and then changed into full strength solution from the next week and renewed every five days. Three-leaf-stage seedlings were subjected to low-K treatment. The potassium concentration was adjusted to 0.01 mM (low-K treatment) and 1 mM as control, respectively.

Biomass and potassium content determination

At 15 d after low-K treatment, the roots of all seedlings were thoroughly rinsed with tap water and dried with tissue papers. Then shoots and roots of seedlings were harvested and separated. All the plant samples were dried at 80°C for 72 h until their weight remained constant for biomass measurement. Dry shoots and roots were ground into powder, and approximately 0.1 g tissue sample was prepared for K content determination using Inductively Coupled Plasma-Optical Emission Spectroscopy (ICP-OES) (Optima 8000DV, PerkinElmer, USA).

Gene expression assay

For time course pre-analysis of the expression of the gene HvHAK1 under low-K stress, seeds of XZ153 were germinated and seedlings were cultivated as described above. All the endosperms were removed away from the seedlings to eliminate any additional supply of nutrition. The plants were incubated with 1/2 strength nutrient solution for 5 d and refreshed with full-strength for another 5 d. Then the two-leaf-stage seedlings were treated under low-K (0.01 mM) and normal K (1 mM) conditions. The roots of XZ153 were sampled with 3 biological replicates at 6 h, 12 h, 24 h, 48 h, 5 d and 7 d after treatment. The root samples were

frozen in liquid nitrogen immediately and stored at −80°C for RNA extraction.

RNA-Seq sampling and RNA isolation

For RNA-Seq sampling, seeds of XZ153 and XZ141 were germinated at the same condition and placed into a plant growth chamber. The two-leaf-stage seedlings were exposed to low-K stress (0.01 mM) for 0 h, 6 h and 48 h, respectively. Roots of 10 seedlings were collected and mixed together at each time point to reduce the differences between plant individuals. There were 6 samples [2 genotypes (XZ153, low-K-tolerant and XZ141, low-K-sensitive) ×3 time periods (0 h, 6 h, 48 h)] in total for further RNA-Seq research. RNA isolation was carried out according to the instructions of miRNeasy mini kit (QIAGEN, Germany). RNA abundances and purity was tested for meeting the requirements.

Library construction, sequencing and data processing

mRNA enrichment was obtained from the total RNA by the magnetic beads with poly-T oligonucleotide. Then mRNA was randomly broken into fragments. Double-stranded cDNA was synthesized using reverse transcriptase combined with random primers and with adapters ligated at both ends. With those adapter sequences, DNA fragments were selectively amplified and enriched. Thus, the cDNA libraries were ready for sequencing. Qubit quantitation, insert size tested by Agilent 2100, and the Q-PCR were also conducted for accurate quantification before sequencing.

PCR products were loaded onto Illumina HiSeq2000 platform for 2×100 bp pair-ends sequencing. Then the RNA-Seq reads were generated via the Illumina data processing pipeline (version 1.8). To obtain the clean data, the raw reads were trimmed by removing empty reads, adaptor sequences and low quality bases at the 3 end. Then all the clean reads were considered for further analysis. The barley genome sequence and annotation data was downloaded, and TopHat (http://tophat.cbcb.umd.edu/) was adopted to align RNA-Seq reads to the barley reference genomes using the ultra high-throughput short read aligner, and then analyzes the mapping results to identify splice junctions between exons.

Identification of the differentially expression genes (DEGs) and quantitative RT-PCR analysis

For gene expression analysis, the expression level of each gene was calculated by quantifying the number of reads. Gene expression counts were normalized by a variation of the FPKM (fragments per kilo-base of exon per million fragments mapped reads) method [26]. To identify differentially expression genes (DEGs) between the two different samples, the software Cufflinks was employed to output the T-statistic and the p-values for each gene [27]. We calculated the expression ratio of 6 h/0 h or 48 h/0 h as fold changes, respectively. Differentially expressed genes (DEGs) were required to have a 2-fold change and p≤0.01. In addition, an FPKM value≥2 in at least one of the samples was applied to genes for statistical analysis [28].

The RNA samples for RNA-Seq were also used for real-time PCR assays to confirm the reliability of the RNA-Seq result. 1 μg total RNA was treated with DNase I to eliminate the genomic DNA contamination, used as a template for reverse transcription (Takara, Japan). First strand cDNA was synthesized with oligo dT primer and Random 6 mers in a 20 μl reaction. Real time PCR was performed on a CFX96 system machine (Bio-Rad, USA). The PCR profiles were as follows: Pre-denaturation at 95°C for 30 s, 40 cycles of denaturation at 95°C for 5 s and annealing at 60°C for 30 s, followed by steps for Melt-Curve analysis (60°C–95°C, 0.5°C increment for 5 s per step). The relative expression of the chosen genes was calculated according to the comparative CT method [29]. In order to normalize all the data, the amplification of *HvGAPDH* sequence was used as endogenous reference. The gene specific primers were designed using primer-blast (http://www.ncbi.nlm.nih.gov/tools/primer-blast/). All the primers were listed in Table S6.

Gene annotation, GO enrichment and KEGG analysis

The Blast2GO program was used to obtain GO annotation for the DEGs, as well as for KEGG analysis (http://www.blast2go.com/b2ghome) [30]. BLASTx was performed to align against NCBI non-redundant (nr) protein database for homology search. Following the mapping step, the gene ontology (GO) annotation, InterProScan annotation and enzyme code annotation steps were conducted in details with default parameters. The GOs distribution associated with DEGs were then obtained from three levels: molecular functions, biological processes and cellular components. The KEGG maps containing the EC numbers and enzymatic functions in the context of the metabolic pathways, in which they participate as well as annotation results can be available in a variety of formats [31].

Statistical analysis

The significance of difference between the two barley genotypes in physiological traits and gene expression was examined using data processing system (DPS) statistical software package, followed by the Duncan's Multiple Range Test (DMRT) and the difference at $P<0.05$ and 0.01 is considered as significant and highly significant, respectively.

Results

Effect of K level on biomass, K concentration and accumulation of two wild barley accessions

A total of 99 barley accessions were evaluated in a previous experiment under low-K (0.01 mM) and normal K (1 mM) conditions [25]. XZ153 and XZ141 were identified as low-K tolerant and sensitive, respectively. Although two genotypes grew worse under low K than normal K (control), XZ153 was obviously less affected than XZ141 (Table 1). Hence, relatively dry weight of shoot (low K/control) was 90% for XZ153 and 64% for XZ141 (Table 1).

Furthermore, the two wild barley accessions differed greatly in K concentration and accumulation (Table 1). There was little difference in both root and shoot concentrations between XZ153 and XZ141 under normal K. However, shoot K concentration of XZ153 was significantly higher than that of XZ141 under low-K (Table 1). K accumulation is a function of plant dry weight and K concentration. As a result, K accumulation of XZ153 was 1.61 times larger than that of XZ141 under low-K (Table 1).

Evaluation of RNA-Seq reads and mapping results

In order to determine suitable time of sampling for RNA-Seq analysis, relative expression of *HvHAK1* at 6 h, 12 h, 24 h, 48 h, 5 d and 7 d after low-K treatment was compared (Figure S1). The results showed that the *HvHAK1* gene was up-regulated at 6 h after low-K stress, then remained little change at 12 h and 24 h. Obviously, roots has already sensed low-K signal and activated relevant signal transduction at 6 h after treatment, resulting in differential expression of the genes so as to cope with low-K stress. Interestingly, the expression level of the *HvHAK1* gene was significantly increased at 48 h in comparison with those at 6 h,

Table 1. K concentration and accumulation of two wild barley genotypes XZ153 (Low-K-tolerant) and XZ141 (Low-K-sensitive) under low and normal K levels.

Trait		Genotype					
		XZ153			XZ141		
		CK	LK	Relative	CK	LK	Relative
Dry weight (mg plant⁻¹ DW)	Root	63.17a	60.33a	0.96	61.33a	46.17b	0.75
	Shoot	207.67a	187.67a	0.90	212.33a	135.50b	0.64
	Total	270.83a	248.00b	0.92	273.67a	181.67c	0.66
K concentration (mg g⁻¹ DW)	Root	65.44a	10.08b	0.15	64.21a	9.70b	0.15
	Shoot	82.33a	31.08b	0.38	81.31a	26.15c	0.32
K accumulation (mg plant⁻¹ DW)	Root	4.13a	0.61b	0.15	3.94a	0.45b	0.11
	Shoot	17.10a	5.83b	0.34	17.27a	3.54c	0.21
	Total	21.23a	6.44b	0.30	21.21a	3.99c	0.19

CK: Normal K level (1 mM K); LK: Low K level (0.01 mM K); Relative: LK/CK. For each line, different lowercase letters indicate significant differences ($P<0.01$) among the treatments and genotypes, n = 3.

12 h or 24 h, and thereafter, remained little change at 5 d and 7 d (Figure S1). Thus, we took the samples at 6 h and 48 h for RNA-Seq analysis.

To obtain an overview of the transcriptome profiling of the early response to low-K stress in the two wild barley accessions, six sequencing cDNA libraries were constructed, and paired-end reads were sequenced using the Illumina platform. We got raw reads with length ranging from lower than 50 bp to as high as 101 bp. A total of 223 112 382 clean reads were generated by sequencing 6 cDNA libraries. All reads were categorized into three classes, including unmapped, multiple mapped and unique mapped reads. Of the 29–41 million clean reads from each library, 77–81% was mapped to unique locations, whereas 6–10% was mapped to multiple locations in the genome (Table S1). Meanwhile, the number of expressing genes found in each sample ranged from 54 322 to 57 516, thus providing massive data for further expression profiling analysis.

Identification of differentially expressed genes (DEGs) and cluster analysis

Gene profiles of wild barley roots under both normal and low-K conditions were analyzed. FPKM (Fragments per kilo-base of exon per million fragments mapped reads) method was employed to normalize gene expression counts for the sequence. Previous studies suggested that sequencing with low FPKM may not provide reliable expression data statistically [28]. To minimize false positives, FPKM≥2 was required at least for one of the samples [28]. Additionally, differentially expressed genes were identified according to fold change and P value [32]. Hence, in this study, we set a screening threshold (FPKM≥2 at least in one of the samples, 2-fold change, P≤0.01) of differentially expressed genes. We used the same criteria for both XZ153 and XZ141 to obtain genes that had a significant response to low-K stress. A total of 692 genes showed differential expression at 6 h and 48 h under low-K stress in the two accessions (Table S2, Table S3). XZ141 had more differentially expressed genes than XZ153 (Figure 1). There were only 137 differentially expressed genes, which were commonly found in both XZ153 and XZ141. There were more DEGs at 6 h than at 48 h in the two accessions (Table S4, Table S5). Nearly same amount of DEGs were up-regulated and down-regulated at 6 h and 48 h in XZ141 (Figure 1). However, the gene expression pattern in XZ153 differed from that in XZ141. The number of up-regulated genes in XZ153 was almost four times as large as that of down-regulated ones at 6 h (Figure 1).

Meanwhile, the 692 DEGs were divided into eight groups, based on their expression pattern by cluster analysis (Genesis 1.7.5) (Figure S2). Clusters 1 and 2 included the genes with continuous positive or negative response along the whole time course; clusters 3 and 4 included the genes with persistent positive or negative response; clusters 5 and 6 included the genes with latent positive or negative response, and cluster 7 and 8 referred to the genes with initial positive or negative response (Figure S2). In view of the differentially expressed gene patterns and DEGs, it can be suggested that XZ153 has a distinct mechanism differing from XZ141 in response to low-K stress. Thus, it is valuable to make further analysis of different responses to low-K stress between the two accessions.

To assess the validity of the RNA-Seq data, 15 DEGs were randomly selected for real-time PCR analysis (Table S6) and they included CBF protein 4 (XLOC_069634), transporter HKT7-like (XLOC_074072), late embryogenesis abundant protein(XLOC_022400), dehydrin 1 (XLOC_090019), kelch repeat-containing F-box family protein (XLOC_069158), chaperone protein DnaJ (XLOC_024068), Ring-h2 finger protein ATL32-like

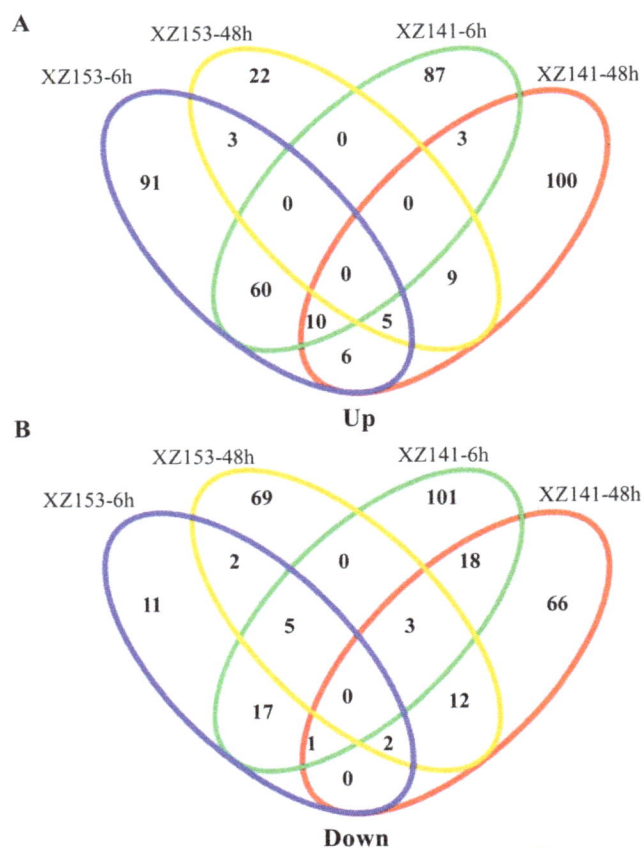

Figure 1. A Venn diagram describing overlaps among differentially expressed genes (DEGs) in XZ153 and XZ141. (A) Up-regulated genes at 6 h and 48 h after low-K treatment. (B) Down-regulated genes at 6 h and 48 h after low-K treatment.

(XLOC_003376), low temperature and salt responsive protein family (XLOC_054609), three pathogenesis-related proteins (XLOC_002529, XLOC_064196, XLOC_067584), response regulator like protein (XLOC_072378), phosphomethylpyrimidine synthase (XLOC_059862) and unknown proteins (XLOC_060141, XLOC_055640) (Table S6). For most of these genes, their expression patterns of the real-time PCR were highly consistent with those shown in the RNA-Seq data (Figure 2).

Transcription factors (TFs) in the differentially expressed genes (DEGs)

Transcription factors are essential for regulation of gene expression. In this study, 46 DEGs encoding transcription factors were identified, and they belonged to diverse families, such as Zinc finger (22), MYB (11), bZIP (3), CBF (2), NF-Y (2), ERF (1), WRKY (1), bHLH (1), MADS-box (1), AP2/EREBP (1), HSF (1) (Figure 3). Interestingly, we found that the proteins with zinc finger domains were the most enriched among the TFs, accounting for 48% of all DEGs (Figure 3). According to the expression patterns, all the TFs could be clustered into several categories. Overall, most TFs in both two accessions displayed a short-term response, and then returned to the original expression level, although a few of them were only differentially expressed at 48 h (Figure 3). Hence, we were focused on these TFs which were significantly up-regulated in XZ153, but remained little change in XZ141, or remained little change in XZ153 but down-regulated in XZ141.

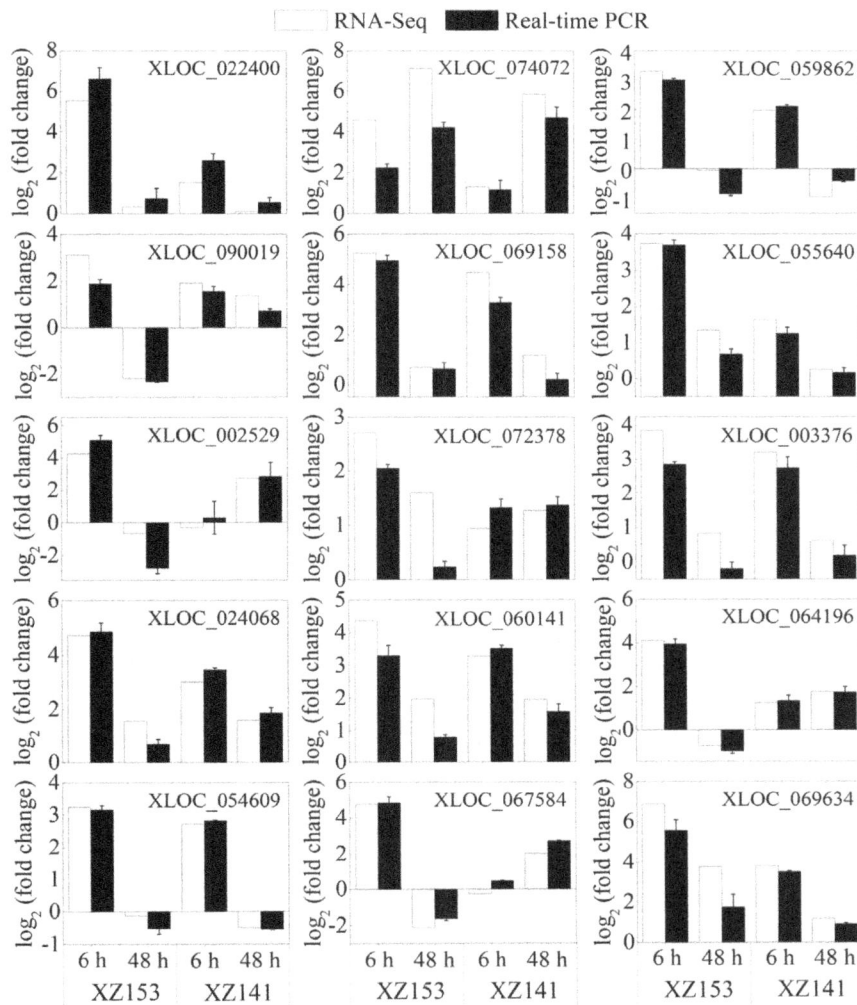

Figure 2. Quantitative real-time PCR validation of 15 differentially expressed genes. Gene-specific primers used for real-time PCR are listed in Table S6.

Transporter and kinase

In the current study, six DEGs encoding K transporters were detected. Surprisingly, expression level of the genes encoding K transporters was not changed obviously until after 48 h stress in XZ141 (Table 2). By contrast, three of them (XLOC_030313, XLOC_086787, XLOC_074072) were up-regulated at 6 h in XZ153, including a HKT7-like gene with higher expression (Table 2). Meanwhile we also found that the expression of the genes encoding nitrate and ammonium transporters changed greatly under low-K stress (Table 2), indicating that K and N uptake is under cross-talking regulation. In addition, the genes associated with ion uptake or translocation were discovered, which is probably involved in ion homeostasis under low-K stress (Table 2). Moreover, some DEGs encoding protein kinases of different groups were also identified (Table 2). Among these genes, there were 4 CBL-interacting protein kinases, 3 leucine-rich repeat receptor-like protein kinases and 2 cysteine-rich receptor-like protein kinases. On the whole, the expression of those groups was mostly down-regulated in XZ141, but up-regulated or unchanged in XZ153.

DEGs related to hormone signaling and oxidative stress

Heatmap clustering analysis was performed to study differentially expressed genes involved in hormone signaling in the two barley accessions under low-K stress (Figure S3). A total of 24 hormone-related DEGs were found, including gibberellin (4), jasmonate (5), cytokinin (3), auxin/IAA (8), ethylene (1) and abscisic (3) (Figure S3). Meanwhile, the key enzymes, such as peroxidase (6), cytochrome P450 (8) and glutathione S-transferase (8) were indentified, and they are critical for the regulation of ROS production and reducing cellular damage in the response to low-K stress (Figure S4). At 6 h after low-K treatment, there were few oxidative stress-related genes (Figure S4). However, the changed genes increased dramatically at 48 h after treatment. Moreover, there were more differentially expressed genes in XZ141 than in XZ153 (Figure S4).

Gene ontology (GO) function and KEGG analysis of low-K tolerance related DEGs

We are focused on those DEGs, whose expression was significantly up-regulated in XZ153 roots, but down-regulated/unchanged in XZ141, or remained little changed in XZ153 but

46

Cereal Grain Crops

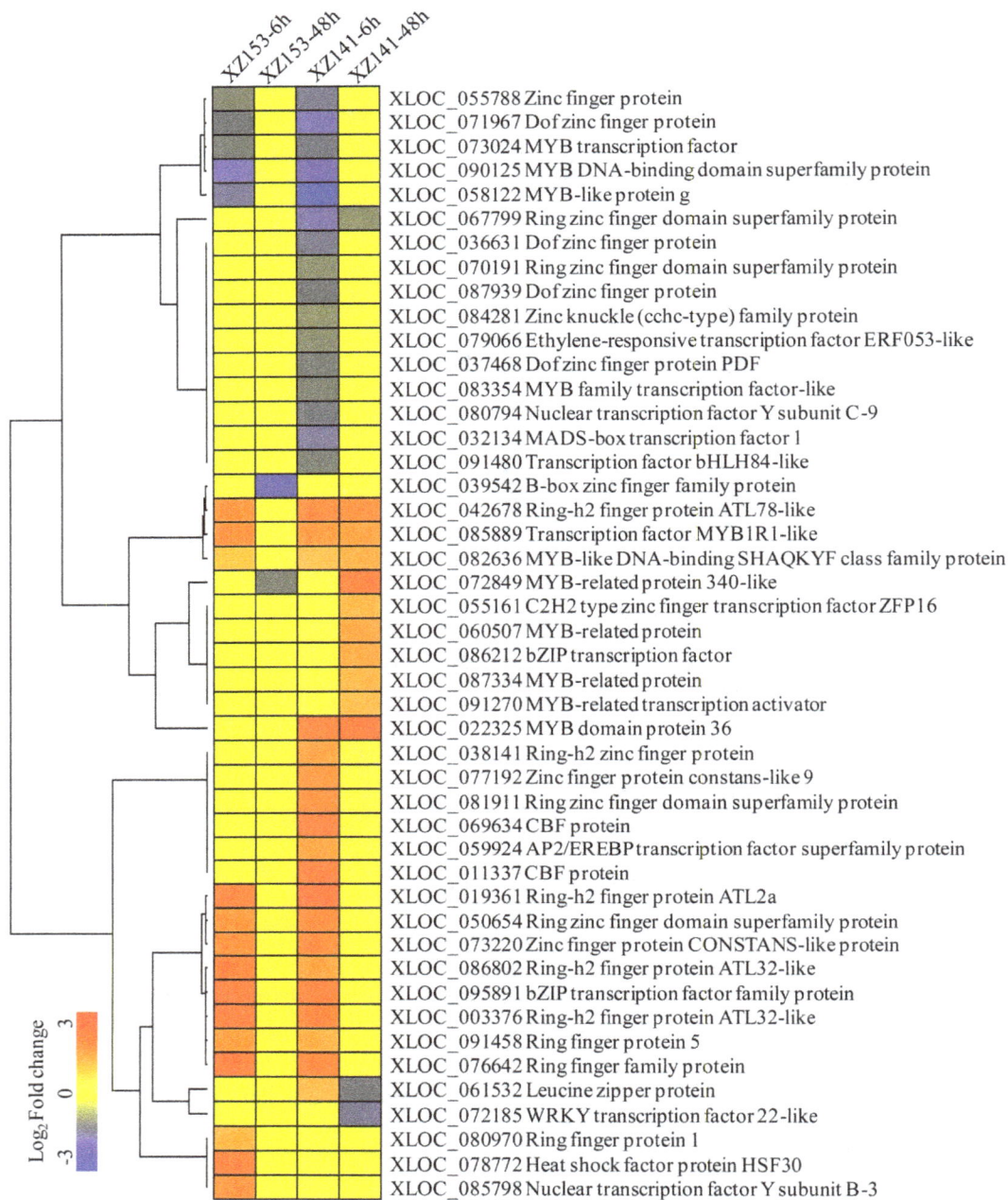

Figure 3. Average linkage hierarchical cluster analysis of Transcription factors (TFs) identified in differentially expressed genes (DEGs). The sample and treatments are displayed above each column. Genes are displayed by different colors. Relative levels of expression are showed by a color gradient from low (blue) to high (red).

down-regulated in XZ141. In this study, a total of 294 DEGs met the above criteria and were further investigated. Hierarchical clustering analysis of those DEGs was performed and they could be mainly grouped into four classes (Figure 4A). GO functional enrichment analysis revealed that the genes associated with binding (GO: 0005488), catalytic activity (GO: 0003824) and transporter activity (GO: 0005215) were significantly enriched, accounting for as much as 89% of molecular function (Figure 4B). The processes represented by the GO terms 'metabolic process', 'response to stimulus', 'single-organism process', 'biological regulation' and 'cellular process', accounted for the majority of

the biological process (Figure 4C). Meanwhile, DEGs related to low-K tolerance also acted as diverse cellular components (Figure 4D).

Totally, 294 DEGs encoding various enzymes were further investigated for KEGG pathway enrichment. Forty-seven enzymes were assigned to 32 KEGG pathways, including amino acid, nucleotide, lipid, carbohydrate, energy and other metabolisms (Figure S5). Based on FPKM value, we found 6 enzymes, which were involved in starch and sucrose metabolisms, and altered markedly (Figure S6). In addition, six pathways, classified as lipid metabolism, including alpha-linolenic acid, fatty acid biosynthesis,

Table 2. Genes encoding protein transporters and kinases showing genotypic difference expression in response to low K stress.

Group	Gene id	Log2(Fold change)				Seq description
		XZ153		XZ141		
		6 h	48 h	6 h	48 h	
Potassium	XLOC_030313	2.62	4.38		4.59	High-affinity potassium transporter
	XLOC_086787	2.36	2.82		3.06	High-affinity potassium transporter
	XLOC_032661		2.01		2.89	High-affinity potassium transporter
	XLOC_033262		1.89			Potassium transporter
	XLOC_035041				3.05	High-affinity potassium transporter
	XLOC_074072	4.59	7.13		5.84	Transporter HKT7-like
Nitrate	XLOC_031163	2.14				Probable nitrate transporter
	XLOC_093217	2.04			2.22	Nitrate transporter -like
	XLOC_082123		−2.61	−1.89	−3.79	Nitrate transporter -like
	XLOC_020339				1.94	High-affinity nitrate transporter -like
Ammonium	XLOC_050205				2.83	Ammonium transporter
	XLOC_027452				1.86	Ammonium transporter
Yellow-strike	XLOC_082435	−3.11	−4.61	−5.10		Metal-nicotianamine transporter
	XLOC_082568	−3.19	−4.05	−4.89		Metal-nicotianamine transporter
	XLOC_082319	−3.19	−4.62	−5.00		Metal-nicotianamine transporter
	XLOC_090161			−2.26		Probable metal-nicotianamine transporter
Mate	XLOC_071812	2.76			2.37	MATE efflux family protein
	XLOC_079551	2.34				MATE efflux family protein
	XLOC_079205				−2.23	MATE efflux family protein, expressed
	XLOC_035193			−4.33	−2.32	MATE efflux family protein
CIPK	XLOC_095894	4.59		4.03		CBL-interacting protein kinase
	XLOC_050803			−2.04		CBL-interacting protein kinase
	XLOC_083910			−2.37		CBL-interacting protein kinase
	XLOC_065318			−2.56		CBL-interacting protein kinase
LRR	XLOC_066629				−3.00	LRR receptor-like protein kinase
	XLOC_081108			3.16		LRR receptor-like protein kinase
	XLOC_087240	2.80				LRR receptor-like protein kinase
	XLOC_084744			−2.61		LRR receptor-like protein kinase
CRK	XLOC_007825			−3.52	−3.99	Cysteine-rich receptor-like protein kinase
	XLOC_071068	5.55				Cysteine-rich receptor-like protein kinase

glycerolipid, glycerophospholipid, linoleic acid and sphingolipid metabolisms, differed in expression patterns between XZ153 and XZ141 (Figure S5). Among them, two DEGs encoding lipase (EC: 3.1.1.3) and 13S-lipoxygenase (EC: 3.2.1.23) showed up-regulation in XZ153, but unchanged in XZ141; whereas three enzymes (EC: 3.2.1.23, EC: 6.3.4.14, EC: 2.3.1.15) were unchanged in XZ153, and down-regulated in XZ141. Meanwhile, the expression of two key enzymes (Homocysteine S-methyltransferase and S-methyl-5-thioribose kinase) participating in S-Adenosyl-L-methionine (SAM) cycle and methionine salvage process, was unchanged in XZ153, but down-regulated in XZ141 (Table S7). Furthermore, one DEGs encoding S-Aminocyclopropane-1-carboxylate synthase (ACS), a key rate-limiting enzyme in ethylene biosynthesis pathway, was also indentified, which showed normal expression in XZ153, and down-regulation in XZ141 (Figure 5, Table S7).

Discussion

Potassium is an essential macronutrient for plant growth and development. However, K deficiency in soil is quite common and becomes more severe in crop production [7]. The most effective approach of overcoming K deficiency is to develop the crop cultivars with high tolerance to low-K stress or high K use efficiency. In the present study, we used the RNA-Seq to reveal the transcriptome profiling of two wild barley accessions differing in low K tolerance. Firstly, phenotypic responses of these two genotypes were compared. It was reported by Hermans *et al* (2006) that biomass was the final result of plant growth and development, so relatively dry weight was often used as indicators of plant tolerance to low nutrition stress [33,34]. Under low K, root and shoot growth was dramatically inhibited for XZ141, whereas remained less effect for XZ153 (Table 1), proving that XZ153 is more tolerant to low-K stress than XZ141. Meanwhile, the reduction of K concentration and accumulation in the shoot under

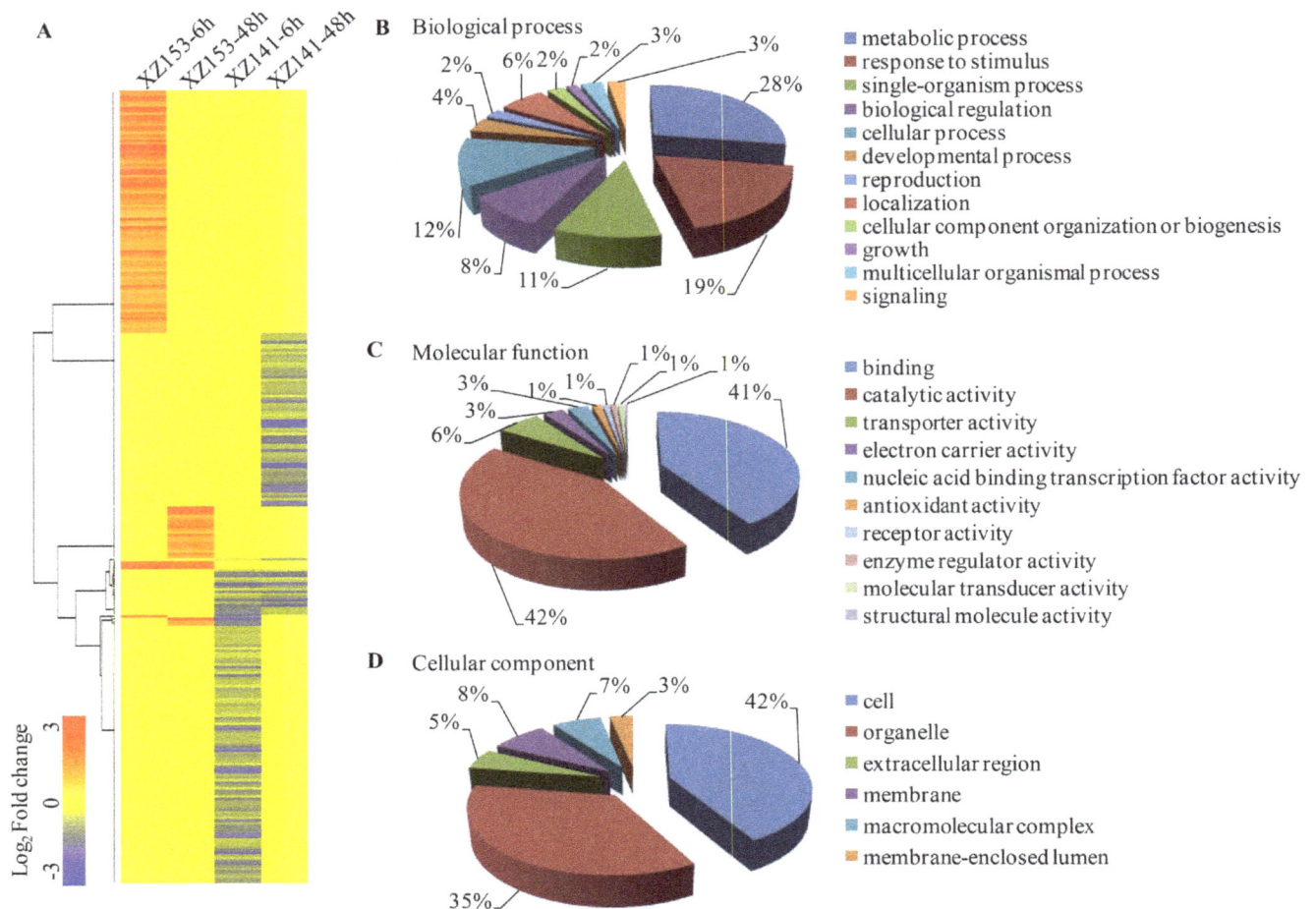

Figure 4. Hierarchical cluster and gene ontology (GO) categories analysis of low-K tolerance related DEGs. A total of 294 low-K tolerance related DEGs were performed on (A) Hierarchical cluster analysis. The samples and treatments are displayed above each column. Genes are displayed by different colors and relative levels of expression are showed by a color gradient from low (blue) to high (red). Gene ontology categories from three levels: (B) Molecular function; (C) Biological process; (D) Cellular component.

low K also differed greatly between the two genotypes, XZ153 being less reduced than XZ141, indicating that XZ153 had the higher capability of K absorption and translocation.

The capacity of K absorption and translocation is related to K transporters and channels, which belong to high- and low-affinity uptake system, respectively. A number of K transporters and channels have been functionally characterized in plants [3,35,36]. It has been demonstrated that *HvHAK1*, as well as *AtHAK5* and *OsHAK1* are all assigned to the KT/KUP/HAK family and could be induced by low external potassium [37–41]. The previous studies reported that there were seven HAK-type transporters in the response to low-K stress in rice; including up- and down-regulated expression patterns [42]. However, in this study, all the HAK transporter proteins were up-regulated in the two genotypes. Interestingly, some of them were up-regulated in XZ153 as early as at 6 h after treatment, but not in XZ141 at that time (Table 2). Thus, it may be assumed that a rapid response to low K stress happened in XZ153 is attributed to its more K uptake and accumulation in plants, resulting in higher low K tolerance.

Transcription factors are a kind of proteins that are bound to *cis*-regulatory elements and can regulate gene expression [43]. The roles of Zinc finger protein and some members of the MYB transcription factor family in abiotic stress tolerance have been

well documented [44–47]. In the present study, 48 DEGs encoding transcription factors were characterized (Figure 3), with Zinc finger and MYB being the largest components, similar to the results that were obtained in rice plants exposed to K starvation [48]. Obviously, some members of those two families are associated with the responses to low-K stress, at least in rice and barley. Nevertheless, nuclear factor Y (NF-Y), a conserved heterotrimeric and CCAAT-specific domain, composed of NF-YA, NF-YB, NF-YC, was only observed in this study, and was not found in rice and soybean subjected to low-K stress [42,49]. Although the members of NF-Y are involved in multiple biological functions, its relation with low-K stress tolerance has been not reported up to data [50,51]. Meanwhile, many TFs, mainly involved in plant growth and development, were repressed in XZ141. It may be assumed that more growth inhibition of XZ141 is described to lower K concentration in plant tissues under low-K stress.

The plant hormone, ethylene is involved in many aspects of the plant life cycle and its production is tightly regulated in response to environmental stimuli from both of biotic and abiotic stresses [52]. Hence, the mechanisms related to biosynthesis of ethylene have been intensively studied. Using chemical and genetic approaches, Jung *et al.*, (2009) found the ethylene-ROS pathway in *Arabidopsis*

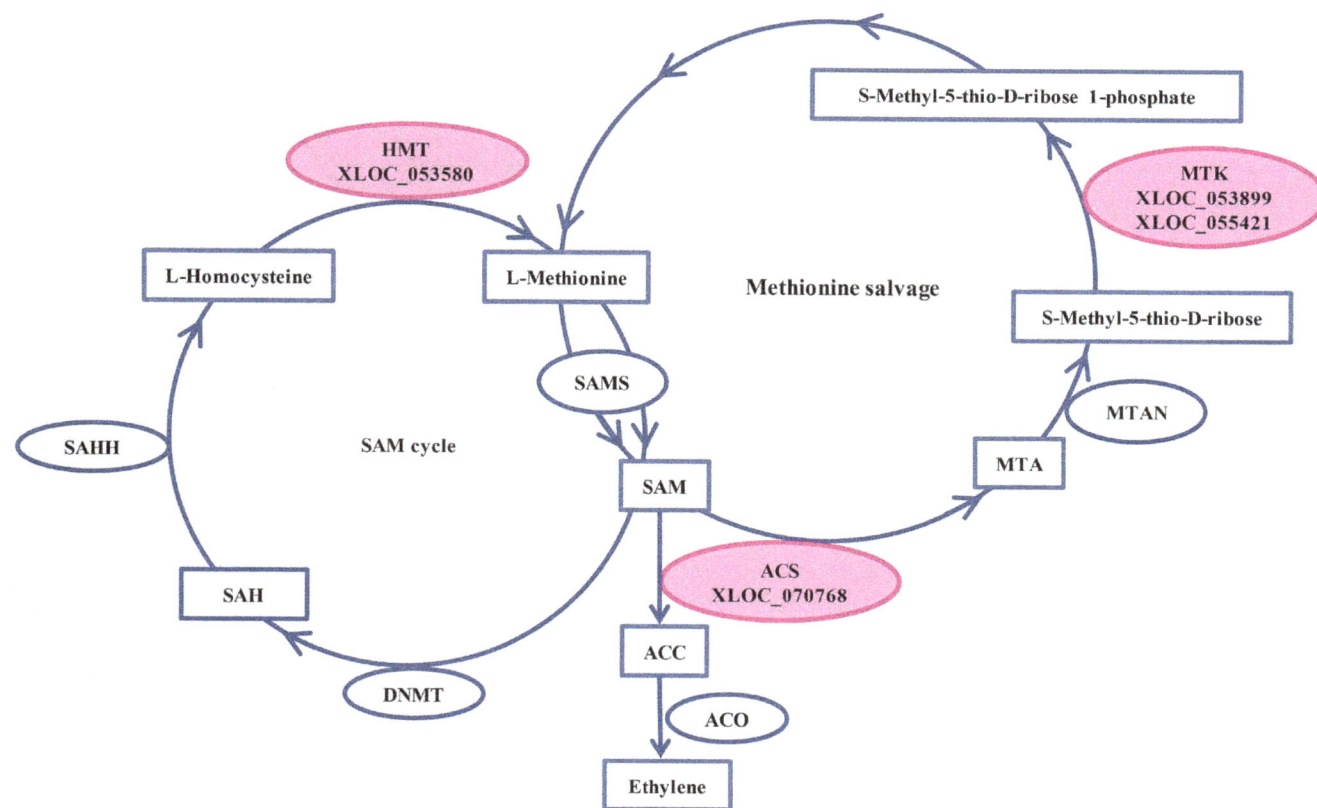

Figure 5. DEGs were mapped to SAM cycle and methionine salvage involving in ethylene biosynthesis process. The three differentially expressed enzymes were colored in pink. SAM: S-Adenosyl-L-methionine; SAMS: S-Adenosyl-L-methionine synthase; ACC: S-Aminocyclopropane-1-carboxylate; ACS: S-Aminocyclopropane-1-carboxylate synthase; ACO: S-Aminocyclopropane-1-carboxylate oxidase; MTA: S-Methyl-5'-thioadenosine; MTAN: 5'-Methylthioadenosine Nucleosidase; MTK: S-methyl-5-thioribose kinase; SAH: S-Adenosyl-L-homocysteine; SAHH: S-Adenosyl-L-homocysteine hydrolase; DNMT: DNA (cytosine-5-)-methyltransferase; HMT: Homocysteine S-methyltransferase.

that ethylene stimulated the production of ROS and played an important role in low-K stress [53]. As S-Aminocyclopropane-1-carboxylate (ACC) is a direct precursor of ethylene biosynthesis, the step from the (S-Adenosyl-L-methionine synthase) SAM to ACC, catalyzed by ACC synthase (ACS) and usually considered as a rate-limiting step in ethylene synthesis, is very important in this process. In addition to the two enzymes involved in SAM cycle and methionine salvage, a key enzyme ACS was also found in the current study (Figure 5). In particular, the DEGs encoding these three enzymes were inhibited in sensitive genotype XZ141 under low-K stress, but unchanged in the tolerant genotype XZ153 (Table S7). Meanwhile, an ethylene-responsive transcription factor ERF053-like was down-regulated simultaneously in XZ141, while little change in XZ153 (Figure 3). The AP2/ERF transcription factor, RAP2.11 in *Arabidopsis*, is a factor in regulating transporter through biding to promoter of *AtHAK5* and has been also identified as a component of ethylene-ROS pathway under low-K stress [54]. Obviously, the ethylene response pathway may account for the genotypic difference in low-K tolerance, which may be reflected by K uptake capability.

KEGG is a major public pathway-related database, which is valuable for research on the complex biological behaviors of genes. In the present study, 294 DEGs should be specially addressed in view of their expression patterns, i.e. the genes being up-regulated in tolerant genotype XZ153 and remained unchanged or down-

regulated in sensitive genotypes XZ141, or unchanged in XZ153, and down-regulated in XZ141 (Figure 4A). KEGG metabolic pathway enrichment of these genes will likely make insight into the mechanism of low-K tolerance. The difference was found in the regulation of some metabolisms between the two genotypes, including carbohydrate, cysteine, methionine and lipid metabolisms (Figure S5). Armengaud *et al* (2009) reported that the genes involved in carbohydrate metabolism played a vital role in the regulation of potassium nutrition in plants [55]. In this context, these differences in metabolic pathways may lead to different energy distribution and capacity of adaptation to low-K stress in the two genotypes.

In conclusion, we applied high-throughput Next Generation Sequencing (NGS) technique to investigate gene expression profiling of two Tibetan wild barley genotypes in response to low-K stress. The results demonstrate that the response of XZ153 and XZ141 to low-K stress differed dramatically in the transcriptional level. Despite the complexity of responses to low-K, a hypothetical model could be suggested for low-K tolerance mechanism in XZ153, based on the available results (Figure 6). While the difference in K absorption and accumulation in plant tissues between the two genotypes may explain the different growth inhibition under low K stress. In our understanding, this is a first comprehensive study of low K tolerance in Tibetan wild barley at transcriptome level. In addition, the current results

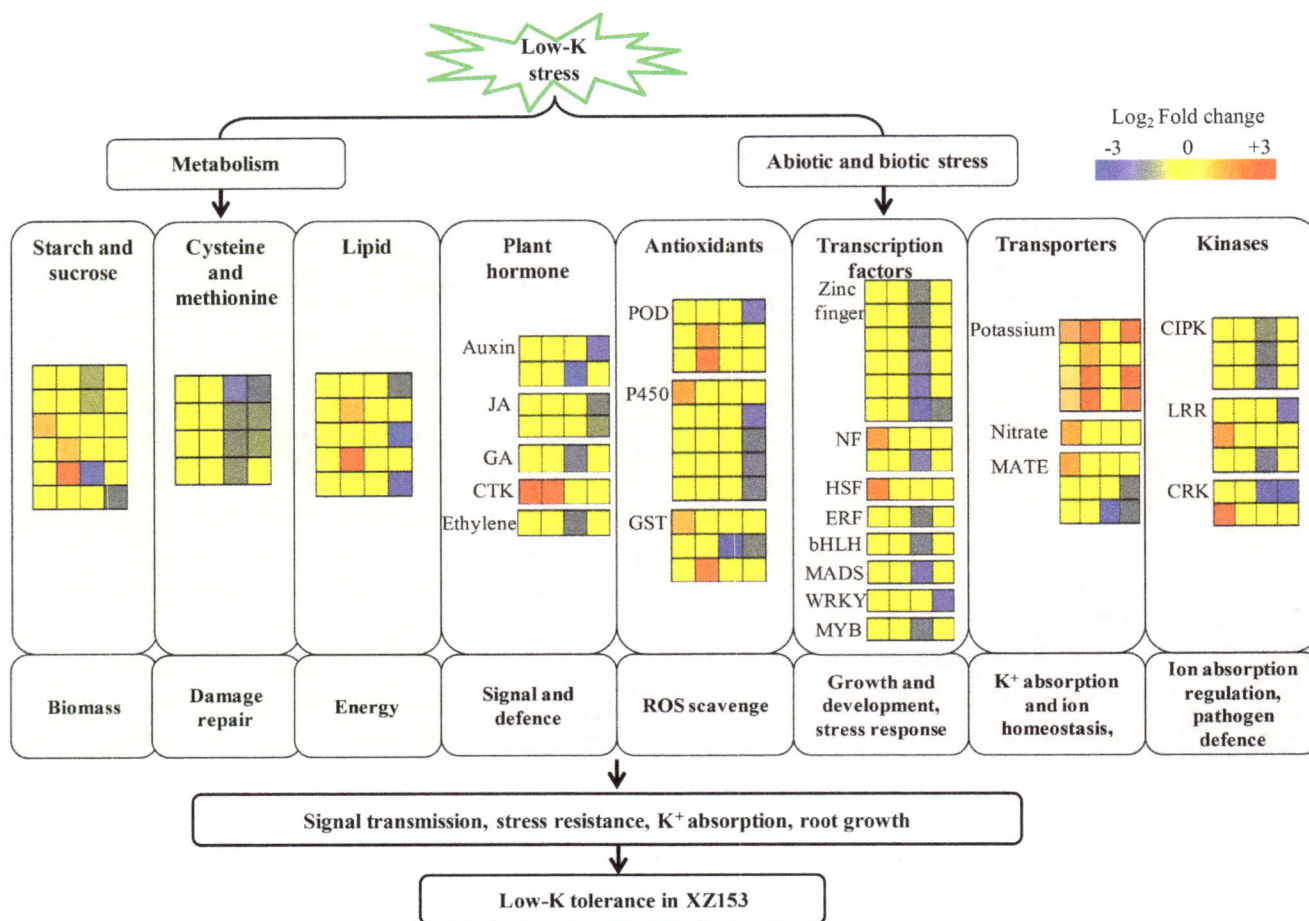

Figure 6. A hypothetical model of low-K tolerance mechanism underlying in XZ153. Genes are shown by different colors and relative expression levels are shown by a color gradient from low (blue) to high (red). For each heatmap from left to right: XZ153-6h (first column), XZ153-48h (second column), XZ141-6h (third column), XZ141-48h (fourth column).

provide some candidate genes, which can be used in barley breeding for improving low-K tolerance.

Supporting Information

Figure S1   Real-time PCR analysis of the *HvHAK1* gene under low K treatment. * represents significant difference according to the Duncan's multiple range, P<0.05, n = 3. Primers of *HvHAK1* and GAPDH for RT-PCR are listed in Table S6. (PDF)

Figure S2   Cluster analysis of the DEGs in the two genotypes. Y-axis represents the gene expression values (FPKM) transformed by logarithms, base 2. The middle white line indicates the gene expression trend of each cluster. (PDF)

Figure S3   Heat Map analysis of DEGs involved in hormone signaling in XZ153 and XZ141. The sample and treatments are displayed above each column. Genes are displayed by different colors. Relative levels of expression are showed by a color gradient from low (blue) to high (red). (PDF)

Figure S4   Heat Map analysis of DEGs involved in oxidative stress-related in XZ153 and XZ141. The samples and treatments are displayed above each column. Genes are displayed by different colors. Relative levels of expression are showed by a color gradient from low (blue) to high (red). (PDF)

Figure S5   KEGG overview of low-K tolerance related DEGs under low K stress. X-axis represents the number of enzymes participating in each pathway; Y- axis depicts the different pathway. (PDF)

Figure S6   DEGs related to low-K tolerance were mapped to the starch and sucrose metabolism pathway. The column marked in color indicates the differentially expressed enzymes. Four small squares from left to right represent different treatment time of the two varieties (XZ153-6h, XZ153-48h, XZ141-6h, XZ141-48h). Gene expression is displayed by different colors and the relative levels of expression are showed by a color gradient from low (blue) to high (red). (PDF)

Table S1   Summary of RNA-seq data and mapping results. (DOC)

Table S2   The FPKM value of 692 DEGs.
(XLS)

Table S3   Gene accession numbers and sequences of 692 DEGs.
(XLS)

Table S4   DEGs at 6 h and 48 h after low K treatment in XZ153.
(XLS)

Table S5   DEGs at 6 h and 48 h after low K treatment in XZ141.
(XLS)

Table S6   The primers used in real-time PCR.
(DOC)

Table S7   DEGs are involved in Starch and Sucrose metabolism or Cysteine and Methionine metabolism.
(DOC)

Acknowledgments

We thank Prof. Dongfa Sun (Huazhong Agricultural University, China) for providing Tibetan wild barley accessions and Dr. Gulei Jin (Hangzhou Guhe Info-technology Co., Ltd.) for excellent technical assistance and valuable suggestions.

Author Contributions

Conceived and designed the experiments: JZ GZ. Performed the experiments: JZ XH DW. Analyzed the data: JZ BZ SC ZJ. Contributed reagents/materials/analysis tools: UN GZ. Contributed to the writing of the manuscript: JZ GZ.

References

1. Amtmann A, Troufflard S, Armengaud P (2008) The effect of potassium nutrition on pest and disease resistance in plants. Physiol Plant 133: 682–691.
2. Britto DT, Kronzucker HJ (2008) Cellular mechanisms of potassium transport in plants. Physiol Plant 133: 637–650.
3. Amtmann A, Armengaud P (2009) Effects of N, P, K and S on metabolism: new knowledge gained from multi-level analysis. Curr Opin Plant Biol 12: 275–283.
4. Maathuis FJM (2009) Physiological functions of mineral macronutrients. Curr Opin Plant Biol 12: 250–258.
5. Szczerba MW, Britto DT, Kronzucker HJ (2009) K$^+$ transport in plants: physiology and molecular biology. J Plant Physiol 166: 447–466.
6. Römheld V, Kirkby EA (2010) Research on potassium in agriculture: needs and prospects. Plant Soil 335: 155–180.
7. Rengel Z, Damon PM (2008) Crops and genotypes differ in efficiency of potassium uptake and use. Physiol Plant 133: 624–636.
8. Fageria NK, Slaton NA, Baligar VC (2003) Nutrient management for improving lowland rice productivity and sustainability. Adv Agron 80: 63–152.
9. Zhang Y, Zhang CC, Yan P, Chen XP, Yang JC, et al. (2013) Potassium requirement in relation to grain yield and genotypic improvement of irrigated lowland rice in China. J Plant Nutr Soil Sci 176: 400–406.
10. Ellis RP, Forster BP, Robinson D, Handley LL, Gordon DC, et al. (2000) Wild barley: a source of genes for crop improvement in the 21st century? J Exp Bot 51: 9–17.
11. Dai F, Nevo E, Wu DZ, Comadran J, Zhou MX, et al. (2012) Tibet is one of the centers of domestication of cultivated barley. Proc Natl Acad Sci USA 109: 16969–16973.
12. Zhao J, Sun HY, Dai HX, Zhang GP, Wu FB (2010) Difference in response to drought stress among Tibet wild barley genotypes. Euphytica 172: 395–403.
13. Qiu L, Wu DZ, Ali S, Cai SG, Dai F, et al. (2011) Evaluation of salinity tolerance and analysis of allelic function of HvHKT1 and HvHKT2 in Tibetan wild barley. Theor Appl Genet 122: 695–703.
14. Wu DZ, Qiu L, Xu LL, Ye LZ, Chen MX, et al. (2011) Genetic variation of HvCBF genes and their association with salinity tolerance in Tibetan annual wild barley. PLoS One 6: e22938.
15. Marioni JC, Mason CE, Mane SM, Stephens M, Gilad Y (2008) RNA-Seq: An assessment of technical reproducibility and comparison with gene expression arrays. Genome Res 18: 1509–1517.
16. Wang Z, Gerstein M, Snyder M (2009) RNA-Seq: a revolutionary tool for transcriptomics. Nat Rev Genet 10: 57–63.
17. Ozsolak F, Milos PM (2010) RNA sequencing: advances, challenges and opportunities. Nat Rev Genet 12: 87–98.
18. Cloonan N, Grimmond SM (2008) Transcriptome content and dynamics at single nucleotide resolution. Genome Biol 9: 234.
19. Mortazavi A, Williams BA, McCue K, Schaeffer L, Wold B (2008) Mapping and quantifying mammalian transcriptomes by RNA-Seq. Nat Methods 5: 621–628.
20. Marguerat S, Bähler J (2010) RNA-seq: from technology to biology. Cell Mol Life Sci 67: 569–579.
21. Oshlack A, Robinson MD, Young MD (2010) From RNA-seq reads to differential expression results. Genome Biol 11: 220.
22. Mochida K, Shinozaki K (2011) Advances in omics and bioinformatics tools for systems analyses of plant functions. Plant Cell Physiol 52: 2017–2038.
23. Wang L, Chen F (2012) Genotypic variation of potassium uptake and use efficiency in cotton Gossypium hirsutum. J Plant Nutr Soil Sci 175: 303–308.
24. Postnikova OA, Shao J, Nemchinov LG (2013) Analysis of the Alfalfa root transcriptome in response to salinity stress. Plant Cell Physiol 54: 1041–1055.
25. Zhu B, Zeng JB, Wu DZ, Cai SG, Yang LN, et al. (2014) Identification and physiological characterization of low potassium tolerant germplasm in Tibetan Plateau annual wild barley. J Zhejiang Univ (Agric & Life Sci) 40: 165–174.
26. Robinson MD, McCarthy DJ, Smyth GK (2010) edgeR: a Bioconductor package for differential expression analysis of digital gene expression data. Bioinformatics 26: 139–140.
27. Trapnell C, Roberts A, Goff L, Pertea G, Kim D, et al. (2012) Differential gene and transcript expression analysis of RNA-Seq experiments with TopHat and Cufflinks. Nat Protoc 7: 562–578.
28. Zenoni S, Ferrarini A, Giacomelli E, Xumerle L, Fasoli M, et al. (2010) Characterization of transcriptional complexity during berry development in Vitis vinifera using RNA-Seq. Plant Physiol 152: 1787–1795.
29. Schmittgen TD, Livak KJ (2008) Analyzing real-time PCR data by the comparative CT method. Nat Protoc 3: 1101–1108.
30. Conesa A, Götz S, García-Gómez JM, Terol J, Talón M, et al. (2005) Blast2GO: a universal tool for annotation, visualization and analysis in functional genomics research. Bioinformatics 21: 3674–3676.
31. Kanehisa M, Goto S (2000) KEGG: Kyoto encyclopedia of genes and genomes. Nucleic Acids Res 28: 27–30.
32. O'Rourke JA, Yang SS, Miller SS, Bucciarelli B, Liu JQ, et al. (2013) An RNA-Seq transcriptome analysis of orthophosphate-deficient white lupin reveals novel insights into phosphorus acclimation in plants. Plant Physiol 161: 705–724.
33. Broadley MR, Bowen HC, Cotterill HL, Hammond JP, Meacham MC, et al. (2004) Phylogenetic variation in the shoot mineral concentration of angiosperms. J Exp Bot 55: 321–336.
34. Hermans C, Hammond JP, White PJ, Verbruggen N (2006) How do plants respond to nutrient shortage by biomass allocation? Trends in Plant Science 11: 610–617.
35. Dreyer I, Uozumi N (2011) Potassium channels in plant cells. FEBS J 278: 4293–4303.
36. Wang Y, Wu WH (2013) Potassium transport and signaling in higher plants. Annu Rev Plant Biol 64: 451–476.
37. Santa-María GE, Rubio F, Dubcovsky J, Rodríguez-Navarro A (1997) The HAK1 gene of barley is a member of a large gene family and encodes a high-affinity potassium transporter. Plant Cell 9: 2281–2289.
38. Bañuelos MA, Garciadeblas B, Cubero B, Rodríguez-Navarro A (2002) Inventory and functional characterization of the HAK potassium transporters of rice. Plant Physiol 130: 784–795.
39. Gierth M, Maser P, Schroeder JI (2005) The potassium transporter AtHAK5 functions in K$^+$ deprivation induced high-affinity K$^+$ uptake and AKT1 K$^+$ channel contribution to K$^+$ uptake kinetics in Arabidopsis roots. Plant Physiol 137: 1105–1114.
40. Gierth M, Mäser P (2007) Potassium transporters in plants-involvement in K$^+$ acquisition, redistribution and homeostasis. FEBS Lett 581: 2348–2356.
41. Hong JP, Takeshi Y, Kondou Y, Schachtman DP, Matsui M, et al. (2013) Identification and characterization of transcription factors regulating Arabidopsis HAK5. Plant Cell Physiol 54: 1478–1490.
42. Ma TL, Wu WH, Wang Y (2012) Transcriptome analysis of rice root responses to potassium deficiency. BMC Plant Biol 12: 161.
43. Lee TI, Young RA (2000) Transcription of eukaryotic protein-coding genes. Annu Rev Genet 34: 77–137.
44. Dubos C, Stracke R, Grotewold E, Weisshaar B, Martin C, et al. (2010) MYB transcription factors in Arabidopsis. Trends Plant Sci 15: 573–581.
45. Lin PC, Pomeranz MC, Jikumaru Y, Kang SG, Hah C, et al. (2011) The Arabidopsis tandem zinc finger protein AtTZF1 affects ABA-and GA-mediated growth, stress and gene expression responses. Plant J 65: 253–268.
46. Zhang X, Ju HW, Chung MS, Huang P, Ahn SJ, et al. (2011) The RR-type MYB-like transcription factor, AtMYBL, is involved in promoting leaf senescence and modulates an abiotic stress response in Arabidopsis. Plant Cell Physiol 52: 138–148.
47. Jan A, Maruyama K, Todaka D, Kidokoro S, Abo M, et al. (2013) OsTZF1, a CCCH-tandem zinc finger protein, confers delayed senescence and stress tolerance in rice by regulating stress-related genes. Plant Physiol 161: 1202–1216.

48. Shankar A, Singh A, Kanwar P, Srivastava AK, Pandey A, et al. (2013) Gene expression analysis of Rice seedling under potassium deprivation reveals major changes in metabolism and signaling components. PLoS One 8: e70321.

49. Wang C, Chen H, Hao Q, Sha A, Shan Z, et al. (2012) Transcript profile of the response of two soybean genotypes to potassium deficiency. PLoS One 7(7): e39856.

50. Nelson DE, Repetti PP, Adams TR, Creelman RA, Wu J, et al. (2007) Plant nuclear factor Y (NF-Y) B subunits confer drought tolerance and lead to improved corn yields on water-limited acres. Proc Natl Acad Sci USA 104: 16450–16455.

51. Kumimoto RW, Zhang Y, Siefers N, Holt BF (2010) NF-YC3, NF-YC4 and NF-YC9 are required for CONSTAN S-mediated, photoperiod-dependent flowering in Arabidopsis thaliana. Plant J 63: 379–391.

52. Merchante C, Alonso JM, Stepanova AN (2013) Ethylene signaling: simple ligand, complex regulation. Curr Opin Plant Biol 16: 554–560.

53. Jung JY, Shin R, Schachtman DP (2009) Ethylene mediates response and tolerance to potassium deprivation in Arabidopsis. Plant Cell 21: 607–621.

54. Kim MJ, Ruzicka D, Shin R, Schachtman DP (2012) The Arabidopsis AP2/ERF transcription factor RAP2.11 modulates plant response to low-potassium conditions. Mol Plant 5: 1042–1057.

55. Armengaud P, Sulpice R, Miller AJ, Stitt M, Amtmann A, et al. 2009. Multi level analysis of primary metabolism provides new insights into the role of potassium nutrition for glycolysis and nitrogen assimilation in Arabidopsis roots. Plant Physiol 1502: 772–785.

5

Effects of Zn Fertilization on Hordein Transcripts at Early Developmental Stage of Barley Grain and Correlation with Increased Zn Concentration in the Mature Grain

Mohammad Nasir Uddin, Agnieszka Kaczmarczyk, Eva Vincze*

Department of Molecular Biology & Genetics, Faculty of Science & Technology, Aarhus University, Slagelse, Denmark

Abstract

Zinc deficiency is causing malnutrition for nearly one third of world populations. It is especially relevant in cereal-based diets in which low amounts of mineral and protein are present. In biological systems, Zn is mainly associated with protein. Cereal grains contain the highest Zn concentration during early developmental stage. Although hordeins are the major storage proteins in the mature barley grain and suggested to be involved in Zn binding, very little information is available regarding the Zn fertilization effects of hordein transcripts at early developmental stage and possible incorporation of Zn with hordein protein of matured grain. Zinc fertilization experiments were conducted in a greenhouse with barley cv. Golden Promise. Zn concentration of the matured grain was measured and the results showed that the increasing Zn fertilization increased grain Zn concentration. Quantitative real time PCR showed increased level of total hordein transcripts upon increasing level of Zn fertilization at 10 days after pollination. Among the hordein transcripts the amount of B-hordeins was highly correlated with the Zn concentration of matured grain. In addition, protein content of the matured grain was analysed and a positive linear relationship was found between the percentage of B-hordein and total grain Zn concentration while C-hordein level decreased. Zn sensing dithizone assay was applied to localize Zn in the matured grain. The Zn distribution was not limited to the embryo and aleurone layer but was also present in the outer part of the endosperm (sub-aleurone layers) which known to be rich in proteins including B-hordeins. Increased Zn fertilization enriched Zn even in the endosperm. Therefore, the increased amount of B-hordein and decreased C-hordein content suggested that B-hordein upregulation or difference between B and C hordein could be one of the key factors for Zn biofortification of cereal grains due to the Zn fertilization.

Editor: Chengdao Li, Department of Agriculture and Food Western Australia, Australia

Funding: Funding provided by The Danish Fødevareforskningsprogrammet 2009; grant #: 3304-FVFP-09-B-004. The funders had no role in study design, data collection and analysis, decision to publish, or preparation of the manuscript.

Competing Interests: The authors have declared that no competing interests exist.

* Email: eva.vincze@agrsci.dk

Introduction

Zinc (Zn) is an essential element for plants and animals. After iron, Zn is the most abundant transition metal in organisms and is also present in all six enzyme classes [1]. Zinc deficiency is considered one of the top priority micronutrient deficiency problems affecting nearly one third of the world population [2–4]. In biological systems Zn is known to be incorporated with protein and prefers tetrahedral coordination by four ligands such as sulphur from cysteine, nitrogen from histidine, oxygen from aspartate and glutamate; much more rarely observed ligands include the hydroxyl of tyrosine, the carbonyl oxygen of the protein backbone and the carbonyl oxygen of either asparagine or glutamine [5,6]. Bioinformatics searches for known zinc binding motifs identified that the human proteome contains 10–15% zinc binding proteins; and in *Arabidopsis* a total of 2367 proteins in 181 gene families are identified as Zn-related [1,7,8]. However, these figures do not reflect the total number of actual zinc binding proteins which might exceed these numbers since a lot of zinc binding motifs are impossible to predict with bioinformatics analyses [9,10]. In addition to undiscovered potential zinc binding motifs, there are intermolecular binding sites (in which Zn ion acts as a bridging ligand between two polypeptides) in the sequences that are extremely difficult to predict in silico [7,11].

Although cereal grains inherently contain lower amounts of proteins and minerals than some legumes [1,12] up to 75% of the daily calorie intake of people living in the rural areas of the developing world comes from cereal-based foods with very low Zn bioavailability and concentrations (www.harvestplus.org).

The cereal grains have several major depositories for nutrients such as testa or pericarp, embryo including the scutellum and endosperm, surrounded by the aleurone layer. Usually the inner part of the endosperm has the lowest concentration of Zn and proteins and higher concentration of starch. The embryo and aleurone layers contain about half of the total Zn of cereal grains but during the milling process they are mostly removed [13]. Therefore, in order to improve the problem of Zn malnutrition, zinc concentration inside the endosperm needs to be increased.

A link among Zn transport and Zn storage proteins as well as nitrogen, sulphur and various amino acids was suggested from a nitrogen fertilization experiment [14]. In wheat, 0.26% increase of grain protein concentration is found with every mg of Zn per kg of top soil [15]. A study of bread wheat (*Triticum aestivum*) with ^{65}Zn application at anthesis shows that the greatest proportion of the [65]Zn is found in the glutenin fraction suggesting that Zn is associated with the grain storage proteins [16]. Furthermore, a strong correlation between wheat gliadin and grain Zn concentration was observed and suggesting that grain Zn in wheat could be bound to sulphur-rich low molecular weight (30–50 kDa) prolamins (γ and α gliadin; and B and C type LMW glutenins) [17]. Increasing Zn application increases the total polymeric glutenin compared to monomeric gliadin but within the glutenin fraction SDS-unextractable large polymeric glutenin is decreased compared to LMW glutenin [18]. It is also suggested that Zn could be primarily bound to protein/peptides in barley grain [19]. Therefore, these facts gave an indication that probably some fraction(s) in barley prolamin that is homologous to γ and α gliadin and LMW glutenin of wheat are capable of binding zinc.

In addition, it is known that Zn has a high affinity to the cysteine groups in disulphide bridges [20] and could be involved in the polymerization processes, which occur late in the grain filling [21,22]. In wheat more than 80% of the total protein fraction (w/w) in the mature endosperm is gluten protein; either as monomeric (gliadins) or polymeric (glutenins) [21,23]. The glutenin polymers are stabilized by inter- and intra-chain disulphide bridges and hydrogen bonds, and a positive relationship between dough strength and the ability of gluten proteins to form glutenin polymers has been observed [21,24–27]. Moreover, increasing of Zn concentration in the grain by foliar Zn fertilization in bread wheat altered the gluten protein composition of the endosperm in favour of the polymeric glutenin. This gave an indication that a high glutenin to gliadin ratio is a trait connected to a high Zn concentration, [18].

In rice (*Oryza sativa*) it was observed that the distribution of Zn changed rapidly during grain development [28]. For instance at 10 days after fertilization (DAF), Zn is abundant in the aleurone layer, thereafter Zn decreases around the aleurone layer and spreads into the inner endosperm adjacent to the aleurone layer; *i.e.*, the sub-aleurone layer [28]. A study using high-definition synchrotron X-ray fluorescence and ICP-MS in matured barley grain showed that in total 58% of zinc ion is present across the testa-aleurone-endosperm gradient [29]. Pearling and immunocytochemical studies of barley also have shown that protein-rich sub-aleurone cells are enriched in B-hordein (S-rich) and C-hordein (S-poor) [27] while D-hordein (the HMW prolamin) is only present in significant amounts in the inner part of the starchy endosperm [22,23,30]. Therefore, it could be assumed that major Zn binding storage proteins in the sub-aleurone layer of barley and wheat belong to proteins homologous to LMW-GS, possibly B-hordeins.

In wheat, the highest Zn concentration in the grain was found at the beginning of grain development such as 10–12 day after pollination (DAP) [31,32]. Therefore, it gives an indication that endosperm proteins expressing in the early developmental stages could be potential sinks of Zn ion, especially those localized in the sub-aleurone layer. A transcriptomics study in barley showed that the proportion of B-hordein among all hordeins are higher at early developmental stages such as 10 DAP [33].

Considering the importance of barley as an ancient cereal grain crop ranking fourth among all crops in dry matter production in the world [34], the aims of the work were (i) to assess the effect of foliar and soil Zn fertilization on hordein transcript of early developmental stages; (ii) to seek a correlation between the proportion of hordein proteins and Zn concentration of the matured grain; and (iii) to detect the hordein fractions with Zn ion binding capabilities.

Materials and Methods

Plant material

Barley (*Hordeum vulgare* cv. Golden Promise) grains were surface sterilized by soaking in 30% H_2O_2 for 10 min, rinsed with distilled water 5 times; 3 grains were planted in a pot containing 200 g soil and grown under greenhouse conditions under a cycle of 16 h illumination and 8 h darkness at 23 and 18°C, respectively, at Research Centre Flakkebjerg, Slagelse, Denmark. After sowing, the pots were watered three times a week. Once germination was completed, plants were thinned to one plant per pot.

Individual spikes were tagged at flowering and harvested in the morning (09.00–11.30) at 10 days after pollination (DAP). The collected spikes were immediately frozen in liquid nitrogen and stored at −80°C until the analysis. The remaining spikes were harvested at maturity.

Soil preparation and fertilizer application

Two hundred g dried PindstrupUnimuld (PindstrupMosebrug A/S, Denmark) soil was put in each plastic pot (size 1 L) without holes. Soil fertilizer was applied during soil preparation by adding 1 L fertilizer solution gradually into the soil until it was absorbed. The fertilization was repeated at 21, 45, 65 and 90 days after sowing the seeds. Foliar application (4 mL for each plant) was done with a hand sprayer twice in a week from 35 to 90 days after sowing the seeds. Three different Zn treatments were applied in the experiment: low, medium and high. In the low Zn treatment fertilizer solution [35,36] containing basic nutrient (composition in mg: $(NH_4)_2SO_4$ 48.2, $MgSO_4$ 65.9, K_2SO_4 15.9, KNO_3 18.5, $Ca(NO_3)_2$ 59.9, KH_2PO_4 24.8, $C_6H_5FeO_7$ 5.0, $MnCl_2 \cdot 4H_2O$ 0.9, $CuSO_4 \cdot 5H_2O$ 0.04, H_3BO_3 2.9, H_2MoO_4 0.01) was added into the soil of each pot and the leaves were sprayed with water. The medium and high Zn treatments were done by adding of basic nutrient solution supplemented with 0.25 mM and 1 mM of $ZnSO_4.7H_2O$ into the soil respectively, plus foliar spraying of $ZnSO_4.7H_2O$ solution on each plant: 1 mM for medium and 10 mM for high treatments.

DNA and RNA extractions, mRNA isolation, cDNA synthesis

DNA coding actin gene (HVSMEi0002G07f) was prepared from plasmid clones using GenElute Plasmid Miniprep kit (Sigma-Aldrich) and DNA quantification was done using HoeferDyNA Quant 200 fluorometer and quantification assay (Sigma-Aldrich) according to manufacturer's protocol.

Total RNA was extracted from two biological samples per treatment representing 2–3 grains taken from the middle part of the spike from each individual plant from different pot. The grains were homogenized in liquid nitrogen using a mortar and pestle, with either RNeasy kit (Qiagen) or TRI Reagent (Sigma-Aldrich) according to the manufacturers' protocols. RNA qualities were checked using an Agilent 2001 Bioanalyzer (Agilent Technologies, Inc.) and mRNA was isolated from total RNA with Dynabeads (Invitrogen) according to manufacturer's protocol. First strand cDNA was synthesized with 10 μL of mRNA, 1 μL of oligodT (Invitrogen), 4 μL of 5Xbuffer (supplied with enzyme), 1 μL RNAsin (Promega), 2 μL of 0.1 M DTT (supplied with enzyme), 1 μL of each dNTP (10 mM each) and 1 μL of Superscript II RT enzyme. The resulting cDNA mixture was diluted to 200 μL by adding 180 μL of MilliQ-water and stored at −20°C.

Quantitative real time PCR (qRT-PCR)

Quantitative RT-PCR was performed in triplicate from each biological samples in 384 well microtiter plates in 7900HT Sequence Detection System (Applied Biosystems). The total reaction volume was 10 μL which contained 1 μL of appropriately diluted template DNA, 2.8 μL of MilliQ water, 0.6 μL of each primer (5 μM) and 5 μL of Power SYBR Green Master Mix (Applied Biosystems). Additional no-template control (NTC) reactions were carried out to check the potential of primer-dimers formation. The thermal profile set up, data analyses and quantification of individual gene of interest using actin standard and reference gene was done according to Kaczmarczyk et al. [33]. Our study conforms to the Minimum Information for Publication of Quantitative Real-Time PCR Experiments (MIQE) [37].

Statistical distributions and interpreting P values were done using online GraphPad software (http://www.graphpad.com/quickcalcs/distMenu/).

Dithizone staining of matured grains

In order to study the localization of Zn in the matured grains, a staining method was developed using dithizone (DTZ), which creates a red/purple Zn-dithizonate complex upon binding with Zn [31,38,39]. Matured barley grains were incubated in milli-Q water for 1 hour and thereafter cut into two pieces longitudinally by scalpel/blade. The half grains were incubated in incubation buffer containing 100 mM Tris-HCl (pH 6.8), 50 mM NaCl and 10 mM DTT for 2 hours followed by half hour incubation with 50% methanol in the incubation buffer but without DTT. After incubation, the grains were stained in 1 mM DTZ (diluted with DMSO from 10 mM DTZ stock solution made with pure acetone) for 30 min. The stained grains were rinsed with buffer containing 50 mM Tris-HCl, pH 7.0 and 50 mM NaCl for enhanced colour reaction, following rinse with water and analysed qualitatively by using a reflectance light microscope (Carl Zeiss Microsystem) with a high-resolution digital camera (AuxioCam MRc5).

Determination of grain Zn concentration

Mature grain Zn concentration (μg/g) was measured from all the biological replicates (plant from individual pots). All of the grains from the individual plants were grinded using a ceramic grain mill (KoMoFidibus 21) and Zn contents were measured from the flour by inductively coupled plasma optical emission spectrometry (ICP-OES) at Department of Plant and Environmental Sciences, Faculty of Science, University of Copenhagen.

Hordein extraction and SDS-PAGE

The milled harvested matured barley grains were used for hordein determination as well. Barley alcohol-soluble proteins (hordeins) were extracted from 50 mg of flour and the isolated proteins were separated on SDS-PAGE according to Uddin et al. [40]. Maltose binding protein (MBP5) (6 mg/mL stock) (New EnglandBioLabs) as a negative control and yeast alcohol dehydrogenase (1 μg/μL stock) (Sigma-Aldrich Inc) as a positive control was used with following dilutions in the sample buffer: maltose binding protein (MBP5) in 1:20; and alcohol dehydrogenase in 1:1. Pre-stained high molecular weight protein standard (HiMark) were purchased from Invitrogen (Life Technologies).

Coomassie staining and calculating the percentage of the band volume

After the SDS-PAGE gel was stained with Coomassie blue [41] and image was taken using BioRadGelDoc. ImageLab 4.01 was used for image analysis and calculating percentage of band volume in each lane. Different hordein bands were assigned according to their approximate molecular weight. Since B- and γ-hordein have similar molecular weight and with 1-D gel, it was not possible to distinguish them, in calculation they were assigned together. The ratio among different hordeins (from the band intensity) was calculated manually in MS excel 2007. Statistical distributions and interpreting P values was done using online GraphPad software (http://www.graphpad.com/quickcalcs/distMenu/).

Blotting, radioactive ^{65}Zn assay

Hordeins extracted from mature barley grain, separated by SDS-PAGE were blotted on membrane, renatured, overlayed and probed with zinc and subsequently zinc binding specificity of certain proteins was detected by autoradiography as described by Uddin et al. [40].

Bioinformatics analyses

Primers for hordein gene families were used according to Kaczmarczyk et al. [33] For multiple sequence alignment of proteins, sequences were collected from uniprotKB (http://www.uniprot.org/help/uniprotkb) database and alignment was done using MEGA5.1 software with clustalW [42] algorithm using the following parameters: Gap penalty = 10; gap extension penalty = 0.2; Protein weight matrix = Gonnet, residues specific penalty = on, hydrophobic penalty = on, gap separation distance = 4, End gap separation off, and delay divergent cut off = 30%.

Results

Effects of Zn fertilization on the Zn concentration of the matured grain

The soil used in our experiments has very little amount of Zn (0.4 g/m^3) and the low Zn treated plants (no additional Zn was applied during the whole growing period) received only slight amount of Zn from soil. We used $ZnSO_4.7H_2O$ as a source of Zn fertilizer in the medium and high Zn treatments. $ZnSO_4$ is the most widely applied fertilizer by the farmer due to high solubility, low cost and availability in the market. Moreover, it is also recommended that in case of biofortification of staple food crops fertilizers with high water-soluble Zn would be the best choice for foliar applications (http://www.zinc.org/crops/resources/publications).

In our experiments, foliar and soil Zn fertilization resulted in an increase of grain Zn concentration. Average matured grain Zn concentration for different Zn treated plants was measured as 65, 151 and 466 μg/g for low, medium and high Zn treatment respectively (Table 1). In comparison to low Zn treated plants the Zn concentration was increased 2.3 and 7.1 folds for the medium and high Zn treated plants respectively (Table 1). Our results are in agreement with previous reports, suggesting that foliar $ZnSO_4$ application together with soil Zn fertilization is an effective way to promote grain Zn concentration in rice and wheat [31,43,44].

Effects of Zn treatment on hordein transcript at 10 DAP

Zn fertilization increased the total amount of hordein transcripts (measured in amole of hordein/amole of actin) (Figure 1). The observed increases compared to the low Zn treatment were 1.3 and 2.5 fold for the medium and high zinc treated group respectively (Figure 1).

In addition, Zn concentration of matured grain also correlated with some of the hordein groups expressed at the early developmental stages (10 DAP). Transcripts of B-hordeins were

Table 1. Biological replicates and Zn concentration of full grain flour from matured barley (*Hordeum vulgare* cv. Golden Promise) measured (by ICP-OES).

Low Zn treated plants*	Grain Zn concentration (µg/g)	Medium Zn treated plants*	Grain Zn concentration (µg/g)	High Zn treated plants*	Grain Zn concentration (µg/g)
P13	56.7	Q11	114.6	R15	239.7
P2	58.2	Q13	119.8	R11	349.2
P7	59.9	Q12	130.5	R14	391.1
P6	64.4	Q9	135.5	R2	395.3
P4	64.5	Q10	147.0	R1	401.2
P11	86.1	Q4	149.7	R6	433.4
		Q5	157.6	R8	525.8
		Q2	165.7	R13	597.2
		Q1	239.6	R7	650.6
				R4	680.7
Average	65.0	Average	151.1	Average	466.4
St. Error	4.4	St. Error	12.4	St. Error	44.9

* P2, P4, P6, P7, P11, P13; Q1, Q2, Q4, Q5, Q9, Q10, Q11, Q12, Q13; and R1, R2, R4, R6, R7, R8, R11, R13, R14, R15 are referring to each biological replicate (one plant in each pot).

found to be highly abundant followed by C-, γ- and D-hordeins in all three Zn treatments (Figure 2).

B-hordeins

Using the primer sets designed to recognize the whole B-hordein gene family [33] we found that the steady state level of B-hordein transcripts increased with the increasing Zn concentrations (Figure 2). Average transcripts of B-hordein were measured as 922, 1254 and 2780 in amole of B-hordein/amole of actin for low, medium and high Zn fertilization/treatment respectively (Figure 2). Also a significant correlation (r = 0.915; DF = 4; the two-tailed P value = 0.0105) was observed between B-hordein expression at 10 DAP and matured grain Zn concentration (Table 2, Figure S1).

Similar trends were observed by using the primers sets [33] recognizing the two major sub-families of B-hordeins: B1 and B3. For instance, average transcript of B1-hordein was found as 262, 454 and 560 amole of B1-hordein/amole of actin for low, medium

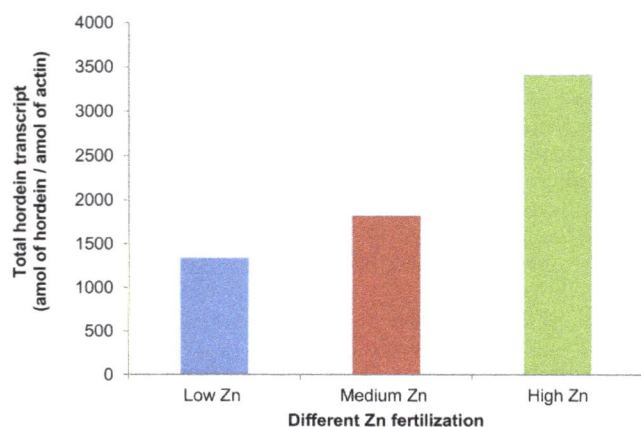

Figure 1. Effects of Zn fertilizations (low - blue; medium - red; high – green) on total amount of hordein transcripts (the sum of all hordeins) measured in amol of hordein/amol of actin.

and high Zn fertilization/treatment respectively. Also a significant correlation (r = 0.89; DF = 4; the two-tailed P value = 0.017) was observed between B1-hordein expression at 10 DAP and matured grain Zn concentration (Table 2, Figure S1). For B3 hordein, average transcript was measured as 254, 402 and 594 amole of B3-hordein/amole of actin for low, medium and high Zn fertilization/treatment respectively. In addition, very significant correlation (r = 0.96; DF = 4; the two-tailed P value = 0.003) was found between B3-hordein expression at 10 DAP and matured grain Zn concentration (Table 2, Figure S1).

C-hordeins

The C-hordein family (MW 55–75 kDa) is a S-poor prolamin group with unique sequences comprising highly conserved tandem repeats [45]. In matured barley grain C-hordeins account for 10–20% of the total hordein protein [25]. It was considered that C-hordein, like most of the S-poor prolamins, are lacking cysteine, and hence unable to form disulphide bonds [25].

In our Zn treatment experiments, on average very slight up-regulation of C-hordein transcript was observed with increasing the Zn concentrations (Figure 2) and average transcript was measured as 228, 246 and 280 amole of C-hordein/amole of actin for low, medium and high Zn fertilization/treatment respectively (Figure 2). Furthermore, no significant correlation (r = 0.71; DF = 4; the two-tailed P value = 0.11) was found between C-hordein expression at 10 DAP and matured grain Zn concentration (Table 2, Figure S1).

D-hordeins

D-hordeins (MW>100 kDa), account for 2–4% of total grain protein in the mature grain and belong to HMW prolamins typified by the HMW subunits of wheat glutenin [25,46].

In our experiment transcripts of D-hordein at 10 DAP were increased upon increasing Zn fertilization. The average transcript of D-hordein was measured as 12, 28 and 38 amole of D-hordein/amole of actin for low, medium and high Zn fertilization/treatment respectively (Figure 2). Furthermore, significant correlation (r = 0.88; DF = 4; the two-tailed P value = 0.021) was

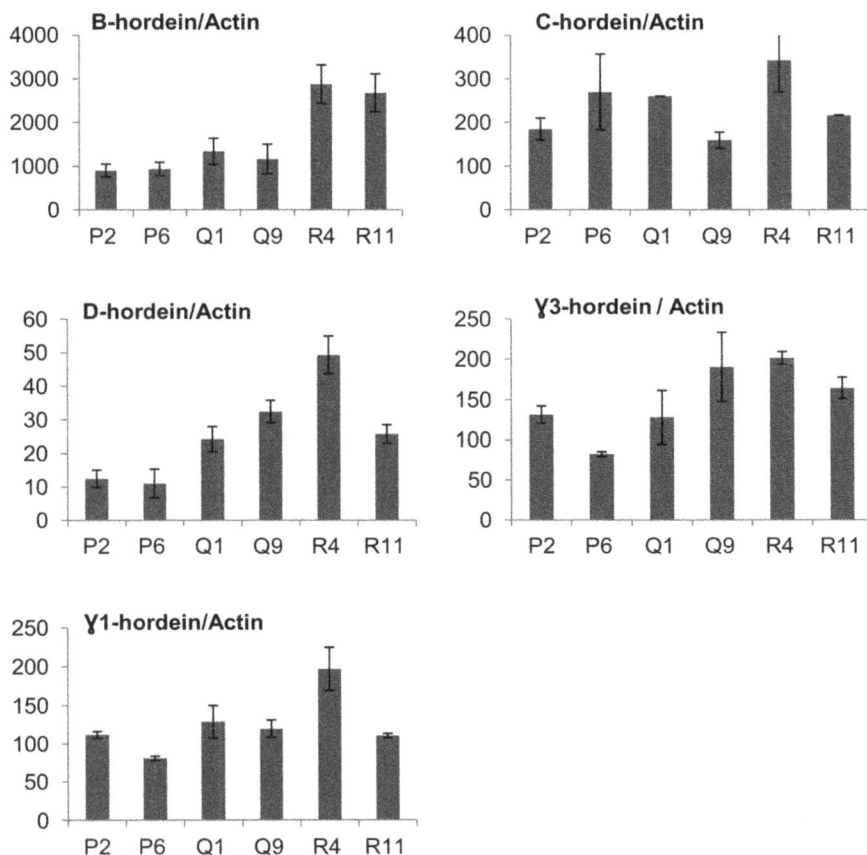

Figure 2. Effects of Zn treatments on the relative expression of different hordein transcripts at 10 DAP measured in amol of hordein/amol of actin. Two biological replicates for each Zn treatment are shown as: Low (P2 & P6), Medium (Q1 & Q9) and High (R4 & R11); and 3 technical replicates presented as means ± SE.

Table 2. Linear correlation between matured grain Zn concentration and hordein transcript at 10 DAP or protein from the matured grain.

Types of hordeins	Correlation coefficient (r)	Degrees of freedom (df)	Two tailed P-value
Transcripts*			
B	0.915	4	0.010
B1	0.891	4	0.017
B3	0.957	4	0.002
γ1	0.654	4	0.158
γ3	0.671	4	0.144
C	0.714	4	0.111
D	0.878	4	0.02
Protein (%)**			
(B+γ)	0.630	23	0.0007
C	0.401	23	0.0464
D	0.106	23	0.6130
(B+γ)–C	0.657	23	0.0004
(B+γ+D)–C	0.697	23	0.0001

* 10 DAP (amol of hordein/amol actin).
** % of protein from matured grain.

observed between D-hordein expression at 10 DAP and matured grain Zn concentration (Table 2, Figure S1).

γ-hordeins

γ-hordeins belong to S-rich prolamin group (MW 36–44 KDa) and are represented by a small group of proteins, their contribution to the total grain protein has not been precisely determined [33,47] although they are thought to be present in very minor amounts [48].

In our experiment, we used the common primers [33] designed for γ1- and γ3- subfamily. Towards higher Zn treatment average γ3-hordein was more responsive than the γ1 hordeins (Figure 2). The average γ3-hordein transcript level was measured as 107, 159 and 183 amole of γ3-hordein/amole of actin for low, medium and high Zn fertilization respectively while the average γ1-hordein transcripts was 96, 124 and 154 amole of γ3-hordein/amole of actin for low, medium and high Zn fertilization respectively (Figure 2). However, none of the γ-hordein showed statistically significant correlation between the expression at 10 DAP and matured grain Zn concentration [γ1 (r = 0.65; DF = 4; the two-tailed P value = 0.15); and γ3 (r = 0.67; DF = 4; the two-tailed P value = 0.14)] (Table 2, Figure S1).

Effect of Zn fertilization on the ratio of different hordein transcripts

Prolamins consist of multiple gene protein families, which are divided into monomeric and polymeric groups. In barley, C-hordein and γ-hordeins are the monomeric prolamins and B- and D-hordeins are considered as polymeric hordeins [49]. In our experiment, although a slight increase was observed in the transcript level of monomeric hordeins at 10 DAP, this increase did not significantly correlate with actual matured grain Zn concentration (Table 2, Figure 2 & Figure S1).

In contrast to monomeric hordeins, polymeric hordeins such as B- and D-hordeins were up regulated at 10 DAP with increasing Zn fertilization and this up regulation also correlated with the actual Zn concentration in matured grain (Table 2, Figure 2 & Figure S1). B-hordeins are S-rich and are the main group of hordeins in barley (70–80% of total prolamin content), and as sub units they are about the same size as γ-hordein [49]. D-hordeins are similar to HMW subunit of wheat [50]. In our experiment, the ratio in the expression data from different hordein transcripts showed an increase of glutenin and gliadin ratio [(B+D):(C+γ)] and the ratio between LMW glutenin and gliadin sub unit [B: (C+γ)] with increasing Zn treatment, whereas ratio between monomeric gliadin and polymeric HMW glutenin (C:D) decreased upon higher zinc treatment (Table 2, Figure 3). In addition, increasing Zn fertilization also increased the percentage of [(B+γ)−C] -hordeins or [(B+γ+D)−C]-hordein (Figure 3).

Effects of Zn fertilization on hordeins of matured grain

The hordeins were isolated from all the biological replicates with known zinc concentration (Table 1) from matured grain and the different hordeins were separated on Coomassie stained SDS-PAGE gel. Image analyses of these protein gels demonstrated that on average the highest Zn treatment decreased the monomeric C-hordein whereas polymeric B- together with γ- and D-hordein was increased (Figure 4). However, plants from medium Zn treated groups showed a slight decrease of B-hordein whereas D-hordein as well as LMW S-rich prolamins such as trypsin/α-amylase inhibitors (known as A- hordeins previously, shown as TI in Figure 4A) was increased (Figure 4A). Although γ- and B-hordein have similar molecular weight we assume that this change of

hordein percentages was due to the B-hordein since the amount of γ-hordein is very low in barley [51]. In addition, increasing Zn treatment also increased the differences between (B+γ+D) and C-hordein [% of (B+γ+D)−C] (Figure 4B).

Furthermore, very significant linear positive correlation (r = 0.63; DF = 23; the two-tailed P value = 0.0007) was observed between (B+γ)-hordein of matured grain and actual grain Zn concentration (Table 2), whereas negative correlation was found in case of C-hordein (Figure S2, Table 2). In contrast to the results from qPCR at 10 DAP, D-hordein did not show any significant correlation between matured grain Zn concentration and percentage of D-hordein present in matured grain (Table 2, Figure S1 & Figure S2).

In addition, increasing polymeric hordein also increased grain zinc concentration and highly significant positive correlation was found between grain Zn concentration and percentage of [(B+γ)−C] or [(B+γ+D)−C] hordeins in the grain (Table 2, Figure S2).

Zinc localization of matured grain with dithizone staining

Dithizone staining was performed for the localization of Zn inside the grain from different Zn treated plants. Regardless of the Zn treatment, embryo and aleurone/sub-aleurone layers showed the highest Zn intensity (Figure 5). However, high zinc application made a slight enrichment of Zn ion inside the endosperm in comparison to low zinc treated groups (Figure 5). This Zn enrichment inside the endosperm shown by DTZ staining was observed up to a certain limit (average total grain Zn concentration 240 μg/g), and beyond this limit colour intensity did not increase further (data not shown).

Zinc blotting and detection of Zinc binding protein

In the [65]Zn assay positive Zn binding was observed for the well-known positive control alcohol dehydrogenase (MW 38 kDa) and the protein bands (MW 35–46 kDa) from our hordein isolates, whereas maltose binding proteins (MBP5) (MW 42.5 kDa) used as negative control did not show any zinc binding (Figure S3). The molecular weights of hordein extracts showing positive bands in [65]Zn autoradiography are about 35–46 kDa representing either as B-hordein or γ-hordein [25,40].

Multiple sequence alignment with Zn binding pumpkin trypsin Inhibitor

From crystallographic studies it was observed that pumpkin trypsin inhibitor, CMTI-I (ITR1_CUCMA) is capable of binding zinc ion in a tetrahedral and symmetric fashion through glutamic acid residue [52]. Like CMTI-I protein, B-hordeins have a trypsin/alpha amylase inhibitor domain in the C-terminal region which might participate in Zn binding [52,53].

Therefore, multiple sequence alignment was done with the full length amino acid sequence from pumpkin trypsin inhibitor (CMTI-I) and different B-hordein protein sequences (Figure 6). The multiple sequence alignment showed presence of 11 conserved sites in C-terminal region of B-hordein and CMTI-I protein, consists of cysteine (C), proline (P), isoleucine (I), leucine (L) and lysine (K) residues (Figure 6).

Zinc binding sites in protein are often distorted tetrahedral or trigonal bipyramidal geometry [6]. Histidine (H), cysteine (C), glutamic acid/glutamate (E/Q) and aspartic acid (D) are the most common amino acids that supply ligands to these sites. Usually in protein zinc-binding sites, zinc ion is coordinated by different combinations of protein side chains: the nitrogen from histidine, oxygen from aspartate or glutamate, and sulphur from cysteine; and among them histidine and cysteine are the most commonly

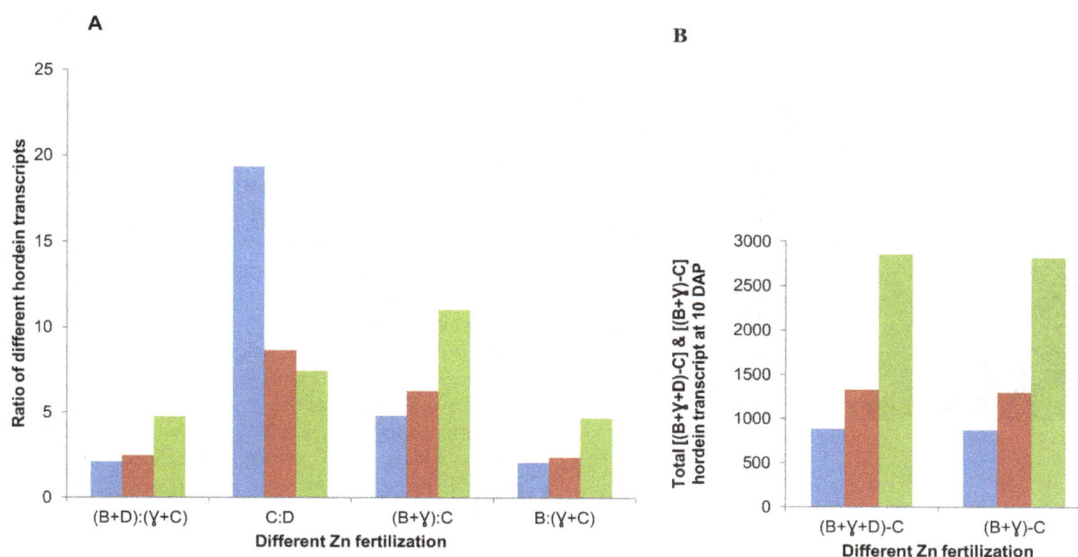

Figure 3. Effects of Zn treatments on ratio of different hordein transcripts measured at 10 DAP. Zn treatment labels: low - blue; medium - red; high – green. A) Ratio of different hordein transcripts measured at 10 DAP from different Zn fertilization; B) Total [(B+γ+D)–C] hordein or [(B+γ)−C] hordein transcripts measured at 10 DAP from different Zn fertilization. In the figure B, C, D and γ refers as B-hordein, C-hordeins, D-hordeins, and γ-hordeins respectively.

observed [6,54]. However, much more rarely found ligands are the hydroxyl of tyrosine, the carbonyl oxygen of protein backbone and the carbonyl oxygen of either asparagine or glutamine [6]. In this alignment, five cysteine residues were found as conserved in B-hordein and CMTI-I (Figure 6). Usually during protein evolution polar amino acids such as asparagine (N), aspartic acid (D), glutamic acid (E), glutamine (Q), serine (S), tyrosine (Y) can be substituted by each other [55]. For instance, the GLU19 residue of CMTI-I, which was found to bind zinc in a crystallographic study [52], was substituted by serine in B-hordein (Figure 6). Furthermore, gluta-

mine (Q) is a very abundant amino acid in B-hordein [49] which can be post translationally modified by deamination and changed to glutamic acid (E) [48,56] that might participate Zn binding [6].

Discussions

Zinc treatment increased the level of hordeins

Zinc exists in soil in various organic and inorganic forms, which affect its bioavailability to plants. More than 90% of zinc in the soil is insoluble and thus unavailable for plants [1]. Therefore, it is very

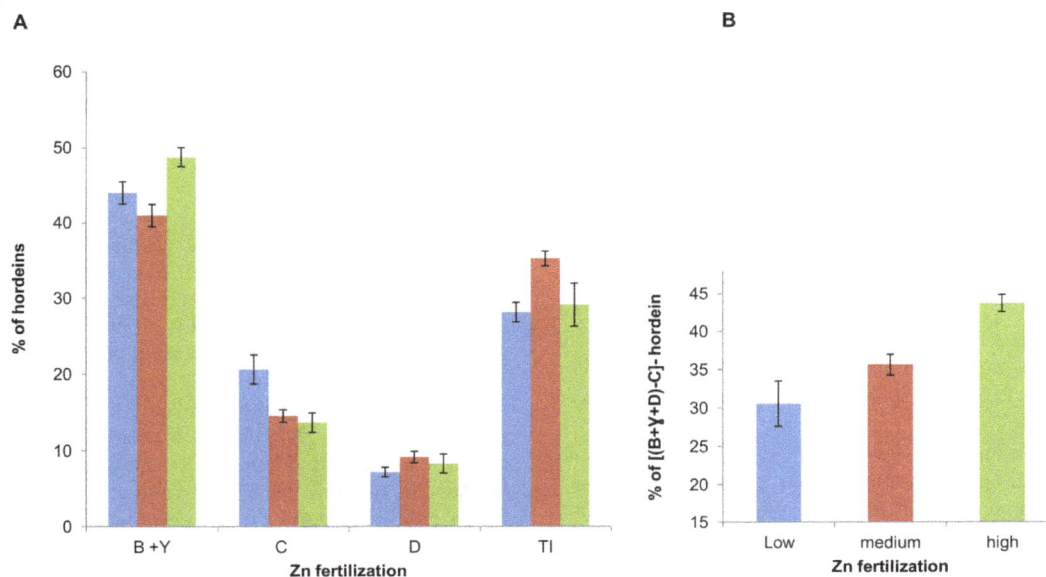

Figure 4. Effects of Zn treatments on percentage of different hordeins of matured grain measured from image analyses of SDS-PAGE gel. In the figure B, C, D, γ and TI refers as B-hordein, C-hordeins, D-hordeins, γ- hordeins and A-hordeins/Trypsin inhibitors/alpha amylase inhibitors respectively. The calculations are based on the biological replicates mentioned in table 1 and presented as means ± SE. A) Percentage of different hordeins after different Zn fertilization; Zn treatment labels: low - blue; medium - red; high - green. B) % of [(B+γ+D)–C]-hordein after different Zn fertilization.

A

B

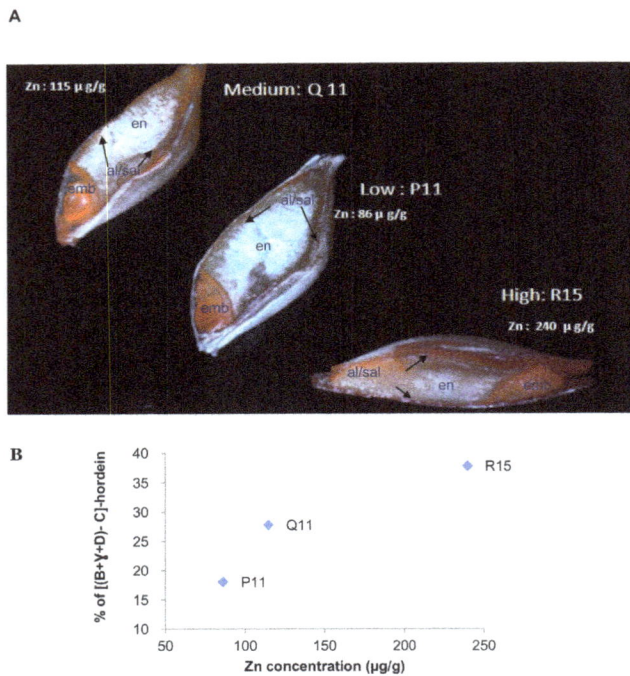

Figure 5. Zinc localization in matured grain. A) Dithizone staining of the mature grain from three different Zn fertilization experiments. Dithizone formed red/pink coloured complexes upon binding Zn ion. En- inner endosperm; emb- embryo surrounding region; al/sal - aleurone and sub-aleurone layers. B) Percentage of [(B+γ+D)-C]-hordein in the mature grains from three different Zn fertilization experiments. P11, Q11 & R15 refers as individual grains from low, medium and high Zn fertilization group.

important to keep sufficient amount of available Zn in soil (by soil Zn applications) and in leaf tissue (by foliar Zn applications) to maintenance of adequate root Zn uptake and transport of Zn from leaf tissue to the seeds during reproductive growth stage [57,58]. The main goals of our experiments were to increase grain Zn ion concentration, especially in the endosperm and to explore the effects of zinc treatments on hordein protein composition of barley grain. Our experiments with combined foliar and soil Zn fertilization showed that by applying $ZnSO_4$ the total hordein transcript levels were significantly increased with increasing Zn fertilization at 10 DAP (Figure 1). It is known that primary control of the synthesis of storage protein in wheat is at the level of transcription [59,60]. Therefore, it is possible that increasing Zn fertilization increases DNA transcription, which results in an increase in all of hordein transcripts at 10 DAP. The major fraction of the hordein transcripts that showed up-regulations were B-hordeins (Figure 2).

In wheat, significant correlations have been observed between concentration of grain Zn and the protein content which showed that increasing supplies of Zn up to an optimum concentration could significantly increase the protein content of wheat grain [61,62]. By examining data from 63 wheat crops in the Mallee region of South-eastern Australia over 3 growing seasons 0.26% increase in grain protein concentration is reported with every mg Zn per kg top soil [15]. Moreover, Zn speciation study by SEC-ICP-MS analyses of barley grain suggests that Zn is mainly binding with protein/peptide [19,29] hence it is possible that increasing Zn concentration could increase total hordein protein or vice versa.

Considering the facts that hordeins are the major storage proteins in barley, our results showed that it was possible to increase matured grain Zn concentration and it was correlated with the increasing protein concentration of the barley hordeins. (Figure 1 & Figure S1).

Association of Zn ion with B-hordein

In this experiment B- and D-hordein which are homologous to the glutenin subunit of wheat, showed significant correlation between grain zinc concentration and expression of transcript at 10 DAP (Figure S1). However, in the matured grain only the percentage of B-hordein showed significant positive correlation with grain Zn concentration, whereas C-hordein showed a negative correlation (Figure S2A). In wheat, labelling studies with [65]Zn have shown that the glutenins have a much greater level of [65]Zn incorporation (47–65%) than the other grain proteins [16]. Similar results were obtained from barley storage protein extracts and it is reported that western immunoblotting assay shows the presence of B-hordein within this MW range [40]. In addition, we assume that some protein bands in the lower molecular weight region (MW<20 kDa) which showed positive [65]Zn binding probably belonged to trypsin/alpha amylase inhibitor families, previously known as A-hordein [40,63] (Figure S3).

Therefore, our results are in agreement with the above mentioned studies that Zn can be incorporated in the glutenin fraction of grain, in this case B-hordein of barley, which is homologous to LMW-GS of wheat.

B-hordein expression at early developmental stage and correlation with Zn concentration of matured grain

After foliar application of $ZnSO_4$ the highest concentration of Zn ion in the grain of bread wheat is at the beginning of the grain development (10–12 DAP) and at later stages to maturity the relative concentration of Zn ion is stable or decreases [31]. In barley the proportion of the different hordein groups during grain development varies slightly but the amount of total hordein transcripts increased from 10 DAP onward and the B-hordein transcripts always appear as the highest proportion (more than 80%) compared to the other hordeins [33]. Zinc uptake rate of the wheat endosperm reaches its maximum level at 14 days after anthesis (DAA). During the most active starch deposition period (14 to 34 DAA) the demand for Zn from the endosperm cells remains stable or is slightly reduced [32]. The lack of Zn demand or low sink capacity of the starchy endosperm are the reasons for the low concentration of zinc in the endosperm, and therefore enhanced levels of Zn can only be reached when additional sink is created in the endosperm [32,64]. Therefore, we assumed that B-hordein transcript at 10 DAP could not vary too much compared to the matured grain and the linear relationship between matured grain Zn concentration and proportion of B-hordein transcripts at 10 DAP could be also true for all stages of grain development. It could be suggested that B-hordeins are incorporating Zn and the percentage of B or [(B+γ+D)-C] is positively correlating with the concentration of Zn. Therefore, decreasing the percentage of C-hordein in the grain achieved by Zn fertilization could help to increase the total grain Zn concentration. Decreasing C-hordein could help to increase B-hordein content of the grain that might participate in Zn binding. Previously, a study with transgenic approach for down-regulation of C-hordein transcripts with RNAi antisense technique showed pleiotropic effects including up-regulation of B-hordein as well as other Zn binding proteins [65,66]. The expression of B-hordein coding gene using D-hordein promoter could be another possible way to enrich the Zn concentration in inner endosperm. However, additional

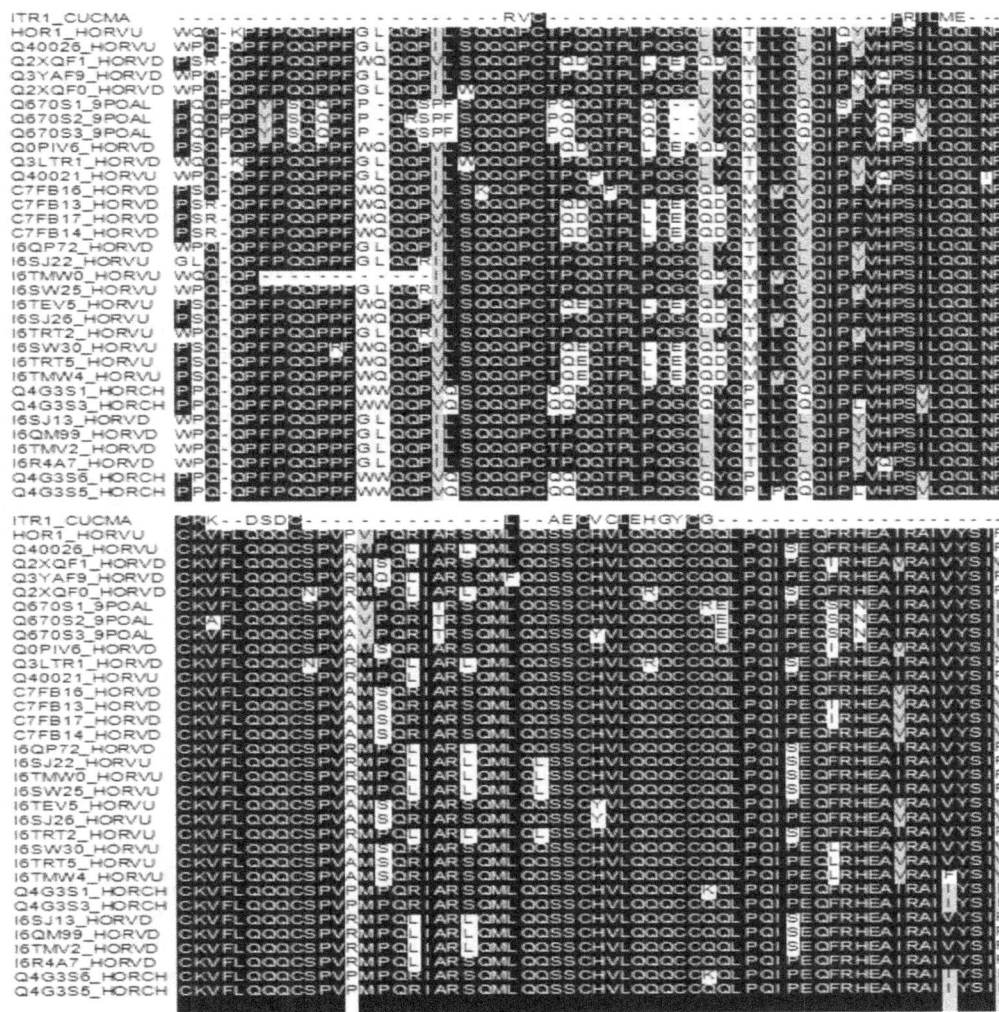

Figure 6. Multiple sequence alignment of zinc binding trypsin inhibitor 1 of *Cucurbita maxima* **(ITR1_CUCMA) and several B hordein of** *Hordeum vulgare.* The alignment was done using MEGA5.1 software with clustalW algorithm. Identical amino acids (conserved sites) are shaded in black.

experiment with transgenic approach is needed to see whether increasing B-hordein could be another biofortification strategy.

Zn could have a role in hordein polymerization

Usually monomeric prolamins contain intramolecular disulphide bonds, whereas polymeric prolamins have inter- and intramolecular disulphide bonds [48].

Foliar Zn ion application increases grain Zn concentration and has a special effect on protein composition by decreasing the proportion of gliadin in the flour, and increasing the ratio of polymeric protein to gliadin [18]. We also observed that increasing Zn application increased the ratio of polymeric to monomeric hordein transcript [(B+D):(γ+C)] (Figure 3A). In addition, transcript of total [(B+γ+D)–C] hordein or [(B+γ)–C] hordein also increased with the increasing Zn fertilization (Figure 3B). Image analyses of SDS-PAGE gel also showed an increase in percentages of [(B+γ+D)–C] hordeins (Figure 4B). In addition, our results showed significant linear correlation between percentage of [(B+γ+D)–C]- or [(B+γ)−C]- hordein and matured grain Zn concentration (Figure S2B) which is in agreement with the suggestion that

sulfhydryl groups of the cysteine residue are involved in this polymerization [18]. Since Zn has a strong affinity to sulfhydryl groups and prevents their peroxidation to form disulphide bonds [20], the concentration of grain Zn could influence the degree of polymerisation of the proteins which could shift the balance to low molecular weight polymers (eg. B-hordein) [18]. In our experiment, this is consistent with the increase in B-hordein with increasing Zn concentration in the grain both at 10 DAP and at the matured stages.

Zinc incorporation with B-hordein in sub-aleurone layer in endosperm

We reported that Zn concentration in matured grain was predominantly high in embryo, aleurone layer and outer part of the endosperm (sub-aleurone layers) (Figure 5A). However, increasing Zn fertilization could move/increase Zn concentration inside the endosperm up to a certain level and the difference of [(B+γ+D)–C]-hordein increased (Figure 5B). Our results are consistent with previous findings in developing wheat grain (subjected to foliar application of Zn) that Zn predominantly is

located in the embryo, the aleurone layer and outer part of the endosperm (sub-aleurone layer) in the matured grain [31,32,58]. Zn distribution in barley grain is not limited to the aleurone layer but is also present in the outer part of the endosperm (sub-aleurone layers) which is known to be rich in proteins [29,31]. In addition, Zn is present in the ventral part of the endosperm but found to be limited to the aleurone layer in the lateral and dorsal parts of the grain [67].

Pearling and immunocytochemical studies of barley have shown that the protein-rich sub-aleurone cells are enriched in B-hordein (S-rich) and C-hordein (S-poor) [27] while D-hordein (the HMW prolamin) is only present in a significant amount in the inner part of the starchy endosperm [68]. A study from scanning electron microscopy and mass spectrometry suggests that the structure of the aleurone layer storing large quantities of protein did not differ between the high- and low-protein forms of barley biotypes. However, the quantity of protein in barley is determined by proteins localized in the sub-aleurone layer [69]. Two proteins, B3-hordein and Z-type serpin present in the sub-aleurone layer make the biggest quantitative difference between high- and low-protein kernels in forms of barley. In our experiment, both B1- and B3-hordeins showed high responses towards Zn treatment at early developmental stage, and significant linear correlation was observed between B1- and B3-hordein transcript and grain Zn concentration of matured stage (Table 2). Therefore, our experiments showing that B-hordeins are capable of binding Zn are in agreement with those previous reports and it could be assumed that major Zn binding storage proteins in the sub-aleurone layer of barley and wheat belong to proteins homologous to LMW-GS.

B-hordein expression at early developmental stage and correlation with Zn concentration of matured grain

Barley storage protein transcripts encoding hordeins (B, C, D & γ) start to appear in the endosperm/sub-aleurone fraction at the early developmental stage (about 10–12 DAP) and reach their highest levels from 16 DAP onward [70]. Experiments with foliar application of $ZnSO_4$ in wheat showed highest concentration of Zn in the grain at the beginning stage (early milk) of the grain development (10–12 DAP) and at later stages to maturity Zn concentration was stable or decreased although the total amount of Zn increased due to the increasing dry mass volume of the grain [31]. Therefore, we choose to study hordein expression at the early stage of grain development (10 DAP). Kaczmarczyk et al. [33] studied expression of different hordein transcripts at different developmental stages and found that the proportion of the different hordein groups during grain developmental varied slightly but the amount of total hordein transcripts increased from 10 DAP onward and B-hordein always appeared as a higher proportion (more than 80%) than any other hordeins [33]. Therefore, from our experiment we assumed that the relative proportion of B-hordein transcript at 10 DAP did not vary too much compared to the matured grain and the linear relationship between matured grain Zn concentration and proportion of B-hordein transcripts at 10 DAP could be also true for all stages of grain development.

The Zn uptake rates of the wheat endosperm reaches its maximum level at 14 DAA and after the Zn concentration in endosperm remains more or less stable [32]. It is suggested that the low sink capacity of the starchy endosperm are the reasons for the low concentration of zinc in the endosperm, and therefore enhanced levels of Zn can be reached only when an additional sink is created [32,64]. The sink capacity could be increased either by decreasing starch content or increasing the proportion of different hordeins that are incorporating Zn. B-hordeins could be that sink protein as we reported that the percentage of B or [(B+γ+D)–C] was correlating with the concentration of Zn. Therefore, a possible approach to enrich the Zn concentration in the inner endosperm could be the expression of B-hordein using D-hordein promoter to allow translocation in the endosperm. Alternatively decreasing C-hordein level could help to increase B-hordein content of the grain as it is shown in C-hordein antisense lines [65,66]. We showed that increasing Zn fertilization could decrease percentage of C-hordein in matured grain (Figure 4), as well as the ratio of C/D transcript at the 10 DAP (Figure 3A). Therefore, decreasing the percentage of C-hordein in the grain could also help to increase the total grain Zn concentration.

Hordein evolution and possible Zn binding sites/domain in B-hordein

Prolamins are thought to have evolved about 100 million years ago through gene duplication and addition/deletion of repetitive domains in the sequences of this ancestral protease inhibitor [49,71–73]. B-hordein and CMTI-I, a small-protein trypsin inhibitor from pumpkin known to bind Zn ion, share a common trypsin inhibitor domain. Therefore, it is possible that one of several potential zinc binding sites/domains is located within this region of the B-hordein sequence (Figure 6). Usually zinc is considered as a "borderline" metal, which does not consistently act either "hard" (not very polarizable) or "soft" (highly polarizable); and hence does not have a strong preference for coordinating with either oxygen or nitrogen or sulphur atoms [5,6]. Speciation of Zn in barley and wheat grain is not well understood although several hypotheses exist regarding the association with S or O atoms in the aleurone layers of the grain [19,74]. B-hordein is known to be localized mainly in the sub-aleurone layer of the grain. Therefore, if S is the main ligand in the sub-aleurone layer of the grain then it is possible that B-hordein is participating in binding Zn through cysteine residues; Zn could either have a role in stabilizing the structure of the protein or be involved in the polymerization of hordein.

Conclusions

Greenhouse experiments showed that combined foliar and soil Zn fertilization could significantly increase the total Zn concentration of barley grain. Although the total hordein transcripts increased in the grain at 10 DAP, the increase of B-hordein transcript level was the most prominent. Significant positive correlation was observed between the amount of B-hordein transcript level at 10 DAP and grain Zn concentration at the matured stage. Furthermore, increasing Zn fertilization also increased the percentage of B hordein and we found significant positive correlations between proportion of [(B+D)–C or (B–C)]-hordein and Zn concentration in the matured grain.

The zinc fertilization increased the total zinc concentration of the grain and the enrichment happened mostly around areas such as embryo, aleurone and sub-aleurone layer. The outer part of the endosperm (sub-aleurone layer) was enriched with Zn due to the increased presence of B-hordein. Considering the obtained Zn concentration of the grain and our protein data we concluded that not all hordein, but B-hordein upregulation or difference between B- and C-hordein is one of the key factors for increased Zn concentration in the mature grain. More experimental evidences are required to see whether down regulation of C-hordein level and up regulation of B-hordein could be a viable option for Zn biofortification.

Supporting Information

Figure S1 Linear correlations between different steady state level of hordein gene at 10 DAP and Zn concentration of the matured grain.
(TIF)

Figure S2 Linear correlation between matured grain Zn concentration and different proportion of hordein measured from SDS-PAGE gel by image analyses. A) linear correlations of grain Zn ion concentration and % of hordeins; B) linear correlations of grain Zn concentration and percentage of [(B+γ+D)−C] or [(B+γ)−C]-hordeins.
(TIF)

Figure S3 Selective binding of zinc ion by alcohol soluble protein from barley (cv. Golden Promise) grain. A) Replica membrane stained with amido black, B) Zinc binding protein specified by autoradiography showing black bands on the membrane. In A & B: 1- Maltose binding protein (MBP5); 2-HiMark prestained protein marker, 3-Hordein extract; 4- Alcohol dehydrogenase. Numbers in vertical axis represent the approximate molecular weight (kDa) of the protein bands.
(TIF)

Acknowledgments

We are grateful to the Department of Plant and Environmental Sciences, University of Copenhagen, Denmark for measuring the Zn concentration of flour. We also like to thank K. B. Nellerup and O. B. Hansen for their technical support.

Author Contributions

Conceived and designed the experiments: MNU EV. Performed the experiments: MNU AK. Analyzed the data: MNU EV. Contributed reagents/materials/analysis tools: MNU AK EV. Wrote the paper: MNU EV.

References

1. Broadley MR, White PJ, Hammond JP, Zelko I, Lux A (2007) Zinc in plants. New Phytologist 173: 677–702.
2. Hotz C, Brown KH (2004) Assessment of the risk of zinc deficiency in populations and options for its control: International nutrition foundation: for UNU.
3. Stein AJ (2010) Global impacts of human mineral malnutrition. Plant and soil 335: 133–154.
4. Corbo MD, Lam J (2013) Zinc deficiency and its management in the pediatric population: A literature review and proposed etiologic classification. Journal of the American Academy of Dermatology.
5. Auld D (2001) Zinc coordination sphere in biochemical zinc sites. Biometals 14: 271–313.
6. McCall KA, Huang C-c, Fierke CA (2000) Function and Mechanism of Zinc Metalloenzymes. The Journal of Nutrition 130: 1437S–1446S.
7. Andreini C, Bertini I, Rosato A (2009) Metalloproteomes: a bioinformatic approach. Accounts of chemical research 42: 1471–1479.
8. White PJ (2011) Physiological limits to zinc biofortification of edible crops. Frontiers in Plant Science 2.
9. Haase H, Rink L (2013) Zinc signals and immune function. BioFactors: n/a-n/a.
10. Andreini C, Bertini I, Cavallaro G (2011) Minimal Functional Sites Allow a Classification of Zinc Sites in Proteins. PLoS ONE 6: e26325.
11. Maret W (2013) Zinc Biochemistry: From a Single Zinc Enzyme to a Key Element of Life. Advances in Nutrition: An International Review Journal 4: 82–91.
12. White PJ, Broadley MR (2009) Biofortification of crops with seven mineral elements often lacking in human diets - iron, zinc, copper, calcium, magnesium, selenium and iodine. New Phytologist 182: 49–84.
13. Hansen T, Laursen K, Persson D, Pedas P, Husted S, et al. (2009) Micro-scaled high-throughput digestion of plant tissue samples for multi-elemental analysis. Plant Methods 5: 12.
14. Kutman UB, Yildiz B, Ozturk L, Cakmak I (2010) Biofortification of Durum Wheat with Zinc Through Soil and Foliar Applications of Nitrogen. Cereal Chemistry Journal 87: 1–9.
15. Sadras V, Roget D, O'Leary G (2002) On-farm assessment of environmental and management constraints to wheat yield and efficiency in the use of rainfall in the Mallee. Crop and Pasture Science 53: 587–598.
16. Starks TL, Johnson PE (1985) Techniques for intrinsically labeling wheat with zinc-65. Journal of Agricultural and Food Chemistry 33: 691–698.
17. Gomez-Becerra HF, Abugalieva A, Morgounov A, Abdullayev K, Bekenova L, et al. (2010) Phenotypic correlations, G x E interactions and broad sense heritability analysis of grain and flour quality characteristics in high latitude spring bread wheats from Kazakhstan and Siberia. Euphytica 171: 23–38.
18. Peck AW, McDonald GK, Graham RD (2008) Zinc nutrition influences the protein composition of flour in bread wheat (Triticum aestivum L.). Journal of Cereal Science 47: 266–274.
19. Persson DP, Hansen TH, Laursen KH, Schjoerring JK, Husted S (2009) Simultaneous iron, zinc, sulphur and phosphorus speciation analysis of barley grain tissues using SEC-ICP-MS and IP-ICP-MS. Metallomics 1: 418–426.
20. Maret W (2012) New perspectives of zinc coordination environments in proteins. Journal of Inorganic Biochemistry 111: 110–116.
21. Shewry PR (2009) Wheat. Journal of Experimental Botany 60: 1537–1553.
22. Shewry PR, Halford NG (2002) Cereal seed storage proteins: structures, properties and role in grain utilization. Journal of Experimental Botany 53: 947–958.
23. Shewry PR, Darlington H (2002) The proteins of the mature barley grain and their role in determining malting performance. Barley Science: Recent advances from molecular biology to agronomy of yield and quality: 503–521.
24. Gupta R, Khan K, Macritchie F (1993) Biochemical basis of flour properties in bread wheats. I. Effects of variation in the quantity and size distribution of polymeric protein. Journal of Cereal Science 18: 23–41.
25. Shewry PR (1993) Barley seed proteins. Barley: Chemistry and Technology: 131–197.
26. Shewry PR (1995) Plant Storage Proteins. Biological Reviews 70: 375–426.
27. Shewry PR (1996) Cereal grain proteins. In: Henry RJ, Kettlewell PS, editors. Cereal Grain Quality: Springer Netherlands. pp. 227–250.
28. Iwai T, Takahashi M, Oda K, Terada Y, Yoshida KT (2012) Dynamic changes in the distribution of minerals in relation to phytic-acid accumulation during rice seed development. PLANT PHYSIOLOGY.
29. Lombi E, Smith E, Hansen TH, Paterson D, de Jonge MD, et al. (2011) Megapixel imaging of (micro)nutrients in mature barley grains. Journal of Experimental Botany 62: 273–282.
30. Molina-Cano JL, Sopena A, Polo JP, Bergareche C, Moralejo MA, et al. (2002) Relationships Between Barley Hordeins and Malting Quality in a Mutant of cv. Triumph. II. Genetic and Environmental Effects on Water Uptake. Journal of Cereal Science 36: 39–50.
31. Ozturk L, Yazici MA, Yucel C, Torun A, Cekic C, et al. (2006) Concentration and localization of zinc during seed development and germination in wheat. Physiologia Plantarum 128: 144–152.
32. Stomph TJ, Choi EY, Stangoulis JCR (2011) Temporal dynamics in wheat grain zinc distribution: is sink limitation the key? Annals of Botany 107: 927–937.
33. Kaczmarczyk A, Bowra S, Elek Z, Vincze E (2012) Quantitative RT-PCR based platform for rapid quantification of the transcripts of highly homologous multigene families and their members during grain development. BMC plant biology 12: 184.
34. Baik B-K, Ullrich SE (2008) Barley for food: characteristics, improvement, and renewed interest. Journal of Cereal Science 48: 233–242.
35. Chen F, Wang F, Zhang G, Wu F (2008) Identification of Barley Varieties Tolerant to Cadmium Toxicity. Biological Trace Element Research 121: 171–179.
36. Zhao J, Sun H, Dai H, Zhang G, Wu F (2010) Difference in response to drought stress among Tibet wild barley genotypes. Euphytica 172: 395–403.
37. Bustin SA, Benes V, Garson JA, Hellemans J, Huggett J, et al. (2009) The MIQE guidelines: minimum information for publication of quantitative real-time PCR experiments. Clinical chemistry 55: 611–622.
38. Mager M, McNary WF, Lionetti F (1953) The histochemical detection of zinc. Journal of Histochemistry & Cytochemistry 1: 493–504.
39. McNary WF (1954) Zinc-Dithizone reaction of pancreatic islets. Journal of Histochemistry & Cytochemistry 2: 185–195.
40. Uddin MN, Nielsen AL-L, Vincze E (2014) Zinc blotting assay for detection of zinc binding prolamin in barley (Hordeum vulgare) grain. Cereal Chemistry 91: 228–232.
41. Schägger H, Von Jagow G (1987) Tricine-sodium dodecyl sulfate-polyacrylamide gel electrophoresis for the separation of proteins in the range from 1 to 100 kDa. Analytical biochemistry 166: 368–379.
42. Thompson JD, Higgins DG, Gibson TJ (1994) CLUSTAL W: improving the sensitivity of progressive multiple sequence alignment through sequence weighting, position-specific gap penalties and weight matrix choice. Nucleic Acids Research 22: 4673–4680.
43. Wei Y, Shohag MJI, Yang X (2012) Biofortification and Bioavailability of Rice Grain Zinc as Affected by Different Forms of Foliar Zinc Fertilization. PLoS ONE 7: e45428.
44. Cakmak I, Kalayci M, Kaya Y, Torun AA, Aydin N, et al. (2010) Biofortification and Localization of Zinc in Wheat Grain. Journal of Agricultural and Food Chemistry 58: 9092–9102.

45. Tatham AS, Shewry PR (2012) The S-poor prolamins of wheat, barley and rye: Revisited. Journal of Cereal Science 55: 79–99.
46. Shewry P, Bunce NAC, Kreis M, Forde B (1985) Polymorphism at the Hor 1 locus of barley (Hordeum vulgare L.). Biochem Genet 23: 391–404.
47. MacGregor AW, Bhatty RS (1993) Barley: chemistry and technology: American Association of Cereal Chemists (AACC).
48. Kanerva P (2011) Immunochemical analysis of prolamins in gluten-free foods.
49. Shewry PR, Tatham AS (1990) The prolamin storage proteins of cereal seeds: structure and evolution. Biochemical Journal 267: 1.
50. Shewry PR, Tatham AS (1999) The characteristics, structures and evolutionary relationships of prolamins. Seed Proteins: Springer. pp. 11–33.
51. Rechinger K, Simpson D, Svendsen I, Cameron-Mills V (1993) A role for gamma3 hordein in the transport and targeting of prolamin polypeptides to the vacuole of developing barley endosperm. Plant J 4: 841–853.
52. Thaimattam R, Tykarska E, Bierzynski A, Sheldrick GM, Jaskolski M (2002) Atomic resolution structure of squash trypsin inhibitor: unexpected metal coordination. Acta Crystallographica Section D 58: 1448–1461.
53. Breiteneder H, Radauer C (2004) A classification of plant food allergens. Journal of Allergy and Clinical Immunology 113: 821–830.
54. Auld DS (2001) Zinc coordination sphere in biochemical zinc sites. Biometals 14: 271–313.
55. Betts MJ, Russell RB (2003) Amino acid properties and consequences of substitutions. 289 p.
56. Vader LW, de Ru A, van der Wal Y, Kooy YM, Benckhuijsen W, et al. (2002) Specificity of tissue transglutaminase explains cereal toxicity in celiac disease. The Journal of experimental medicine 195: 643–649.
57. Kutman U, Kutman B, Ceylan Y, Ova E, Cakmak I (2012) Contributions of root uptake and remobilization to grain zinc accumulation in wheat depending on post-anthesis zinc availability and nitrogen nutrition. Plant and Soil 361: 177–187.
58. Cakmak I (2008) Enrichment of cereal grains with zinc: Agronomic or genetic biofortification? 302: 1–17.
59. Bartels D, Thompson RD (1986) Synthesis of mRNAs coding for abundant endosperm proteins during wheat grain development. Plant Science 46: 117–125.
60. Triboï E, Martre P, Triboï-Blondel AM (2003) Environmentally-induced changes in protein composition in developing grains of wheat are related to changes in total protein content. Journal of Experimental Botany 54: 1731–1742.
61. Lorenz K, Loewe R (1977) Mineral composition of US and Canadian wheats and wheat blends. Journal of agricultural and food chemistry 25: 806–809.
62. Dikeman E, Pomeranz Y, Lai F (1982) Minerals and protein contents in hard red winter wheat. Cereal Chemistry 59.
63. Shewry PR, Ellis JR, Pratt HM, Miflin BJ (1978) A comparison of methods for the extraction and separation of hordein fractions from 29 barley varieties. Journal of the Science of Food and Agriculture 29: 433–441.
64. Stomph TJ, Jiang W, Struik PC (2009) Zinc biofortification of cereals: rice differs from wheat and barley. Trends in Plant Science 14: 123–124.
65. Hansen M, Lange M, Friis C, Dionisio G, Holm PB, et al. (2007) Antisense-mediated suppression of C-hordein biosynthesis in the barley grain results in correlated changes in the transcriptome, protein profile, and amino acid composition. Journal of experimental botany 58: 3987–3995.
66. Lange M, Vincze E, Wieser H, Schjoerring JK, Holm PB (2007) Suppression of C-Hordein Synthesis in Barley by Antisense Constructs Results in a More Balanced Amino Acid Composition. Journal of Agricultural and Food Chemistry 55: 6074–6081.
67. Lombi E, Scheckel KG, Kempson IM (2011) In situ analysis of metal(loid)s in plants: State of the art and artefacts. Environmental and Experimental Botany 72: 3–17.
68. Tosi P, Gritsch CS, He J, Shewry PR (2011) Distribution of gluten proteins in bread wheat (Triticum aestivum) grain. Annals of Botany 108: 23–35.
69. Macewicz J, Orzechowski S, Dobrzy ska U, Haebel S (2006) Is quantity of protein in barley forms determined by proteins localized in the subaleurone layer? Acta Physiologiae Plantarum 28: 409–416.
70. Sreenivasulu N, Usadel B, Winter A, Radchuk V, Scholz U, et al. (2008) Barley grain maturation and germination: metabolic pathway and regulatory network commonalities and differences highlighted by new MapMan/PageMan profiling tools. Plant Physiol 146: 1738–1758.
71. Kreis M, Forde BG, Rahman S, Miflin BJ, Shewry PR (1985) Molecular evolution of the seed storage proteins of barley, rye and wheat. Journal of Molecular Biology 183: 499–502.
72. Holding D, Messing J (2013) Evolution, Structure, and Function of Prolamin Storage Proteins. Seed Genomics: 138–158.
73. Onda Y, Kawagoe Y (2011) Oxidative protein folding: Selective pressure for prolamin evolution in rice. Plant Signaling & Behavior 6: 1966–1972.
74. Neal A, Geraki K, Borg S, Quinn P, Mosselmans JF, et al. (2013) Iron and zinc complexation in wild-type and ferritin-expressing wheat grain: implications for mineral transport into developing grain. JBIC Journal of Biological Inorganic Chemistry 18: 557–570.

6

Genetic Basis and Selection for Life-History Trait Plasticity on Alternative Host Plants for the Cereal Aphid *Sitobion avenae*

Xinjia Dai[1,2,3], **Suxia Gao**[1,2,3], **Deguang Liu**[1,2,3]*

1 State Key Laboratory of Crop Stress Biology for Arid Areas (Northwest A&F University), Yangling, Shaanxi Province, China, **2** Key Laboratory of Integrated Pest Management on Crops in Northwestern Loess Plateau, Ministry of Agriculture, Yangling, Shaanxi Province, China, **3** College of Plant Protection, Northwest A&F University, Yangling, Shaanxi Province, China

Abstract

Sitobion avenae (F.) can survive on various plants in the Poaceae, which may select for highly plastic genotypes. But phenotypic plasticity was often thought to be non-genetic, and of little evolutionary significance historically, and many problems related to adaptive plasticity, its genetic basis and natural selection for plasticity have not been well documented. To address these questions, clones of *S. avenae* were collected from three plants, and their phenotypic plasticity under alternative environments was evaluated. Our results demonstrated that nearly all tested life-history traits showed significant plastic changes for certain *S. avenae* clones with the total developmental time of nymphs and fecundity tending to have relatively higher plasticity for most clones. Overall, the level of plasticity for *S. avenae* clones' life-history traits was unexpectedly low. The factor 'clone' alone explained 27.7–62.3% of the total variance for trait plasticities. The heritability of plasticity was shown to be significant in nearly all the cases. Many significant genetic correlations were found between trait plasticities with a majority of them being positive. Therefore, it is evident that life-history trait plasticity involved was genetically based. There was a high degree of variation in selection coefficients for life-history trait plasticity of different *S. avenae* clones. Phenotypic plasticity for barley clones, but not for oat or wheat clones, was frequently found to be under significant selection. The directional selection of alternative environments appeared to act to decrease the plasticity of *S. avenae* clones in most cases. G-matrix comparisons showed significant differences between *S. avenae* clones, as well as quite a few negative covariances (i.e., trade-offs) between trait plasticities. Genetic basis and evolutionary significance of life-history trait plasticity were discussed.

Editor: Ulrich Melcher, Oklahoma State, United States of America

Funding: This research was supported by the Specialized Research Fund for the Doctoral Program of Higher Education from Ministry of Education of China (No. 20110204120001) (http://www.moe.edu.cn/), and a grant from Northwest A&F University (No. QN2011059) (http://www.nwsuaf.edu.cn). DL received the funding. The funders had no role in study design, data collection and analysis, decision to publish, or preparation of the manuscript.

Competing Interests: The authors have declared that no competing interests exist.

* Email: dgliu@nwsuaf.edu.cn

Introduction

All organisms live in spatially and temporally variable and sometimes predictable environments. Organisms may cope with the highly variable environments by adaptation through genetic modifications under natural selection [1]. Phenotypic plasticity may also facilitate successful use of changing environments by an organism, and thus is considered to be another mechanism for adaptation [2]. However, the optimal strategy of organisms is assumed to be non-plastic and maximal in fitness traits for all environments, so one would rarely expect traits tightly linked to fitness to be plastic [3]. In addition, phenotypic plasticity was often considered to buffer the impact of natural selection, and thus act to slow evolutionary changes [4]. Therefore, phenotypic plasticity of an organism in response to variable environments had long been considered to be non-genetic and of little evolutionary importance until late 1980s [5]. Bradshaw coined the term 'phenotypic plasticity' to describe environmentally contingent morphological expression when developmental variability was considered to be uninteresting noise by most scientists [6–7]. But now it has been broadly used to describe all phenotypic responses to environmental changes [8]. Following Sultan [9], we define plasticity as variation in phenotypic expression of a genotype that occurs in response to particular environmental conditions. Plasticity of a particular trait may have positive (i.e., adaptive), negative (i.e., maladaptive) or no (i.e., neutral) consequences for a genotype's fitness under different environments, so phenotypic plasticity can either retard or accelerate rates of phenotypic evolution based on relative fitness of the new phenotype [10–13]. Over the past two decades, interest in phenotypic plasticity has grown exponentially, and the change of interest reflects the new understanding that plasticity could be a powerful means of adaptation [7,14–15]. Despite the large volume of work in this field, problems related to adaptive plasticity, genetic basis and natural selection for plasticity have not been well documented [14,16]. All such problems are conceptually crucial for our understanding not only of evolutionary consequences of plasticity, but also of phenotypic evolution in general [16–17].

There exist fundamental differences in biological features among various organisms (e.g., plants and animals), which can have important implications for the evolution of phenotypic plasticity. For example, plants (sessile organisms) usually have greater plasticity than animals (mobile organisms) in morphological and developmental responses to changes in their biotic and abiotic environments, probably because animals can often move away from unfavorable environments [5,18]. As a large and abundant group of organisms, http://en.wikipedia.org/wiki/Phenotypic_plasticity - cite_note-3#cite_note-3insect herbivores (many of them are highly mobile) have their own biological properties, and can often utilize a variety of host plant species. Significant variation in morphology, physiology and chemistry can occur among these plants [19]. Different host plants of insects often exist in temporally and spatially discrete patches that act as differential selective environments. Therefore, insect herbivores were also shown to be plastic in morphology, physiology, behavior or life-history in response to different host plants [4,15,20–21]. Surprisingly, studies on phenotypic plasticity of insects' life-history traits on different host plants have been rare.

Aphids' success in a wide diversity of ecosystems is partially attributed to their broad phenotypic plasticity in color, wing production and reproduction, although many of them are specialized on particular host plants [22–23]. The cotton aphid (*Aphis gossypii* Glover) or black bean aphid (*Aphis fabae*) was shown to be plastic in morphology [20], insecticide susceptibility [24], host choice behavior [21] or life-history traits (e.g., developmental times) [4]. Host plants (i.e., different environments) were found to be an important factor in inducing aphids' plastic changes in phenotypes [4,20–21]. Host plants also showed conditioning effects on the tested aphids, which is another piece of evidence for plasticity that happens without substantial modifications on the insect genome [21,25–26]. Plasticity resulting from different host plants can play significant roles influencing the evolutionary trajectory of aphids. Surprisingly, studies on phenotypic plasticity of aphids' life-history traits have been rare.

The cereal aphid, *Sitobion avenae* (Fabricius), can survive on numerous species in the Poaceae [27], and provides a good model to study life-history trait plasticity for insects that can disperse long distances. Several studies characterized key life-history traits (e.g., developmental times and fecundity) of *S. avenae* clones on wheat, barley and oat [25,28–31]. In our previous study, we compared barley clones to oat clones from Shaanxi province, and found that barley and oat clones differentiated significantly in life-history traits, heritabilities of those traits, and the extent of specialization on a particular host plant; however, divergent selection on both host plants did not result in the formation of highly specialized clones or host races [23]. Therefore, differential adaptation of *S. avenae* clones to barley or oat in our previous study might result from their phenotypic plasticity. Since spatial heterogeneity and dispersal of organisms can have significant consequences for the evolution of plasticity [32], the study mentioned above had limitations in the number of host plants tested (i.e., two) and the source of tested *S. avenae* clones (from a single location) for analyzing the effects of plasticity. So, we collected *S. avenae* clones on three cereal crops from two provinces of China, and tested them in the laboratory. We hypothesize that key life-history traits of *S. avenae* was highly plastic in alternative environments, and the trait plasticity involved is genetically based and evolutionarily important. The aims of this study were to: 1) characterize phenotypic plasticity of key life-history traits for *S. avenae* clones in alternative environments (i.e., on alternative host plants); 2) assess the underlying genetic basis and natural selection for life-

history trait plasticity of *S. avenae* clones; and 3) evaluate whether the observed plastic responses of *S. avenae* are adaptive.

Materials and Methods

Aphids and Plants

In order to increase genetic variability, *S. avenae* clonal genotypes were sampled from three host species and distinct locations from two provinces of China. Individual clones of *S. avenae* were collected in May of 2013 in Shaanxi Province, and in August of 2013 in Qinghai province. These clones came from barley, oat and wheat fields in the Shaanxi area (collected at three sites: 34°17′21.64″N, 108°4′10.09″E; 34°18′6.35″N, 108°5′20.54″E; 34°18′35.73″N, 107°57′42.20″E) and Qinghai area (collected at three sites: 36°48′82.33″N, 101°59′89.94″E; 37°04′87.06″N, 101°90′00.31″E; 37°13′04.89″N, 101°28′78.21″E) (no specific permissions were required for the sample collecting activities at all the abovementioned sites, and no endangered or protected species were involved in the collecting activities). In order to limit the chance of re-sampling individuals from the same parthenogenetic mother, an individual wingless adult aphid was collected from a plant separated by at least 10 m from other samples [25]. At least 20 different clones were collected for each plant species in each area, and they were used to start separate colonies in the laboratory. Aphid clones were cultured on the species of plants from which they were originally collected (i.e., barley, oat or wheat). Wheat (*Triticum aestivum* L. cv. Aikang 58), oat (*Avena sativa* L. cv. Sandle) and barley (*Hordeum vulgare* L. cv. Xian 91–2) seeds were planted in plastic pots (6 cm in diameter) containing turfy soil, vermiculite and perlite (4:3:1, v/v/v). Plants with collected clones were enclosed with a transparent cylinder (5.5 cm in diameter, 15 cm in height) which had a Terylene mesh top for ventilation, and maintained in rearing rooms (at 20±2°C and a photoperiod of 16: 8 (L: D)). Plants were watered or replenished as needed. Prior to the bioassays, aphid clones were reared for two generations under common conditions in the lab to minimize the effects of confounding factors (e.g., weather conditions) according to Pitchers et al. [33]. After that, aphid clones were randomly selected from the colony for use in the following tests.

Life History Data Collection

Barley, oat and wheat seedlings (one per pot) used in life-history tests were planted as described above. When they reached one- to two-leaf stage, seedlings then received aphids that were transferred from rearing plants. To have a cohort of first instar nymphs with the same age, young wingless female adults of 16 different clones (10 from Shaanxi and 6 from Qinghai) for each plant species were transferred to test plants (one individual per plant). After 2–3 h, plants were checked and all aphids except one newborn nymph in each pot were removed by a fine paint brush. To prevent the aphids from escaping, each pot of plant was enclosed with a transparent plastic cylinder described above. Test plants were maintained in environmental growth chambers at 20±1°C, a relative humidity of 65±2%, and a photoperiod of 16:8 (L:D). Four to six replicates were established on each test plant species for each clone. All aphid clones were tested on both original and alternate host species (i.e., barley, oat and barley). Each test aphid individual was observed twice daily from birth until the onset of reproduction, and molting and mortality were recorded at about the same time each day. After the initiation of reproduction, which usually occurred 1 d after the fourth molt, mortality and fecundity of aphids were recorded daily, and their offspring were then removed from each test plant daily for 7 d.

Statistical Analysis

Developmental times of first to fourth nymphal instars (hereafter referred to as DT1 to DT4), the total developmental time of nymphs (from birth to adult emergence) (hereafter referred to as DT5), and 7-d fecundity (nymphs born in the first 7 d after the initiation of reproduction) were calculated. The abovementioned fitness traits were analyzed with three way nested analysis of variance (ANOVA), which was conducted with clones nested in plant origin (i.e., barley, oat and wheat) in SAS [34]. The main effects of location, plant origin and test plant were analyzed, and the interaction between the latter two was also considered. When the overall variation in ANOVA was significant, post-hoc comparisons among means were carried out by using Tukey tests at $\alpha = 0.05$.

The amount of plasticity was evaluated by calculating the coefficient of variation ($CV = SD/\overline{x} \times 100$; SD, standard deviation of treatments; $\overline{x}$, mean of treatments) for each trait in different environments (i.e., on different hosts). Another nested ANOVA with abovementioned factors was performed to analyze the phenotypic plasticity of life-history traits and genetic variation underlying phenotypic plasticity. Data were log-transformed if needed to meet the assumptions of normality and homoscedasticity required for these analyses.

Our experimental design with clonal genotypes allows us to estimate the total variance of a particular trait (V_P), which includes among-clone genetic components V_G, the (broad sense) genetic variance, and within-clone components V_E (i.e., environmental variance or residual variance). Broad-sense heritabilities were calculated as the proportion of the total variance accounted for by the among-clone variance component ($H^2 = V_G/V_P$). The statistical significance of heritabilities was assessed by using likelihood-ratio tests (LRTs) following Carter et al. [35].

Genetic variance and covariance estimates for life-history trait plasticities were obtained with the restricted maximum likelihood (REML) method implemented in the software VCE 6.0.2 [36]. The genetic correlation between traits x and y was calculated from the genetic covariance estimate (cov[x, y]) and their additive variances as $r = \text{cov}(x,y)/\left[(v_x) \times (v_y)\right]^{0.5}$. The resulting G matrices were compared using the Flury hierarchical method, using the software CPCrand [37]. Based on maximum likelihood, this method can analyze structural differences among G matrices by comparing their eigenvectors and eigenvalues. Specifically, the method can test the following models in order: (1) unrelated structure (meaning that matrices do not share any eigenvector), (2) partial common principal component (matrices sharing some eigenvectors), (3) common principal components (matrices sharing all eigenvectors but not eigenvalues), (4) proportionality (matrices with same eigenvectors and proportional eigenvalues), and (5) equality (matrices with same eigenvectors and eigenvalues) (see also in [35]). The significance of genetic covariances and correlations between trait plasticities were evaluated using LRTs following Carter et al. [35].

In this study, we used 7 d fecundity as the fitness estimate [23,31]. Relative fitness of an aphid clone was calculated by dividing the clone's 7 d fecundity by the mean of all clones in each treatment. All traits were standardized to mean zero and unit variance. We then calculated univariate standardized selection differentials using parametric regression analysis following Lande and Arnold [38] to quantify the strength of selection for S. avenae on the three cereals. Selection differentials can estimate the total strength of selection on a trait, and thus include both direct selection and indirect selection arising through covariances with other traits [39]. To separate the effects of direct selection on focal traits from the effects of indirect selection on other traits, standard linear selection gradients were estimated by performing multiple regressions following Lande and Arnold [38]. Regression analyses were performed using the PROC REG procedure in SAS [34].

Principal component analysis (PCA) (Proc PRINCOMP) was performed with plasticities of all life-history traits measured above [34]. The factor weightings for each replicate in the PCA mentioned above were calculated and the resulting values were used as a composite plasticity factor in subsequent regression analyses. The PROC REG procedure in SAS [34] was used to identify the relationships between the relative fitness of S. avenae clones on test plants and the level of life-history plasticity (i.e., the composite plasticity factor calculated as the first component extracted from PCA).

Results

Life-history trait plasticity

After transfer to alternative environments (i.e., alternative host plants), all tested clones from both areas showed non-significant changes in DT1 except the barley clones from the Qinghai area (Fig. 1A). DT1 of barley clones from Qinghai was significantly reduced in alternative environments, showing higher mean plasticity of these clones in comparison to other clones (Fig. 1A). DT2 presented non-significant changes for all clones (i.e., barley, oat and wheat) from both areas, indicating similar levels of mean plasticity for them (Fig. 1B). Barley clones from the Shaanxi area showed significantly lower DT3 in alternative environments (meaning relatively higher mean plasticity), but all other clones presented relatively lower mean plasticity in DT3, indicated by non-significant changes in DT3 in different environments (Fig. 1C). After they were transferred from plants of origin to alternative ones, significant changes in DT4 were found for oat clones from Shaanxi and wheat clones from Qinghai; non-significant changes were found for all other clones tested (Fig. 1D). After switching environments, all clones presented significant changes in DT5 but oat clones from Qinghai, indicating that DT5's mean plasticity was relatively higher for S. avenae clones in comparison to other trait plasticities (Fig. 1E). Significant changes in 7-d fecundity were identified for all clones of both areas but wheat clones of Shaanxi and barley clones from Qinghai, showing relatively higher mean plasticity of 7-d fecundity for the majority of S. avenae clones (Fig. 1F).

Genetic variation of trait plasticity

The plasticity for the developmental time of 1st instar nymphs was significantly influenced by 'location', 'origin', and interaction between 'origin' and 'test', as well as 'clone' (nested in plant origin) (Table 1). All factors (i.e., 'location', 'origin', 'test', 'origin x test', and 'clone' nested in plant origin) showed significant effects on the plasticity for the developmental times of 2nd to 4th instar nymphs. Similar results were found for the plasticities for the total developmental time of nymphs and 7-d fecundity. The significant interactions between 'plant origin' and 'test plant' for the plasticity of all tested life-history traits indicated differences in host adaptation among different clones. The factor 'clone' alone accounted for a significant proportion (i.e., 27.7–62.3%) of the total variance for the plasticity of life-history traits mentioned above. 'Clone' and 'origin' together explained 48.6–68.6% of the total variance, and 'location' accounted for 1.3–19.3%, whereas 'test' and 'origin x test' contributed relatively little (i.e., 0.7–3.7% and 1.1–6.5%, respectively) to the total variance. So, genetic effects were evident for the plasticity of all the fitness traits tested.

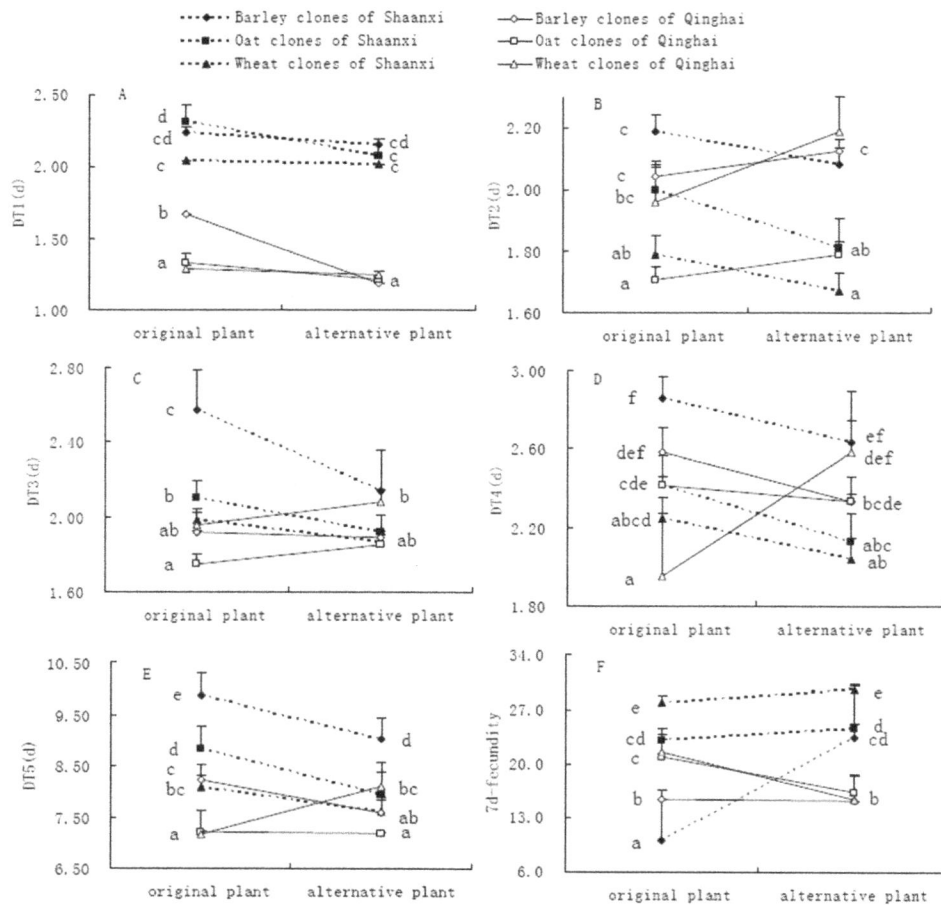

Figure 1. Comparisons of life-history traits for barley, oat and wheat clones of *Sitobion avenae* from two areas on original and alternative host plants, showing mean plasticity of tested clones (A-D for DT1-DT4, the developmental time of 1st to 4th instar nymphs; E for DT5, the total developmental time of nymphs; F for 7-d fecundity; data for a particular trait with different letters were significantly different at the $P<0.05$ level, ANOVA followed by Tukey tests).

Sitobion avenae clones from barley, oat and wheat presented little differentiation in broad-sense heritability of life-history trait plasticities (Table 2). The broad-sense heritabilities for all trait plasticities of all clones were high and significant except for DT1 of barley clones from both areas. Heritabilities were particularly high for the plasticities of DT3, DT4, DT5 and 7-d fecundity for barley clones from Shaanxi, and for those of DT5 for barley clones from Qinghai and 7-d fecundity for oat or wheat clones from Qinghai.

Significant genetic correlations were found between plasticities of life-history traits for all clones from both areas (Table 3). All trait-plasticity pairs for barley clones from the Qinghai area presented significant genetic correlations but DT1-DT3, DT2-DT3, DT2-DT5, and DT2-7d fecundity, and all significant correlations for these clones were positive except the pair of DT1-DT2. For barley clones from Shaanxi, Plasticity of DT1was correlated with none of all other trait plasticities, which were significantly correlated with one another; the only significantly negative correlation (i.e., trade-off) for these clones was found between plasticities of DT2 and 7-d fecundity. Similar patterns were found for oat and wheat clones from both Qinghai and Shaanxi, where the majority of trait-plasticity pairs showed significant correlations, and all significant correlations were positive with few exceptions. The only negative correlations of

oat clones were found for the pairs of DT2-DT4 and DT2-DT5 of clones from Qinghai, and for DT1-DT3 and DT1-7 d fecundity from Shaanxi. The only significantly negative correlation of wheat clones was between plasticities of DT3 and 7 d fecundity for clones from Qinghai.

G-matrix comparisons by Flury's method and jump-up approach (that is, at each step in the hierarchy, the hypothesis is tested against the hypothesis of unrelated structure) showed significant differences between paired matrices of life-history trait plasticities (Table 4). The difference between G matrices for barley and oat clones from both areas was best explained by the full CPC model (i.e., all principal components shared in common), but the matrices were not equal (for Shaanxi: LRT = 90.5, $P<0.001$; for Qinghai: LRT = 332.6, $P<0.001$). The CPC(4) model best explained the differences between matrices for barley and wheat clones (LRT = 128.1, $P<0.001$), and between those for oat and wheat clones (LRT = 47.7, $P<0.001$) from the Shaanxi area, in other words, matrices shared four of the six possible principal components. However, unrelated structures were found between matrices for barley and wheat clones (LRT = 362.7, $P<0.001$), and between those for oat and wheat clones (LRT = 52.6, $P<0.001$) from the Qinghai area.

Table 1. Estimates of variance components for trait plasticities of *Sitobion avenae* clones showing main effects of collecting locations (location), plant origin (origin), test plant (test), clone nested in origin and interactions (significant effects highlighted in boldface).

Trait	Variance source	df	F	P	% total
Developmental time of 1st instar nymphs	Location	1	10.74	**0.001**	**1.3**
	Origin	2	24.20	**<0.001**	**6.0**
	Test	2	2.79	0.063	0.7
	Origin×test	1	27.98	**<0.001**	**3.5**
	Clone(origin)	38	10.98	**<0.001**	**52.1**
	Error	291	–	–	36.3
Developmental time of 2nd instar nymphs	Location	1	92.47	**<0.001**	**10.6**
	Origin	2	3.52	**0.03**	**0.8**
	Test	2	3.76	**0.02**	**0.9**
	Origin×test	1	56.59	**<0.001**	**6.5**
	Clone(origin)	38	10.97	**<0.001**	**47.8**
	Error	291	–	–	33.4
Developmental time of 3rd instar nymphs	Location	1	68.40	**<0.001**	**6.2**
	Origin	2	20.91	**<0.001**	**3.8**
	Test	2	17.30	**<0.001**	**3.1**
	Origin×test	1	16.33	**<0.001**	**1.5**
	Clone(origin)	38	17.10	**<0.001**	**59.0**
	Error	291	–	–	26.4
Developmental time of 4th instar nymphs	Location	1	72.26	**<0.001**	**5.8**
	Origin	2	39.82	**<0.001**	**6.3**
	Test	2	8.24	**<0.001**	**1.3**
	Origin×test	1	14.36	**<0.001**	**1.1**
	Clone(origin)	38	20.56	**<0.001**	**62.3**
	Error	291	–	–	23.2
Total Developmental time of nymphs	Location	1	99.72	**<0.001**	**10.3**
	Origin	2	42.58	**<0.001**	**8.8**
	Test	2	17.64	**<0.001**	**3.7**
	Origin×test	1	62.29	**<0.001**	**6.5**
	Clone(origin)	38	10.31	**<0.001**	**40.6**
	Error	291	–	–	30.1
7-d fecundity	Location	1	472.05	**<0.001**	**19.3**
	Origin	2	442.41	**<0.001**	**36.2**
	Test	2	43.62	**<0.001**	**3.6**
	Origin×test	1	30.64	**<0.001**	**1.3**
	Clone(origin)	38	17.81	**<0.001**	**27.7**
	Error	291	–	–	11.9

Selection of alternative environments on trait plasticity

Directional selection differentials and gradients were estimated for barley, oat and wheat clones from both areas (Table 5). Barley clones from the Qinghai area presented significantly negative differentials for the plasticity of all life-history traits tested but DT2 and DT3. For barley clones from Qinghai, the directional selection gradients of plasticity for DT1 and DT4 were significantly negative, those for DT2 and DT3 significantly positive, and those for DT5 and 7-d fecundity non-significant. All selection coefficients for barley clones from the Shaanxi area were significant except the selection gradient of plasticity for DT2 and

selection differentials of plasticity for DT1 and DT2, and all the significant selection coefficients were negative except the selection gradient of plasticity for DT5. Oat clones from Qinghai showed significantly negative selection coefficients (i.e., differential and gradient) for plasticity of DT4 and 7-d fecundity, the selection differential of DT5 plasticity was also significantly negative, and all other selection coefficients for these clones were non-significant. The only significant coefficient for oat clones from Shaanxi was the selection gradient of DT2 plasticity. Wheat clones from Qinghai had a significantly negative differential and gradient for DT5 plasticity, and they also had a significantly negative selection

Table 2. Broad-sense heritabilities (SE) of life history trait plasticities for different *Sitobion avenae* clones from barley, oat and wheat on alternative host plants.

| Traits | Clone sources | | | | | |
| | Shaanxi area | | | Qinghai area | | |
	Barley	Oat	Wheat	Barley	Oat	Wheat
DT1	0.0402n (0.102)	0.6398*** (0.121)	0.6566*** (0.102)	0.3123n (0.201)	0.6529* (0.187)	0.4998* (0.203)
DT2	0.5908* (0.221)	0.5286** (0.194)	0.5291*** (0.053)	0.6117* (0.197)	0.4333** (0.167)	0.4932*** (0.083)
DT3	0.8759*** (0.053)	0.5273*** (0.063)	0.4037** (0.179)	0.6817*** (0.128)	0.3852** (0.122)	0.3265* (0.167)
DT4	0.8907*** (0.033)	0.7834*** (0.086)	0.7852*** (0.085)	0.5077** (0.107)	0.7726** (0.173)	0.6165*** (0.093)
DT5	0.8088*** (0.082)	0.7628*** (0.095)	0.7886*** (0.080)	0.8156*** (0.065)	0.7315*** (0.096)	0.7955*** (0.105)
7 d fecundity	0.9432*** (0.039)	0.7016*** (0.105)	0.7651*** (0.088)	0.7469*** (0.088)	0.8026** (0.120)	0.8251*** (0.073)

Note: DT1-DT4, the developmental time of 1st to 4th instar nymphs; DT5, the total developmental time of nymphs; statistical significance of heritability for a trait within clones (i.e. barley, oat, wheat) evaluated using likelihood-ratio tests; *, $P<0.05$; **, $P<0.01$; ***, $P<0.001$.

differential for DT2 plasticity. Alternative host plants also showed little selection for life-history trait plasticity of wheat clones from Shaanxi, and the only significant selection coefficients were the selection differential of plasticity for DT4, and the selection differential and gradient of plasticity for 7-d fecundity.

Relationship between plasticity and fitness

The relative fitness of *S. avenae* clones was regressed against the first factor (PC1) extracted from PCA of all life-history trait plasticities. The results of PCA for all clones showed the first three components explaining 83.2% (45.9% for PC1) of the total data variability. Barley clones with higher plasticity tended to have

lower fitness, whereas the fitness of wheat clones tended to rise with increasing plasticity (Fig. 2). The linear relationship between relative fitness and plasticity was found to be significant for barley clones (Fig. 2A, $R^2 = 0.25$, $P<0.001$) or wheat clones (Fig. 2C, $R^2 = 0.06$, $P<0.01$). But the linear relationship between fitness and plasticity was not significant for oat clones (Fig. 2B).

Discussion

Genetic basis of phenotypic plasticity

Phenotypic plasticity is considered a universal quality of life, but it's often neglected and thought to be non-genetic historically (until

Table 3. Genetic correlations among life history trait plasticities for different *Sitobion avenae* clones from barley, oat and wheat in two areas.

| Trait-plasticity pairs | Clone sources | | | | | |
| | The Qinghai area | | | The Shaanxi area | | |
	Barley	Oat	Wheat	Barley	Oat	Wheat
DT1-DT2	−0.9079**	0.0492	0.3974*	−0.0298	0.2332*	0.3634*
DT1-DT3	0.0480	0.5457**	−0.0545	0.0658	−0.2651*	0.1720*
DT1-DT4	0.5144**	0.0407	0.2627*	0.0308	0.0602	0.2832*
DT1-DT5	0.3264*	0.3767*	0.1778*	0.0958	0.3304*	0.1812*
DT1-7 d fecundity	0.3816*	0.3511*	0.0926	−0.0489	−0.2455*	−0.0846
DT2-DT3	0.1221	−0.1156	−0.1479	0.1964*	−0.0943	0.0999
DT2-DT4	0.3761*	−0.1558*	0.3252*	0.1614*	−0.0985	0.3825**
DT2-DT5	−0.1324	−0.1924*	0.2670*	0.2843*	0.3874**	−0.0046
DT2-7 d fecundity	−0.1279	0.0425	0.1822*	−0.3327*	−0.0464	0.1560*
DT3-DT4	0.2112*	0.1041	0.1560	0.8745***	0.2259*	0.3118*
DT3-DT5	0.6200**	0.2999*	0.0703	0.8429***	0.1953*	0.3133*
DT3- 7 d fecundity	0.2548*	0.2546*	−0.3055*	0.2577*	0.3048*	0.0046
DT4-DT5	0.3724**	0.4431**	0.4312**	0.8271***	0.5573**	0.5580**
DT4-7 d fecundity	0.4789**	0.4762**	0.1668*	0.4057**	0.2568*	0.5298**
DT5-7 d fecundity	0.4740**	0.4965**	0.5531**	0.4536**	0.2848*	0.1787*

Note: DT1-DT4, the developmental time of 1st to 4th instar nymphs; DT5, the total developmental time of nymphs; statistical significance of genetic correlations evaluated using likelihood-ratio tests; *, $P<0.05$; **, $P<0.01$; ***, $P<0.001$.

Table 4. Comparisons of G-matrices for life-history trait plasticities of barley, oat and wheat clones of *Sitobion avenae* in two areas.

Clone source	G matrices	Flury hierarchy		
		LRT	*P*-value	Verdict
The Shaanxi area	Barley vs. oat	90.5	<0.001	Full CPC
	Barley vs. wheat	128.1	<0.001	CPC(4)
	Oat vs. wheat	47.7	<0.001	CPC(4)
The Qinghai area	Barley vs. oat	332.6	<0.001	CPC
	Barley vs. wheat	362.7	<0.001	Unrelated
	Oat vs. wheat	52.6	<0.001	Unrelated

Note: The verdict of the Flury hierarchy is the model shown to be the best in explaining the difference between paired matrices; the *P*-values are for the test of equality of two matrices; full CPC, all principal components shared in common; CPC(4), four of the six possible components shared in common; unrelated, no relations between the matrices.

late 1980s) [4–7,40]. Therefore, studies on the genetic basis of phenotypic plasticity in insects have been rare, especially for their key life-history traits. Our tests of *S. avenae* clones on barley, oat and wheat showed that the factor 'clone' accounted for a significant proportion (i.e., 27.7–62.3%) of the total variance for the life-history trait plasticities, indicating that the divergence of plasticity among *S. avenae* clones had a genetic basis. Another piece of evidence for the genetic basis of plasticity is that nearly all plasticities of life-history traits for all tested clones showed significant heritability with very few exceptions. Significant genetic correlations were also found between plasticities of life-history traits for all tested *S. avenae* clones, which provided an additional piece of information regarding the genetic basis of phenotypic plasticity. Three genetic models have been proposed to explain phenotypic plasticity, and they are overdominance, pleiotropy, and epistasis [3]. The overdominance model states that plasticity decreases with increasing heterozygosity (i.e., the more heterozy-

gous a clone, the less plastic it will be), and this model can be important for explaining plasticity in aphids that frequently show high heterozygote excess due to parthenogenesis [41–42]. The genetic mechanisms that underlie plastic response are still poorly understood [43]. Further studies with tested clones using microsatellites can clarify the relationship between heterozygosity and plasticity in aphids, and improve our understanding of genetic basis for plasticity in aphids.

Selection on life-history trait plasticity

Alternative environments can impose natural selection not only on life-history traits, but on their plasticity. Questions related to how (and how frequently) natural selection acts on plasticity are conceptually crucial for our understanding not only of genotype-by-environment interactions, but also of phenotypic evolution in general [16]. In our study, there was substantial selection directly on the plasticity of developmental times for barley clones, which

Table 5. Selection differentials and gradients for life-history trait plasticities of *Sitobion avenae* clones collected from barley, oat and wheat in two areas.

Traits	Barley clones		Oat clones		Wheat clones	
	Differential	Gradient	Differential	Gradient	Differential	Gradient
For clones from the Qinghai area						
DT1	−0.1425***	−0.0659*	−0.0655	−0.0090	0.0048	0.0631
DT2	0.0495	0.0763*	−0.0288	−0.0286	−0.0860*	−0.0775
DT3	0.0091	0.0815*	−0.0328	0.0294	0.0024	0.0350
DT4	−0.1368***	−0.1045**	−0.1991***	−0.1161***	−0.0568	0.0366
DT5	−0.1104**	−0.0743	−0.1259**	0.0212	−0.1584***	−0.1999***
7-d fecundity	−0.1341***	−0.0333	−0.2480***	−0.2057***	−0.0916*	0.0487
For clones from the Shaanxi area						
DT1	−0.0342	−0.0384**	−0.0201	−0.0215	−0.0264	−0.0257
DT2	0.0262	−0.0042	0.0158	0.0290*	−0.0260	−0.0026
DT3	−0.0881***	−0.0895**	0.0174	0.0208	0.0075	0.0220
DT4	−0.0986***	−0.0597*	−0.0030	0.0094	−0.0499***	−0.0245
DT5	−0.0784***	0.0959**	−0.0063	−0.0149	−0.0219	−0.0027
7-d fecundity	−0.0927***	−0.0919***	−0.0090	−0.0176	−0.0531***	−0.0413*

Note: DT1-DT4, the developmental time of 1st to 4th instar nymphs; DT5, the total developmental time of nymphs; *, *P*<0.05; **, *P*<0.01; ***, *P*<0.001.

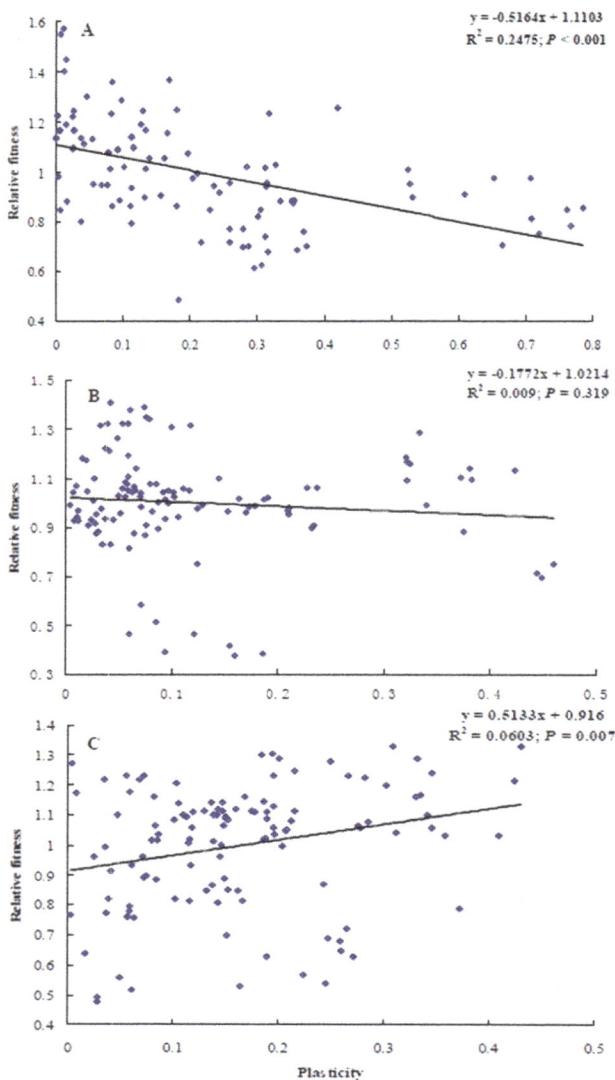

$$y = -0.5164x + 1.1103$$
$$R^2 = 0.2475; P < 0.001$$

$$y = -0.1772x + 1.0214$$
$$R^2 = 0.009; P = 0.319$$

$$y = 0.5133x + 0.916$$
$$R^2 = 0.0603; P = 0.007$$

Figure 2. Relationship between total life-history trait plasticity and relative fitness of *Sitobion avenae* clones collected from barley (A), oat (B) and wheat (C) (the amount of plasticity evaluated by calculating the coefficient of variation for each trait in different environments; total life-history trait plasticity calculated from the first factor extracted from the principal component analysis of all life-history trait plasticities).

was indicated by their significant selection gradients. However, there was little such selection for oat or wheat clones. Substantial direct selection on the plasticity of fecundity was also found for certain clones of barley, oat and wheat. The direct selection of alternative environments acted to decrease the plasticity of *S. avenae*'s life-history traits in most cases except that it increased the plasticity of developmental times of certain nymphal instars for barley clones. Alternative environments produced positive direct selection on the plasticity of total developmental time of nymphs for barley clones from Shaanxi area, but this was masked in the selection differential by indirect selection through some correlated characters. However, differing signs on the coefficients for the directional selection gradient of the developmental times for barley clones were found, so selection could act separately on the trait

plasticities involved. Therefore, our study revealed significant directional selections on the developmental times and fecundity of *S. avenae*, but the identified directional selection acted to decrease the plasticity in most cases. The results appeared to be in agreement with the findings that some *S. avenae* clones were specialized to a certain extent on different hosts [23,25,44], since highly plastic genotypes may lower their plasticity to become relatively specialized to certain environment. Other relatively generalized clones may be sufficiently plastic to survive well on the three cereals (i.e., barley, oat and wheat), so that natural selection will not occur to produce specialized ecotypes. Although phenotypic plasticity is a common phenomenon in aphids, it appears that highly plastic clones of *S. avenae* have been removed from the population by natural selection. It is assumed that traits that are tightly linked to fitness should be more strongly canalized as a result of past stabilizing selection [3,45]. Our finding is consistent with the abovementioned assumption, since developmental times and fecundity of an organism are both key fitness components.

Evolutionary significance of plasticity

After cereal crops are harvested in the summer, some individuals of *S. avenae* may disperse a short distance to find wild grass hosts, others may have to move long distance northward to find other cereal crop fields. So, a *S. avenae* clone may experience several host plant species in a single year. This pattern might lead to the maintenance of moderate phenotypic plasticity in response to changes in host plant species. It makes sense to assume that a clone with higher plasticity can become established more easily in alternative environments than that with low plasticity. Adaptive phenotypic plasticity can evolve in natural populations, which is suggested by the frequent observation of genetic variation for plasticity, but direct experimental evidence is rare due to logistical reasons [16]. In our study, some wheat clones of *S. avenae* with higher level of life-history trait plasticity tended to have higher fitness, indicating that phenotypic plasticity can be adaptive for these clones. Positively affected clones may reinforce the relationship between plasticity and fitness during feeding on alternative host plants. Significant heritability and genetic correlations for *S. avenae*'s life-history trait plasticities identified in our study also indicated the evolutionary potential of adaptive plasticity. However, evidence for adaptive plasticity was not found for oat or barley clones. Cost of plasticity was an important factor influencing evolution of adaptive plasticity for insect populations experiencing heterogeneous environments [46]. So, the cost of being plastic could be high for barley or oat clones. Another mutually non-exclusive explanation is that a population that inhabits heterogeneous environments may be selected to evolve a genetic constitution that allows different levels of phenotypic plasticity to adjust to different environments so as to increase its overall fitness [47].

The evolution of adaptive plasticity can also be influenced by the structure of G-matrix for *S. avenae*'s life-history trait plasticities. Interestingly, quite a few negative covariances (i.e., trade-offs) were found between trait plasticities. These trade-offs (also shown by negative genetic correlations) may also playing a role in slowing the evolution of highly plastic *S. avenae* clones. Significant differences between matrices for barley, oat and wheat clones were found, which have important implications for the overall direction of plasticity evolution in *S. avenae*. The G-matrix structure of different traits for aphid clones have complex relationships with factors like clone specialization, trade-offs, and genotype-by-environment interactions [48]. It may be interesting to determine the stability of the G-matrix for life-history trait plasticity of *S. avenae* over time and further explore its evolutionary implications in future studies.

The selection of clones with relatively low life-history trait plasticity was found for barley clones in our study. This may facilitate the process where *S. avenae* clones become specialized to a certain host plant. Recently, plasticity has been widely recognized as a significant mode of phenotypic diversity and hence as an important aspect of how organisms evolve in different environments [43]. It has been pointed out that speciation in herbivore species can start with phenotypic plasticity, not necessarily with reproductive isolation, therefore, the presence of sufficient variation in phenotypic plasticity of aphids may facilitate host race formation and sympatric speciation [4]. A recent study showed that five unique *S. avenae* biotypes might have developed on commonly planted wheat varieties in China [49]. It appears that perfect plasticity is hard to evolve for *S. avenae*, but plasticity may be important for the evolution of specialized genotypes. This leads to the challenging question of whether phenotypic plasticity is the raw material for speciation and biodiversity [45]. Further studies with costs of life-history trait plasticity may provide insights into its evolution, as well as its significance in host race formation and sympatric speciation for *S. avenae*.

Acknowledgments

We appreciate the laboratory and field assistance of X.-L. Huang, P. Dai, and D. Wang (Northwest A&F University, China). We want to thank Y.-G. Hu and X.-S. Hu (Northwest A&F University, China) for providing plant seeds used in this study.

Author Contributions

Conceived and designed the experiments: SG DL. Performed the experiments: SG XD. Analyzed the data: DL XD. Contributed to the writing of the manuscript: DL XD.

References

1. Orr HA (2005) The genetic theory of adaptation: a brief history. Nature Reviews Genetics, 6: 119–127.
2. Sultan SE (1995) Phenotypic plasticity and plant adaptation. Acta Botanica Neerlandica, 44: 363–383.
3. Scheiner SM (1993) Genetics and evolution of phenotypic plasticity. Annual Review of Ecology and Systematics 24: 35–68.
4. Gorur G, Lomonaco C, Mackenzie A (2005) Phenotypic plasticity in host-plant specialization in *Aphis fabae*. Ecological Entomology 30: 657–664.
5. Schlichting CD (1986) The evolution of phenotypic plasticity in plants. Annual Review of Ecology and Systematics 17: 667–693.
6. Bradshaw AD (1965) Evolutionary significance of phenotypic plasticity in plants. Advances in Genetics 13: 115–155.
7. Sultan SE (2007) Development in context: the timely emergence of eco-devo. Trends in Ecology and Evolution 22: 575–582.
8. Kelly SA, Panhuis TM, Stoehr AM (2012). Phenotypic Plasticity: Molecular Mechanisms and Adaptive Significance. Comprehensive Physiology 2: 1417–1439.
9. Sultan SE (1987) Evolutionary Implications of Phenotypic Plasticity in Plants. In: Hecht M, Wallace B, Prance G, editors. Evolutionary Biology. New York, NY: Springer US. 127–178 p.
10. Price TD, Qvarnstrom A, Irwin DE (2003) The role of phenotypic plasticity in driving genetic evolution. Proceedings of the Royal Society of London Series B 270: 1433–1440.
11. Ghalambor CK, McKay JK, Carroll SP, Reznick DN (2007) Adaptive versus non-adaptive phenotypic plasticity and the potential for contemporary adaptation in new environments. Functional Ecology 21: 394–407.
12. Lande R (2009) Adaptation to an extraordinary environment by evolution of phenotypic plasticity and genetic assimilation. Journal of Evolutionary Biology 22: 1435–1446.
13. Chevin L-M, Lande R, Mace GM (2010) Adaptation, plasticity, and extinction in a changing environment: towards a predictive theory. PLoS Biology 8: e1000357.
14. DeWitt TJ, Scheiner SM (2004) Phenotypic variation from single genotypes: a primer. In: DeWitt TJ, Scheiner SM, editors. Phenotypic plasticity: Functional and Conceptual Approaches. New York, NY: Oxford University Press, Inc. 1–9 p.
15. Whitman DW, Agrawal A (2009) What is phenotypic plasticity and why is it important? In: Whitman DW, Ananthakrishnan TN, editors. Phenotypic plasticity of insects: mechanisms and consequences. Enfield, NH: Science Publishers Inc. 1–63 p.
16. Pigliucci M (2005) Evolution of phenotypic plasticity: where are we going now? Trends in Ecology & Evolution 20: 481–486.
17. Bradshaw AD (2006) Unraveling phenotypic plasticity - why should we bother? New Phytologist 170: 644–648.
18. Borges RM (2008) Plasticity comparisons between plants and animals. Plant Signaling and Behavior 3: 367–375.
19. Fordyce JA (2005) Clutch size plasticity in the Lepidoptera. In: Ananthakrishnan TN, Whitman DW, editors. Insects and Phenotypic Plasticity. Enfield, NH: Science Publishers Inc. 125–144 p.
20. Wool D, Hales DF (1997) Phenotypic plasticity in Australian cotton aphid (Homoptera: Aphididae): host plant effects on morphological variation. Annals of the Entomological Society of America 90: 316–328.
21. Gorur G, Lomonaco C, Mackenzie A (2007) Phenotypic plasticity in host choice behavior in black bean aphid, *Aphis fabae* (Homoptera: Aphididae). Arthropod-Plant Interactions 1: 187–194.
22. Agarwala BK (2007) Phenotypic plasticity in aphids (Homoptera: Insecta): components of variation and causative factors. Current Science 93: 308–313.
23. Gao S-X, Liu D-G, Chen H, Meng X-X (2014) Fitness traits and underlying genetic variation related to host plant specialization in the aphid *Sitobion avenae*. Insect Science 21: 352–362.
24. Godfrey LD, Fuson FJ (2001) Environmental and host plant effects on insecticide susceptibility of the cotton aphid (Homoptera: Aphididae). Journal of Cotton Science 5: 22–29.
25. Gao S-X, Liu D-G (2013) Differential performance of *Sitobion avenae* clones from wheat and barley with implications for its management through alternative cultural practices. Journal of Economic Entomology 106: 1294–1301.
26. Via S (1991) Specialized host plant performance of pea aphid clones is not altered by experience. Ecology 72: 1420–1427.
27. Blackman R, Eastop VF (2006) *Aphids on the world's herbaceous plants and shrubs*. Available: http://www.aphidsonworldsplants.info/index.htm.
28. Adams JB, Drew ME (1964) Grain aphids in New Brunswick II. Comparative development in the greenhouse of three aphid species on four kinds of grasses. Canadian Journal of Zoology 42: 741–744.
29. Dean GJW (1973) Bionomics of aphids reared on cereals and some Gramineae. Annals of Applied Biology, 73: 127–135.
30. Watson SJ, Dixon AFG (1984) Ear structure and the resistance of cereals to aphids. Crop Protection 3: 67–76.
31. Huang X-L, Liu D-G, Gao S-X, Chen H (2013) Differential performance of *Sitobion avenae* populations from both sides of the Qinling Mountains under common garden conditions. Environmental Entomology, 42: 1174–1183.
32. Via S, Lande R (1985) Genotype-environment interaction and the evolution of phenotypic plasticity. Evolution 39: 505–522.
33. Pitchers WR, Brooks R, Jennions MD, Tregenza T, Dworkin I, et al. (2013) Limited plasticity in the phenotypic variance-covariance matrix for male advertisement calls in the black field cricket, *Teleogryllus commodus*. Journal of Evolutionary Biology 26: 1060–1078.
34. SAS (1998) SAS System, version 6.12, SAS Institute Inc., Cary, NC, USA.
35. Carter MJ, Simon J-C, Nespolo RF (2012) The effects of reproductive specialization on energy costs and fitness genetic variances in cyclical and obligate parthenogenetic aphids. Ecology and Evolution 2: 1414–1425.
36. Neumaier A, Groeneveld E (1998) Restriced maximum likelihood estimation of covariances in sparse linear models. Genetics Selection Evolution 30: 3–26.
37. Phillips PC, Arnold DE (1999) Hierarchical comparison of genetic variance-covariance matrices. I. Using the Flury hierarchy. Evolution 53: 1506–1515.
38. Lande R, Arnold SJ (1983) The measurement of selection on correlated characters. Evolution 37: 1210–1226.
39. Svensson EI, Kristoffersen L, Oskarsson K, Bensch S (2004). Molecular population divergence and sexual selection on morphology in the banded demoiselle (*Calopteryx splendens*). Heredity 93: 423–433.
40. West-Eberhard MJ (1989) Phenotypic plasticity and the origins of diversity. Annual Review of Ecology and Systematics 20: 249–278.
41. Papura D, Simon JC, Halkett F, Delmotte F, Le Gallic JF, et al. (2003) Predominance of sexual reproduction in Romanian populations of the aphid Sitobion avenae inferred from phenotypic and genetic structure. Heredity 90: 397–404.
42. Simon J-C, Baumann S, Sunnucks P, Hebert PDN, Pierre J-S, et al. (1999) Reproductive mode and population genetic structure of the cereal aphid *Sitobion avenae* studied using phenotypic and microsatellite markers. Molecular Ecology 8: 531–545.
43. Sultan SE (2000) Phenotypic plasticity for plant development, function and life history. Trends in Plant Science 5: 537–542.
44. Sunnucks P, De Barro PJ, Lushai G, Maclean N, Hales DF (1997) Genetic structure of an aphid studied using microsatellite: cyclic parthenogenesis, differentiated lineages, and host specialization. Molecular Ecology, 6: 1059–1073.

45. Nylin S, Gotthard K (1998) Plasticity in life-history traits. Annual Review of Entomology 43: 63–83.

46. Relyea RA (2002) Costs of phenotypic plasticity. The American Naturalist 159: 272–282.

47. West-Eberhard MJ (2003) Developmental plasticity and evolution. New York: Oxford University Press. 794 p.

48. Nespolo RF, Halkett F, Figueroa CC, Plantegenest M, Simon J-C (2009) Evolution of Trade-Offs between Sexual and Asexual Phases and the Role of Reproductive Plasticity in the Genetic Architecture of Aphid Life Histories. Evolution 63: 2402–2412.

49. Xu Z-H, Chen J-L, Cheng D-F, Sun J-R, Liu Y, et al. (2011) Discovery of English grain aphid (Hemiptera: Aphididae) biotypes in China. Journal of Economic Entomology 104: 1080–1086.

7

Subcutaneous Adipose Fatty Acid Profiles and Related Rumen Bacterial Populations of Steers Fed Red Clover or Grass Hay Diets Containing Flax or Sunflower-Seed

Renee M. Petri[1,9], **Cletos Mapiye**[2,3,9], **Mike E. R. Dugan**[2], **Tim A. McAllister**[1*]

1 Lethbridge Research Centre, Agriculture and Agri-Food Canada, Lethbridge, Alberta, Canada, **2** Lacombe Research Centre, Agriculture and Agri-Food Canada, Lacombe, Alberta, Canada, **3** Department of Animal Sciences, Faculty of AgriSciences, Stellenbosch University, Matieland, Western Cape, South Africa

Abstract

Steers were fed 70:30 forage:concentrate diets for 205 days, with either grass hay (GH) or red clover silage (RC), and either sunflower-seed (SS) or flaxseed (FS), providing 5.4% oil in the diets. Compared to diets containing SS, FS diets had elevated ($P<0.05$) subcutaneous *trans* (*t*)-18:1 isomers, conjugated linoleic acids and *n*-6 polyunsaturated fatty acid (PUFA). Forage and oilseed type influenced total *n*-3 PUFA, especially α-linolenic acid (ALA) and total non-conjugated diene biohydrogenation (BH) in subcutaneous fat with proportions being greater ($P<0.05$) for FS or GH as compared to SS or RC. Of the 25 bacterial genera impacted by diet, 19 correlated with fatty acids (FA) profile. *Clostridium* were most abundant when levels of conjugated linolenic acids, and *n*-3 PUFA's were found to be the lowest in subcutaneous fat, suggestive of their role in BH. *Anerophaga, Fibrobacter, Guggenheimella, Paludibacter* and *Pseudozobellia* were more abundant in the rumen when the levels of VA in subcutaneous fat were low. This study clearly shows the impact of oilseeds and forage source on the deposition of subcutaneous FA in beef cattle. Significant correlations between rumen bacterial genera and the levels of specific FA in subcutaneous fat maybe indicative of their role in determining the FA profile of adipose tissue. However, despite numerous correlations, the dynamics of rumen bacteria in the BH of unsaturated fatty acid and synthesis of PUFA and FA tissue profiles require further experimentation to determine if these correlations are consistent over a range of diets of differing composition. Present results demonstrate that in order to achieve targeted FA profiles in beef, a multifactorial approach will be required that takes into consideration not only the PUFA profile of the diet, but also the non-oil fraction of the diet, type and level of feed processing, and the role of rumen microbes in the BH of unsaturated fatty acid.

Editor: Ayyalasomayajula Vajreswari, National Institute of Nutrition, India

Funding: This work was financed by the Alberta Meat and Livestock Agency (ALMA). Dr. C. Mapiye acknowledges the receipt of NSERC Fellowships funded through ALMA (alma.alberta.ca). The funders had no role in study design, data collection and analysis, decision to publish, or preparation of the manuscript.

Competing Interests: The authors have declared that no competing interests exist.

* Email: tim.mcallister@agr.gc.ca

9 These authors contributed equally to this work.

Introduction

The healthfulness of beef has been challenged because of its relatively high concentrations of saturated fatty acids (SFA [1]), including myristic (14:0) and palmitic (16:0) acids which have been shown to raise serum levels of low-density lipoproteins, a risk factor for cardiovascular disease in humans [2]. However, meat also contains essential fatty acids (EFA) such as α-linolenic acid (18:3*n*-3, ALA) and its elongation and desaturation products including eicosapentaenoic acid (20:5*n*-3, EPA), docosapentenoic acid (22:5*n*-3) and docosahexaenoic acid (22:6*n*-3; DHA) and rumen biohydrogenation (BH) products including rumenic acid (*c*9,*t*11-18:2, RA) and its precursor vaccenic acid (*t*11-18:1, VA) which have purported human health-promoting properties [3–6]. In this context, current research efforts have been directed at finding dietary strategies that facilitate higher fore-stomach bypass of *n*-3 polyunsaturated fatty acids (PUFA) and specific PUFA BH products (*i.e*, VA and RA) for absorption and incorporation into adipose tissue.

Our previous studies have evaluated the effects of feeding 10 to 15% flaxseed (FS, a rich source of ALA) in the diet on ALA and its

BH products in beef [7–9]. In an initial examination, feeding steers FS in a barley grain-based (73% of dry matter [DM]) diet resulted in limited absolute increases *n*-3 PUFA and their BH products in beef. In this research, the BH pathway promoted tissue accumulation of *t*13-/*t*14-18:1 rather than VA, and non-conjugated diene BH products (*i.e.*, atypical dienes, AD) in place of conjugated linoleic acids (CLA) [8]. In a second investigation, we fed cull cows grass hay (GH) or barley silage-based (50% of DM) diets supplemented with FS for 20 weeks and found that cows fed GH had a greater concentration of *n*-3 PUFA and BH products, with VA as the main *t*-18:1 isomer, as well as greater levels of AD instead of CLA in adipose tissue as compared to those fed barley silage [7,9].

Recent research in our laboratory has shown that supplementing a red clover silage (RC, 70% of DM) diet with FS for 215 days resulted in greater levels of *n*-3 PUFA and its BH products as a proportion of total fatty acids (FA) when compared to feeding other silages. As much as 2.9% of RA was found in the subcutaneous fat of cattle fed a RC diet supplemented with FS [10]. In this research, the increased amounts of PUFA BH products were in part associated with the amount and duration of FS feeding and the presence of relatively increased levels of polyphenol oxidase (PPO) activity which can produce quinones in RC, reducing the rate of PUFA lipolysis and BH in the rumen [11]. Consequently, adipose deposition of PUFA and their BH products was increased [12]. To our knowledge, no studies have directly compared RC with other forages for its effect on the concentrations of PUFA and BH intermediaries in beef adipose tissue [7].

Results from comparisons of different oilseeds suggest that those rich in linoleic acid (LA, 18:2n-6) such as sunflower-seed (SS) may be more effective at increasing VA and CLA in beef [13–15], while those rich in ALA increase ALA and its specific BH products. In this regard, it may be important to investigate how the deposition of VA, CLA and n-3 PUFA in beef differs when feeding RC as opposed to GH supplemented with SS or FS as PUFA sources.

Previously we have observed large coefficients of variation in concentrations of PUFA BH products among cattle, especially for VA, when feeding FS in high-forage diets [7,16], but the source of this inter-animal variation remains uncertain. Overall, differences in FA composition have been reported to originate from inter-animal variation in the rumen environment, including the kinds and numbers of rumen microbes present as well as other factors [17]. In this respect, determining an individual animal's rumen microbial profile with current metagenomic technologies could be useful in minimizing FA variation amongst animals consuming the same diet by designing management strategies that enhance the levels of beneficial FA in beef through microbial manipulation. It may also help to identify those microbial populations involved in BH that are inhibited by plant secondary compounds such as the quinones in RC, an outcome that results in a more favorable FA profile in beef. The objectives of the current research were to compare the effects on rumen bacterial populations and related subcutaneous FA profiles when steers were fed GH or RC diets supplemented with FS or SS. Subcutaneous fat was chosen as a representative tissue as it is used to make hamburger, which is the most consumed beef product in North America, and it is considered a more representative indicator of rumen FA metabolism than muscle [18].

Materials and Methods

Animals and diets

Animal management and diets were previously described by Mapiye et al. [19] with ethical experimental practices reviewed and approved (Protocol #201102) by Lacombe Research Centre Animal Care Committee using guidelines which are accredited by the national Canadian Council of Animal Care [20]. Briefly, 64 British × Continental crossbred steers were stratified by weight to four experimental diets, with two pens of eight steers per diet. The four diets were GH-FS, GH-SS, RC-FS and RC-SS. On a dry matter (DM) basis, diets contained 70% forage and 30% concentrate with sunflower-seed (SS) or flaxseed (FS) at a level that resulted in the addition of 5.4% oil to the diets (Table 1). In an attempt to equalize the digestible energy of the diets, additional ground-barley grain was also included in diets containing SS, and additional barley straw was added to diets containing FS. Flaxseed was triple rolled, while SS was fed whole. Nutrient and FA composition of the experimental diets are also shown in Table 1.

Sample collection procedures

Subcutaneous fat thickness was measured monthly by a certified ultrasound technician using an Aloka 500V diagnostic real-time

ultrasound with a 17 cm 3.5 Mhz linear array transducer (Overseas Monitor Corporation Ltd., Richmond, B.C., Canada) following procedures of Brethour, [21]. Steers were slaughtered at the Lacombe Research Centre abattoir over four slaughter dates in November 2011 (two steers/pen/diet/slaughter day) at an average of 205 d on feed corresponding to subcutaneous fat depths of 6–8 mm between the 12th and 13th rib over the right *longissimus thoracis* muscle of each animal.

On the morning of slaughter, steers were transported 2 km to the Lacombe Research Centre abbatoir for immediate slaughter. At slaughter, final live weights were recorded and steers were stunned, exsanguinated and dressed in a commercial manner. At approximately 20 min *post-mortem*, samples of subcutaneous fat adjacent to the 12th rib were collected and stored at −80°C until analysed for FA. About 30 min *post-mortem*, the rumen was opened and the ruminal contents collected and thoroughly mixed. Thereafter, samples of ruminal contents (solids and fluid) were taken from mid-ventral region of the rumen (250 g) by the same researcher throughout the trial and placed into an open 2 L plastic container. Samples were then hand-mixed, subsampled and put in 2×50 ml plastic culture tubes. The tubes were immediately flash-frozen in liquid nitrogen and stored in a −80 freezer until DNA was extracted.

Bacterial DNA extraction, sequencing and quantification

From the 64 steers, 24 were selected for rumen bacterial analysis, based on which six animals in each dietary treatment had the lowest (n = 3) or the highest (n = 3) vaccenic acid in their subcutaneous fat. Subsamples of ruminal contents (25–40 ml) were lyophilized in a VirTis Freezemobile 25 freeze dryer (SP Industries Warminste, PA, USA), and 2–4 g of dried material was ground for 5 minutes at a frequency of 30 cycles/s in 10 ml grinding jars with a 20 mm stainless steel ball using a Qiagen TissueLyser II (Qiagen, Toronto, ON). Total DNA was extracted from dried, ground samples (30 mg) in two parallel procedures: one from the initial lysis supernatant, and the other from the initial lysis pelleted fraction of the QIAamp DNA Stool Kit as described by Narvaez et al. [22]. Final DNA elution volume was 150 µl for the supernatant fraction, and 100 µl for the pelleted fraction. Concentrations of DNA were determined spectrophotometrically using a NanoDrop 2000 (ThermoScientific, Wilmington, DE, USA), and supernatant and pellet elutions were pooled 1:1 (v/v).

Bacterial tag-encoded FLX-Titanium amplicon pyrosequencing (bTEFAP) of the pooled DNA, [23], was performed at MR DNA (Shallowater, TX, USA). Bacterial 16S primers 530F 5′-GTGCCAGCMGCNGCGG-3′ and 1100R 5′-GGGTTNCG-NTCGTTG-3′ were used in PCR amplification of the V4–V6 hyper-variable regions of 16S rRNA gene. Sequencing primers and barcodes were trimmed from the DNA sequences and quality control measures using Mothur [24] were used to exclude sequences <200 bp or those containing homopolymers longer than 8 base pairs. Pyrosequencing errors were minimized in the dataset, using the pre-cluster algorithm in Mothur [25], whereby rare sequences highly similar to abundant sequences were re-classified as their abundant homologue. Chimeras were removed from the samples, using the sequence collection as its own reference database [24]. Clean reads were submitted to EBI European Nucleotide Archive (ENA) database (http://www.ebi.ac.uk/ena, accession number PRJEB6402). Calculation of treatment based rarefaction curves, using the Mothur pipeline provided a way of comparing the phylogenetic richness among samples and determining the extent of sequencing relative to sampling needed to accurately describe the microbial community. While the total number of sequences obtained was decreased in the RC-FS

Table 1. Ingredient, nutrient and fatty acid composition of the experimental diets.

Variable	Diet[1]			
	GH-FS	GH-SS	RC-FS	RC-SS
Ingredient (% DM basis)				
Red clover silage	0.0	0.0	70.0	70.0
Grass hay	70.0	70.0	0.0	0.0
Barley straw	11.5	0.0	11.5	0.0
Sunflower-seed	0.0	18.4	0.0	18.4
Flaxseed	14.3	0.0	14.3	0.0
Vitamin/mineral supplement[2]	4.2	4.2	4.2	4.2
Barley grain	0.0	7.4	0.0	7.4
Nutrient (DM basis)				
Dry matter (%)	93.1	93.0	46.9	46.9
Crude protein (%)	13.3	13.4	14.2	14.0
Crude fat (%)	6.4	6.6	8.2	8.4
Calcium (%)	1.1	1.1	1.1	1.2
Phosphorus (%)	0.3	0.3	0.3	0.2
ADF (%)	44.3	45.4	43.0	44.0
NDF (%)	53.2	57.6	55.5	61.6
Digestible Energy[3] (Mcal/kg)	2.08	2.02	2.16	2.10
Fatty acid (% of total fatty acids)				
14:0	0.2	0.2	0.1	0.1
16:0	8.6	10.2	7.5	8.4
18:0	3.0	4.1	2.9	4.2
20:0	0.4	0.5	0.3	0.4
22:0	0.7	0.9	0.4	0.8
24:0	0.6	0.5	0.4	0.4
c9-18:1	11.6	11.3	11.6	11.7
c11-18:1	0.8	0.9	0.8	0.7
18:2n-6	23.4	66.0	21.4	70.4
18:3n-3	50.7	5.3	54.6	2.8

[1]GH-FS, grass hay + flaxseed; GH-SS, grass hay + sunflower-seed, RC-FS, red clover silage + flaxseed; RC-SS, red clover silage + sunflower-seed.
[2]Vitamin/mineral supplement per kg DM contained 1.86% calcium, 0.93% phosphorous, 0.56% potassium, 0.21% sulphur, 0.33% magnesium 0.92% sodium, 265 ppm iron, 314 ppm manganese, 156 ppm copper, 517 ppm zinc, 10.05 ppm iodine, 5.04 ppm cobalt, 2.98 ppm selenium, 49722 IU/kg vitamin A, 9944 IU/kg vitamin D3, and 3222 IU/kg vitamin E.
[3]Digestible energy was calculated according to Bull, [72].

compared to the other diets, the degree of coverage was similar across diets (Fig. S1). However, none of the curves reached a plateau, indicating that the observed level of richness (unique operational taxonomic unit), as determined by the unique sequences and overall sampling intensity was insufficient to fully describe the richness of rumen bacterial communities.

Sequences were then grouped according to diet in order to determine the effect of forage, oilseed and the forage by oilseed interaction and to account for low sequence abundance in some individual samples. A distance matrix was constructed using the average neighbour algorithm at 0.05 (genus) and 0.25 (phylum) phylogenetic distances to determine the most accurate phylogenetic tree structure. Pairwise distances between aligned sequences were calculated at a 0.97% similarity cut off and then clustered into unique OTUs (operational taxonomic unit). Any sequences aligning for more than 97% of the sequence were considered to be from the same bacterial species (OTU). In total, there were 64,396 quality reads with an average of 16,099±6731 reads and an

average of 364 unique OTUs per diet. Mothur was also used to calculate the coverage for each treatment (Fig. S1), and to create a dendrogram based on treatment differences using OTU dissimilarity between the structures of two communities [26]. Calculations of percentage of sequences within taxonomic classifications at the genus level were performed using a custom summation script [27].

Subcutaneous fatty acid analysis

Subcutaneous fat samples (50 mg) were freeze-dried and directly methylated with sodium methoxide [28]. As an internal standard, 1 ml of 1 mg c10-17:1 methyl ester/ml toluene (standard no. U-42M form Nu-Check Prep Inc., Elysian, MN, USA) was added prior to addition of methylating reagents. Fatty acid methyl esters (FAME) were analysed by gas chromatography using a CP-Sil88 column (100 m, 25 μm ID, 0.2 μm film thickness) in a CP-3800 gas chromatograph equipped with an 8600-series autosampler (Varian Inc., Walnut Creek, CA, USA).

Table 2. Effect of forage type and oilseed interaction on fatty acid profiles of subcutaneous fat from beef steers.

Variable	Grass hay		Red clover		SEM	P-value		
	Flax	Sunflower	Flax	Sunflower		Oilseed	Forage	O*F[1]
$\sum$ PUFA	1.97	2.10	1.80	2.11	0.07	0.001	0.25	0.19
$\sum$ n-6	1.40	1.75	1.29	1.82	0.06	<0.001	0.78	0.15
18:2n-6	1.30	1.63	1.20	1.70	0.06	<0.001	0.81	0.17
20:2n-6	0.03	0.03	0.03	0.03	0.00	0.10	0.18	0.50
20:3n-6	0.04	0.05	0.03	0.05	0.00	<0.001	0.26	0.27
20:4n-6	0.03	0.04	0.03	0.05	0.00	<0.001	0.65	0.18
$\sum$ n-3	0.58	0.35	0.51	0.29	0.02	<0.001	0.001	0.90
18:3n-3	0.50	0.30	0.44	0.24	0.02	<0.001	0.001	0.88
20:3n-3	0.02	0.01	0.02	0.01	0.00	<0.001	0.25	0.72
22:5n-3	0.06	0.04	0.05	0.04	0.00	<0.001	0.03	0.81
$\sum$ CLNA	0.29[a]	0.08[b]	0.32[a]	0.05[b]	0.01	<0.001	0.90	0.03
c9,t11,t15-18:3	0.10	0.03	0.11	0.02	0.01	<0.001	0.66	0.16
c9,t11,c15-18:3	0.19[a]	0.05[b]	0.21[a]	0.03[b]	0.01	<0.001	0.71	0.04
$\sum$ AD	2.61	1.68	2.35	1.17	0.09	<0.001	<0.001	0.17
$\sum$ CLA	2.10	2.47	2.07	2.34	0.18	0.02	0.55	0.72
$\sum$ c,t-CLA	1.91	2.37	1.90	2.26	0.18	0.001	0.65	0.73
$\sum$ t,t-CLA	0.19	0.10	0.17	0.09	0.01	<0.001	0.06	0.95
$\sum$ t-18:1	6.14	8.20	5.51	7.66	0.38	<0.001	0.09	0.90
$\sum$ c-MUFA	45.2	44.7	46.0	43.5	1.09	0.16	0.85	0.35
c9-14:1	1.47	1.55	1.75	1.34	0.14	0.25	0.79	0.09
c7-16:1	0.20	0.19	0.22	0.22	0.01	0.42	0.001	0.74
c9-16:1	5.12	4.77	5.95	4.68	0.32	0.02	0.26	0.16
c9-17:1	0.78	0.68	0.83	0.71	0.03	0.001	0.27	0.76
c9-18:1	34.3	34.0	34.3	33.7	0.73	0.45	0.80	0.82
c11-18:1	1.33	1.11	1.32	1.12	0.09	0.02	0.97	0.88
c12-18:1	0.62[c]	1.38[a]	0.36[d]	0.81[b]	0.05	<0.001	<0.001	0.001
c13-18:1	0.44	0.38	0.47	0.34	0.03	0.01	0.90	0.29
c14-18:1	0.09	0.09	0.08	0.07	0.00	0.54	<0.001	0.43
c15-18:1	0.49	0.23	0.39	0.18	0.02	<0.001	<0.001	0.16
c16-18:1	0.08	0.08	0.07	0.07	0.00	0.10	0.06	0.70
c10-19:1	0.08	0.04	0.08	0.04	0.00	<0.001	0.19	0.53
c9-20:1	0.12	0.11	0.12	0.11	0.00	0.01	0.70	0.47
c11-20:1	0.15	0.13	0.15	0.14	0.01	0.23	0.67	0.40
$\sum$ BCFA	2.01	1.83	1.93	1.90	0.04	0.01	0.86	0.07

Table 2. Cont.

Variable	Grass hay		Red clover		SEM	P-value		
	Flax	Sunflower	Flax	Sunflower		Oilseed	Forage	O*F[1]
iso-14:0	0.07	0.06	0.07	0.07	0.001	0.75	0.45	0.38
iso-15:0	0.24	0.22	0.22	0.22	0.01	0.07	0.16	0.07
anteiso-15:0	0.28	0.27	0.28	0.29	0.01	0.83	0.79	0.28
iso-16:0	0.29	0.27	0.29	0.30	0.01	0.80	0.16	0.24
iso-17:0	0.34	0.31	0.32	0.31	0.01	0.01	0.05	0.12
anteiso-17:0	0.64	0.57	0.61	0.59	0.01	0.001	0.88	0.07
iso-18:0	0.15	0.12	0.15	0.13	0.00	<0.001	0.89	0.32
∑ SFA	38.6	37.9	38.9	40.3	1.07	0.75	0.19	0.28
14:0	3.45	3.24	3.51	3.36	0.14	0.17	0.47	0.83
15:0	0.59	0.55	0.63	0.61	0.02	0.06	0.001	0.61
16:0	22.8	22.1	23.8	23.1	0.50	0.06	0.02	0.97
17:0	0.75[a]	0.68[c]	0.70[b]	0.75[a]	0.03	0.72	0.60	0.03
18:0	10.8	11.1	10.1	12.4	0.71	0.08	0.74	0.17
19:0	0.05	0.05	0.04	0.05	0.00	0.68	0.42	0.39
20:0	0.08	0.09	0.08	0.10	0.01	0.12	0.28	0.38
22:0	0.02	0.02	0.02	0.02	0.00	0.10	0.54	0.08

[a,b,c,d]Means with different superscripts for a particular fatty acid profile have a significant oilseed × forage interaction($P<0.05$); SEM, standard error of mean;

[1]Oilseed type × forage type interaction; c, cis; t, $trans$;

$\sum$ PUFA, sum of polyunsaturated fatty acids $= \sum n\text{-}6 + \sum n\text{-}3$; $\sum n\text{-}6 =$ sum of 18:2n-6, 20:2n-6, 20:3n-6, 20:4n-6; $\sum n\text{-}3$ sum of 18:3n-3, 20:3n-3, 22:5n-3; $\sum$CLNA, sum of conjugated α-linolenic acid $= c9,t11,t15$-, $c9,t11,c15$-; $\sum$AD, total atypical dienes $=$ sum of $t11,t15$-, $c9,t13$-/$t8,c12$-, $t8,c13$-, $c9,t12$-/$c16$-18:1, $t9,c12$-, $t11,c15$-, $c9,c15$-, $c12,c15$-; $\sum$CLA, conjugated linoleic acid $=$ sum of c,t-CLA $+$ sum of c,t-CLA; $\sum$ trans-trans-CLA $=$ sum of $t12,t14$-, $t11,t13$-, $t10,t12$-, $t9,t11$-, $t8,t10$-, $t7,t9$-$t6,t8$-; $\sum$ cis-/trans-CLA $=$ sum of $c9,t11$-, $t7,c9$-, $t11,c13$-, $t12,c14$-, $c11,t13$-, $t10,c12$-, $t8,c10$-, $t9,c11$-; $\sum t$-18:1, sum of trans-18:1 isomers $= t6,t7,t8$-, $t9$-, $t10$-, $t11$-, $t12$-, $t13,t14$-, $t15$-, $t16$-; $\sum c$-MUFA $=$ sum of $c9$-14:1, $c7$-16:1, $c9$-16:1, $c11$-16:1, $c9$-17:1, $c9$-18:1, $c11$-18:1, $c12$-18:1, $c13$-18:1, $c14$-18:1, $c15$-18:1, $c9$-20:1, $c11$-20:1; $\sum$ BCFA, branched chain fatty acids $=$ sum of iso-15:0, anteiso15:0, so16, iso17:0, anteiso17:0, iso18:0; $\sum$ SFA $=$ sum of 14:0, 15:0, 16:0, 17:0, 18:0, 19:0, 20:0, 22:0.

Two gas chromatography (GC) analyses were conducted per sample using complementary temperature programs with 150°C and 175°C plateaus according to Kramer, et al. [29]. CLA isomers not separated by GC were further analysed using Ag$^+$-HPLC as described by Cruz-Hernandez et al. [30].

For the identification of FAME by GC, the reference standard no. 601 from Nu-Check Prep Inc, Elysian, MN, USA was used. Branched-chain FAME were identified using a GC reference standard BC-Mix1 purchased previously from Applied Science (State College, PA, USA). For CLA isomers, the UC-59M standard from Nu-Chek Prep Inc. was used which contains all four positional CLA isomers. Trans-18:1, CLA isomers and other BH products not included in the standard mixtures were identified by their retention times and elution orders as reported in literature [29–31]. The FAME were quantified, using chromatographic peak area and internal standard based calculations. Only FAME representing more than 0.01% of total FAME were included in tables and figures with the exception of BH products, where all the quantified isomers were reported.

Statistical analysis

All data were analysed, using the PROC MIXED procedure of SAS [32]. The statistical model for FA profiles and genus percent abundance included the fixed effects of oilseed, forage and oilseed × forage interaction,with slaughter date and pen considered as random effects and animal as the experimental unit. Since the random effect of pen nested within the oilseed × forage interaction was not significant, it was removed from the model. Treatment means were generated and separated, using the LSMEANS and PDIFF options respectively [32].

For the comparison of high and low levels of VA within diet to bacterial abundance using 2×2×2 factorial analysis, there was a significant FA level by oilseed type interaction. Therefore, data were reanalyzed as 2×4 factorial ANOVA comparing forage to high and low levels of a FA for each oilseed. For the analysis, treatment means were generated and separated, using Tukey-Kramer and PDIFF options [32]. To relate rumen bacterial profiles to FA profiles, percent abundance of genus level taxa were additionally analyzed in pairwise Pearson correlation to FA data, using the PROC CORR procedure of SAS [32]. The significance threshold for all statistical analyses was set at $P<0.05$.

Results

Animal performance

Data on animal performance are detailed in a companion paper by Mapiye et al [19]. In summary, steers fed SS diets (13.2±0.37) had higher ($P<0.05$) DM intake than those fed FS (12.1±0.37), and steers fed GH consumed more (13.3±0.37; $P<0.05$) feed than those fed RC (12.1±0.37). As a result, average daily gain and final live weights were higher ($P<0.05$) in steers fed SS (0.7 kg/d±0.08 and 550 kg±9.55) compared to FS (0.51 kg/d±0.08 and 503 kg±9.55) and final live weights were also higher ($P<0.05$) in steers fed GH (551 kg±9.55), as compared to those fed RC (517 kg±9.55). Steers fed SS diets (7.66±0.38 mm) had a tendency to have thicker (P = 0.08) final subcutaneous fat depth than steers fed FS diets (6.69±0.38 mm), but forage type and its interaction with oilseed had no effect ($P>0.05$) on final subcutaneous fat thickness.

Subcutaneous fatty acid profiles

The proportions of total PUFA in subcutaneous fat were affected by oilseed type with steers fed SS diets having greater ($P<0.05$) proportions than those fed FS (Table 2). Oilseed type also

influenced the proportions of total and individual n-6 PUFA (18:2n-6, 20:3n-6, 20:4n-6), with steers fed SS having greater ($P<0.05$) proportions in subcutaneous fat than steers fed FS. For all diets, LA was the most prominent n-6 PUFA, accounting for more than 90% of total n-6 PUFA.

For n-3 PUFA, diets containing FS as opposed to SS, and GH as opposed to RC exhibited elevated ($P<0.05$) proportions of total n-3 PUFA, 18:3n-3 (ALA) and 22:5n-3 (Table 2), but in general, increases were influenced more by oilseed than forage type. Alpha-linolenic acid was the most abundant n-3 PUFA, making up over 80% of total n-3 PUFA in all diets.

An oilseed × forage type interaction was detected for total conjugated linolenic acids (CLNA) and c9,t11,c15-18:3, but upon means separation only an effect of oilseed was detected with FS diets resulting in greater ($P<0.05$) proportions than SS diets (Table 2). The proportions of c9,t11,t15-18:3 were also only influenced by oilseed type, with steers fed FS having greater ($P<0.05$) proportions than those fed SS.

Based on the type of oilseed and forage fed, two AD isomer patterns were observed (Fig. 1A). The proportions of AD isomers likely largely derived from ALA (t8,c13-; t9,t12-/c9,t13-; t11,c15-; t11,t15-; c12,c15-18:2) were increased by FS (P<0.05). Of the isomers in this group, minor forage effects were found for t8,c13- and t9,t12-/c9,t13-18:2 (P<0.05), with diets containing GH resulting in slightly increased ($P<0.05$) proportions than diets containing RC. A small but significant forage × oilseed type interaction (P<0.05) was found for c12,c15-18:2 with the GH-FS diet resulting in the greatest proportions followed by RC-FS, GH-SS and RC-SS diets, respectively. For AD likely largely derived from LA (t8, c12-; c9,t12-; t9,c12-18:2), forage × oilseed type interactions were found ($P<0.05$) with amounts being greatest for the GH-SS diet. The dominant AD isomer irrespective of diet was t11, c15-18:2 accounting for up to 25% and 35% of total AD in steers fed SS and FS, respectively.

For CLA, two clear isomer patterns were found based on the type of oilseed fed (Fig. 1B). The proportions of CLA isomers with the first double bond at carbon 10 or closer to the carboxyl end (t7,c9-; t8,c10-; c9,t11- (RA); t9,c11-; t10,c12-; t10,t12-18:2) were more elevated ($P<0.05$) when feeding SS as compared to FS. Some minor forage effects were also found for t7,c9-; t8,c10-; t10,c12- and t10,t12-18:2 with diets containing GH vs. RC yielding slightly greater proportions. The proportions of CLA isomers with the first double bond at carbon 11 or further from the carboxyl end (c11,t13-; t11,c13-; t11,t13-; c12,t14-; t12,t14-; t12,c14-18:2) were mostly increased ($P<0.05$) by inclusion of FS. Minor forage effects were also noted for c12, t14-18:2 ($P<0.05$) while forage × oilseed type interactions were noted for t11,c13- and t12, c14-18:2 ($P<0.05$). For all diets, RA accounted for over 70% of total CLA (Fig. 1B).

For t-18:1, an isomer pattern was found based on the type of oilseed and forage fed (Fig. 1C). The proportions of t-18:1 isomers with double bonds from carbon 6 to 12 were primarily greater ($P<0.05$) with SS than FS, and for the majority of these isomers, feeding GH vs. RC diets also led to increases ($P<0.05$), but at a reduced magnitude as compared to oilseeds. For t-18:1 isomers with double bonds from carbon 13 to 16, the pattern of differences was less strongly linked to forage or oilseed effects (Fig. 1C). Feeding diets containing FS as opposed to SS increased ($P<0.05$) the proportions t15- and t16-18:1 (Fig. 1C). The proportions of t13-/t14- and t16-18:1 were elevated ($P<0.05$) by diets containing GH as opposed to RC. Vaccenic acid was the predominant t-18:1 isomer and accounted for 45% and 50% of total t-18:1 isomers in the subcutaneous fat of steers fed FS and SS diets, respectively (Fig. 1C).

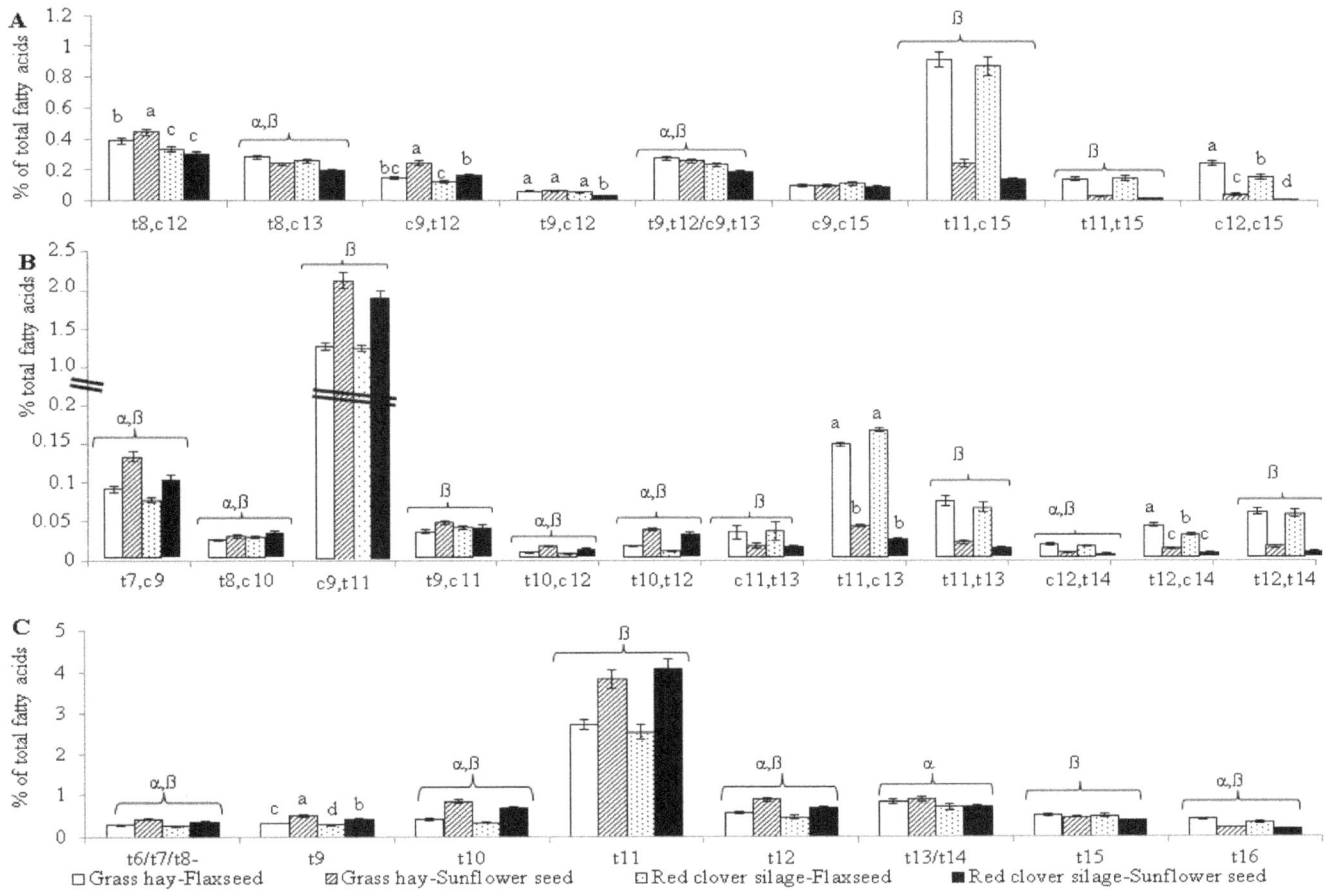

Figure 1. Effect of forage type and oilseed supplementation on atypical dienes (A), conjugated linoleic acid (B) and *trans*-18:1 isomers (C) in subcutaneous fat of beef steers. [a,b,c,d] Means (± standard error) with different superscripts for a particular fatty acid profiles are significantly different ($P<0.05$). α: Significant forage effect ($P<0.05$); β: Significant oilseed effect ($P<0.05$).

Overall, feeding diets containing FS as opposed to SS elevated ($P<0.05$) the proportions of several individual *c*-monounsaturated fatty acids (MUFA) isomers (*c*9-16:1, *c*9-17:1, *c*11-18:1, *c*13-18:1, *c*15-18:1, *c*10-19:1, *c*9-20:1). An oilseed type × forage type interaction influenced the proportions of *c*12-18:1, with steers fed GH-SS having the largest proportions followed by those fed RC-SS, GH-FS and RC-FS, respectively ($P<0.05$; Table 2). Relative to feeding RC, GH reduced ($P<0.05$) the proportions of *c*7-16:1 and increased ($P<0.05$) the proportions of *c*14- and *c*15-18:1 (Table 2). Oleic acid (*c*9-18:1) was the dominant MUFA isomer, accounting for over 75% of total *c*-MUFA in all diets (Table 2).

Feeding diets containing FS compared to SS increased total branched-chain fatty acids (BCFA; $P<0.05$) as a result of increases in the proportions of *iso*-17:0, *anteiso*-17:0 and *iso*-18:0 ($P<0.05$) (Table 2). Neither oilseed nor forage type had any effect on total SFA ($P<0.05$), but proportions of 15:0, 16:0 and *iso*-17:0 were influenced by forage type, with steers fed RC having higher ($P<0.05$), 15:0, 16:0 and lower ($P<0.05$) *iso*-17:0 proportions than those fed GH. An oilseed type × forage type interaction was observed for steers fed GH-FS and RC-SS,,with steers fed RC-FS having intermediate and steers fed GH-SS having the lowest proportions of 17:0 ($P<0.05$). Palmitic acid (16:0) was the most abundant saturated FA (SFA), constituting about 60% of the total SFA in subcutaneous fat of steers across all diets.

Rumen bacterial profiles

The Yue and Clayton [26] measure of dissimilarity among communities was used to create a dendrogram showing the separation of OTU in samples from individual diets (Fig. 2). Despite clustering of diets based on type of forage, diets were not found to differ ($P = 1.0$). The OTU's calculated for each diet were used to construct a Venn diagram (Fig. 3), which identified a total of 558 OTU's across the 4 diets. The number of OTU's that were associated with each diet ranged from 326 to 427. Of these OTU's, 217 were shared by all 4 diets. Using non-parametric estimators in Mothur, it was predicted that the core microbiome was composed of 256 OTU's.

Using a summation script, the percent abundance of each of the 87 classified genera were determined and those which differed (P< 0.05) by forage or oilseed type are listed in Table 3. Of these genera, 12 were impacted by forage type, 9 were impacted by oilseed type and 5 exhibited a forage × oilseed interaction. The taxa *Butyrivibrio*, and *Syntrophococcus* were higher ($P<0.05$) in the GH diets, whereas *Fibrobacter was* higher ($P<0.01$) in RC diets. *Blautia*, *Eubacterium*, and *Olsenella* were higher in the GH-FS diet and *Anaerophaga* was lowest in GH fed cattle. *Johnsonella* was highest with GH-SS, whereas *Mogibacterium* and *Wandonia* were highest in cattle fed RC-FS. When comparing oilseed supplementation, *Acidaminobacter* was the only taxon that was consistently higher in cattle fed FS. *Barnesiella* was abundant

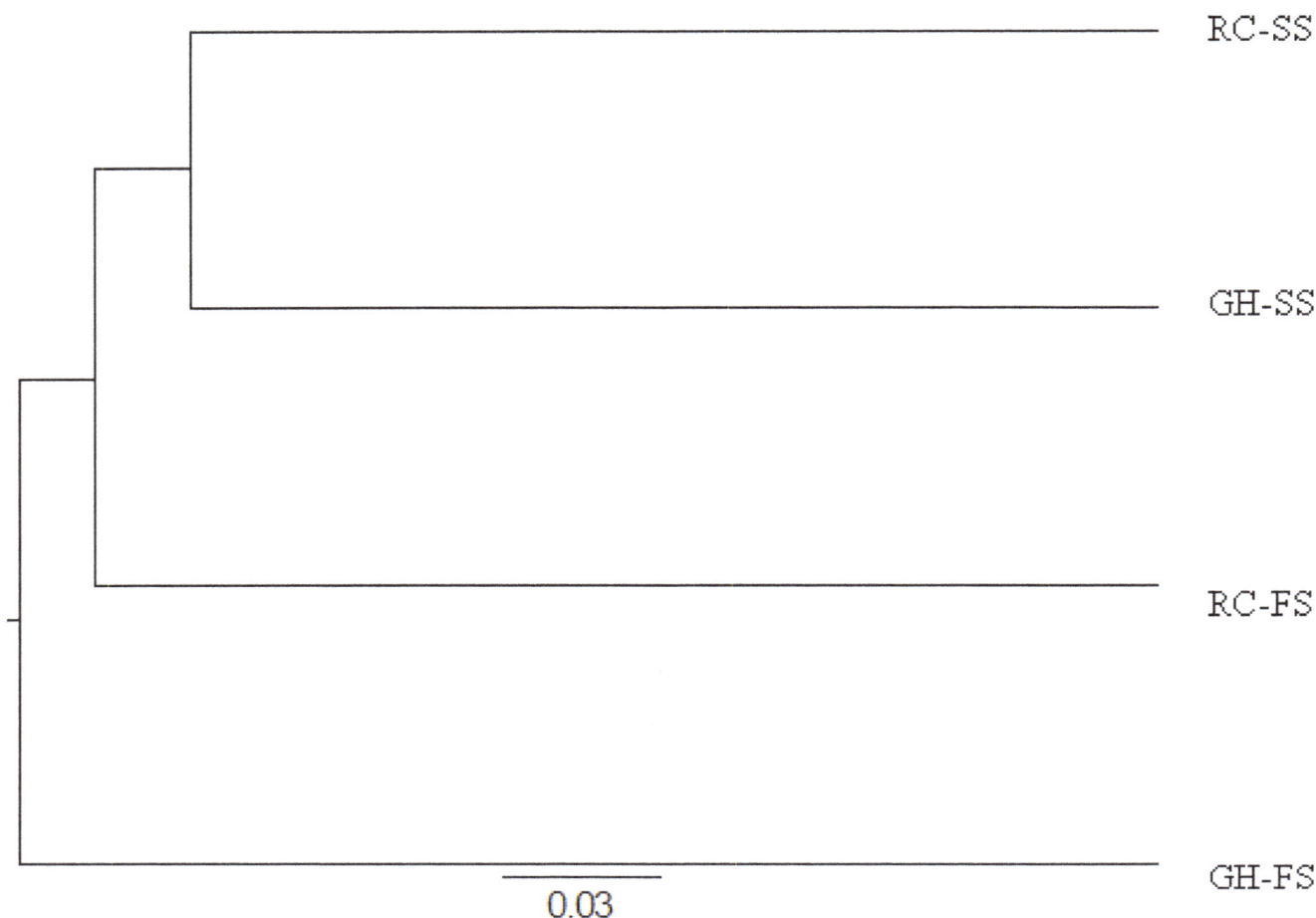

Figure 2. Dendrogram showing distances between dietary treatments based on similarity of sequences using Jaccard analysis (GH-SS: hay-sunflower seed, GH-FS: hay-flax seed, RC-SS: red clover-sunflower seed, RC-FS: red clover-flax seed).

across all diets, but it was higher (*P*<0.05) in the GH-FS diet and lowest with GH-SS. *Marvinbryantia* exhibited the opposite response being more abundant (*P*<0.02) with GH-SS than with GH-FS. *Bellilinea, Blautia, Guggenheimella, Nubsella* and *Sporobacter* were all influenced by an oilseed × forage interaction. Of these, *Bellilnea, Guggenheimella,* and *Nubsella* were present in greater abundance (*P*<0.05) in the RC-FS fed cattle.

Additionally, Pearson correlation was used to relate the percent abundance of measured bacterial taxa to measured subcutaneous FA. The highest correlation (*P*<0.004) between percent abundance and FA profile was for the genus *Clostridium IV* (Table 4). *Clostridium IV* was highly correlated to total *n*-3 PUFA (*P*<0.004; $R^2 = 0.32$), total CLNA (*P*<0.001; $R^2 = 0.38$) and total atypical dienes (*P*<0.001; $R^2 = 0.40$). The only other genera, which was highly correlated with specific FA profiles, was *Fibrobacter*, which correlated to total CLA (*P*<0.001; $R^2 = 0.40$) and *Marvinbryantia*, which correlated to total *t*MUFA (*P*<0.002; $R^2 = 0.36$). Additionally, two bacterial genera, *Acidaminobacter* and *Asteroleplasma* were highly (*P*<0.001) correlated to specific individual FA. *Acidaminobacter* was correlated to FA *iso*-18:0 with a $R^2 = 0.39$ and *Asteroleplasma* was correlated to *c*12-18:1 with a $R^2 = 0.44$.

This difference in VA levels coincided with changes in the percent abundance of several genera including *Anaerophaga, Asaccharobacter, Fibrobacter, Guggenheimella, Marvinbryantia,*

Paludibacter, Pseudosphingobacterium, Pseudozobellia and *Syntrophococcus* (Table 5).

Discussion

Animal performance

The finding that steers fed diets containing FS grew slightly slower and had lower final live weight than steers fed diets containing SS may be related to palatability issues resulting from the addition of FS [33] and/or the inclusion of ground straw [34] in FS diets to balance for digestible energy across oilseed diets. The explanations for the lower final live weight observed, when feeding RC as opposed to GH are less clear, but the quality of silage fed and higher crude fat content of RC diets may have been partially responsible. Silage quality (dry matter content and fermentation quality) was not assessed in this experiment, but fat levels greater than 5% have been reported to inhibit ruminal fiber digestion, increasing rumen fill and reducing DM intake [35].

Subcutaneous fatty acid profiles

The elevated levels of total PUFA in the subcutaneous fat of steers fed diets containing SS was related to the increase in *n*-6 PUFA, especially LA. This resulted from the large proportion of LA in diets containing SS. In most diets, the extent of LA BH is lower than that of ALA [36] because greater levels of LA in the

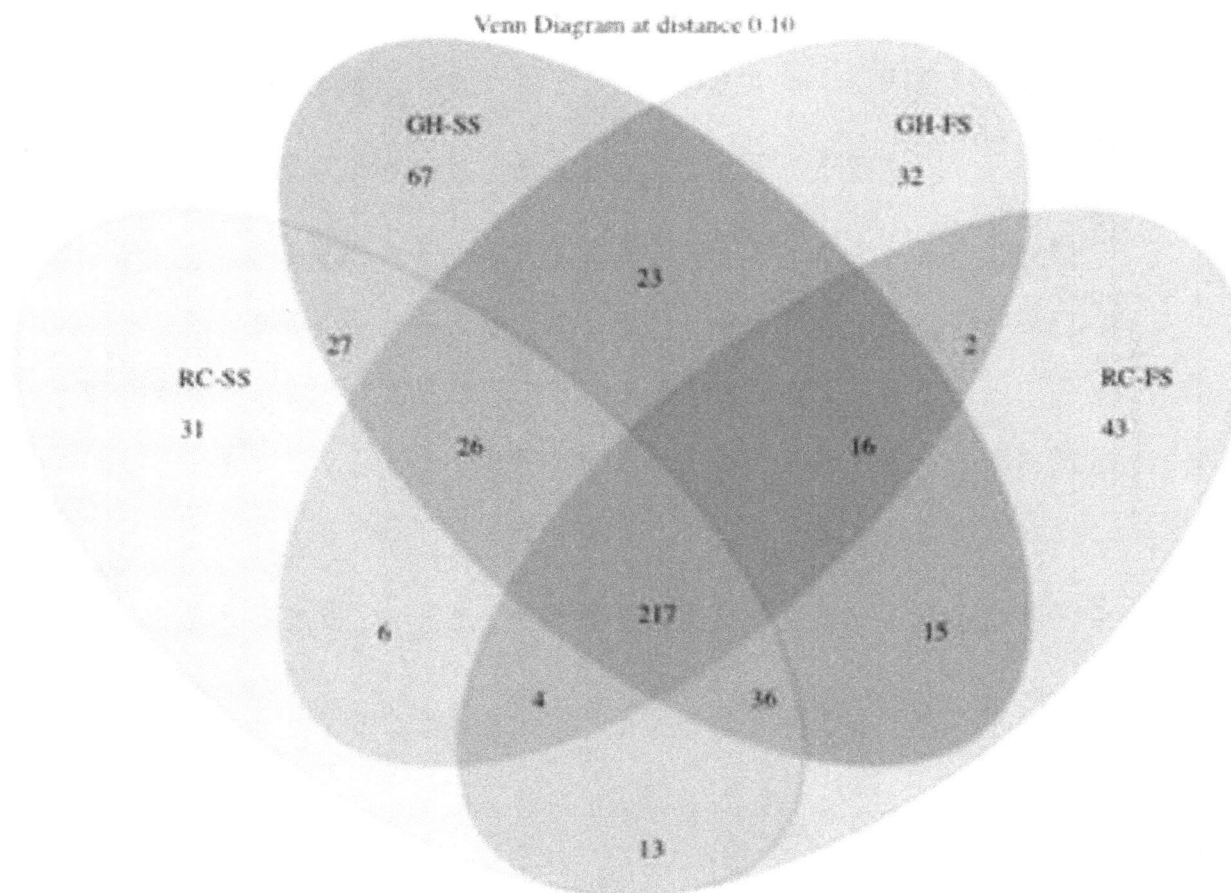

Figure 3. Venn diagram showing OTU's unique to and shared by each of the dietary treatments (GH-SS: hay-sunflower seed, GH-FS: hay-flax seed, RC-SS: red clover-sunflower seed, RC-FS: red clover-flax seed).

rumen inhibits the growth of many of the bacterial species involved in BH [37]. This inhibitory effect results in increased tissue deposition of LA and its long chain n-6 PUFA derivatives when compared to ALA and its long chain n-3 PUFA derivatives [38]. Additionally, the physical form of the oilseed fed (whole SS vs. triple rolled FS) might have also influenced BH in the rumen. Rolling has been reported to enhance the availability of PUFA in the rumen, thereby increasing the degree to which they are BH [39]. The increase in surface area as a result of rolling would have increased microbial attachment and therefore potentially increased the rate of lipolysis and BH. The effect of forage on FA profiles was confounded by the physical form of the feed. The lack of accumulation of PUFA in cattle fed RC diets was unexpected as it is well known that RC has increased levels of poly-phenol oxidase (PPO) which increases PUFA levels in meat and milk [11]. It is possible that the activity of PPO was reduced during ensiling.

As anticipated, feeding diets containing FS, a rich source of ALA, increased the deposition of ALA and its long chain derivatives, 20:3n-3 and 20:5n-3 in subcutaneous fat. This is likely due to part of dietary ALA escaping ruminal BH and therefore being available for direct deposition in the tissue. The reason for the increased n-3 PUFA proportions observed in steers fed diets containing GH diets remains uncertain, but may be partly associated with the higher feed intake previously reported for this diet [19] or the preservation method of the forage (i.e., hay vs. silage). It has been reported that ALA content in tissues is often

higher, when hay as opposed to silage is fed due to a higher efficiency of transfer from the rumen to the tissue [40]. This could be related to changes in forage lipids, during preservation and/or ruminal lipid metabolism [41]. The process of ensiling has been shown to increase the levels of PUFA that are BH in the rumen due to lipolysis in the silo, which occurs as a result of increased moisture content during fermentation [42]. Biohydrogenation of PUFA's in the rumen has been shown to be lower in hay compared to ensiled forages, and therefore with hay a greater percentage of PUFA bypass the rumen and are available for uptake in the duodenum [40,43]. This may be a result of differences in physical properties of the forage impacting microbial attachment [43] or shifts in microbial communities that result in differences in lipid metabolism and outflow of FA [44]. In this investigation, the mean percentage concentration of ALA in subcutaneous fat (0.44–0.50% of total FA) of steers fed diets containing FS was higher than the mean percentage concentration of ALA in the subcutaneous fat of animals fed FS in high-grain diets (0.09–0.12%, [33]), but lower than those found, when feeding FS in 50:50 forage:grain (0.73–0.79%, [7]) or high-forage diets (0.87%, [13]; 1.1%, [9]). However, the differences in ALA across these trials could be related to a number of factors within these experiments including feeding management, forage type and quality or animal variability.

The finding that CLNA proportions in the subcutaneous fat were accentuated by feeding diets containing FS, as opposed to

Table 3. Percent abundance of genera impacted by forage type (F), oilseed type included in the diet (O), level of vaccenic acid (VA) in backfat or any interaction of these factors.

| | Hay | | | | Red Clover | | | | | P-value for indicated factor | | | | | | |
| | Flax | | Sunflower | | Flax | | Sunflower | | | | | | | | | |
Bacterial Genera	High VA[1]	Low VA	High VA	Low VA	High VA	Low VA	High VA	Low VA	SEM	F	O	VA	F×O	F×VA	O×VA	F×O×VA
Acidaminobacter	0.00	0.37	0.00	0.00	0.17	0.17	0.00	0.00	0.11	0.93	0.03	0.24	0.93	0.24	0.24	0.24
Anaerophaga	0.76	1.86	0.92	1.82	1.99	2.21	1.90	2.66	0.40	0.01	0.68	0.02	0.85	0.38	0.77	0.52
Anaerosporobacter	0.61	0.42	0.61	0.71	0.20	0.54	0.37	0.00	0.19	0.04	0.91	0.83	0.24	0.90	0.47	0.09
Anaerostipes	0.18	0.22	0.18	0.00	0.47	0.73	0.00	0.23	0.17	0.10	0.03	0.49	0.15	0.21	0.63	0.70
Asaccharobacter	0.46	0.00	0.17	0.00	0.00	0.00	0.00	0.00	0.10	0.05	0.35	0.05	0.35	0.05	0.35	0.35
Asteroleplasma	0.00[b]	0.00[b]	0.83[a]	0.21[ab]	0.20[ab]	0.00[b]	0.00[b]	0.43[ab]	0.15	0.34	0.01	0.37	0.07	0.06	0.98	0.01
Blautia	1.48	1.31	1.19	0.99	0.47	0.66	1.34	0.89	0.23	0.02	0.45	0.35	0.02	0.87	0.31	0.36
Butyrivibrio	2.75	2.64	2.56	2.20	1.26	1.54	1.68	1.46	0.29	0.00	0.73	0.62	0.25	0.54	0.38	0.77
Clostridium_IV	0.88	0.70	1.09	1.13	0.77	0.83	1.47	1.50	0.25	0.30	0.01	0.96	0.33	0.76	0.79	0.73
Eubacterium	1.02	0.76	0.56	0.61	0.23	0.40	0.28	0.23	0.21	0.01	0.25	0.88	0.42	0.59	0.89	0.39
Faecalibacterium	0.00[b]	0.26[ab]	0.47[ab]	0.00[b]	0.68[a]	0.23[ab]	0.35[ab]	0.60[a]	0.23	0.10	0.70	0.55	0.82	0.99	0.96	0.04
Fibrobacter	0.59	1.89	0.37	0.98	2.06	2.18	1.48	2.85	0.38	0.00	0.36	0.01	0.28	0.70	0.62	0.09
Galbibacter	0.4[c]	2.55[c]	2.6[c]	2.8[bc]	3.41[a]	3.71[bc]	1.99[c]	3.74[ab]	0.11	0.00	0.52	0.44	0.10	0.77	0.00	0.00
Guggenheimella	0.40	2.55	2.60	2.63	2.61	3.49	1.99	3.09	0.50	0.05	0.38	0.01	0.03	0.89	0.20	0.12
Hallella	2.18[a]	1.47[ab]	1.19[ab]	1.22[ab]	1.54[ab]	2.25[a]	1.39[ab]	1.08[b]	0.30	0.82	0.01	0.73	0.93	0.22	0.73	0.05
Johnsonella	0.83	0.37	0.98	0.72	0.37	0.00	0.33	0.23	0.30	0.02	0.37	0.13	0.69	0.74	0.53	0.92
Marvinbryantia	1.43	0.55	1.88	1.30	1.38	1.12	1.66	1.21	0.24	0.78	0.03	0.01	0.24	0.28	0.88	0.49
Mogibacterium	0.17	0.18	0.40	0.00	0.88	0.71	0.70	0.50	0.17	0.00	0.48	0.13	0.35	0.95	0.37	0.43
Nubsella	0.27[b]	1.45[ab]	1.25[ab]	1.13[ab]	2.16[a]	1.45[ab]	0.85[ab]	1.50[ab]	0.28	0.03	0.44	0.22	0.03	0.17	0.93	0.00
Olsenella	0.48	0.20	0.18	0.00	0.00	0.00	0.00	0.00	0.10	0.03	0.19	0.22	0.19	0.22	0.77	0.77
Oscillibacter	0.00	0.61	0.60	0.76	0.54	0.53	0.80	0.79	0.19	0.21	0.03	0.18	0.68	0.16	0.43	0.42
Paludibacter	0.00	0.79	0.24	0.75	1.04	0.79	0.56	1.12	0.20	0.01	0.94	0.01	0.55	0.10	0.36	0.08
Phocaeicola	2.33[a]	1.23[ab]	1.06[ab]	1.57[ab]	1.05[ab]	1.88[ab]	1.39[ab]	0.62[b]	0.31	0.17	0.05	0.55	0.99	0.46	0.99	0.00
Pseudo-sphingobacterium	1.69[b]	5.94[a]	1.99[b]	4.38[ab]	6.7[a]	5.64[a]	4.52[ab]	6.14[a]	0.70	0.00	0.16	0.00	0.84	0.01	0.68	0.04
Pseudozobellia	0.18	0.54	0.00	0.00	0.00	0.41	0.00	0.19	0.15	0.80	0.05	0.04	0.26	0.57	0.21	0.76
Sporobacter	1.21	0.76	1.12	1.49	1.16	1.30	0.97	0.90	0.86	0.42	0.66	0.57	0.05	0.51	0.49	0.12
Syntrophococcus	2.31	1.50	2.28	1.55	1.36	1.35	1.61	1.08	0.23	0.00	0.99	0.01	0.94	0.14	0.51	0.37
Treponema	1.95[b]	2.62[ab]	2.56[ab]	2.11[ab]	2.52[ab]	2.14[ab]	2.06[ab]	2.77[a]	0.23	0.82	0.79	0.60	0.95	0.92	0.98	0.05
Wandonia	0.00	0.29	0.19	0.00	0.32	0.70	0.34	0.51	0.23	0.05	0.70	0.33	0.92	0.49	0.30	0.67

a,b,cMeans with different superscripts within a row are significantly different (P<0.05) based on the three-way interaction.
[1]VA, Vaccenic acid; F, Forage; O, Oilseed; SEM, standard error of the mean.

Table 4. Correlation of bacterial genera to subcutaneous fatty acid groupings.

Genera Taxa		Fatty Acid Group						
		Total n-6 and n-3	Total n-6	Total n-3	Total CLNA	Total AD	Total CLA	Total trans MUFA
Acidaminobacter	P-value	0.75	0.31	0.05	0.07	0.26	0.33	0.03
	R^2	0.00	0.05	0.17	0.14	0.06	0.04	0.19
	Corr.[1]			+				–
Anaerostipes	P-value	0.03	0.02	0.57	0.16	0.56	0.36	0.03
	R^2	0.20	0.22	0.01	0.09	0.02	0.04	0.19
	Corr.	–	–					
Asteroleplasma	P-value	0.20	0.11	0.32	0.17	0.53	0.14	0.05
	R^2	0.07	0.11	0.04	0.08	0.02	0.09	0.17
	Corr.							+
Clostridium_IV	P-value	0.39	0.06	0.004	0.001	0.001	1.00	0.38
	R^2	0.03	0.15	0.32	0.38	0.40	0.00	0.04
	Corr.			–	–	–		
Ethanoligenens	P-value	0.97	0.43	0.03	0.04	0.09	0.48	0.33
	R^2	0.00	0.03	0.19	0.19	0.12	0.02	0.04
	Corr.			–	–			
Fibrobacter	P-value	0.37	0.54	0.50	0.87	0.33	0.001	0.06
	R^2	0.04	0.02	0.02	0.00	0.04	0.40	0.15
	Corr.							
Galbibacter	P-value	0.20	0.25	0.83	0.52	0.66	0.01	0.31
	R^2	0.07	0.06	0.00	0.02	0.01	0.25	0.05
	Corr.						–	
Marvinbryantia	P-value	0.30	0.18	0.39	0.30	0.45	0.08	0.002
	R^2	0.05	0.08	0.03	0.05	0.03	0.13	0.36
	Corr.							+
Olsenella	P-value	0.54	0.28	0.17	0.26	0.04	0.26	0.49
	R^2	0.02	0.05	0.08	0.06	0.18	0.06	0.02
	Corr.					+		
Pseudo-butyrivibrio	P-value	0.42	0.43	0.97	0.74	0.64	0.03	0.38
	R^2	0.03	0.03	0.00	0.01	0.01	0.20	0.03
	Corr.						+	

[1]Positive (+) or negative (–) correlation (Corr.) for $P < 0.05$.

Table 5. Changes in composition of bacterial abundance in animals with high vs.low vaccenic acid in the backfat.

| | Vaccenic Acid within Treatment | | | |
	High (n = 12)	Low (n = 12)-	SEM	P-value
Vaccenic Acid (% of total FAME)[1]	4.44	2.44	0.09	
Bacterial Abundance (% of total)				
Anaerophaga	1.39[b]	2.14[a]	0.02	0.02
Asaccharobacter[2]	0.02	0.00	0.01	0.05
Fibrobacter	1.13[b]	1.97[a]	0.02	0.01
Guggenheimella	1.90[b]	2.94[a]	0.02	0.01
Marvinbryantia	1.59[a]	1.04[b]	0.01	0.01
Paludibacter	0.46[b]	0.87[a]	0.01	0.01
Pseudosphingobacterium[2,3]	3.73	5.53	0.04	<0.01
Pseudozobellia	0.05[b]	0.28[a]	0.01	0.04
Syntrophococcus	1.89[a]	1.37[b]	0.01	0.01

[1]Significant two-way interaction of vaccenic acid level × oilseed (P<0.05).
[2]Significant two-way interaction of vaccenic acid level × forage (P<0.05).
[3]Significant three-way interaction of vaccenic acid level × oilseed × forage (P<0.05).
[a,b]Means with different superscripts within a row are significantly different (P<0.05).

feeding SS, was anticipated as ALA is known to initially get isomerised to CLNA as it undergoes BH in the rumen [45]. Current findings agree with our earlier results, when feeding GH or barley silage diets containing FS [7] and FS combined with RC [10]. Previous cell culture and animal model studies have demonstrated that plant-derived CLNA can improve biomarkers of cardiovascular health and has anti-inflammatory, immune-modulatory, anti-obesity and anti-carcinogenic properties [46]. Therefore, addition of FS to high-forage diets to achieve greater levels of CLNA in beef fat may be perceived as a positive outcome from a human health perspective.

The increase in AD isomers largely derived from ALA, including t11,c15-18:2 the major AD isomer in steers fed diets containing FS was expected as ALA is isomerised to CLNA, before BH [45]. These findings agree with previous reports with diets containing FS [7,10,47]. Some AD isomers (t8,c12-; c9,t12-; c9,c15-18:2) reported by Mapiye et al. [10] and Nassu et al. [7] were increased by FS, but in the present study these isomers were preferentially enriched by diets containing SS as reported by Bessa et al. [47]. This could imply that these isomers are found in common with BH pathways for both LA and ALA. The inconsistencies in AD proportions seen between steers fed GH vs. RC, may also reflect changes in the physical state of the RC as a result of ensiling increasing the degree of BH of PUFAs in the rumen [40,41], as enhanced microbial attachment to silage may have decreased the rumen bypass of PUFAs [43]. The observation that the proportions of AD likely largely derived from ALA were higher, when feeding diets containing FS as opposed to SS supports previous observations by Jenkins et al. [48] and Doreau et al. [49]. It was found that BH of ALA proceeds mostly through pathways incorporating AD vs. CLA, whereas BH of LA proceeds mostly through CLA. Definitive experiments on the impact of AD isomers on human health are required before recommendations can be made about including SS or FS in ruminant diets.

As found previously in ruminants [13,47,50], supplementing high-forage diets with SS as opposed to FS enhanced the proportions of CLA isomers with the first double bond from carbon 7 to 10 and t-18:1 isomers with double bonds between

carbons 6 and 12. This is because of the common precursor LA, which is abundant in SS containing diets. Feeding cattle, a blend of oils or oilseeds rich in ALA and LA has previously been shown to result in a synergistic accumulation of VA [51] and CLA in the tissue [52]. Therefore, further investigations regarding whether blending SS and FS in high-forage diets would simultaneously enhance concentrations of VA, RA and n-3 PUFA in beef tissues are necessary. Furthermore, the blending of ALA and LA with long chain PUFA such as EPA and DHA, which are known inhibitors of BH of 18:1 to 18:0 in the rumen [53,54] merits investigation.

In this investigation, proportions of RA and VA in subcutaneous fat (2.1% and 2.6%, respectively) of steers fed diets containing SS were greater than those previously reported, when high-forage diets were supplemented with oils or oilseeds [7,55,56]. Conversely, mean percentage concentrations of RA and VA in subcutaneous fat was determined to be less than that observed, when grass pastures were grazed by cattle supplemented with FS or sunflower oil (2.7–3.9% and 7.5–9.7%, respectively; [13,56]) or RC with FS (2.9% and 5.9%; [10]). Variation in mean percentage concentrations of RA and VA across studies could be due to the effects of non-oil components of the diet on rumen BH. In the current study, the sunflower seeds were relatively low in oil (29.5%) compared to normal values of ≈140% [57]. As a result, the amount of SS that had to be added to the diet to get 5.4% oil was relatively high (18.4%), necessitating the addition of hulls that have a low energy value. Consequently, to balance the digestible energy content across diets, straw had to be added to flaxseed diets, leading to a diminished overall quality of forage in these diets. Therefore, the present findings strongly emphasize the importance of the non-oil components of the diet when trying to enrich for BH intermediates, particularly RA and VA, in beef. Overall, direct comparisons among studies are often difficult due to RA commonly being combined with other CLA isomers such as t7,c9-18:2, and VA containing more than one t-18:1 isomer.

The observation that feeding diets containing SS as compared to FS increased t10-18:1 confirms earlier reports by AbuGhazaleh and Jacobson [58] and Martínez Marín [59] that the BH of ALA as opposed to LA is less prone to promote *t*10-18:1 formation. The relative amounts of *t*10- and *t*13/*t*14-18:1 in the current study, are lower than those found by Juárez, et al., [8], whereas the relative amounts of *t*15- and *t*16-18:1 are less than those obtained by Nassu, et al., [7] and Mapiye et al [10]. These inconsistences could be attributed to differences in time on feed or levels, types and forms of forage fed in these experiments.

The increase in the proportions of minor CLA and *t*-18:1 isomers observed when feeding GH as opposed to RC may be partly associated with changes in the composition of ruminal microbes [60] or differences in ruminal and duodenal flow rates between these two forages [61]. For CLA and *t*-18:1 isomers, effects of oilseed type on these FA were greater than that of forage type.

This experiment showed that measured concentrations of *c*-MUFA isomers in subcutaneous fat were decreased when feeding SS as opposed to FS, which concurs with previous findings [62,63] when feeding LA- and ALA-rich diets to dairy cows and goats, respectively. Overall, the levels of *c*-MUFA in subcutaneous fat when feeding PUFA are indicative of their dietary levels, the level of PUFA BH to *c*-MUFA in the rumen and the capacity of PUFA/ or its BH products to inhibit stearoyl-CoA desaturase (SCD) activity, which converts 18:0 to *c*9-18:1 (oleic acid) in subcutaneous fat [64]. This further suggests that either LA has a slower BH rate and extent of ruminal BH in the rumen than ALA [36] or is more effective in down-regulating SCD activity than ALA [62]. In the present study, reductions in the proportions of total and individual *c*-MUFA isomers in steers fed SS diets could also relate to displacement of *c*-MUFA due to increased formation of *t*-18:1 during BH of PUFA in the rumen, as reflected by the elevated levels of total and individual *t*-18:1 isomers in subcutaneous fat.

Steers fed diets containing FS had a greater percentage of total BCFA as well as individual BCFA and odd-chain FA in subcutaneous fat than steers fed SS. This suggests that ALA did not inhibit rumen microbial FA synthesis as much as LA. In general, the decrease in BCFA and odd-chain FA in ruminant products is related to a direct inhibition of PUFA on microbial FA synthesis [65]. Given that BCFA have the potential to reduce necrotizing enterocolitis [66] and cancer [67,68] in humans, diets that increase levels of BCFA in beef are desirable from a human health perspective.

Rumen bacterial profiles

The fatty acid results of the present experiment demonstrate that feeding oilseeds in combination with diets containing a greater proportion of forage directly effects the mean percentage concentration of PUFA and their BH products in beef fat. However, there are many factors that simultaneously affect the amount and variation of PUFA and their BH products, making ad hoc estimation of fatty acid profiles extremely challenging. In order to perform such calculations a multifactorial approach is required, and among the potential variables, the bacterial ecosystem of the rumen will undoubtedly have an overriding influence. Where historically others have been able to culture isolated bacterial species which metabolize dietary FA [53,69], the global perspective of how the entire population may be changing in response to diet and correlate with differences in FA profiles is lacking. This prompted our study taking advantage of current technologies to more fully characterize differences in rumen bacterial populations and their relationship to diet and FA deposited in beef. Previous research has cultured individual bacteria to examine their role in BH [69,70]. However, the sensitivity of rumen bacteria to temperature, pH, osmolality, and the bacterial competition, quorum sensing, nutrient limitations

and syntrophism within the rumen, makes it difficult to replicate this ecosystem in the laboratory [71,72]. The use of pyrosequencing of the rumen ecosystem creates the opportunity to gain an overview of rumen bacterial populations simultaneously at any one point in time and compare it to other points in time. By looking at changes in the overall ecosystem within a variety of diets and animal hosts, one may be able to predict which bacterial populations are involved in rumen BH. This approach has been previously used to document the role of rumen bacteria in acidosis [73]. Although the ability of pyrosequencing to identify individual species is limited, the data generated could be used as a guideline for further research and when combined with culture techniques provide insight into the manipulation of individual rumen species involved in BH.

While overall sequence data was unable to fully describe the rumen bacterial ecosystem in cattle fed each diet due to limited coverage (Fig. S1), it was able to show that there were a large number of OTU's shared regardless of forage or oilseed type (Fig. 3). Therefore, despite dietary differences, there was a core group of microbes that was similar among all diets. However, changes in both ingredient and nutrient composition can impact the prevalence and density of specific groups or species without impacting the overall bacterial community structure [74]. While a number of genera were clearly impacted by diet (Table 3), variations in diet quality (*i.e.*, ADF, NDF) and composition (*i.e.*, inclusion of barley straw *vs.* barley grain) confounded the impact of oilseed or forage source on bacterial abundance. Regardless of diet, more than half the genera in Table 3 were members of the phylum Firmicutes, class Clostridia, order Clostridiales. This order is known to be Gram-negative, obligate anaerobes with a large representation in the rumen ecosystem [75]. However, despite the extensive diversity of this genus, previous research has indicated a possible link between some of these species and rumen BH [76].

When the 25 bacterial genera impacted by diet where compared in a correlation analysis to individual and groups of FA, 19 showed a significant correlation to one or more of the FA's (Table 4). Of these 19, five genera including *Acidaminobacter*, *Clostrium IV*, *Fibrobacter*, *Marvinbryantia* and *Olsenella* showed correlation to 4 or more individual FA's. *Clostridium IV* were highly negatively correlated to levels of CLNAisomers *c*9,*t*11,*t*15-18:3 and *c*9,*t*11,*c*15-18:3, as well as ALA. Increased levels of ALA and its long chain derivatives were found in the subcutaneous fat of cattle fed diets containing FS. Therefore, the increased percent abundance of *Clostridium IV* and *Ethanoligenens* in diets containing SS and not FS indicates that these two genera may be negatively associated with ALA and other n-3 PUFA in the rumen. However, several other genera including *Barnesiella*, *Butyrivibrio*, *Eubacterium* and *Olsenella* all showed increased abundance in diets with increased levels of at least one n-3 PUFA. Since ALA is considered a beneficial FA for human consumption [3–6], further research into the role of these organisms in rumen BH of ALA could elucidate how to increase ruminal bypass of ALA to adipose tissue.

Anaerostipes was the only genus that was found to be negatively correlated (*P*<0.001) to n-6 PUFA, which is known to inhibit BH in the rumen. Interestingly, n-6 PUFA were found to be highest in diets containing SS and *Anaerostipes* was also most abundant in RC-FS diets. This bacterial group may be also be involved in the rumen BH process and could be an indicator of inhibition of BH.

Post hoc analysis comparing cattle with high *vs.* low levels of VA in subcutaneous fat showed a significant separation of means (*P*< 0.001; Table 5). Individuals with the lowest levels of VA in subcutaneous fat were found to have rumen populations with a greater abundance of *Anaerophaga*, *Fibrobacter*, *Guggenheimella*,

Paludibacter and *Pseudozobellia*. Increasing amounts of VA in adipose tissue has been a target for ongoing research as it is a precursor for RA and may have beneficial health effects on its own [77]. However, some rumen bacteria are able to hydrogenate VA to steric acid [48,69]. Therefore, bacteria found at significantly increased levels in cattle with low as compared to high VA in tissue may have greater VA BH activity in the rumen.

The main bacterial species within the rumen which is known to impact FA metabolism is *Butyrivibrio proteoclasticus* [69]. Similar to previous research, a relation between FA metabolism and *Butyrivibrio* was noted in the current experiment. However, increased levels of *Butyrivibrio* were related to increased levels of *n*-3 PUFA in the tissue and not increased levels of SFA as has been previously noted [69]. However, the current study was unable to identify changes in percent abundance of bacterial groups at a species level and therefore other *Butyrivibrio* spp may also be involved in rumen BH. Previous research has also indicated that *Propionibacteria acnes*, *Selenomonas ruminantium*, *Enterococcus faecium*, *Staphylococcus sp.*, *Flavobacterium sp.* and *Streptococcus bovis* may also play a role in the hydrogenation of *trans* FA to SFA [69]. While none of these bacterial groups were noted as being significantly different in this study, a number of previously unreported bacterial genera were identified as being correlated, either positively or negatively, to the levels of various FA in subcutaneous fat. These additional bacterial genera may also play a role in ruminal BH that has not yet been elucidated using traditional microbial culturing techniques. Further research is required to clarify the role of not only these bacterial groups but also of ciliate protozoa in relation to dietary changes impacting the FA profiles of adipose tissue in ruminants.

Conclusions

The present study showed that cattle fed high-forage diets supplemented with SS had increased proportions of VA, RA and LA in subcutaneous fat, whereas supplementing with FS increased *n*-3 PUFA, CLNA, and AD. Present results demonstrate that dietary changes have the potential to produce beef with enhanced levels of PUFA BH products. However, in order to produce beef with consistently increased levels of PUFAs, further knowledge

regarding how the type of PUFA, composition of the non-oil fraction of the diet and the type and level of feed processing impact PUFA deposition in the subcutaneous fat is required. Optimizing ruminal outflows of BH products will also require a better understanding of the precise role of rumen microbes and how the individual animal's genetics contribute to inter-animal variation. While some bacterial genera in this experiment were found to be correlated to specific FAs, the effect of diet on lipolytic rumen microorganisms and preferential pathways for BH requires greater insight by combining molecular techniques with classical culture techniques. Bacterial genera positively correlated to increasing BH intermediates in the present experiment need to be further researched in order to confirm whether these correlations are consistent over a range of diet types.

Acknowledgments

Special thanks are extended to staff at the Lacombe Research Centre (LRC) Beef Unit of AAFC for animal care, animal management and sample collection. The slaughter and processing of the cattle by the LRC abattoir staff is gratefully acknowledged. Contributions of the meat grading and quality staff at the LRC to the result are appreciated. Ms. I.L. Larsen and Dr. D. Petri are acknowledged for their valuable assistance in statistical analysis.

Supporting Information

Figure S1 Rarefaction curves representing estimated coverage of bacterial sequences representing unique taxon for each of the diets (GH-SS: hay-sunflower seed, GH-FS: hay-flax seed, RC-SS: red clover-sunflower seed, RC-FS: red clover-flax seed).
(DOCX)

Author Contributions

Conceived and designed the experiments: MERD TAM. Performed the experiments: RMP CM. Analyzed the data: RMO CM MERD. Contributed reagents/materials/analysis tools: MERD TAM. Wrote the paper: RMP CM MERD TAM.

References

1. Raes K, De Smet S, Demeyer D (2004) Effect of dietary fatty acids on incorporation of long chain polyunsaturated fatty acids and conjugated linoleic acid in lamb, beef and pork meat: a review. Animal Feed Science and Technology 113: 199–221.
2. Salter AM (2014) Dietary fatty acids and cardiovascular disease. Animal doi:10.1017/S1751731111002023. In press.
3. Dilzer A, Park Y (2012) Implication of conjugated linoleic acid (CLA) in human health. Critical Reviews in Food Science and Nutrition 52: 488–513.
4. Jaudszus A, Jahreis G, Schlörmann W, Fischer J, Kramer R, et al. (2012) Vaccenic acid-mediated reduction in cytokine production is independent of c9,t11-CLA in human peripheral blood mononuclear cells. Biochimica et Biophysica Acta - Molecular and Cell Biology of Lipids 1821: 1316–1322.
5. McNeill S, Van Elswyk ME (2012) Red meat in global nutrition. Meat Science 92: 166–173.
6. Sofi F, Buccioni A, Cesari F, Gori AM, Minieri S, et al. (2010) Effects of a dairy product (pecorino cheese) naturally rich in cis-9, trans-11 conjugated linoleic acid on lipid, inflammatory and haemorheological variables: A dietary intervention study. Nutrition, Metabolism and Cardiovascular Diseases 20: 117–124.
7. Nassu RT, Dugan MER, He ML, McAllister TA, Aalhus JL, et al. (2011) The effects of feeding flaxseed to beef cows given forage based diets on fatty acids of longissimus thoracis muscle and backfat. Meat Science 89: 469–477.
8. Juárez M, Dugan MER, Aalhus JL, Aldai N, Basarab JA, et al. (2011) Effects of vitamin E and flaxseed on rumen-derived fatty acid intermediates in beef intramuscular fat. Meat Science 88: 434–440.
9. He ML, McAllister TA, Kastelic JP, Mir PS, Aalhus JL, et al. (2012) Feeding flaxseed in grass hay and barley silage diets to beef cows increases alpha-linolenic

acid and its biohydrogenation intermediates in subcutaneous fat. Journal of Animal Science 90: 592–604.
10. Mapiye C, Turner TD, Rolland DC, Basarab JA, Baron VS, et al. (2013) Adipose tissue and muscle fatty acid profiles of steers fed red clover silage with and without flaxseed. Livestock Science 151: 11–20.
11. Van Ranst G, Lee MRF, Fievez V (2011) Red clover polyphenol oxidase and lipid metabolism. Animal 5: 512–521.
12. Lee MRF, Evans PR, Nute GR, Richardson RI, Scollan ND (2009) A comparison between red clover silage and grass silage feeding on fatty acid composition, meat stability and sensory quality of the M. Longissimus muscle of dairy cull cows. Meat Science 81: 738–744.
13. Noci F, French P, Monahan FJ, Moloney AP (2007) The fatty acid composition of muscle fat and subcutaneous adipose tissue of grazing heifers supplemented with plant oil-enriched concentrates. Journal of Animal Science 85: 1062–1073.
14. Schmid A, Collomb M, Sieber R, Bee G (2006) Conjugated linoleic acid in meat and meat products: A review. Meat Science 73: 29–41.
15. Crumb DJ (2011) Conjugated linoleic acid (CLA)-An overview. International Journal of Applied Research in Natural Products 4: 12–15.
16. Mapiye C, Dugan MER, Turner TD, Rolland DC, Basarab JA, et al. (2013) Erythrocytes assayed early ante-mortem can predict adipose tissue and muscle trans-18:1 isomeric profiles of steers fed red clover silage with and without flaxseed. Canadian Journal of Animal Science 93: 1–5.
17. Hegarty RS (2004) Genotype differences and their impact on digestive tract function of ruminants: A review. Australian Journal of Experimental Agriculture 44: 459–467.
18. Dugan MER, Rolland DC, Aalhus JL, Aldai N, Kramer JKG (2008) Subcutaneous fat composition of youthful and mature Canadian beef: emphasis

on individual conjugated linoleic acid and trans-18:1 isomers. Canadian Journal of Animal Science 88: 591–599.

19. Mapiye C, Turner TD, Rolland DC, Basarab JA, Baron VS, et al. (2013) Effects of feeding flaxseed or sunflower-seed in high-forage diets on beef production, quality and fatty acid composition Meat Science 95: 98–109.

20. Candian Council of Animal Care (CCAC) (1993) In Olfert ED, Cross BM, McWilliams AA, (Eds.). (2nd ed.) Guide to the care and use of experimental animals, vol. 1, Ottawa, Ontario.

21. Brethour JR (1992) The repeatability and accuracy of ultrasound in measuring backfat of cattle. Journal of Animal Science 70: 1039–1044.

22. Narvaez N, Wang Y, Xu Z, Alexander T, Garden S, et al. (2013). Effects of hop varieties on ruminal fermentation and bacterial community in an artificial rumen (rusitec). Journal of the Science of Food and Agriculture 93: 45–52.

23. Dowd SE, Callaway TR, Wolcott RD, Sun Y, McKeehan T, et al. (2008). Evaluation of the bacterial diversity in the feces of cattle using 16S rDNA bacterial tag-encoded FLX amplicon pyrosequencing (bTEFAP). BMC Microbiology 8: 125.

24. Schloss PD, Westcott SL, Ryabin T, Hall JR, Hartmann M, et al. (2009). Introducing mothur: open-source, platform-independent, community-supported software for describing and comparing microbial communities. Applied and Environmental Microbiology 75: 7537–7541.

25. Schloss PD, Gevers D, Westcott SL (2011). Reducing the effects of PCR amplification and sequencing artifacts on 16S rRNA-based studies. PLOS ONE 6:e27310.

26. Huse SM, Welch DM, Morrison HG, Sogin ML (2010). Ironing out the wrinkles in the rare biosphere through improved OTU clustering. Environmental Microbiology 12: 1889–1898.

27. Petri RM, Schwaiger T, Penner GB, Beauchemin KA, Forster RJ, et al. (2013). Changes in the rumen epimural bacterial diversity of beef cattle as affected by diet and induced ruminal acidosis. Applied and Environmental Microbiology 79: 3744–3755.

28. Aldai N, Dugan MER, Rolland DC, Kramer JKG (2009) Survey of the fatty acid composition of Canadian beef: Backfat and longissimus lumborum muscle. Canadian Journal of Animal Science 89: 315–329

29. Kramer JKG, Hernandez M, Cruz-Hernandez C, Kraft J, Dugan MER (2008) Combining results of two GC separations partly achieves determination of all cis and trans 16:1, 18:1, 18:2 and 18:3 except CLA isomers of milk fat as demonstrated using ag-ion SPE fractionation. Lipids 43: 259–273

30. Cruz-Hernandez C, Deng Z, Zhou J, Hill AR, Yurawecz MP, et al. (2004) Methods for analysis of conjugated linoleic acids and trans-18:1 isomers in dairy fats by using a combination of gas chromatography, silver-ion thin-layer chromatography/gas chromatography, silver-ion liquid chrmatography. Journal of AOAC International 87: 545–562.

31. Gómez-Cortés P, Bach A, Luna P, Juárez M, de la Fuente MA (2009) Effects of extruded linseed supplementation on n-3 fatty acids and conjugated linoleic acid in milk and cheese from ewes. Journal of Dairy Science 92: 4122–4134.

32. SAS (2009) SAS user's guide: Statistics. SAS for windows. Release 9.2. Cary NC: SAS Institute Inc.

33. Kim CM, Kim JH, Oh YK, Park EK, Ahn GC, et al. (2009) Effects of flaxseed diets on performance, carcass characteristics and fatty acid composition of Hanwoo steers. Asian-Australasian Journal of Animal Sciences 22: 1151–1159.

34. Sarnklong C, Coneja JW, Pellikaan W, Hendriks WH (2010) Utilization of rice straw and different treatments to improve its feed value for ruminants: A review. Asian-Australasian Journal of Animal Sciences 23: 680–692.

35. Jenkins TC (1993) Lipid metabolism in the rumen. Journal of Dairy Science 76: 3851–3863.

36. Doreau M, Ferlay A (1994) Digestion and utilisation of fatty acids by ruminants. Animal Feed Science and Technology 45: 379–396.

37. Harfoot CG, Noble RC, Moore JH (1973) Factors influencing the extent of biohydrogenation of linoleic acid by Rumen micro-organisms in vitro. Journal of the Science of Food and Agriculture 24: 961–970.

38. Geay Y, Bauchart D, Hocquette JF, Culioli J (2001) Effect of nutritional factors on biochemical, structural and metabolic characteristics of muscles in ruminants, consequences on dietetic value and sensorial qualities of meat. Reproduction Nutrtion and Development 41: 1–26.

39. Doreau M, Aurousseau E, Martin C (2009) Effects of linseed lipids fed as rolled seeds, extruded seeds or oil on organic matter and crude protein digestion in cows. Animal Feed Science and Technology 150: 187–196.

40. Shingfield KJ, Salo-Väänänen P, Pahkala E, Toivonen V, Jaakkola S, et al. (2005) Effect of forage conservation method, concentrate level and propylene glycol on the fatty acid composition and vitamin content of cows' milk. Journal of Dairy Research 72: 349–361.

41. Shingfield KJ, Chilliard Y, Toivonen V, Kairenius P, Givens DI (2008) Fatty acids and bioactive lipids in ruminant milk in: Bioactive components of milk, Advances in experimental medicine and biology. In: Bösze Z (ed), Springer New York.

42. Khan NA, Cone JW, Hendriks WH (2009) Stability of fatty acids in grass and maize silages after exposure to air during the feed out period. Animal Feed Science and Technology 154: 183–192.

43. Scollan ND, Richardson I, De Smet S, Moloney AP, Doreau M, et al. (2005) Enhancing the content of beneficial fatty acids in beef and consequences for meat quality. Indicators of milk and beef quality, EAAP Publications: 151–162.

44. Halmemies-Beauchet-Filleau A, Kairenius P, Ahvenjarvi S, Crosley LK, Muetzel S, et al. (2013) Effect of forage conservation method on ruminal lipid

metabolism and microbial ecology in lactating cows fed diets containing a 60:40 forage-to-concentrate ratio. Journal of Dairy Science, 96: 2428–2447.

45. Lee YJ, Jenkins TC (2011) Biohydrogenation of linolenic acid to stearic acid by the rumen microbial population yields multiple intermediate conjugated diene isomers. Journal of Nutrition 141: 1445–1450.

46. Hennessy AA, Ross RP, Devery R, Stanton C (2011) The health promoting properties of the conjugated isomers of α-linolenic acid. Lipids 46: 105–119.

47. Bessa RJB, Alves SP, Jerónimo E, Alfaia CM, Prates JAM, et al. (2007) Effect of lipid supplements on ruminal biohydrogenation intermediates and muscle fatty acids in lambs. European Journal of Lipid Science and Technology 109: 868–878.

48. Jenkins TC, Wallace RJ, Moate PJ, Mosley EE (2008) Board-Invited Review: Recent advances in biohydrogenation of unsaturated fatty acids within the rumen microbial ecosystem. Journal of Animal Science 86: 397–412.

49. Doreau M, Laverroux S, Normand J, Chesneau G, Glasser F (2009) Effect of linseed fed as rolled seeds, extruded seeds or oil on fatty acid rumen metabolism and intestinal digestibility in cows. Lipids 44: 53–62.

50. Jerónimo E, Alves SP, Alfaia CM, Prates JAM, Santos-Silva J, et al. (2011) Biohydrogenation intermediates are differentially deposited between polar and neutral intramuscular lipids of lambs. European Journal of Lipid Science and Technology 113: 924–934

51. AbuGhazaleh AA, Schingoethe DJ, Hippen AR, Kalscheur KF, Whitlock LA (2002) Fatty Acid profiles of milk and rumen digesta from cows fed fish oil, extruded soybeans or their blend. Journal of Dairy Science 85: 2266–2276.

52. Lock AL, Garnsworthy PC (2002) Independent effects of dietary linoleic and linolenic fatty acids on the conjugated linoleic acid content of cows' milk. Animal Science 74: 163–176

53. Maia MRG, Chaudhary LC, Figueres L, Wallace RJ (2007) Metabolism of polyunsaturated fatty acids and their toxicity to the microflora of the rumen. Antonie van Leeuwenhoek 91: 303–314.

54. Chow TT, Fievez V, Moloney AP, Raes K, Demeyer D, et al. (2004) Effect of fish oil on in vitro rumen lipolysis, apparent biohydrogenation of linoleic and linolenic acid and accumulation of biohydrogenation intermediates. Animal Feed Science and Technology 117: 1–12.

55. Basarab JA, Mir PS, Aalhus JL, Shah MA, Baron VS, et al. (2007) Effect of sunflower seed supplementation on the fatty acid composition of muscle and adipose tissue of pasture-fed and feedlot finished beef. Canadian Journal of Animal Science 87: 71–86.

56. Raes K, De Smet S, Balcaen A, Claeys E, Demeyer D (2003) Effect of diets rich in N-3 polyunsaturated fatty acids on muscle lipids and fatty acids in Belgian Blue double-muscled young bulls. Reproduction Nutrition Development 43: 331–345.

57. Gunstone FD, Harwood JL (1994) Occurrence and characterization of oils and fats. In Gunstone, F. D., Harwood, J. L., Padley, F. B., (Eds.), The Lipid Handbook (pp 101). New York: Chapman & Hall.

58. AbuGhazaleh AA, Jacobson BN (2007) The effect of pH and polyunsaturated C18 fatty acid source on the production of vaccenic acid and conjugated linoleic acids in ruminal cultures incubated with docosahexaenoic acid. Animal Feed Science and Technology 136: 11–22.

59. Martínez Marín AL, Gómez-Cortés P, Gómez Castro G, Juárez M, Pérez Alba L, et al. (2012) Effects of feeding increasing dietary levels of high oleic or regular sunflower or linseed oil on fatty acid profile of goat milk. Journal of Dairy Science 95: 1942–1955.

60. Kong Y, He M, McAlister T, Seviour R, Forster R (2010) Quantitative fluorescence in situ hybridization of microbial communities in the rumens of cattle fed different diets. Applied and Environmental Microbiology 76: 6933–6938.

61. Lee MRF, Harris LJ, Dewhurst RJ, Merry RJ, Scollan ND (2003) The effect of clover silages on long chain fatty acid rumen transformations and digestion in beef steers. Animal Science 76: 491–501.

62. Jacobs AAA, van Baal J, Smits MA, Taweel HZH, Hendriks WH, et al. (2011) Effects of feeding rapeseed oil, soybean oil, or linseed oil on stearoyl-CoA desaturase expression in the mammary gland of dairy cows. Journal of Dairy Science 94: 874–887.

63. Bernard L, Bonnet M, Leroux C, Shingfield KJ, Chilliard Y (2009) Effect of sunflower-seed oil and linseed oil on tissue lipid metabolism, gene expression, milk fatty acid secretion in Alpine goats fed maize silage–based diets. Journal of Dairy Science 92: 6083–6094.

64. Nakamura MT, Nara TY, (2004) Structure, function, dietary regulation of Δ6, Δ5, Δ9 desaturases, 2004. p. 345–376.

65. Vlaeminck B, Fievez V, Cabrita ARJ, Fonseca AJM, Dewhurst RJ (2006) Factors affecting odd- and branched-chain fatty acids in milk: A review. Animal Feed Science and Technology 131: 389–417.

66. Ran-Ressler RR, Khailova L, Arganbright KM, Adkins-Rieck CK, Jouni ZE, et al. (2011) Branched chain fatty acids reduce the incidence of necrotizing enterocolitis and alter gastrointestinal microbial ecology in a neonatal rat model. PLOS ONE 6: e29032

67. Wongtangtintharn S, Oku H, Iwasaki H, Toda T (2004) Effect of branched-chain fatty acids on fatty acid biosynthesis of human breast cancer cells. Journal of Nutritional Science and Vitaminology 50: 137–143.

68. Yang Z, Liu S, Chen X, Chen H, Huang M, et al. (2000) Induction of apoptotic cell death and in vivo growth inhibition of human cancer cells by a saturated branched-chain fatty acid, 13-methyltetradecanoic acid. Cancer Research 60: 505–509.

69. McKain N, Shingfield KJ, Wallace RJ (2010) Metabolism of conjugated linoleic acids and 18: 1 fatty acids by ruminal bacteria: products and mechanisms. Microbiology 156: 579–588.

70. Maia MR, Chaudhary LC, Figueres L, Wallace RJ (2007) Metabolism of polyunsaturated fatty acids and their toxicity to the microflora of the rumen. Antonie van Leeuwenhoek 91: 303–314.

71. Hobson PN (1965) Continuous Culture of some anaerobic and facultatively anaerobic rumen bacteria. Journal of General Microbiology 38: 167–180.

72. Russell JB, Baldwin RL (1979) Comparison of substrate affinities among several rumen bacteria: a possible determinant of rumen bacterial competition. Applied and Environmental Microbiology 37: 531–536.

73. Petri RM, Schwaiger T, Penner GB, Beauchemin KA, Forster RJ, et al. (2013) Characterization of the core rumen microbiome in cattle during transition from forage to concentrate as well as during and after an acidotic challenge. PLOS ONE 8: e83424.

74. Petri RM, Forster RJ, Yang W, McKinnon JJ, McAllister TA (2012) Characterization of rumen bacterial diversity and fermentation parameters in concentrate fed cattle with and without forage. Journal of Applied Microbiology 112: 1152–1162.

75. Yutin N, Galperin MY (2013) A genomic update on clostridial phylogeny: Gram-negative spore formers and other misplaced clostridia. Environmental Microbiology 15: 2631–2641.

76. Huws SA, Kim EJ, Lee MR, Scott MB, Tweed JK, et al. (2011) As yet uncultured bacteria phylogenetically classified as Prevotella, Lachnospiraceae incertae sedis and unclassified Bacteroidales, Clostridiales and Ruminococcaceae may play a predominant role in ruminal biohydrogenation. Environmental Microbiology 13: 1500–1512.

77. Field CJ, Blewett HH, Proctor S, Vine D (2009) Human health benefits of vaccenic acid. Applied Physiology, Nutrition, Metabolism 34: 979–991.

8

Novel Bioassay for the Discovery of Inhibitors of the 2-*C*-Methyl-*D*-erythritol 4-Phosphate (MEP) and Terpenoid Pathways Leading to Carotenoid Biosynthesis

Natália Corniani[1], Edivaldo D. Velini[1], Ferdinando M. L. Silva[1], N. P. Dhammika Nanayakkara[2], Matthias Witschel[3], Franck E. Dayan[4]*

1 São Paulo State University, Faculty of Agronomic Sciences, Botucatu, SP, Brazil, **2** National Center for Natural Products Research, School of Pharmacy, University of Mississippi, University, MS, United States of America, **3** BASF SE, GVA/HC-B009, Ludwigshafen, Germany, **4** USDA-ARS Natural Products Utilization Research Unit, University, MS, United States of America

Abstract

The 2-*C*-methyl-*D*-erythritol 4-phosphate (MEP) pathway leads to the synthesis of isopentenyl diphosphate in plastids. It is a major branch point providing precursors for the synthesis of carotenoids, tocopherols, plastoquinone and the phytyl chain of chlorophylls, as well as the hormones abscisic acid and gibberellins. Consequently, disruption of this pathway is harmful to plants. We developed an *in vivo* bioassay that can measure the carbon flow through the carotenoid pathway. Leaf cuttings are incubated in the presence of a phytoene desaturase inhibitor to induce phytoene accumulation. Any compound reducing the level of phytoene accumulation is likely to interfere with either one of the steps in the MEP pathway or the synthesis of geranylgeranyl diphosphate. This concept was tested with known inhibitors of steps of the MEP pathway. The specificity of this *in vivo* bioassay was also verified by testing representative herbicides known to target processes outside of the MEP and carotenoid pathways. This assay enables the rapid screen of new inhibitors of enzymes preceding the synthesis of phytoene, though there are some limitations related to the non-specific effect of some inhibitors on this assay.

Editor: Manfred Jung, Albert-Ludwigs-University, Germany

Funding: Grant# BEX 0226/12-2 from the Brazilian Agency CAPES (Coordination for the Improvement of Higher Education Personnel) that supported the research of N. Corniani in a USDA laboratory. The funders had no role in study design, data collection and analysis, decision to publish, or preparation of the manuscript.

Competing Interests: The authors from BASF SE, the University of Sao Paulo, The University of Mississippi and the USDA ARS declare that no competing interests exist.

* Email: franck.dayan@ars.usda.gov

Introduction

The terms isoprenoid, terpenoid, and terpene are used interchangeably in the literature to refer to a broad class of natural products derived from C5 isopentenyl diphosphate (IPP) [1,2]. Plants produce a myriad of isoprenoids that are functionally important in many physiological and biochemical processes [3,4]. Carotenoids comprise a large isoprenoid family that are derived from the C40 tetraterpenoid phytoene [5] and produced by all photosynthetic organisms (plants, algae and cyanobacteria) as well as certain non-photosynthetic bacteria and fungi [6]. In plants, carotenoids participate in photosynthetic processes, including light harvesting, energy conversion, electron transfer, and quenching of excited chlorophyll triplets [7] in addition to a number of other functions.

Through evolution, two independent biosynthetic routes have been selected for the synthesis of these two basic building blocks [8]. In the cytosol and mitochondria, IPP and dimethylallyl diphosphate (DMAPP) are assembled from three molecules of acetyl-CoA by the mevalonate (MVA) pathway. This pathway was first described in the early work of Bloch and Lynen [9,10], and

was thought to be the sole source of all terpenoids. However, it is now known that it is responsible for the synthesis of sterols and ubiquinone. The MVA pathway is the subject of several reviews [5,11], and is not the focus of this paper.

The existence of an alternative pathway was suggested based on the observation that genes encoding enzymes catalyzing the late steps of the MVA pathway are absent in some archaeal genomes [12]. Furthermore, plants treated with the herbicide clomazone had reduced carotenoid levels but their levels of sterols were not affected [13–15]. This plastid-localized independent pathway, called the 2-*C*-methyl-*D*-erythritol 4-phosphate (MEP) pathway (also non-mevalonate or 1-deoxy-*D*-xylulose 5-phosphate (DOXP) pathway), was reported in 1997 (Figure 1) [9,16–19]. The MEP pathway is a major branch point providing precursors for the synthesis of plastidic monoterpenes, diterpenes, carotenoids, the phytyl chain of chlorophylls, tocopherols, plastoquinone as well as the hormones abscisic acid and gibberellins. There is limited crossover between the two pathways [20,21].

The MEP and carotenoid pathways are well characterized and have been reviewed extensively [1,5,12,22]. Briefly, carotenoid

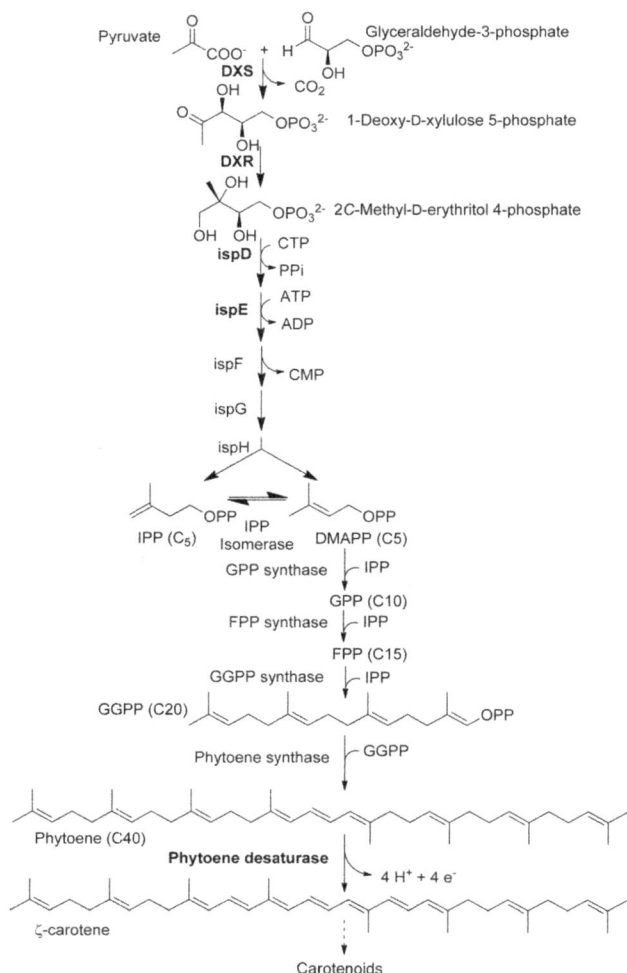

Figure 1. Biosynthesis of carotenoids starts with the 2-*C*-methyl-*D*-erythritol 4-phosphate (MEP) pathway leading to the formation of IPP, continues with the isoprenoid pathway to obtain GGPP. The first committed step to the synthesis of carotenoids consists of the head to head condensation of 2 GGPP to form phytoene. The enzymes in bold letters denote enzyme targets that were tested in this study.

biosynthesis can be divided into three phases (Figure 1). Phase I includes the formation of IPP and DMAPP via the plastid-localized MEP pathway. The first step of the MEP pathway is catalyzed by 1-deoxy-*D*-xylulose 5-phosphate synthase (DXS, EC 2.2.1.7), converting pyruvate and glyceraldehyde-3-phosphate to 1-deoxy-*D*-xylulose 5-phosphate (DOXP). The intramolecular rearrangement and reduction of DOXP to 2-*C*-methyl-*D*-erythritol 4-phosphate (MEP) is catalyzed by 1-deoxy-*D*-xylulose 5-phosphate reductoisomerase (DXR, EC 1.1.1.267). Diverse experimental evidence demonstrates that DXS and DXR represent potential regulatory control points in the MEP pathway [23,24]. Phase I concludes with the formation of the C5 building blocks IPP and DMAPP. In Phase II, a single DMAPP serves as the substrate for successive head-to-tail condensations of IPP units to ultimately form the C20 geranylgeranyl diphosphate (GGPP) [25]. Phase III begins with the head to head condensation of two GGPP molecules to produce phytoene (Figure 1) by the enzyme phytoene synthase (PSY, EC 2.5.1.32). Subsequently, phytoene desaturase (PDS, EC 1.3.5.5) and ζ-carotene desaturase (ZDS, EC 1.3.5.6)

catalyze similar dehydrogenation reactions introducing four double bonds in phytoene to form lycopene. Desaturation requires a plastid-localized terminal oxidase and plastoquinone in photosynthetic tissues [26,27].

Carotenoids are important for plant survival, especially in their role as protection from photooxidation [4,28]. Several important bleaching herbicides inhibit carotenoid synthesis [29]. PDS is the target site for several herbicides such as norflurazon, fluridone and flurochloridone [30]. When sensitive plants are exposed to these herbicides, PDS activity is inhibited, resulting in a rapid accumulation of phytoene and cessation of carotenoid biosynthesis [31]. In plants, inhibition of *p*-hydroxyphenylpyruvate dioxygenase (HPPD, EC 1.13.11.27) by triketone, isoxazole and pyrazole herbicides affects the formation of homogentisic acid [32], which is a key precursor for the biosynthesis of plastoquinone, a critical cofactor of PDS [26] and leads to inhibition of its enzymatic activity. This class of herbicides represents the last herbicide mode of action to have been commercialized in the last twenty years [33].

Blockage of any of the steps preceding the formation of lycopene inhibits carotenoid synthesis. However, only the herbicide clomazone targets the MEP pathway by inhibiting DXS [34], although it does so indirectly [35]. It has been postulated that, once absorbed into the plant, clomazone is oxidized by cytochrome P450 monooxygenases (P450s) to form ketoclomazone (Figure 2), which is the putative active herbicidal form [34]. Early evidences of this requirement for metabolic activation was observed in plants treated with phorate or other P450s inhibitors being protected from the herbicidal effect of clomazone [36,37]. The antibiotic fosmidomycin is not used as an herbicide, but this compound inhibits DXR, the second step of the MEP pathway [38,39].

The MEP and carotenoid pathways are not present in animals, and thus, their enzymes are preferred targets for new herbicides [40]. In spite of the potential relevance of all other enzymes of this pathway, no herbicide targeting the early steps of carotenoid synthesis, other than clomazone, has been developed [41].

The emergence of resistance to herbicides is an increasing problem facing agriculture [42,43], and there do not appear to be any herbicides with novel mechanisms of action being developed [33]. The pathways leading to carotenoids (MEP and isoprenoid pathways) offer several attractive targets for new molecules discovery efforts [44]. Indeed, the unique target sites inhibited by clomazone [37] and fosmidomycin [45] illustrate the potential benefits of developing new herbicides which interfere with the early steps of carotenoid synthesis.

One approach to discover novel inhibitors has been high throughput *in vitro* assays that are most likely to identify

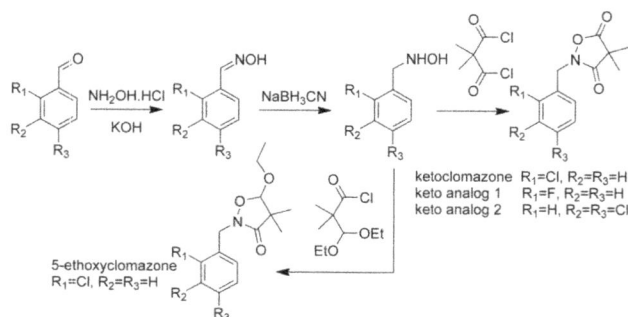

Figure 2. Schematics of the synthesis of 5-ethoxyclomazone, ketoclomazone and keto analogs 1 and 2.

mechanism-based inhibitors of the various steps in the MEP pathway [46]. The aim of our research was to develop a simple, fast, and inexpensive, *in vivo* assay to identify inhibitors of the early steps in carotenoid synthesis by measuring the carbon flux through the MEP and isoprenoid pathways using phytoene as a biomarker.

Materials and Methods

Chemicals and supplies

Phorate, O,O-diethyl S-[(ethylthio]methyl) phosphorodithioate; fosmidomycin (3-(formylhydroxyamino)propyl)-phosphonic acid sodium salt; FR-900098, p-[3-(acetylhydroxyamino)propyl]-phosphonic acid; amitrol, 1,2,4-triazol-3-amine; dinoterb, 2-(1,1-dimethylethyl)-4,6-dinitro-phenol; endothall monohydrate, 7- oxabicyclo[2.2.1] heptane-2, 3-dicarboxylic acid monohydrate; oryzalin, 4-(dipropylamino)-3,5-dinitro-benzenesulfonamide; sulcotrione, 2-[2-chloro-4-(methylsulfonyl)benzoyl]-1,3-cyclohexanedione were purchased from Sigma-Aldrich (St. Louis, MO 63103) and dichlobenil (2,6-dichlorobenzonitrile); glufosinate-ammonium, 2-amino-4-(hydroxymethylphosphinyl)butyric acid ammonium salt were purchased from Allied-Signal Inc., (Morristown, NJ 07960).

Clomazone, 2-[(2-chlorophenyl)methyl]-4,4-dimethyl-3-isoxazolidinone; imazapyr, 2-[4,5-dihydro-4-methyl-4-(1-methylethyl)-5-oxo-1H-imidazol-2-yl]-3-pyridinecarboxylic acid; quinclorac, 3,7-dichloro-8-quinolinecarboxylic acid; atrazine, 6-chloro-$N2$-ethyl-$N4$-(1-methylethyl)-1,3,5-triazine-2,4-diamine; paraquat CL tetrahydrate, 1,1'-dimethyl-4,4'-bipyridinium dichloride; imazethapyr, 2-(4,5-dihydro-4-methyl-4-(1-methylethyl)-5-oxo-1H-imidazol-2-yl)-5-ethyl-3-pyridinecarboxylic acid; alachlor, 2-chloro-N-(2,6-diethylphenyl)-N-(methoxymethyl)-acetamide; asulam, N-[(4-aminophenyl)sulfonyl]-carbamic acid methyl ester; sulfentrazone, N-[2,4-dichloro-5-[4-(difluoromethyl)-4,5-dihydro-3-methyl-5-oxo-1H-1,2,4-triazol-1-yl]-phenyl] methanesulfonamide; glyphosate-isopropylammonium, isopropylammonium N-(phosphonomethyl) glycine; diclofop-methyl, 2-[4-(2,4-dichlorophenoxy)phenoxy]-propanoic acid methyl ester were purchased from Chem-Service (West Chester, PA 19381).

Norflurazon, 4-chloro-5-(methylamino)-2-[3-(trifluoromethyl)-phenyl]-3(2H)-pyridazinone was provided by Sandoz, Inc. Crop Protection (now Syngenta, Greensboro, NC 27419), and experimental inhibitors for some of the enzymes of the MEP pathway were provided by BASF-SE (Ludwigshafen, Germany) [41].

Synthesis of halogenated analogs

Reagents and solvents were purchased from Sigma-Aldrich Chemical Co. (St Louis, MO, USA) and Fisher Scientific (Pittsburgh, PA, USA). NMR spectra were recorded on a Varian-Mercury-plus-400 or Varian Unity-Inova-600 spectrometer using $CDCl_3$ and methanol-d_4 unless otherwise stated. MS data were obtained from an Agilent Series 1100 SL equipped with an ESI source (Agilent Technologies, Palo Alto, CA, USA). Column chromatography and preparative TLC were performed on Merck silica gel 60 (230–400 mesh) and silica gel GF plates (20×20 cm, thickness 0.25 mm), respectively. General synthesis of 5-ethoxyclomazone, ketoclomazone and analogs is shown in Figure 2.

Synthesis of halogenated-benzaldehyde oximes

Oximes of 2-chloro-, 2-fluoro, and 3,4-dichlorobenzaldehyde were prepared by the procedure reported by Zamponi et al. [47].

2-Chlorobenzaldehyde oxime: ^{1}H NMR δ(CDCl3): 7.26 (1H, brt, J = 7.2 Hz), 7.31 (1H, td, J = 7.6, 1.6 Hz), 7.38 (1H, dd, J = 8.0, 1.2 Hz), 8.62 (1H, s).

2-Fluorobenzaldehyde oxime: ^{1}H NMR δ(CDCl3): 7.10 (1H, brt, J = 9.2 Hz), 7.66 (1H, brt, J = 7.6 Hz), 7.37 (1H, brq, J = 7.2 Hz), 7.71 (1H, t, J = 7.6 Hz), 8.38 (1H, s).

3,4-Dichlorobenzaldehyde oxime: ^{1}H NMR δ(CDCl3): 7.40 (1H, dd, J = 8.4, 2.0 Hz), 7.46 (1H, d, J = 8.4 Hz), 7.67 (1H, d, J = 2.0 Hz), 8.01 (1H, s).

Synthesis of halogenated N-(benzyl)hydroxylamine

A mixture of oxime (500 mg) in glacial acetic acid (10 ml) was treated with sodium cyanoborohydride (NaCNBH$_3$) portion wise under stirring while maintaining the temperature below 20°C until the reaction was complete as evidenced by TLC. The reaction mixture was basified with ice-cold NaOH and extracted with ethyl acetate. The organic layer was washed with water, dried and evaporated to afford a white solid. This product was chromatographed over silica gel and eluted with ethyl acetate:hexane 3:7 to yield a pure product which was crystallized from CHCl$_2$/hexanes.

N(2-Chlorobenzyl)hydroxylamine: ^{1}H NMR δ(CDCl3): 4.1 (2H, s), 7.21–7.26 (2H, m), 7.35–7.39 (2H, m).

N(2-Fluorobenzyl)hydroxylamine: ^{1}H NMR δ(CDCl3): 4.06 (2H, s, CH$_2$), 7.05 (1H, brt, J = 8.8 Hz), 7.13 (1H, brt, J = 7.2 Hz), 7.27 (1H, brq, J = 6.8 Hz), 7.33 (1H, brt, J = 7.6 Hz).

N(3,4-Dichlorobenzyl)hydroxylamine (**6**): ^{1}H NMR δ(CDCl3): 3.93 (2H, s), 7.15 (1H, dd, J = 8.0, 2.0 Hz), 7.40 (1H, d, J = 8.0 Hz), 7.44 (1H, d, J = 2.0 Hz).

Synthesis of halogenated ketoclomazone

Dimethylmalonyl dichloride (1.25 mM) in CH$_2$Cl$_2$ (1 ml) was slowly added to a solution of hydroxylamine (1 mM) and triethylamine (0.3 ml) in CH$_2$Cl$_2$ (5 ml) at 10°C. After addition was complete, the reaction mixture was stirred for 30 min poured onto ice and extracted with CH$_2$Cl$_2$. The organic layer was washed with aqueous Na$_2$CO$_3$, 1 N HCl and saturated aqueous NaCl. The resulting solution was then dried and the gummy product obtained was chromatographed over silica gel. The product was eluted with ethyl acetate:hexane 5:95 and crystallized in CH$_2$Cl$_2$/hexanes.

Ketoclomazone: ^{1}H NMR δ(CDCl$_3$): 1.44 (6H, s), 5.05 (2H, s), 7.24–7.31 (2H, m), 7.34 (1H, m), 7.39 (1H, m). 13C NMR δ(CDCl3): 21.4 (CH$_3$), 41.9 (C), 47.3 (CH$_2$), 127.3 (CH), 130.0 (CH), 130.1 (CH), 130.3 (CH), 131.4 (C), 133.8 (C), 172.1 (C), 173.8 (C); HRESIMS [M+H]$^+$ m/z 254.0600 (calcd for (C$_{12}$H$_{12}$ClNO$_3$+H)$^+$254.0584).

2-(2-Fluorobenzyl)-4,4-dimethylisoxazolidine-3,5-dione (keto analog **1**): ^{1}H NMR δ(CDCl$_3$): 1.41 (6H, s), 4.98 (2H, s), 7.07 (1H, brt, J = 9.2 Hz), 7.13 (1H, brt, J = 7.6 Hz), 7.25–7.34 (2H, m). 13C NMR δ(CDCl3): 21.2 (CH$_3$), 41.8 (C), 47.3 (CH$_2$, J_{CF} = 4.4 Hz), 115.8 (CH, J_{CF} = 21.3 Hz), 120.7 (CH, J_{CF} = 14.0 Hz), 124.5 (CH, J_{CF} = 3.6 Hz), 130.5 (CH), 130.6 (CH, J_{CF} = 5.9 Hz), 160.9 (C, J_{CF} = 248 Hz), 172.4 (C), 173.6 (C); HRESIMS [M+H]$^+$ m/z 237.0811 (calcd for (C$_{12}$H$_{12}$FNO$_3$+ H)$^+$237.0801).

2-(3,4-Dichlorobenzyl)-4,4-dimethylisoxazolidine-3,5-dione (keto analog **2**): ^{1}H NMR δ(CDCl3): 1.40 (6H, s), 4.83 (2H, s), 7.17 (1H, dd, J = 8.0, 2.0 Hz), 7.41 (1H, brs), 7.42 (1H, d, J = 2 Hz), 7.43 (1H, d, J = 8.0 Hz). 13C NMR δ(CDCl3): 21.3 (CH$_3$), 42.0 (C), 48.8 (CH$_2$), 128.0 (CH), 130.6 (CH), 131.1 (CH), 133.1 (C), 133.2 (C), 134.0 (C), 172.9 (C), 173.5 (C); HRESIMS [M+H]$^+$ m/z 288.0207 (calcd for (C$_{12}$H$_{11}$Cl$_2$NO$_3$+H)$^+$288.0194).

Synthesis of 5-ethoxyclomazone

A mixture of ethyl 3,3-diethoxy-2,2-dimethylpropionate [48] and KOH (2.2 g) in 10% aqueous ethanol (40 ml) was refluxed for 2 hours and the solvent was evaporated under vacuum. The residue was dissolved in water (40 ml), neutralized with succinic acid and extracted with CH_2Cl_2. The organic layer was washed with saturated aqueous NaCl, dried over anhydrous Na_2SO_4, and evaporated to give 3-diethoxy-2-dimethylpropanoic acid as a thick oil. Upon storage at 4°C this oil formed a white crystalline solid.

1H-NMR (400 MHz) δ 1.17 (6H, s, CH_3), 1.17 (3H, t, $J = 7.0$ Hz, CH3), 3.55 (2H, dq, $J = 16.0$, 7.2 Hz, CH_2), 3.82 (2H, dq, $J = 16.0$, 7.2 Hz, CH_2), 4.54 (1H, s, CH); 13C-NMR (100 MHz) δ 15.3 (CH_3), 19.6 (CH_3), 48.4 (C), 66.5 (CH_2), 107.3 (CH), 181.9 (CO); HRESIMS $[M+Na]^+$ m/z 213.1099 (calcd for $(C_9H_{18}O_4+Na)^+$213.1103).

Oxalyl chloride (260 mg, 4 mm) was added to a solution of 3-diethoxy-2-dimethylpropanoic acid (380 mg) in toluene (3 ml) and the reaction mixture was stirred at 60°C for 30 minutes. The solvent was evaporated under vacuum to afford oil. This oil was dissolved in toluene 5 ml and evaporated under vacuum. This product was used in the next reaction immediately without further purification.

Triethylamine (0.5 ml) and 2-chloro-N-hydroxybenzylamine (200 mg 1.27 mm) were added sequentially to a solution of 3-diethoxy-2-dimethylpropanoic acid chloride (2 mm) in CH_2Cl_2 (4 ml) at 0°C under stirring. The reaction mixture was stirred over night at room temperature and partitioned between water and CH_2Cl_2. The organic layer was washed with water, dried over Na_2SO_4, and evaporated. The products were chromatographed on silica gel and elution with ethyl acetate/hexane (1:99) gave the least polar product, 5-ethoxyclomazone, as an colorless oil (32 mg).

1H-NMR (400 MHz) δ 1.06 (3H, t, $J = 7.2$ Hz, CH_3), 1.16 (3H, s, CH_3), 1.16 (3H, s, CH_3), 1.24 (3H, s, CH_3), 3.38 (1H, dq, $J = 16.0$, 7.2 Hz, CH) 3.53 (1H, dq, $J = 16.0$, 7.2 Hz, CH), 4.73 (1H, d, $J = 16.8$ Hz, CH_2), 4.83 (1H, s, CH), 4.73 (1H, d, $J = 16.8$ Hz, CH), 4.89 (1H, d, $J = 16.8$ Hz, CH), 7.16–7.25 (2H, m, CH), 7.29–7.35 (2H, m, CH); 13C-NMR (100 MHz) δ 14.8 (CH_3), 16.6 (CH_3), 22.4 (CH_3), 46.1 (CH_2), 46.3 (C), 64.1 (CH_2), 108.0 (CH), 126.8 (CH), 129.0 (CH), 129.2 (CH), 129.4 (CH), 132.9 (C), 133.2 (C), 172.8 (CO); HRESIMS $[M+H]^+$ m/z 283.0969 (calcd for $(C_{14}H_{18}ClNO_3+H)^+$283.0975).

Plant material

Barley (Hordeum vulgare L.) seeds were purchased from Johnny's Selected Seeds (Waterville, Maine 04903). Seeds were sown in moist commercial Metromix potting soil and grown either in a dark growth chamber set at 25°C or in the greenhouse under natural light for 4 days.

Bioassays

Approximately 0.1 g of fresh young barley leaves were weighed, cut in 3 mm sections with a razor blade, and incubated (60×15 mm Petri dishes) in 5 ml of 5 mM 2-[N-morpholino]etha-nesulfonic acid buffer (MES, pH 6.5) containing 200 μM of norflurazon for 24 h in a growth chamber with the 16/8 light/dark cycle at 25°C. The herbicides and other test compounds (see section Chemicals and Supplies) were tested either at a fixed concentration or at different concentrations(dose-response curves ranged from 0.1 to 100 μM inhibitor in acetone). Control tissues were exposed to the same amount of acetone as the treated tissues but without the test compounds. All experiments had 3–5 replicates and were repeated in time. The effect of 50 μM phorate, a cytochrome P450 monooxygenase inhibitor, was tested in some of the assays with clomazone to determine the requirement for the metabolic activation of this herbicide.

The usefulness of this bioassay in identifying potential novel inhibitors of the early steps of carotenoid biosynthesis was tested with a number of experimental compounds provided by BASF or synthesized in our laboratory. These compounds were tested at a 100 μM final concentration on greening etiolated barley leaf cuttings and their activity is expressed as inhibition of phytoene accumulation relative to the amount of phytoene accumulating in the norflurazon alone treatment. Finally, the specificity of this assay was evaluated by testing representative compounds inhibiting all the known target sites of commercial herbicides. These compounds were tested at a 100 μM final concentration on etiolated barley leaf cutting either exposed to light or maintained in total darkness for 24 h. Herbicides causing a 10% or less reduction of phytoene level were considered not active. Those causing 10 to 50% inhibition and those causing more than 50% inhibition were considered slightly and highly active, respectively.

Phytoene extraction and determination

Phytoene was extracted and quantified according to a protocol modified from Sprecher et al. [49] as described in Dayan et al. [50]. After 24 h incubation, the barley leaf samples were homogenized (Polytron, PT 3300) in 3 ml of 6% KOH in methanol (w/v), stored for 15 min at room temperature (RT) and then centrifuged for 5 min at 1,300 g (Sorvall Swinging Bucket SH-3000 rotor). The supernatant was transferred into new tubes and mixed with 3 ml of petroleum ether (Acros, Fair Lawn, NJ, boiling range 80–110°C). Saturated NaCl solution was added (1.5 ml), mixed and centrifuged for 10 min at 1,300 g. A clear partition is formed by centrifugation and an aliquot (1.250 ml) of the epiphase was transferred to disposable cuvettes (methacrylate) followed by UV-spectrophotometric (Shimadzu, UV-3101PC) measurements at 287 nm. All manipulations were performed in a dark room under green light. The phytoene content was calculated by its extinction coefficient (ε) of 1108 mM cm^{-1} and expressed as μg g^{-1} fresh weight (FW) according to equation 1:

$$\mu g\, g^{-1} FW = \frac{(A/\varepsilon)\times 0.03}{g\,FW} 10^6 \qquad (1)$$

Statistical Analysis

Phytoene accumulation was plotted against inhibitor concentrations to generate dose-response curves. Data were analyzed by a four-parameters log-logistic model [51] using R software (version 2.15.2, R Foundation for Statistical Computing, Vienna, Austria) with the drc module [52]. Means and standard deviations were obtained using the raw data and the half-maximal inhibitory response (I_{50}) was defined as the concentration at which this accumulation was inhibited by 50% compared with controls. I_{50} values were obtained from the parameters in the regression curves. Graphs were generated with Sigma Plot (version 11, Systat Software Inc., San Jose, CA, USA). Means were separated with the Duncan multiple range test at $P = 0.05$ using the Agricolae module [53].

The quality of the assay was determined by calculating the Z' factor using equation 2 according to Zhang et al. [54]

$$Z' = 1 \frac{(3\sigma_{c+} + 3\sigma_{c-})}{|\mu_{c+} - \mu_{c-}|} \qquad (2)$$

where $3\sigma_{c+} = 3$ standard deviations of positive control, $3\sigma_{c-} = 3$ standard deviations of negative control, $\mu_{c+} =$ mean of positive control and $\mu_{c-} =$ mean of negative control. The utilization of the 3 standard deviations ensures 99.73% confidence limit. The Z' of the phytoene accumulation assay is based on the measurements of 33 positive controls and 33 negative controls.

Results and Discussion

Accumulation of phytoene over time

This simple assay relies on the premise that the accumulation of phytoene resulting from the inhibition of phytoene desaturase by norflurazon is a reflection of the carbon flow through the MEP pathway. Incubation of barley leaves floating on a medium supplemented with 200 µM norflurazon for 24 h caused a time-dependent linear accumulation of phytoene (Figure 3). Therefore, any compound reducing the level of phytoene accumulation during this assay is likely to inhibit one of the many enzymatic steps leading to phytoene synthesis (Figure 1). It has been suggested that inhibiting the MEP pathway of plants could be useful in the search of novel herbicides [55] and the bioassay described herein may be a useful new tool to discover such compounds.

Barley was selected because its small seeds germinate quickly in the dark and contain highly active MEP and carotenoid pathways during its light-induced thylakoid formation, in the transition of etioplasts to chloroplasts during the greening process [56]. Additionally, it is highly sensitive to clomazone [57], which is important because some plants do not metabolize clomazone to ketoclomazone (the putative active form) very rapidly, and their inhibitory effects may not be detected during the time-span of this experiment.

Effect of clomazone, ketoclomazone, and 5-ethoxyclomazone

Clomazone is the only commercial herbicide known to inhibit carotenoid synthesis upstream from phytoene desaturase. Actually, clomazone is inactive, but its metabolite ketoclomazone inhibits DXS [34,56], the thiamine diphosphate-dependent enzyme that catalyzes the first step in the MEP pathway [58].

In our simple barley leaf cutting assay, clomazone inhibited phytoene accumulation in a dose-dependent manner that illustrates the inhibition of carbon flow into the MEP pathway in both green (Figure 4A) and greening etiolated tissues (Figure 4B). Clomazone had an I_{50} for inhibition of phytoene accumulation of 0.6 ± 0.16 and 0.33 ± 0.05 µM in green and greening tissues, respectively. Sandmann and Böger [30] reported an I_{50} value for clomazone of less than 15 µM for inhibition of phytoene and phytol biosynthesis in spinach extracts, suggesting that our *in vivo* assay may be more sensitive as it allows for the metabolic activation of clomazone [36,37].

The requirement for metabolic activation of clomazone was confirmed by repeating the same dose-response curves in the presence of 50 µM phorate. Phorate is an organophosphate insecticide that inhibits cytochrome P450 monooxygenases in plants [59]. In some species, clomazone is rapidly metabolized by P450s [60–62] and phorate can protect plants against the

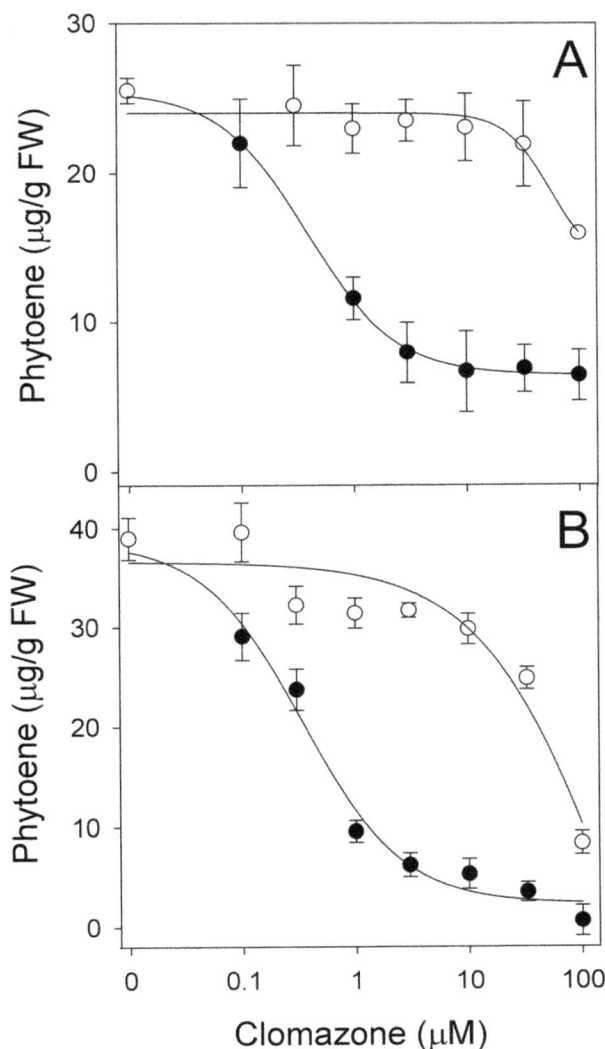

Figure 4. **Dose-response curves showing the effect of the herbicide clomazone with (○) and without (●) phorate on phytoene accumulation induced by 200 µM norflurazon.** (A) Green and (B) greening etiolated young barley leaves. Data represent means of three replications with standard deviation.

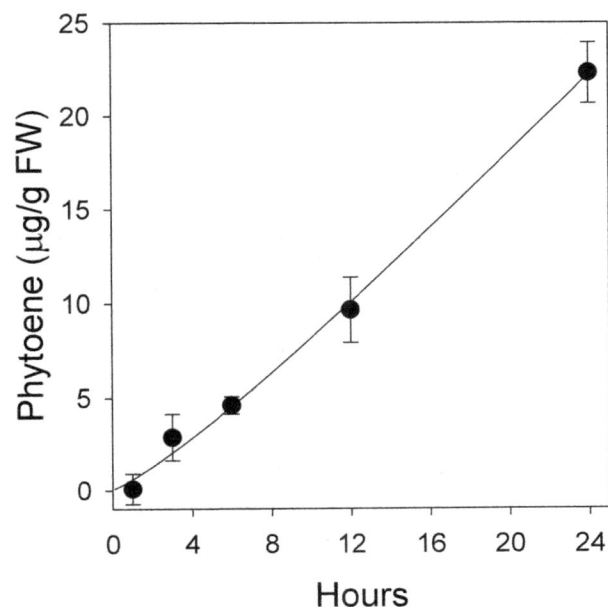

Figure 3. **Time-dependent phytoene accumulation in barley (*Hordeum vulgare* L.) exposed to 200 µM norflurazon.** Data represent means of three replications with standard deviation.

phytotoxic effect of clomazone by preventing this metabolic activation [36,63].

The addition of 50 μM phorate to the solution prevented clomazone from inhibiting phytoene accumulation in both green and greening etiolated, with $I_{50} > 100$ μM (Figures 4 A and B). These results are consistent with previous studies indicating that clomazone must be metabolically activated by P450s and that phorate can abolish its herbicidal activity [36,63].

In plants, metabolism of clomazone follows a couple of pathways. One route involves a N-dealkylation step yielding 2-chlorobenzyl alcohol. The other route follows the progressive oxidation of carbon 5 to yield 5-hydroxyclomazone and ultimately ketoclomazone [60,62]. In both of these studies, 5-hydroxyclomazone and ketoclomazone were present in equivalent amounts, suggesting that the conversion of clomazone to ketoclomazone is not instantaneous. While ketoclomazone has been proposed as the active form responsible for the herbicidal activity of clomazone [34,56], early work on the development of clomazone by FMC (Philadelphia, PA USA) also reported that 5-hydroxyclomazone and several 5-alkoxy derivatives were also potent herbicides [64]. It is postulated that all these 5-hydroxy derivatives are further metabolized into ketoclomazone.

When tested in our bioassay, 5-ethoxyclomazone was as active as clomazone, and phorate did not negatively affect that activity (Figure 5A). Although still very active, ketoclomazone was 16 times less active in greening tissues than clomazone, with I_{50} values of 5 μM (Figure 5B). This was unexpected since clomazone is a proherbicide and appears to be bioactivated by P450s (Figure 4). Phorate did not affect the activity of ketoclomazone, suggesting that it is not subject to further oxidation by P450s (Figure 5B). Similar results were obtained in green tissues (Figure S1).

From a metabolic perspective, the pattern observed with clomazone, 5-ethoxyclomazone and ketoclomazone, and the effect of phorate on the activity of these compounds is particularly informative. Clomazone is a potent *in vivo* inhibitor of phytoene accumulation but phorate greatly reduces this activity, confirming that clomazone is not herbicidal and that one of its oxidized metabolites must be very active, causing inhibition of the carbon flow at submicromolar concentrations. 5-Ethoxyclomazone is as active as the commercial herbicide. On the other hand, the activity of ketoclomazone was lower than that of clomazone, and phorate had no effect on the herbicidal activity of ketoclomazone, suggesting that either the uptake of ketoclomazone was limiting, or that another intermediate is the active form of the herbicide.

Clomazone, 5-hydroxyclomazone and a series of acylated 5-hydroxy derivatives do not inhibit purified plant DXS, whereas ketoclomazone is highly active, with an I_{50} value of 80 nM [41]. Additionally, the *seco*-ketoclomazone analog with a 3-[methyl(hydroxy)amino]-2,2-dimethyl-3-oxopropanoic acid side chain, which is a known plant metabolite of clomazone [60,62], was recently reported as a potent inhibitor of *Haemophilus influenza* DXS [65]. Analysis of the structure-activity relationship of the compounds tested in that study highlighted the requirement for the hydroxamate, the 2,2-dimethylmethylidene, and the carboxylic functional groups. While this *seco*-ketoclomazone analog has structural features reminiscent of a putative transition state intermediate analog, neither the exact form of the clomazone pharmacophore responsible for inhibition of DXS nor its binding to the enzyme is known.

Validity of the bioassay with inhibitors of early steps of the MEP pathway

Keto analogs **1** and **2** are structurally related to ketoclomazone with simple changes in the halogen substitutions on the phenyl

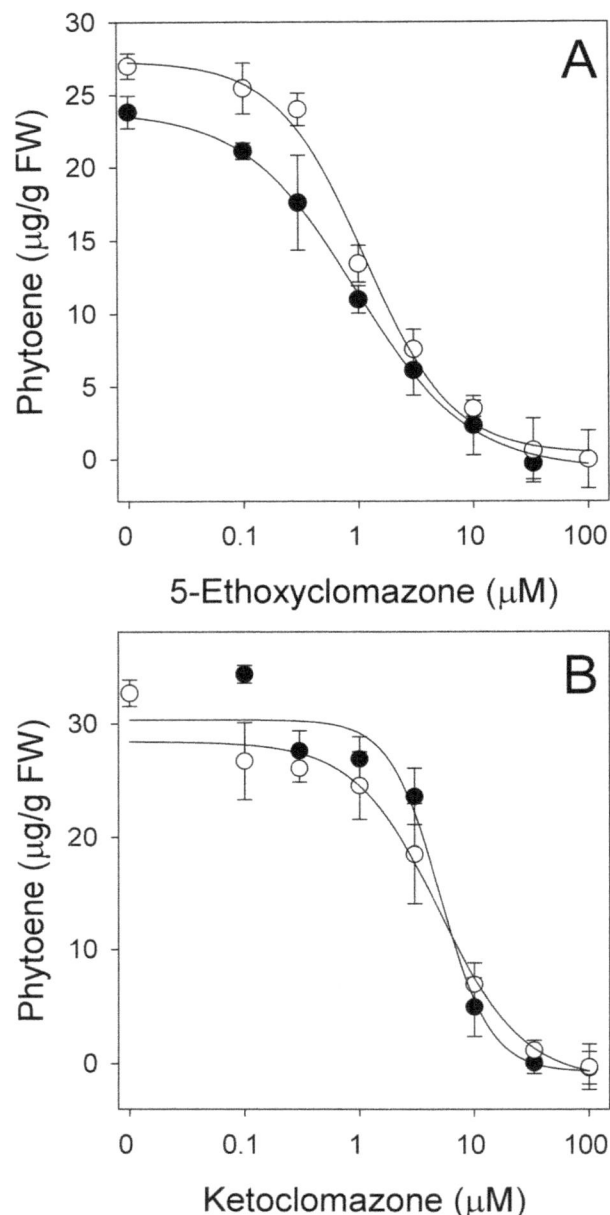

Figure 5. Dose-response curves of (A) 5-ethoxyclomazone and (B) ketoclomazone with (○) and without (●) phorate on greening etiolated barley leaves. Phytoene was caused to accumulate by the presence of 200 μM norflurazon. Data represent means of three replications with standard deviation.

ring (Figure 2). In this study, keto analog **1** was a weak inhibitor of phytoene accumulation, with an I_{50} value of 22 μM on greening etiolated tissue (Figure S2A). Its activity on green tissue was even lower, with an I_{50} value of 52 μM (Figure S2B). However, the structural similarity between ketoclomazone and keto analog **1** suggests that this new compound may inhibit DXS. Keto analog **2** did not affect phytoene accumulation (data not shown), suggesting that the position of the halogen on the phenyl ring has a significant effect on the activity of ketoclomazone.

The first compound described to inhibit the MEP pathway was fosmidomycin, also known as FR-31564 and 3-(N-formyl-N-hydroxyamino) propylphosphonic acid [66,67]. This compound prevents the conversion of labeled DOXP into carotenoids in the *Capsicum* chromoplast system [68] by inhibiting DXR (Figure 1) [66]. Since then, a considerable number of fosmidomycin derivatives have been synthesized to identify new DXR inhibitors [69,70].

In our study, fosmidomycin and its structural analog FR-900098 inhibited phytoene accumulation in a dose-dependent manner with I_{50} values of 5 and 5.5 μM for greening etiolated, respectively (Figure 6). Similar activity was observed on green barley leaves, with I_{50} values of 11 and 17.5 μM for fosmidomycin and FR-900098, respectively (Figure S3).

Interest in inhibitors of the MEP pathway has led a research group at BASF-SE (Ludwigshafen, Germany) to screen for new herbicides targeting this pathway in target-based high throughput *in vitro* assays [41]. Eleven of these experimental compounds were tested in our *in vivo* bioassay.

None of these compounds were very active in our *in vivo* bioassay (Table 1). The most active cluster were **BASF 11** and **BASF 12**, the azolopyrimidine inhibitors of 2C-methyl-D-erythritol 4-phosphate cytidyltransferase (IspD), the third step on the MEP pathway (Figure 1), with 63.6% and 61.3% inhibition of phytoene accumulation, respectively (Table 1). Interestingly, these compounds were the most potent MEP inhibitors generated by BASF with *in vitro* I_{50} values against IspD activity at 140 nM and 35 nM, respectively [41]. IspD activity has been validated by antisense experiments to be essential to plant survival and inhibitors of this enzyme were anticipated to be potentially herbicidal [71,72]. All of the other BASF compounds, including the putative inhibitors of 4-diphosphocytidyl-2C-methyl-D-erythritol kinase (ispE), had *in vitro* activity in the micromolar range [41] and were not very active in the barley bioassay (Table 1). The difference between the activity reported on the target sites and in the barley leaf bioassay may be due to a number of factors. There is often no correlation between the *in vitro* activity of experimental compounds and their performance as herbicides *in vivo* because their physicochemical properties may be less than ideal for uptake

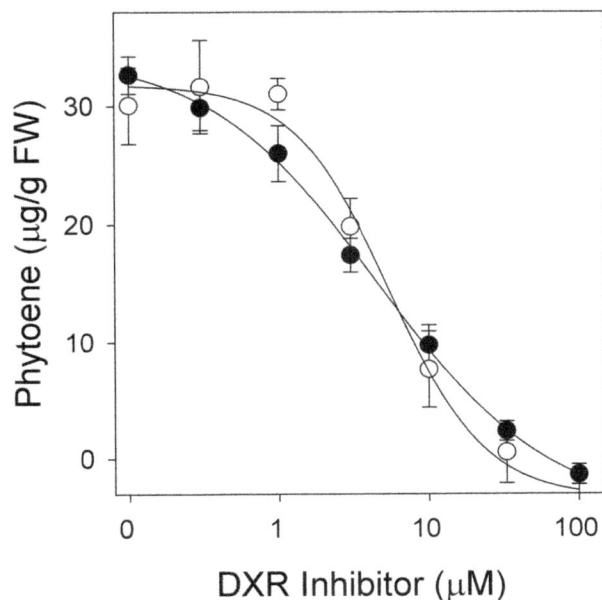

Figure 6. Dose-response curves of the DXR inhibitors fosmidomycin (●)and FR-900098 (○) on greening etiolated barley leaves. Phytoene accumulation was induced in the presence of 200 μM norflurazon. Data represent means of three replications with standard deviation.

and translocation [33]. Metabolic degradation of the compounds may also play a role.

Selectivity of the bioassay

Herbicides have unique affinities for their respective molecular target sites within important plant biochemical pathways and/or physiological processes [35,73]. To test the selectivity of the bioassay to identify inhibitors of the early steps of carotenoid synthesis, herbicides representative of all known modes of action were selected to survey their effects on phytoene accumulation (Table 2). All of the compounds were tested at 100 μM to determine whether they would be detected as false positives in a screening program.

Most of the herbicides targeting enzymes in pathways unrelated to carotenoid synthesis had either little or no effect on the carbon flow into phytoene (−/+). The slight inhibition caused by glyphosate may be a reflection of the secondary effect of this herbicide on the carotenoid pathway. Studies to understand the mechanism of vacuolar sequestration involved in resistance to glyphosate revealed that 2-C-methyl-D-erythritol-2,4-cyclopyrophosphate, the last intermediate in the MEP pathway, accumulated in the presence of this herbicide [74,75].

Activities of a large number of herbicide families are directly or indirectly influenced by light. Herbicide mechanisms that are light-dependent or enhanced by light fall into several different categories [76]. Herbicides with the strongest indirect inhibitory effects (++) on phytoene accumulation were those generating reactive oxygen species (ROS) that lead to lipid peroxidation via light-dependent processes (e.g., inhibitors of PPO, electron diverters from photosystem I) (Table 2). This effect is most likely due to membrane degradation affecting all the biochemical process in the damaged tissues. In most cases, this interference could be alleviated by performing the assays in the dark.

To explore the impact of oxidative stress on the expression of MEP-pathway enzymes, Chang [12] exposed *Catharanthus roseus* leaf discs to a 0.5 μM paraquat solution. This treatment resulted in the bleaching within 10 h of exposure, whereas control treatment did not show any bleaching over a 30 h period. The paraquat treatment also caused a strong induction of *DXS* transcripts, suggesting deregulation of the MEP pathway.

The precise relationship between the effect of the herbicide dinoterb, a synthetic phenol that is no longer used as a herbicide, on phytoene accumulation and its mechanism of action is not well understood but this compound is a strong generator of ROS that destabilizes membranes under both light and dark conditions. Dinoterb also uncouples oxidative phosphorylation, which may reduce the endogenous levels of ATP, thereby inhibiting some of the steps in the MEP pathway (Figure 1). Endothall, an inhibitor of serine/threonine protein phosphatases, also caused a reduction of phytoene accumulation under both light and dark conditions, but the biochemical basis for this effect is unknown.

Conclusions

The simple *in vivo* bioassay developed in this study proved to be an efficient and inexpensive screening method for putative novel inhibitors of the MEP and terpenoid pathways preceding carotenoid synthesis. The method permitted determination of carbon flow through this pathway with accuracy, reproducibility, and minimal sample consumption. The assay appears to be robust in terms of detecting the inhibitory activity of compounds that target the early steps of carotenoid synthesis. The quality of an assay is most commonly assessed by calculating the Z' factor [77]. It is generally accepted that assays with 0.5<Z'<1 have good separation of the

Table 1. Effect of BASF experimental compounds on phytoene accumulation in the presence of 200 µM norflurazon.

MEP target site	BASF #[a]	Inhibition[b] (%)	
DXS			
1-Deoxyxylulose-5-phosphate synthase			
	3	11.1	BC
	4	15.6	BC
	5	27.2	CD
	6	44.4	DE
DXR			
1-Deoxyxylulose-5-phosphate reductoisomerase			
	7	(117.7)[c]	A
	8	40.3	DE
IspD			
2C–Methyl-D-erythritol 4-phosphate cytidyltransferase			
	9	20.6	BCD
	10	26.0	CD
	11	63.6	E
	12	61.3	E
IspE			
4-Diphosphocytidyl-2C-methyl-D-erythritol kinase			
	13	28.8	CD

Inhibition of phytoene accumulation by compound treatment was expressed in percentage of inhibition related to maximum accumulation in control assays. All compounds were tested at 100 µM.
[a]Numbering corresponds to structures in Witschel et al. (2013) [41].
[b]Means values followed by the same letter do not differ significantly at the 5% level by Duncan's multiple range test.
[c]Caused greater phytoene accumulation.

Table 2. Effect of herbicides with different modes of action on phytoene accumulation in the presence of 200 µM norflurazon.

WSSA Class	Compound[a]	MOA	Inhibition[b] Light	Dark
A	Diclofop	Acetyl-CoA carboxylase	+	–
B	Imazethapyr	Acetolactate synthase	–	–
C1	Atrazine	Photosystem II	–	–
D	Paraquat	Photosystem I electron diverter	++	–
E	Sulfentrazone	Protoporphyrinogen oxidase	++	+
F2	Sulcotrione	p-Hydroxyphenylpyruvate dioxygenase	+	–
F4	Clomazone	Deoxyxylulose-5-phosphate synthase	++	++
G	Glyphosate	Enolpyruvylshikimate synthase	+	–
H	Glufosinate	Glutamine synthetase	–	–
I	Asulam	Dihydropteroate synthase	+	–
K1	Oryzalin	Tubulin	–	–
K3	Alachlor	Very long chain fatty acid elongases	–	–
L	Dichlobenil	Cellulose synthase	–	–
M	Dinoterb	Oxidative phosphorylation uncoupler	++	++
NC	Endothall	Serine/threonine protein phosphatase	++	+
O	Quinclorac	Synthetic auxin	–	–

The bioassay was performed in three replications of each treatment.
[a]All compounds were tested at 100 µM to determine whether they would be detected as false positives in a high throughput screening program.
[b]no inhibition (0–10%) (–), slight inhibition (10 to 50%) (+) and strong inhibition (50 to 100%) (++).

distributions and are excellent assays. The assay developed in this study has a Z' factor of 0.584 (Figure S4), which indicates that is likely to successfully identify active compounds [54]. Additionally, it would easily complement enzyme-based high-throughput screenings and evaluate the *in vivo* activity potential performance of inhibitors identified in the *in vitro* assay, without requiring greenhouse tests. The sensitivity of the bioassay to compounds producing reactive oxygen species may be a limitation.

Supporting Information

Figure S1 Dose-response curves of ketoclomazone (a clomazone metabolite) with ($\bigcirc$) and without ($\bullet$) phorate in green barley leaves. Phytoene was caused to accumulate by the presence of 200 µM norflurazon. Data represent means of three replications with standard deviation.
(DOC)

Figure S2 Dose-response to the keto analog 1 on greening etiolated (A) and green (B) young barley leaves ($\bullet$). Data represent means of three replications with standard deviation.
(DOC)

Figure S3 Dose-response curves of the DXR inhibitors fosmidomycin ($\bigcirc$) and FR-900098 ($\bullet$) on green young barley leaves. Phytoene accumulation was induced in the presence of 200 µM norflurazon. Data represent means of three replications with standard deviation.
(DOC)

Figure S4 Calculation of the Z' factor of the bioassay.
(DOCX)

Acknowledgments

We are grateful to J'Lynn Howell, Susan B. Watson and Robert Johnson for their excellent technical assistance. We also thank BASF-SE (Ludwigshafen, Germany) for providing their experimental compounds.

Author Contributions

Conceived and designed the experiments: FED NC. Performed the experiments: NC FMLS NPDN FED. Analyzed the data: NC MW EDV FED. Contributed reagents/materials/analysis tools: MW FED. Contributed to the writing of the manuscript: NC MW FED.

References

1. Vranová E, Coman D, Gruissem W (2013) Network analysis of the MVA and MEP pathways for isoprenoid synthesis. Ann Rev Plant Biol 64: 665–700.
2. Croteau RB, Kutchan TM, Lewis NG (2000) Natural products (secondary metabolites). In: Buchanan BB, Gruissem W, Jones RL, editors. Biochemistry and Molecular Biology of Plants. Rockville, MD: American Society of Plant Biologists. pp. 1250–1268.
3. Cordoba E, Porta H, Arroyo A, San Román C, Medina L, et al. (2011) Functional characterization of the three genes encoding 1-deoxy-D-xylose 5-phosphate synthase in maize. J Exp Bot 62: 2023–2038.
4. Bartley GE, Scolnik PA (1995) Plant carotenoids: Pigments for photoprotection, visual attraction, and human health. Plant Cell 7: 1027–1038.
5. DellaPenna D, Pogson BJ (2006) Vitamin synthesis in plants: Tocopherols and carotenoids. Ann Rev Plant Biol 57: 711–738.
6. Botella-Pavía P, Besumbes Ó, Phillips MA, Carretero-Paulet L, Boronat A, et al. (2004) Regulation of carotenoid biosynthesis in plants: evidence for a key role of hydroxymethylbutenyl diphosphate reductase in controlling the supply of plastidial isoprenoid precursors. Plant J 40: 188–199.
7. Malkin R, Niyogi K (2000) Photosynthesis. In: Buchanan BB, Gruissem W, Jones RL, editors. Biochemistry and Molecular Biology of Plants. Rockville, MD: American Society of Plant Biologists. pp. 568–628.
8. Lichtenthaler HK (2010) The non-mevalonate DOXP/MEP (deoxyxylulose 5-phosphate/methylerythritol 4-phosphate) pathway of chloroplast isoprenoid and pigment biosynthesis. In: Rebeiz CA, Benning C, Bohnert HJ, Daniell H, Hoober JK et al., editors. The Chloroplast: Basics and Applications: Springer Science+Business Media B.V. pp. 95–118.
9. Lichtenthaler HK, Rohmer M, Schwender J (1997) Two independent biochemical pathways for isopentenyl diphosphate and isoprenoid biosynthesis in higher plants. Physiol Plant 101: 643–652.
10. Lichtenthaler HK (2000) Sterols and isoprenoids: Non-mevalonate isoprenoid biosynthesis: enzymes, genes and inhibitors. Biochem Soc Trans 28: 785–789.
11. Cunningham FX Jr., Gantt E (1998) Genes and enzymes of carotenoid biosynthesis in plants. Ann Rev Plant Physiol Plant Mol Biol 49: 557–583.
12. Chang W-C, Song H, Liu H-W, Liu P (2013) Current development in isoprenoid precursor biosynthesis and regulation. Curr Opin Chem Biol 17: 571–579.
13. Weimer MR, Balke NE, Buhler DD (1992) Herbicide clomazone does not inhibit *in vitro* geranylgeranyl synthesis from mevalonate. Plant Physiol 1992: 427–432.
14. Duke SO, Paul RN, Becerril JM, Schmidt JH (1991) Clomazone causes accumulation of sesquiterpenoids in cotton (*Gossypium hirsutum* L.). Weed Sci 39: 339–346.
15. Croteau RB (1992) Clomazone does not inhibit the conversion of isopentenyl pyrophosphate to geranyl, farnesyl, or geranylgeranyl pyrophosphate *in vitro*. Plant Physiol 98: 1515–1517.
16. Rohmer M (1999) The discovery of a mevalonate-independent pathway for isoprenoid biosynthesis in bacteria, algae and higher plants. Nat Prod Rep 16: 565–574.
17. Eisenreich W, Bacher A, Arigoni D, Rohdich F (2004) Biosynthesis of isoprenoids via the non-mevalonate pathway. Cell Mol Life Sci 61: 1401–1426.
18. Lichtenthaler HK, Schwender J, Disch A, Rohmer M (1997) Biosynthesis of isoprenoids in higher plant chloroplasts proceeds via a mevalonate-independent pathway. FEBS Lett 400: 271–274.
19. Schwender J, Zeidler J, Groner R, Müller C, Focke M, et al. (1997) Incorporation of 1-deoxy-D-xylulose into isoprene and phytol by higher plants and algae. FEBS Lett 414: 129–134.
20. Schuhr C, Radykewicz T, Sagner S, Latzel C, Zenk M, et al. (2003) Quantitative assessment of crosstalk between the two isoprenoid biosynthesis pathways in plants by NMR spectroscopy. Phytochem Rev 2: 3–16.
21. Laule O, Fürholz A, Chang H-S, Zhu T, Wang X, et al. (2003) Crosstalk between cytosolic and plastidial pathways of isoprenoid biosynthesis in *Arabidopsis thaliana*. Proc Natl Acad Sci USA 100: 6866–6871.
22. Lichtenthaler HK (1999) The 1-deoxy-D-xylulose-5-phosphate pathway of isoprenoid biosynthesis in plants. Ann Rev Plant Physiol Plant Mol Biol 50: 47–65.
23. Gong YF, Liao ZH, Guo BH, Sun XF, Tang KX (2006) Molecular cloning and expression profile analysis of *Ginkgo biloba* DXS gene encoding 1-deoxy-D-xylulose 5-phosphate synthase, the first committed enzyme of the 2-C-methyl-D-erythritol 4-phosphate pathway. Planta Med 72: 329–355.
24. Cordoba E, Salmi M, León P (2009) Unravelling the regulatory mechanisms that modulate the MEP pathway in higher plants. J Exp Bot 60: 2933–2943.
25. Koyama T, Ogura K (1999) Isopentenyl diphosphate isomerase and prenyltransferases. In: Cane DE, editor. Comprehensive Natural Products Chemistry: Isoprenoids Including Carotenoids and Steroids. Oxford: Pergamon Press. pp. 69–96.
26. Norris SR, Barrette TR, DellaPenna D (1995) Genetic dissection of carotenoid synthesis in *Arabidopsis* defines plastoquinone as an essential component of phytoene desaturation. Plant Cell 7: 2139–2149.
27. Carol P, Stevenson D, Bisanz C, Breitenbach J, Sandmann G, et al. (1999) Mutations in the Arabidopsis gene IMMUTANS cause a variegated phenotype by inactivating a chloroplast terminal oxidase associated with phytoene desaturation. Plant Cell 11: 57–68.
28. Duke SO, Kenyon WH, Paul RN (1985) FMC 57020 effects on chloroplast development in pitted morningglory (*Ipomoea lacunosa*) cotyledons. Weed Sci 33: 786–794.
29. Dayan FE, Duke SO (2003) Herbicides: Carotenoid biosynthesis inhibitors. In: Plimmer JR, Gammon DW, Ragsdale NN, editors. Encyclopedia of Agrochemicals. New York, NY: John Wiley & Sons. pp. 744–749.
30. Sandmann G, Böger P (1989) Inhibition of carotenoid biosynthesis by herbicides. In: Sandmann G, Böger P, editors. Target Sites of Herbicide Action. Boca Raton, FL: CRC Press. pp. 25–44.
31. Weinberg T, Lalazar A, Rubin B (2003) Effects of bleaching herbicides on field dodder (*Cuscuta campestris*). Weed Sci 51: 663–670.
32. Lee DL, Prisbylla MP, Cromartie TH, Dagarin DP, Howard SW, et al. (1997) The discovery and structural requirements of inhibitors of *p*-hydroxyphenylpyruvate dioxygenase. Weed Sci 45: 601–609.
33. Duke SO (2012) Why have no new herbicide modes of action appeared in recent years? Pest Manag Sci 68: 505–512.
34. Müller C, Schwender J, Zeidler J, Lichtenthaler HK (2000) Properties and inhibition of the first two enzymes of the non-mevalonate pathway of isoprenoid biosynthesis. Biochem Soc Trans 28: 792–793.
35. Dayan FE, Duke SO, Grossmann K (2010) Herbicides as probes in plant biology. Weed Sci 58: 340–350.

36. Ferhatoglu Y, Avdiushko S, Barrett M (2005) The basis for the safening of clomazone by phorate insecticide in cotton and inhibitors of cytochrome P450s. Pestic Biochem Physiol 81: 59–70.

37. Ferhatoglu Y, Barrett M (2006) Studies of clomazone mode of action. Pestic Biochem Physiol 85: 7–14.

38. Rohmer M (1998) Isoprenoid biosynthesis via the mevalonate-independent route, a novel target for antibacterial drugs? Progr Drug Res 50: 135–154.

39. Rohmer M, Grosdemange-Billiard C, Seemann M, Tritsch D (2004) Isoprenoid biosynthesis as a novel target for antibacterial and antiparasitic drugs. Curr Opin Investig Drugs 5: 154–162.

40. Withers S, Keasling J (2007) Biosynthesis and engineering of isoprenoid small molecules. Appl Microbiol Biotechnol 73: 980–990.

41. Witschel M, Röhl F, Niggeweg R, Newton T (2013) In search of new herbicidal inhibitors of the non-mevalonate pathway. Pest Manag Sci 69: 559–563.

42. Service RF (2007) A growing threat down on the farm. Science 316: 1114–1117.

43. Service RF (2013) What happens when weed killers stop killing? Science 341: 1329.

44. Hale I, O'Neill PM, Berry NG, Odom A, Sharma R (2012) The MEP pathway and the development of inhibitors as potential anti-infective agents. Med Chem Comm 3: 418–433.

45. Singh N, Cheve G, Avery MA, McCurdy CR (2007) Targeting the methyl erythritol phosphate (MEP) pathway for novel antimalarial, antibacterial and herbicidal drug discovery: Inhibition of 1-deoxy-D-xylulose-5-phosphate reductoisomerase (DXR) enzyme. Curr Pharm Des 13: 1161–1177.

46. Zhao L, Chang W-c, Xiao Y, Liu H-w, Liu P (2013) Methylerythritol phosphate pathway of isoprenoid biosynthesis. Ann Rev Biochem 82: 497–530.

47. Zamponi GW, Stotz SC, Staples RJ, Andro TM, Nelson JK, et al. (2002) Unique structure–activity relationship for 4-isoxazolyl-1,4-dihydropyridines. J Med Chem 46: 87–96.

48. Deno NC (1947) Diethyl acetals of α-formyl esters. J Am Chem Soc 69: 2233–2234.

49. Sprecher SL, Netherland MD, Stewart AB (1998) Phytoene and carotene response of aquatic plants to fluridone under laboratory conditions. J Aquat Plant Manage 36: 111–120.

50. Dayan FE, Owens DK, Corniani N, Silva FML, Watson SB, et al. (2014) Biochemical markers and enzyme assays for herbicide mode of action and resistance studies. Weed Sci DOI:10.1614/WS-D-1613-00063.00061.

51. Seefeldt SS, Jensen JE, Fuerst EP (1995) Log-logistic analysis of herbicide dose-response relationships. Weed Technol 9: 218–227.

52. Ritz C, Streibig JC (2005) Bioassay analysis using R. J Statist Soft 12: 1–22.

53. de Mendiburu F (2014) Agricolae Version 1.1–4. Practical Manual: 1–60.

54. Zhang J-H, Chung TDY, Oldenburg KR (1999) A simple statistical parameter for use in evaluation and validation of high throughput screening assays. J Biomol Screen 4: 67–73.

55. Lichtenthaler HK, Zeidler J, Schwender J, Müller C (2000) The non-mevalonate isoprenoid biosynthesis of plants as a test system for new herbicides and drugs against pathogenic bacteria and the malaria parasite. Z Naturforsch 55c: 305–313.

56. Zeidler J, Schwender J, Mueller C, Lichtenthaler HK (2000) The non-mevalonate isoprenoid biosynthesis of plants as a test system for drugs against malaria and pathogenic bacteria. Biochem Soc Trans 28: 796–798.

57. Anderson RL (1990) Tolerance of safflower (Carthamus tinctorius), corn (Zea mays), and proso millet (Panicum miliaceum) to clomazone. Weed Technol 4: 606–611.

58. Lois LM, Campos N, Putra SR, Danielsen K, Rohmer M, et al. (1998) Cloning and characterization of a gene from Escherichia coli encoding a transketolase-like enzyme that catalyzes the synthesis of D-1-deoxyxylulose 5-phosphate, a common precursor for isoprenoid, thiamin, and pyridoxol biosynthesis. Proc Natl Acad Sci USA 95: 2105–2110.

59. Baerg RJ, Barrett M, Polge ND (1996) Insecticide and insecticide metabolite interactions with cytochrome P450 mediated activities in maize. Pestic Biochem Physiol 55: 10–20.

60. Yasuor H, Zou W, Tolstikov VV, Tjeerdema RS, Fischer AJ (2010) Differential oxidative metabolism and 5-ketoclomazone accumulation are involved in Echinochloa phyllopogon resistance to clomazone. Plant Physiol 153: 319–326.

61. Weimer MR, Buhler DD, Balke NE (1991) Clomazone selectivity: Absence of differential uptake, translocation, or detoxication. Weed Sci 39: 529–534.

62. ElNaggar SF, Creekmore RW, Shcocken MJ, Rosen RT, Robinson RA (1992) Metabolism of clomazone herbicide in soybean. J Agric Food Chem 40: 880–883.

63. Culpepper AS, York AC, Marth JL, Corbin FT (2001) Effect of insecticides on clomazone absorption, translocation, and metabolism in cotton. Weed Sci 49: 613–616.

64. Chang JH, Konz MJ, Aly EA, Sticker RE, Wilson KR, et al. (1987) 3-Isoxazolidinones and related compounds a new class of herbicides. In: Baker DR, Fenyes JG, Moberg WK, Cross B, editors. Synthesis and Chemistry of Agrochemicals. Washington, DC: American Chemical Society. pp. 10–23.

65. Hayashi D, Kuzuyama N, Kato T, Sato Y, Ohkanda J (2013) Antimicrobial N-(2-chlorobenzyl)-substituted hydroxamate is an inhibitor of 1-deoxy-D-xylulose 5-phosphate synthase. Chem Comm 49: 5535–5537.

66. Zeidler J, Schwender J, Müller C, Weisner J, Weidemeyer C, et al. (1998) Inhibition of the non-mevalonate 1-deoxy-D-xylulose-5-phosphate pathway of plant isoprenoid biosynthesis by fosmidomycin. Z Naturforsch 54c: 980–986.

67. Rodríguez-Concepción M (2004) The MEP pathway: A new target for the development of herbicides, antibiotics and antimalarial drugs. Curr Pharm Des 10: 2391–2400.

68. Fellermeier M, Kis K, Sagner S, Maier U, Bacher A, et al. (1999) Cell-free conversion of 1-deoxy-D-xylulose 5-phosphate and 2-C-methyl-D-erythritol 4-phosphate into β-carotene in higher plants and its inhibition by fosmidomycin. Tetrahed Lett 40: 2743–2746.

69. Ershov YV (2007) 2-C-methylerythritol phosphate pathway of isoprenoid biosynthesis as a target in identifying new antibiotics, herbicides, and immunomodulators: A review. Appl Biochem Microbiol 43: 115–138.

70. Jomaa H, Wiesner J, Sanderbrand S, Altincicek B, Weidemeyer C, et al. (1999) Inhibitors of the non-mevalonate pathway of isoprenoid biosynthesis as antimalarial drugs. Science 285: 1573–1576.

71. Fellermeier M, Raschke M, Sagner S, Wungsintaweekul J, Schuhr CA, et al. (2001) Studies on the nonmevalonate pathway of terpene biosynthesis. Eur J Biochem 268: 6302–6310.

72. Rohdich F, Wungsintaweekul J, Eisenreich W, Richter G, Schuhr CA, et al. (2000) Biosynthesis of terpenoids: 4-Diphosphocytidyl-2C-methyl-D-erythritol synthase of Arabidopsis thaliana. Proc Natl Acad Sci USA 97: 6451–6456.

73. Duke SO, Dayan FE (2011) Bioactivity of Herbicides. In: Murray M-Y, editor. Comprehensive Biotechnology. 2 ed. Amsterdam: Elsevier. pp. 23–35.

74. Ge X, d'Avignon DA, Ackerman JJH, Duncan B, Spaur MB, et al. (2011) Glyphosate-resistant horseweed made sensitive to glyphosate: low-temperature suppression of glyphosate vacuolar sequestration revealed by ^{31}P NMR. Pest Manag Sci 67: 1215–1221.

75. Ge X, d'Avignon DA, Ackerman JJH, Sammons RD (2012) Observation and identification of 2-C-methyl-D-erythritol-2,4-cyclopyrophosphate in horseweed and ryegrass treated with glyphosate. Pestic Biochem Physiol 104: 187–191.

76. Hess FD (2000) Light-dependent herbicides: An overview. Weed Sci 48: 160–170.

77. Sui Y, Wu Z (2007) Alternative statistical parameter for high-throughput screening assay quality assessment. J Biomol Screen 12: 229–234.

9

Caspase-Like Activities Accompany Programmed Cell Death Events in Developing Barley Grains

Van Tran, Diana Weier, Ruslana Radchuk, Johannes Thiel, Volodymyr Radchuk*

Institute of Plant Genetics and Crop Plant Research (IPK), Gatersleben, Germany

Abstract

Programmed cell death is essential part of development and cell homeostasis of any multicellular organism. We have analyzed programmed cell death in developing barley caryopsis at histological, biochemical and molecular level. Caspase-1, -3, -4, -6 and -8-like activities increased with aging of pericarp coinciding with abundance of TUNEL positive nuclei and expression of *HvVPE4* and *HvPhS2* genes in the tissue. TUNEL-positive nuclei were also detected in nucellus and nucellar projection as well as in embryo surrounding region during early caryopsis development. Quantitative RT-PCR analysis of micro-dissected grain tissues revealed the expression of *HvVPE2a*, *HvVPE2b*, *HvVPE2d*, *HvPhS2* and *HvPhS3* genes exclusively in the nucellus/nucellar projection. The first increase in cascade of caspase-1, -3, -4, -6 and -8-like activities in the endosperm fraction may be related to programmed cell death in the nucellus and nucellar projection. The second increase of all above caspase-like activities including of caspase-9-like was detected in the maturing endosperm and coincided with expression of *HvVPE1* and *HvPhS1* genes as well as with degeneration of nuclei in starchy endosperm and transfer cells. The distribution of the TUNEL-positive nuclei, tissues-specific expression of genes encoding proteases with potential caspase activities and cascades of caspase-like activities suggest that each seed tissue follows individual pattern of development and disintegration, which however harmonizes with growth of the other tissues in order to achieve proper caryopsis development.

Editor: Srinivasa M. Srinivasula, IISER-TVM, India

Funding: This work was supported in part by the Deutsche Forschungsgemeinschaft (DFG grants WE1608/2-1 for Volodymyr Radchuk and RA2061/3-1 for Ruslana Radchuk) and a grant of Vietnam Ministry of Education and Training for Van Tran. The funders had no role in study design, data collection and analysis, decision to publish, or preparation of the manuscript.

Competing Interests: The authors have declared that no competing interests exist.

* Email: radchukv@ipk-gatersleben.de

Introduction

Programmed cell death (PCD) is a highly regulated cellular suicide process essential for growth, development and survival of all eukaryotic organisms. In plants, developmental PCD accompanies the entire life cycle: seed germination [1], aerenchyma formation [2], tracheary and sieve element differentiation [3,4], leaf shape formation [5], reproduction [6,55], somatic embryogenesis [54,56], senescence [7] and responses against abiotic stresses and pathogens [8].

Development of cereal seeds, including barley grains, is largely accompanied by regular cell death. Mature cereal grains, a main source for human food, domestic animal feed and many industrial applications, consist mainly of dead material. Only the relatively small embryo and aleurone layer are still alive in ripe grains. The regular cell degeneration in cereal caryopses starts soon after fertilization with disintegration of antipodal and synergid cells. Embryo and endosperm develop within the maternal tissues nucellus, inner and outer integuments, and pericarp, which represent the bulk of the early grain. The pericarp can be divided in exocarp or epidermis, mesocarp (representing the majority of pericarp cells) and endocarp or chlorenchyma [9]. The nucellus degenerates within several days after flowering (DAF) providing space and nutrients for the early endosperm [10–12]. Only the

nucellar region opposite to the main vascular bundle stays alive and differentiates into the nucellar projection, which functions as a transfer tissue to deliver the assimilates to the endosperm [13]. The assimilate release from the nucellar projection requires PCD of the tissue [11,14]. The growth of the endosperm takes place at the expense of pericarp which largely degenerates till 12 DAF [12] with the exception of the region surrounding the main vascular bundle. Also cells of the starchy endosperm undergo PCD during later development [15,16].

Little is known about molecular mechanisms underlying PCD in plants. In animals, classical PCD is executed by specific proteases, called caspases, with characteristic cysteines in the catalytic domain. Caspases cleave target peptides at C-terminal after aspartate [17,18] and are involved in apoptosis and development [17]. PCD execution in plants is also often associated with caspase-like activities [19]. Caspase-1, caspase-3 and caspase-6-like activities were detected in the degenerating nucellus of *Sechium edule* [20]. In the developing barley grains, several caspase-like activities were measured at 10 and 30 days after flowering [21]. Taking into account that diverse and often contradictory processes happens simultaneously (i.g., degeneration of pericarp coincides with endosperm expansion) in the caryopses, measurements of caspase activities in distinct tissues over whole development are necessary to detect PCD processes in the developing grain. While caspase activities have been detected in plants, sequences similar to animal caspases are not present in plant genomes. The metacaspases with weak structural similarity to caspases are likely involved in PCD [22], but do not execute caspase-specific proteolytic activity recognizing substrates with either lysine or arginine instead of aspartate [23,24]. Other plant proteases with limited similarity to animal caspases display caspase-like activities and are involved in diverse types of PCD. In particular, vacuolar processing enzyme (VPE), also called legumain, is responsible for caspase-1 activity in plants [25–27]. The 20S proteasome, composed of many α and β subunits, executes caspase-3 activity during xylem development [3] and in response to biotic stress [28]. It has been also shown that the β1 subunit (PBA) and, possibly, the β2 subunit (PBB) provide caspase-3 activity whereas the β5 subunit of the 20S proteasome does not [3]. The subtilisin-like protease called phytaspase has been found to exhibit caspase-6 activity in tobacco and rice [29]. The saspase from *Avena sativa*, which is very similar to phytaspase, is also uses caspase-6 substrates [30]. The caspase-2 and caspase-4 like activities have not been reported in plants so far [31]. With exclusion of VPE genes [12,27], other genes encoding proteases with the respective caspase-like activity have not been described so far in barley.

While the PCD events have been well documented in maternal seed parts of early developing barley grain [12], there is no information about timing and localization of PCD during later seed development. Here we have revealed temporal and spatial PCD patterns over whole barley grain development using the TUNEL assay. Caspase-like activities in separated pericarp and endosperm fractions have been investigated and expression of candidate genes potentially responsible for these activities was studied. The course of PCD events in the different tissues of the developing grain in combination with possible executors of PCD will be delineated.

Materials and Methods

Plant material

Hordeum vulgare cv. Barke plants were grown in greenhouses (18°C and 16/8 h light/dark regime). Caryopses were harvested in two-day interval and hand-separated into the pericarp and endosperm fractions as described previously [12]. For micro-dissections, whole caryopses were collected and kept at −80°C until use.

TUNEL assay

TUNEL assay was performed as described [12]. Both negative and positive controls were performed only at 10 DAF. For negative control, TdT was omitted in the reaction. For positive control of the reaction, the sections were treated with DNase (1500 U ml^{-1}) prior to labelling with the TUNEL mix (Fig. S1).

Caspase assay

The samples for caspase assays were homogenized in liquid nitrogen and re-suspended in 2xCASPB buffer (100 mM HEPES, 0.1% CHAPS, 1 M DTT, pH 7.0) at 4°C. Cell debris was separated by centrifugation at 13000 rpm for 10 min at 4°C and the supernatant was used for the reactions or stored at −70°C. Protein concentration in the extracts was estimated by Bradford assay (BioRad, Hercules, CA, USA). Caspase-like activities were measured in 150 μl reaction mixtures containing 25 μg of protein sample and 10 μM of caspase substrate. Caspase-like activities were detected using the following substrates: acetyl-Tyr-Val-Ala-Asp-7-amido-4-methyl coumarin (Ac-YVAD-AMC) for caspase-1 activity; acetyl-Asp-Glu-Val-Asp-7-amido-4-methyl coumarin (Ac-

DEVD-AMC) for caspase-3 activity; acetyl-Leu-Glu-Val-Asp-7-amido-4-methyl coumarin (Ac-LEVD-AMD) for caspase-4 activity; acetyl-Val-Glu-Ile-Asp-7-amido-4-methyl coumarin (Ac-VEID-AMC) for caspase-6 activity; acetyl-Ile-Glu-Thr-Asp-7-amido-4-methyl coumarin (Ac-IETD-AMD) for caspase-8 activity; and acetyl-Leu-Glu-His-Asp-7-amido-4-methyl coumarin (Ac-LEHD-AMC) for caspase-9 activity. Emitted fluorescence was measured after one hour incubation at room temperature with a 360 nm excitation wave length filter and 460 nm emission wave length filter in a spectrofluorometer (Spectra Max Gemini, Molecular Devices, U.S.A). Four repetitions were performed for determination of each value and standard deviations were calculated. The system was calibrated with known amounts of AMC hydrolysis product in a standard reaction mixture. Blanks were used to account for the spontaneous breakdown of the substrates. The data were analyzed by one-way analysis of variance (ANOVA) followed by a posthoc-test after Holm-Sidak using Microsoft Excel version 2010, with Daniel's XL toolbox version 6.10 [57].

To check the specificity of the caspase assays, the specific protease inhibitors were used to suppress the respective caspase-like activity. The following inhibitors were used: Ac-YVAD-CHO to suppress caspase-1 activity, Ac-DEVD-CHO to suppress caspase-3 activity, Ac-LEVD-CHO to suppress caspase-4 activity, Ac-VEID-CHO to suppress caspase-6 activity, Ac-IETD-CHO to suppress caspase-8 activity and Ac-LEHD-CHO to suppress caspase-9 activity. All caspase substrate and inhibitors were purchased from Enzo Life Sciences (Germany). Assays were performed as described above with the addition of the respective inhibitors (20 μM) to the reaction mixture.

Identification of protease genes with potential caspase-like activity

To identify genes potentially encoding proteases with caspase-like activity, barley full length cDNA data base [32] was screened by BLASTX using already described gene sequences encoding proteases with proven caspase-like activity. The corresponding barley sequences were PCR amplified from a cDNA library of developing grains and re-sequenced using gene-specific primers (Metabion, Germany). Sequence data were processed using the Lasergene software (DNAstar, USA). The phylogenetic trees were built using ClustalW software.

Tissue preparation for laser micro-dissection and pressure catapulting (LMPC)

Frozen caryopses were transferred to a cryostat kept at −20°C. Using a razor blade, the middle part of the caryopses was cut out and glued onto the sample plate by using O.C.T compound. Sections of 20 μm thickness were cut and immediately mounted on PEN membrane slides (PALM). PEN membrane slides were stored for 7 days in the cryostat at −20°C until complete dryness. Prior to laser-assisted micro-dissection, dry cryo-sections were adapted to room temperature for several minutes. LMPC procedure for isolation of specific grain tissues using the PALM® MicroBeam laser system (PALM) has been performed as described in Thiel et al. [33].

RNA processing and qRT-PCR

For each sample, RNA was extracted from 30 to 50 sections of isolated tissues using the Absolutely RNA Nanoprep Kit (Stratagene). Total RNA was amplified by one round of T7-based mRNA amplification using the MessageAmp aRNA Kit (Ambion) to generate tissue-specific antisense RNA (aRNA). After quality

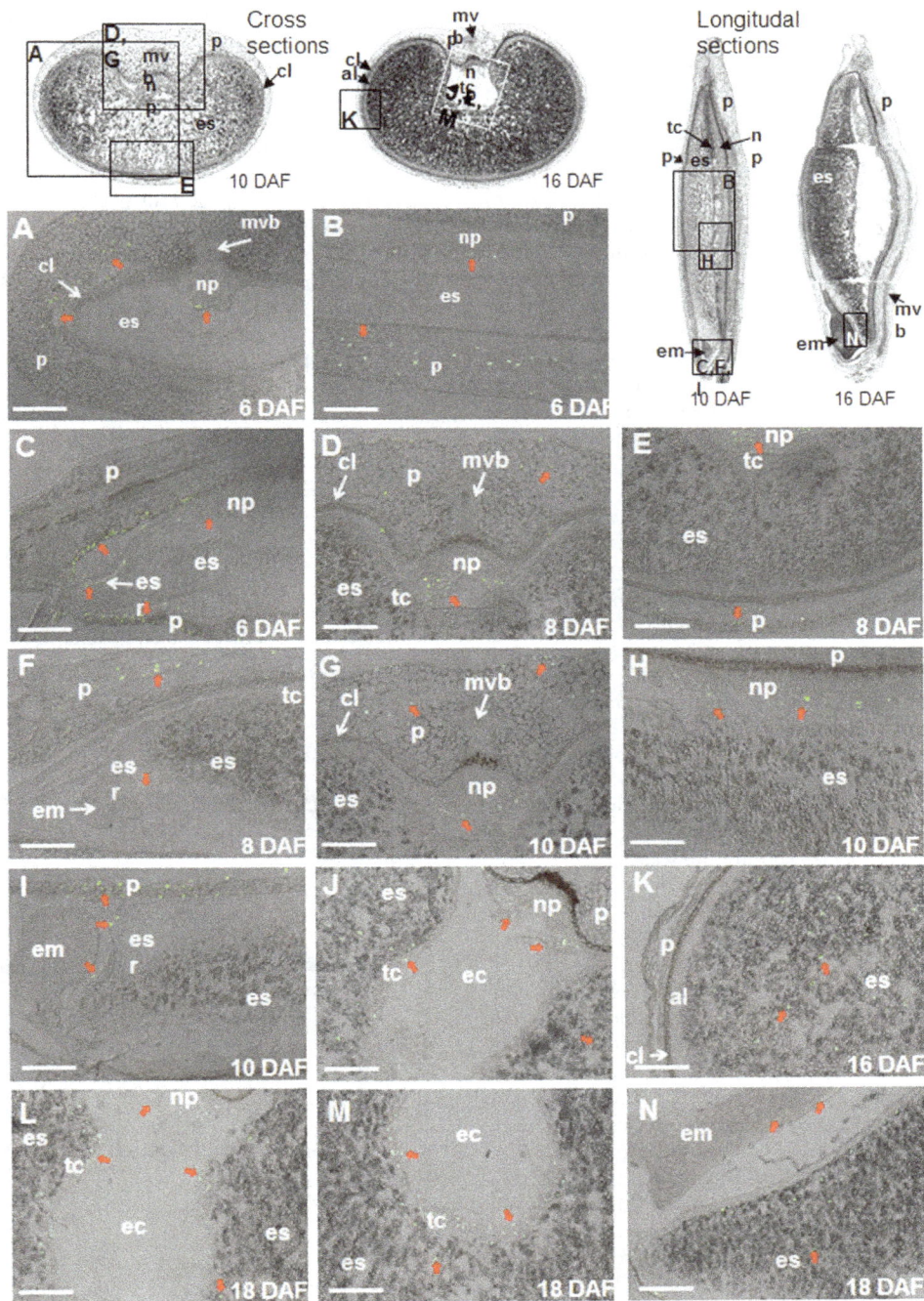

Figure 1. Localization of nuclear DNA fragmentation detected by the TUNEL assay at 6 (A–C), 8 (D–F), 10 (G–I), 16 (J, K), and 18 DAF (L–N). TUNEL-positive nuclei are visualized as green signals and indicated by red arrows. Upper panel demonstrates positions of histological sections used for TUNEL assay at the reconstructed cross and longitudal views of a barley grain. al, aleurone; cl, chlorenchyma, em, embryo; es, endosperm; esr, embryo surrounding region; mvb, main vascular bundle; np, nucellar projection; nu, nucellus; p, pericarp, tc, transfer cells. Bars = 200 μm.

assessment of aRNA populations first strand cDNA was synthesized using SuperScript III (Invitrogen) with random priming according to the manufacturers instructions. The Power SYBR Green PCR master mix was used to perform reactions in an ABI 7900 HT Real-Time PCR system (Applied Biosystems). Data were analyzed using SDS 2.2.1 software (Applied Biosystems). Three replications were conducted for each transcript. The data were analyzed by ANOVA followed by a posthoc-test using Microsoft Excel with Daniel's XL toolbox version 6.10 [57].

The highest relative expression in the group of genes was taken to 100% and expression of the other genes and stages was re-calculated to that value. Primers used for qRT-PCR are listed in Table S1.

Results

Detection of PCD in the developing barley grains by TUNEL assay

Degradation of DNA and disintegration of nuclei are common features of PCD that can be detected by TUNEL assay. Here, we have analyzed PCD pattern during whole development of the barley grain. Only nuclei of the nucellar cells facing to endosperm were TUNEL-labeled between anthesis and three days after flowering (DAF) (Fig. S1) coinciding with endosperm growth [12]. The nucellus is degenerated around 4 DAF, and further endosperm expansion occurs at the expense of pericarp. Coinciding with this, the first TUNEL-labeling nuclei were visible in the innermost cells of the lateral mesocarp region (Fig. S1). The other tissues, including endosperm and nucellar projection, were free of label. Beginning at 6 DAF, the TUNEL-positive nuclei spread throughout the whole mesocarp layer being especially abundant in lateral and dorsal regions (Fig. 1A, B). The chlorenchyma (endocarp) however did not show any labeled nuclei remaining alive till grain maturation (Fig. 1A, D, G, K). The first labeled nuclei were visible at margins of the nucellar projection (Fig. 1A). Numerous labeled nuclei were also detected in close vicinity to the embryo but not in the embryo itself (Fig. 1C). In this region, large number of TUNEL-positive nuclei at margins of the nucellar projection and the pericarp facing the embryo as well as many nuclei of the embryo-surrounding region (ESR) were labeled (Fig. 1C). ESR is the part of the endosperm, which cellularizes first in development [46]. The other endosperm regions were completely free of label (Figs. 1A–C). With ongoing caryopsis development, the dorsal region of the pericarp becomes largely disintegrated and only few labeled nuclei are visible (Fig. 1E). In contrary, the ventral region of pericarp starts to disintegrate and is filled with numerous TUNEL-positive nuclei (Fig. 1D, G). Disintegrating nuclei were also observed at the margins of the nucellar projection, but not in main vascular bundle, chlorenchyma and starchy endosperm (Fig. 1D–H). However, nuclei of endosperm cells close to the embryo were labeled at 8 DAF as well as 10 DAF (Figs. 1F, I) but not cells of embryo itself. In the late grain filling phase (16 DAF), TUNEL-positive nuclei were still detected in the nucellar projection and the ventral parts of pericarp (Fig. 1J). Numerous labeled nuclei were also visible in different regions of starchy endosperm but not the aleurone layer (Fig. 1K). In addition, some nuclei of the transfer cell layer were TUNEL-positive at 16 DAF (Fig. 1J). Two days later, labeling of nuclei spreads to almost all cells of the transfer cell layer (Fig. 1L, M) besides being also detectable in the starchy endosperm and nucellar projection. At 18 DAF, TUNEL-positive nuclei appeared in the embryo. Especially, almost all nuclei of two cell rows in the scutellum were TUNEL-labeled. Many other TUNEL-positive nuclei were also found in other parts of the embryo (Fig. 1N). Because the embryo cells are small in size, the TUNEL-labeled nuclei in the embryo appear to be smaller compared to other tissues.

No TUNEL labeling was detected in control sections when the TdT enzyme had been omitted. Almost all nuclei were labeled in positive controls, treated with DNase prior to TUNEL assay, demonstrating the validity of the procedure (Fig. S1).

Caspase-like activities in the pericarp and endosperm fractions of developing barley grains

As plant PCD has been shown to be associated with caspase-like activities, the profiles of YVADase (caspase-1-like), DEVDase (caspase-3-like), LEVDase (caspase-4-like), VEIDase (caspase-6-like), IETDase (caspase-8-like) and LEHDase (caspase-9-like) activities in pericarp and endosperm fractions of the developing caryopses were investigated.

In the pericarp, the highest activity was detected with the caspase-6 substrate, Ac-VEID-AMC, followed by caspase-3 (Ac-DEVD-AMC), caspase-4 (Ac-LEVD-AMC), caspase-1 (Ac-YVAD-AMC), and caspase-8 (Ac-IETD-AMC) substrates (Fig. 2). Cleavage activities using all substrates generally increased with aging of pericarp and all peaked at 10 DAF (Fig. 2). The caspase-9-like (Ac-LEHD-AMC substrate) activity was overall low in the pericarp and did not show any developmental pattern. Protease inhibitors, specific for each caspase, strongly inhibited the corresponding caspase-like activity, validating these activities in the pericarp (Fig. 2).

In the developing endosperm fraction, two peaks of caspase-like activities were detected for caspase-1, caspase-3, caspase-4, caspase-6 and caspase-8 substrates. The first prolonged increase

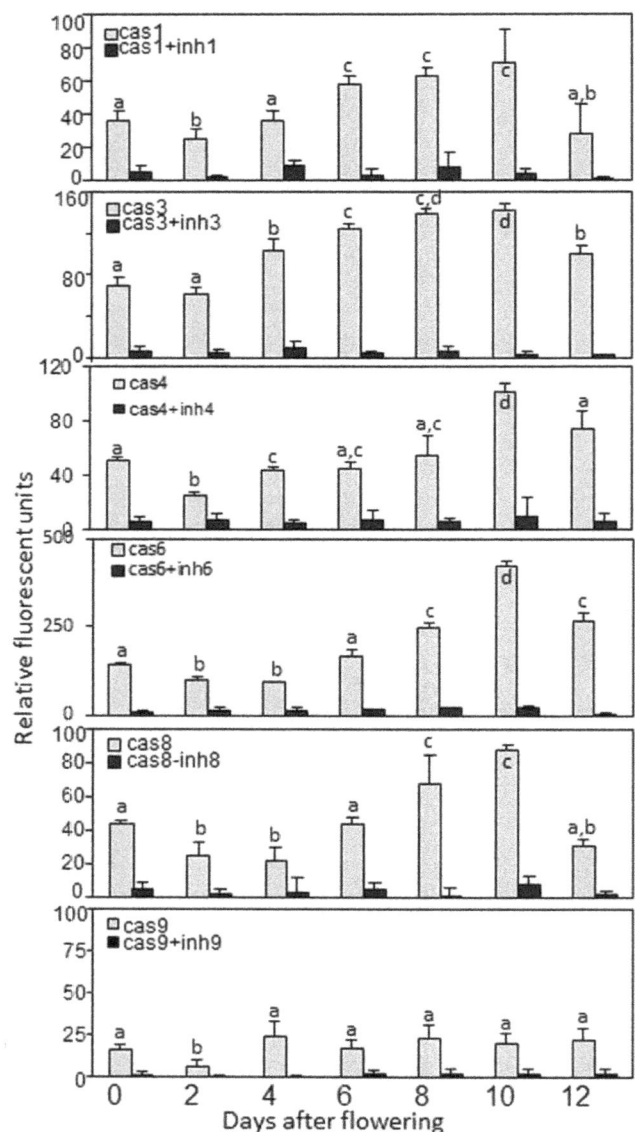

Figure 2. Caspase-like activities and effect of specific caspase inhibitors on corresponding caspase-like activity in barley pericarp. Data are means ± SD, $n = 4$, values followed by the same letter do not differ significantly at $p > 0.05$.

in activity was measured between anthesis and 12 DAF which quickly declined thereafter. The second increase in activity was observed during grain maturation starting from 20–22 DAF (Fig. 3). The second increase in caspase-1-like and caspase-6-like activities was not strongly pronounced (Fig. 3). The activities with caspase-9 substrate were barely detectable in the endosperm fraction throughout development except of grain maturation where the strong increase in the activity was detected after 20 DAF (Fig. 3). The specific caspase inhibitors showed inhibitory effects for either caspase-like activity during endosperm development.

To conclude, caspase-1-like, caspase-3-like, caspase-4-like, caspase-6-like and caspase-8-like activities were detected in the developing pericarp, all with increasing profiles towards 10 DAF.

Two increases in caspase-like activities were measured in the endosperm fraction with majority of caspase substrates. The increase in caspase-9-like activity was only detected during grain maturation.

Identification and expression analysis of genes with potential caspase-like activities in barley grains

We described recently seven genes encoding vacuolar processing enzyme (VPE) with potential caspase-1-like activity and found that *HvVPE4* is exclusively expressed in the deteriorating pericarp, *HvVPE2a* (and possibly *HvVPE2b–HvVPE2d*) transcripts are specific for nucellus/nucellar projection and *HvVPE1*

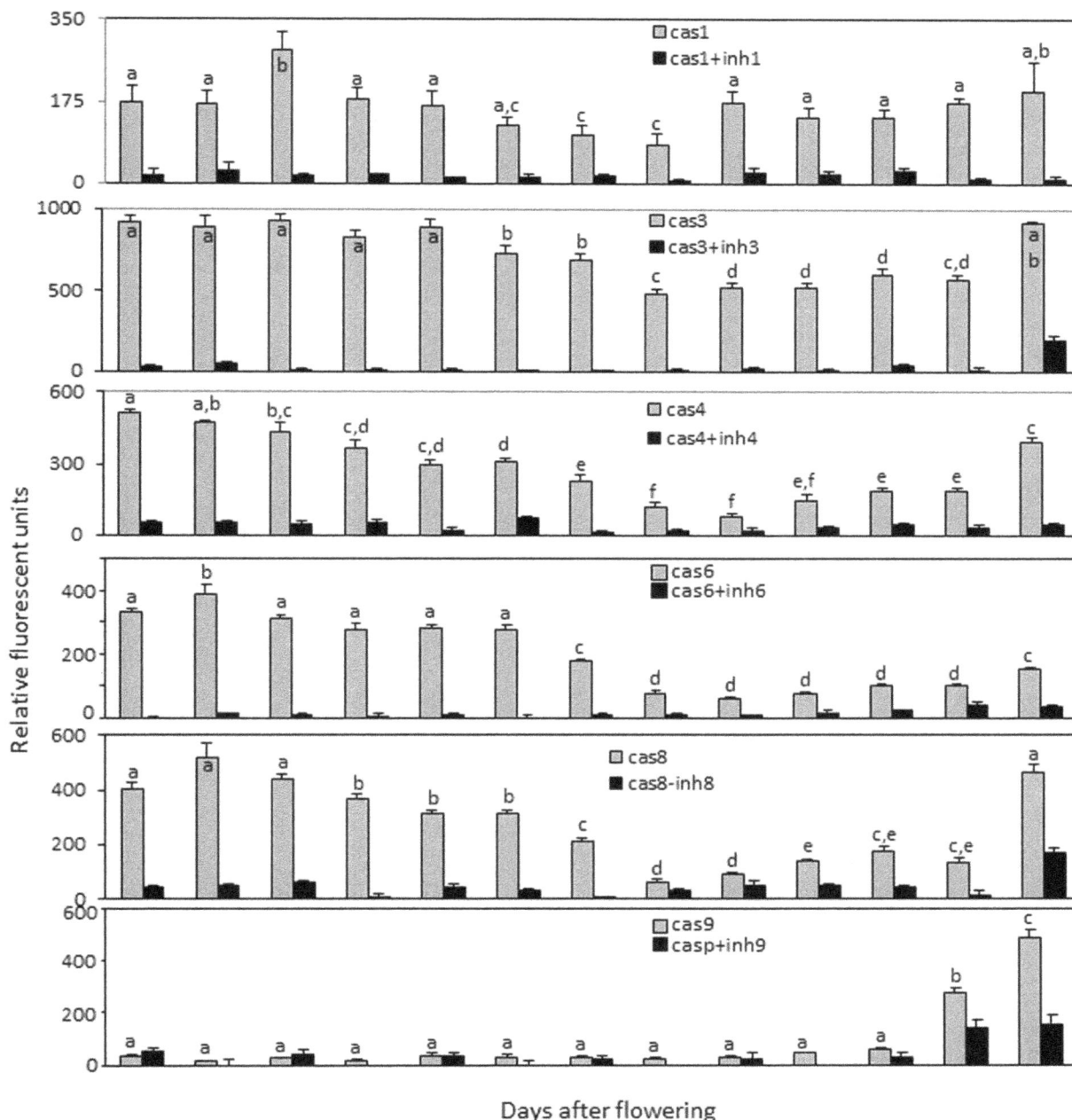

Figure 3. Caspase-like activities and effect of specific caspase inhibitors on corresponding caspase-like activity in the developing endosperm. Data are means ± SD, *n* = 4, values followed by the same letter do not differ significantly at *p* > 0.05.

is transcribes in late endosperm [12]. Here we analyzed barley genes encoding proteases with potential caspase-3 and caspase-6 activities: β1 and β2 subunits of the 20S proteasome with caspase-3 activity and phytaspase with caspase-6 activity.

In Arabidopsis, the 20S proteasome consists of seven α subunits encoded by 12 genes and seven β subunits encoded by 11 genes [35]. However, only β1 (*PBA* gene) and possibly β2 (*PBB* gene) subunits have been shown to exhibit caspase-3 like activity [3,28]. Therefore, we have searched the barley full length cDNA data base [32] for barley *PBA* and *PBB* genes using homologous poplar sequences [3] as queries. Two genes encoding the putative β1 subunit were found in the barley cDNA data base, the same number as found in poplar and rice while only one gene encodes PBA in Arabidopsis (Fig. 4A). HvPBA1 and HvPBA2 are almost identical at amino acid level (95.5% identity) and very similar to other known PBAs sharing 75.5–76.4% identity to AtPBA, 74.8–75.6% to both poplar PBA proteins and 85.7–89.5% identity to putative OsPBA1 and OsPBA2 sequences indicating that the genes encoding PBA are very conserved in plants.

Only one gene encoding β2 subunit was found in barley as well as in Arabidopsis, poplar and rice (Fig. 4A). The deduced HvPBB amino acid sequence is very similar to the other plant counterparts with an identity ranging from 75.7% (Arabidopsis) to 87.9% (rice).

Some members of huge family of subtilisin-like proteases have been shown to possess caspase-6-like activity [29] and are called phytaspases (PhS). Using rice phytaspase [29] as a reference, three

putative barley phytaspase genes have been selected from the full length cDNA database [32]. Barley phytaspases share 77.5–90.2% identity to each other and 68.0–73.7% identity to OsPhS at the amino acid level. The deduced HvPhS1–HvPhS3 proteins group together with tobacco and rice phytaspases and all belong to the subgroup 1 of the subtilisin-like proteases (Fig. 4B).

Gene expression patterns were determined in manually isolated pericarp and endosperm fractions of barley grains between anthesis and 24 DAF by quantitative reverse transcription PCR (qRT-PCR). Both *HvPBA1* and *HvPBA2* as well as *HvPBB* genes did not show developmental expression profile in the pericarp (Fig. 5A). In endosperm fraction, *HvPBA1* ubiquitously expressed while *HvPBA2* and *HvPB* display weak increase in transcription during grain filling (Fig. 5A). The *HvPHS1* gene was also ubiquitously expressed in the pericarp (Fig. 5B) while *HvPhS3* transcripts were barely detected in the tissue. Solely the *HvPhS2* transcripts accumulated in the pericarp with increasing abundance towards 10 DAF and declining afterwards (Fig. 5B). In the endosperm fraction, the *HvPhS1* transcripts were detected at low levels during early grain development but exhibited increase in expression during later grain filling phase starting from 16 DAF (Fig. 5B). The relative expression of *HvPhS2* was the highest in the endosperm fraction among the three barley phytaspases and peaks between 4 and 12 DAF decreasing thereafter (Fig. 5B). Contents of *HvPhS3* mRNA were low in the endosperm fraction

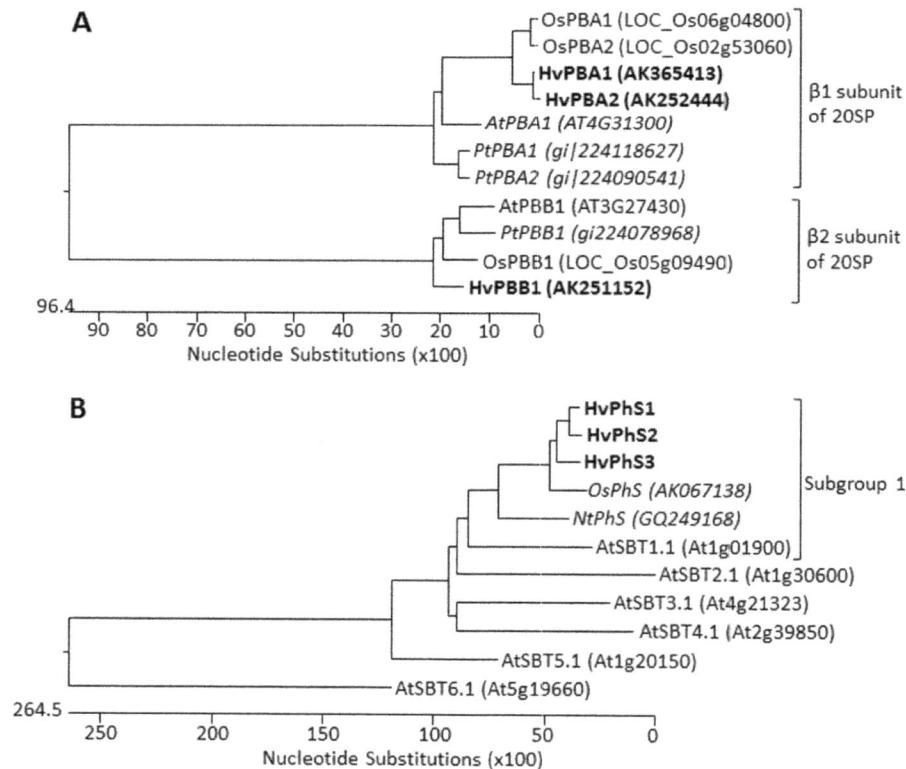

Figure 4. Phylogenetic trees of proteasome subunits PBA and PBB (A) and phytaspases (B) drawn with the ClustalW software. The horizontal scale represents the evolutionary distance expressed as a number of substitutions per amino acid. The putative phytaspases and proteasome subunits PBA1 of barley are shown in bolt. (A) Putative barley (Hv) proteasome subunits PBA and PBB are similar to the corresponding genes from *Arabidopsis thaliana* (At), *Populus tomentosa* (Pt) and *Oryza sativa* (Os). (B) Putative barley (Hv) phytaspases together with phytaspases from *Nicotiana tabacum* (Nt) and *Oryza sativa* (Os) belong to the subgroup 1 of subtilase-like proteases. The phytaspases with proven caspase-6 activity are shown in italic. Only one Arabidopsis subtilase-like protease per subgroup is shown (Rautengarten et al., 2008) in order to simplify the figure.

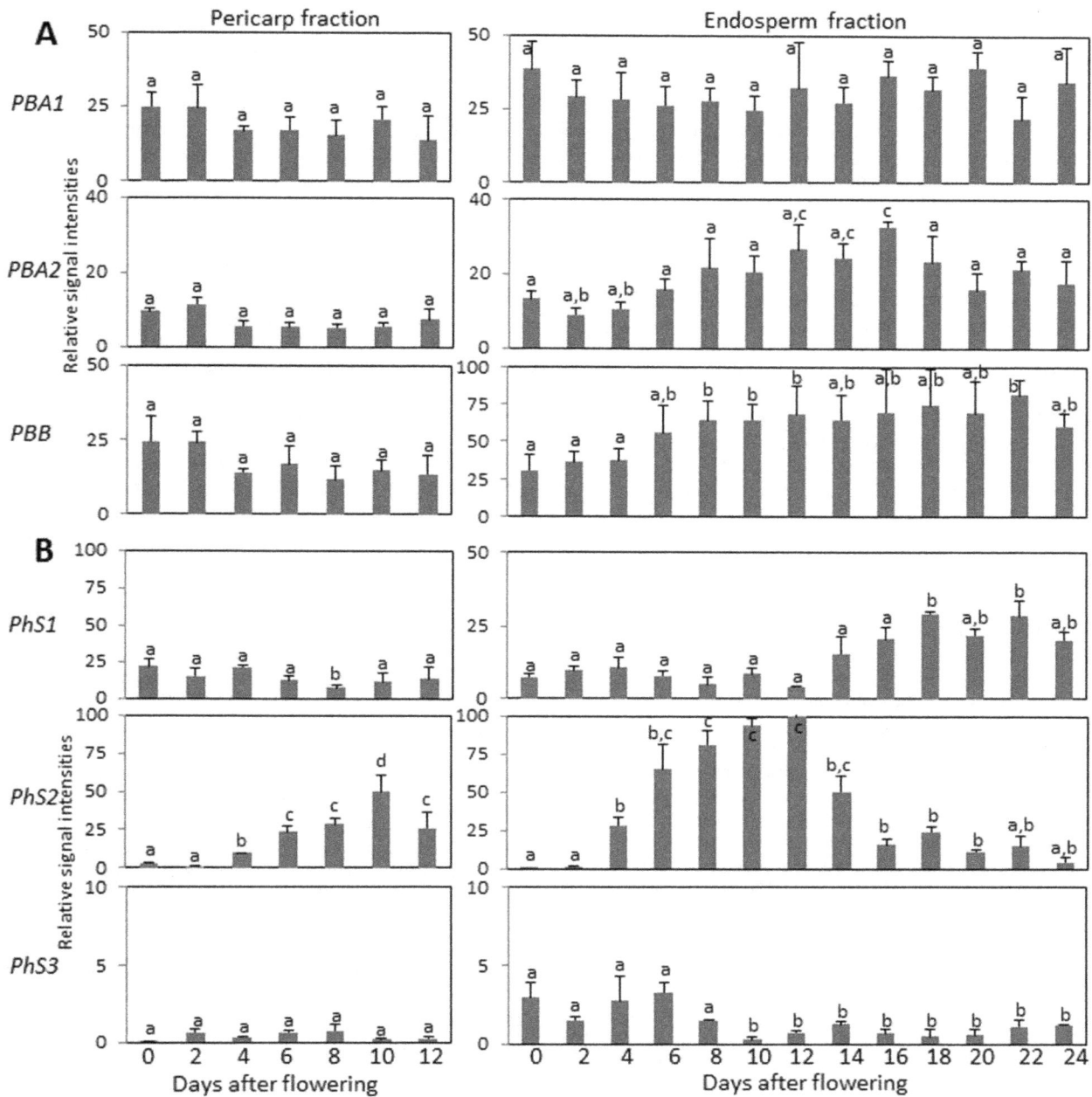

Figure 5. Transcript profiling of the proteasome subunits *PBA* and *PBB* (A) and phytaspase (*PhS*) genes (B) in pericarp (left) and endosperm fractions (right) of the developing barley grains determined by real-time quantitative RT-PCR analysis. Data are means ± SD, $n = 3$, values followed by the same letter do not differ significantly at $p > 0.05$.

throughout development however with weak increase at early developmental stages (Fig. 5B).

The pericarp fraction used for qRT-PCR encloses only maternal tissues consisting predominantly of mesocarp and epidermis (exocarp). Endosperm fraction, however, represents a complex sample consisting of filial and maternal tissues in changing proportions at different developmental stages and encompasses the filial endosperm itself, endosperm transfer cells, aleurone and embryo surrounding region but also maternal nucellus/nucellar projection and chlorenchyma (endocarp). Therefore, to study the tissue-specific gene expression profiles we used micro-dissected samples of these tissues from grains at

different developmental stages. Because the gene expression of the vacuole processing enzymes *HvVPE2b–HvVPE2d* in nucellus/ nucellar projection was not experimentally proven [12], we analyzed also their transcript abundances in micro-dissected tissues. As expected, the expression of *HvVPE2a*, *HvVPE2b* and *HvVPE2d* was found exclusively in nucellus and nucellar projection with a maximum between 7 and 10 DAF (Fig. 6A). *HvVPE2b* gene activity was the highest among all VPEs expressed in these tissues. Accumulation of *HvVPE2a* transcripts was two-fold lower as of *HvVPE2b*, and transcript levels of *HvVPE2d* reached only one tenth of those for *HvVPE2b*. Expression of *HvVPE2c* was detected only at basic level (less than 1% of

HvVPE2b) in all studied tissues (Fig. 6A) confirming previous data [12]. Expression of *HvPBA1*, *HvPBA2* and *HvPBB* genes was detected in all micro-dissected tissues analyzed showing neither preference for any tissue nor characteristic developmental profile (Fig. 6B). Among the three phytaspase genes, *HvPhS2* expression was the highest and found exclusively in the nucellar projection depicting a maximum of expression at 10 DAF (Fig. 6C). *HvPhS3* transcripts were also detected specifically in the nucellar projection peaking around 10 DAF albeit at lower expression level (Fig. 6C). Expression of *HvPhS1* was observed at relatively low levels in all analyzed micro-dissected tissues without clear developmental profile (Fig. 6C).

To conclude, the *HvPBA1*, *HvPBA2*, *HvPBB*, *HvPhS1* and *HvPhS3* are expressed without certain developmental patterns in pericarp. The expression of *HvPhS2* is increased at later stages of pericarp development. Abundance of *HvPBA2* and *HvPBB* mRNAs are weakly increased during grain filling. The transcripts of *HvVPE2a*, *HvVPE2b*, *HvVPE2d*, *HvPhS2* and *HvPhS3* are detected exclusively in the nucellar tissues of the developing barley grains. The *HvPhS1* mRNA abundances increase in the maturing endosperm.

Discussion

Programmed cell death (PCD) is an essential part of the life of any multicellular organism. PCD plays a crucial role in tissue and organ development and in the maintenance of the cellular homeostasis of a tissue. In this work we analyzed PCD events in the developing barley caryopsis. Activation of caspases is a hallmark of apoptosis and inflammatory response in animals [17,36]. Caspase-like activities become also markedly enhanced upon induction of PCD in plants [19,23]. In both pericarp and endosperm fractions of developing grains, distinct caspase-like activities showed similar profiles albeit their relative activity levels were different. Activities with all tested caspase substrates excluding LEHD (caspase-9 substrate) increase during pericarp development (Fig. 2) coinciding with the abundance of TUNEL-positive nuclei (Fig. 1) and degradation of the pericarp tissue [12]. Increased activities with almost all caspase substrates except of caspase-9 were detected in the endosperm fraction during early development (Fig. 3). The second increase in all caspase-like activities including caspase-9-like was found during grain maturation (Fig. 3). Based on these observations we tend to conclude that coaction of caspase-like protease activities may execute and regulate PCD processes in plant tissues similar to that occurring in animal cells [17,37]. In animals, the caspases are classified into inflammatory, apoptotic initiator and apoptotic effector groups [36]. The latter group is processed and activated by upstream caspases and performs downstream steps cleaving multiple cellular substrates. The effector caspases are usually more abundant and active than initiator caspases [36]. In the barley grains, caspase-6-like activity is highest in both pericarp and early endosperm fractions followed by the caspase-3-like activity (both effector activities in animals) while caspase-8-like and especially caspase-9-like activities were substantially lower (Figs. 2, 3). Referring to the animal model, it is tempting to speculate that proteases with caspase-6-like and caspase-3-like activities may fulfill effector role in plant PCD while proteases with caspase-8-like and caspase-9-like activities are PCD initiators. Caspase-like proteases executing PCD may differ among distinct plant tissues. For instance, the caspase cascade in the pericarp and early endosperm fractions may not include caspase-9-like activity, because the latter has been barely measurable in these tissues. In contrast, the potential caspase coaction in the maturing endosperm includes caspase-9-

like activity but caspase-6-like activity may play minor role (Fig. 3). The possible coaction of proteases with caspase-like activities in acquisition and execution of plant PCD needs further experimental confirmation.

We detected the caspase-4-like activity in plants for the first time. Its patterns of activity in both pericarp and endosperm (Figs. 2, 3) coincide with the degeneration processes in the respective tissue (Fig. 1). The specific protease inhibitor could strongly inhibit the caspase-4-like activity. The specific protease responsible for the newly detected caspase-4-like activity remains to be detected.

The expression of the *HvVPE* and *HvPhS* genes largely coincides with PCD of the respective tissue (Fig. 7) (see also below). However, none of the genes encoding β1 or β2 subunits of the 20S proteasome shows a specific developmental profile (Figs. 5, 6). However, caspase-3-like activity, potentially mediated by the corresponding proteins [3,28], displays developmental pattern of the activity in barley grains (Figs. 2, 3). It might be possible that plant β1 or β2 subunits of 20S proteasome subunit are post-translationally regulated to control the PCD. The 20S proteasome subunit as part of the ubiquitin/26S proteasome complex plays a role in nearly all processes of plant development by selectively eliminating regulatory proteins [41] and, therefore, its activity has to be fine controlled. It is also possible that other proteases display caspase-3-like activity in barley grains.

PCD processes in the distinct grain tissues are summarized in Fig. 7 and discussed below in more details.

Programmed cell death in the nucellus and nucellar projection

Nucellus is the first tissue undergoing PCD after beginning of caryopsis development (besides the antipodals and synergid cells, which however belong to gametophyte). The first TUNEL-labeled nuclei are visible at the margins of nucellus facing developing endosperm very soon after fertilization in both barley (Fig. S1) and wheat [10]. With the endosperm growth, PCD in nucellus expands to outward cell layers finally resulting in complete disappearing of the nucellus till 4–5 DAF except the cells adjacent to the main vascular bundle, which develop to nucellar projection [12,14]. The nucellar projection together with the opposite endosperm transfer cells operate as a main conduit for nutrient supply from the main vascular bundle to endosperm [11,13]. The first TUNEL-positive nuclei appear at margins of the nucellar projection facing the endosperm transfer cells around 6 DAF (Fig. 1). Thereafter the degenerating nuclei at margins of the nucellar projection are detectable till late grain maturation (Fig. 1). Permanent cell turnover seems to occur in the nucellar projection. New cells are produced in the mitotic region, then cells elongate, produce thick cell walls [41] and become functionally active before they degenerate and thereby direct cell content and cell remnants into the apoplastic space. This mechanism of nutrient delivery is not fully understood, despite its importance for endosperm filling and grain yield [42]. There are no symplastic connections between nucellar projection and endosperm transfer cells, and nutrient transport across maternal/filial border occurs apoplastically [41]. As deduced from thick cell walls of elongating cells [41] and expression of many transporters [14], the nutrient transfer through nucellar projection involves both symplastic and apoplastic pathways and evidently requires PCD at the site of nucellar projection [11,14]. Disruption of PCD in nucellar tissues affects endosperm development and grain weight in barley [11] and rice [43].

Because hand-isolated endosperm fraction always includes nucellar projection [12,34], the first increase in almost all caspase

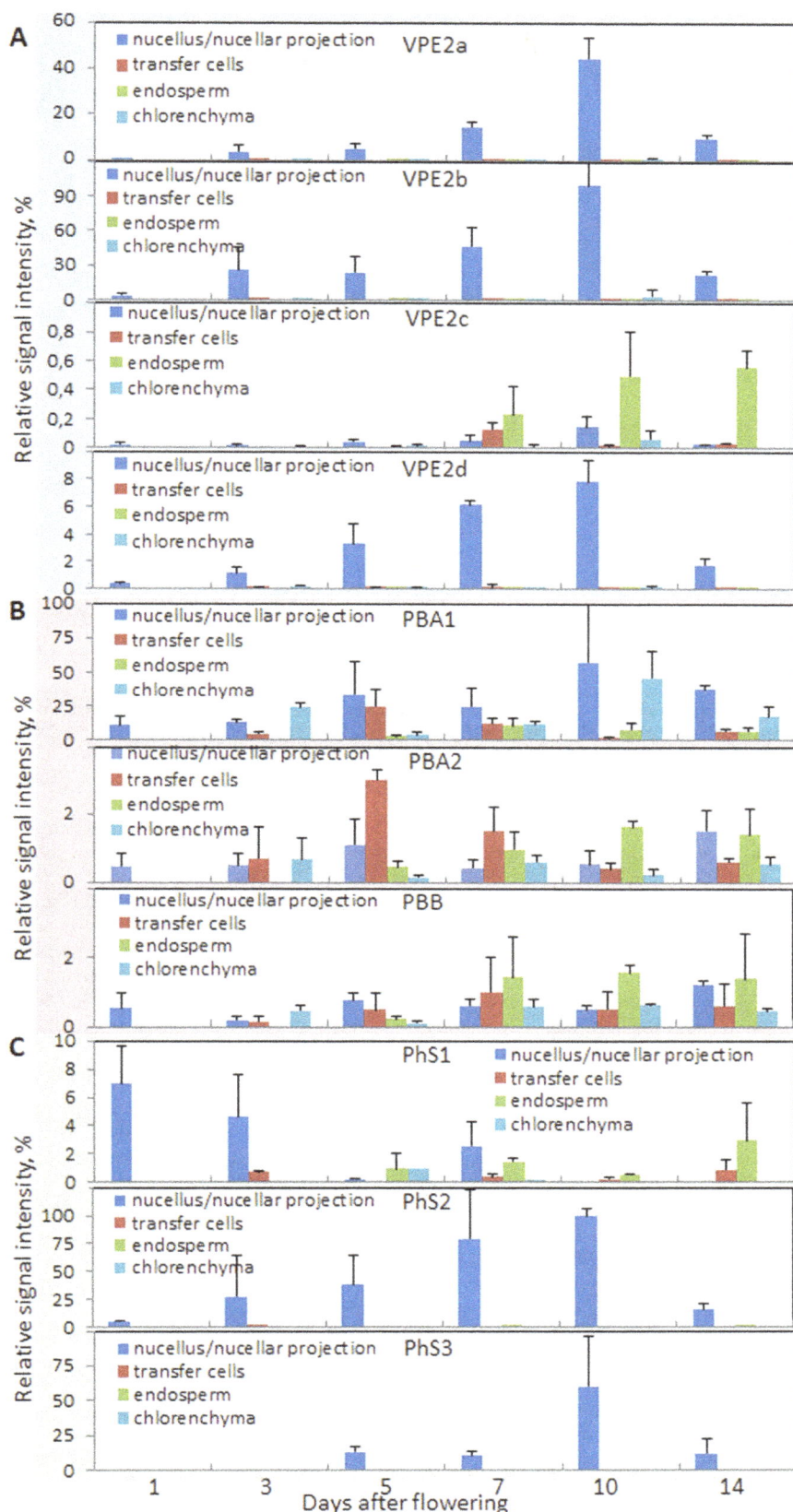

Figure 6. Expression profiles of the vacuolar processing enzymes VPE2a-VPE2d (A), proteasome subunits PBA and PBB (B) and phytaspase (PhS) genes (C) in the different tissues micro-dissected from the developing barley grains.

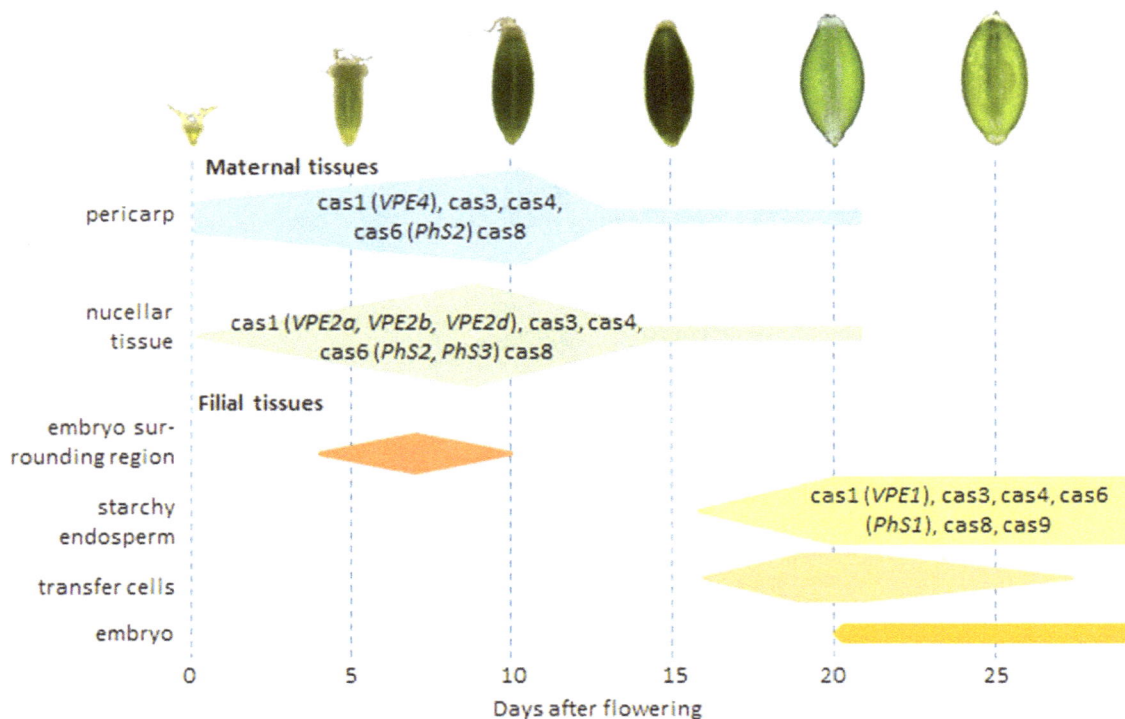

Figure 7. Scheme illustrating programmed cell death processes together with potentially involved activities and genes (in brackets) in distinct tissues of the developing barley grains. Activities: cas1, caspase-1-like; cas3, caspase-3-like; cas4, caspase-4-like; cas6, caspase-6-like; cas8, caspase-8-like; cas9, caspase-9-like. Genes: VPE, vacuolar processing enzyme; PhS, phytaspase.

activities in the endosperm fraction (Fig. 3) may be related at least in part to PCD of nucellus and nucellar projection. The increase in caspase-1-like activity may be acquired by HvVPE2a, HvVPE2b and HvVPE2d proteases which exclusively expressed in nucellus and nucellar projection (Fig. 6). The caspase-1-like activity for HvVPE2b (HvLeg2) has been already proven [27]. The expression patterns of the *HvPhS2* and *HvPhS3*, which are exclusively active in the nucellar projection (Fig. 6), coincide to the caspase-6-like activity profile in the early endosperm fraction (Fig. 3) indicating that HvPhS1 and HvPhS2 may be responsible for the caspase-6-like activity.

Programmed cell death in pericarp

After nucellus degeneration, the endosperm enlarges by cost of pericarp cells which undergo PCD starting from the innermost cell layer of mesocarp between 4 and 5 DAF, as seen from distribution of TUNEL-positive nuclei [12]. The lateral and dorsal parts of the mesocarp disintegrate already till 10–12 DAF (Fig. 1E) whereas the ventral region around the main vascular bundle persists undergoing later a gradual degeneration until grain maturation (Fig. 1J). The green and photosynthetically active chlorenchyma layer [44] however does not show any TUNEL-positive signals during observation time (Fig. 1). Probably this layer disintegrates during desiccation when the maturing grain turns from green to yellow. Obviously, the chlorenchyma plays important role for caryopsis development. Perception of light by photosynthetically active seed layer is thought to represent a strategy to sense environment and provide a means of tuning grain metabolism according to the changing conditions [45].

Coinciding with PCD progression in the pericarp, we detected increase of the caspase-1-like, caspase-3-like, caspase-4-like, caspase-6-like and caspase-8-like but not caspase-9-like activities

towards 10 DAF and their decline thereafter (Fig. 2). The transcript profile of early described mesocarp-expressed *HvVPE4* gene [12] coincides with pattern of caspase-1-like activity (Fig. 2) further supporting that HvVPE4 may be responsible for the activity. The profile of caspase-6-like activity (Fig. 2), expression of *HvPhS2* gene (Fig. 5) and the patterns of TUNEL-positive nuclei (Fig. 1) also coincide indicating that HvPhS2 may be involved in PCD as protease with caspase-6-like activity in the pericarp.

Programmed cell death in the endosperm

Early endosperm develops by divisions of nuclei without cytokinesis resulting in the endosperm coenocyte [46]. Coencyte begins to cellularize around 4 DAF in the embryo surrounding region (ESR) [46,47]. Transfer cell layer is also formed at this time [33]. TUNEL-labeling nuclei in the endosperm are absent between anthesis and 6 DAF indicating that cell degradation processes do not occur during early endosperm development. No genes potentially encoding proteases with caspase-1-like and caspase-6-like activities are expressed in the early developing endosperm (Fig. 6). Transfer cells are also free of the corresponding transcripts (Fig. 6). Therefore, the increase in almost all caspase-like activities in early developing endosperm fraction (Fig. 3) is likely not due to PCD processes in endosperm but may be related to PCD in the nucellar tissues as described above. The first degenerating nuclei appeared in cellularized ESR already 6 DAF (Fig. 1C). Therefore, the high value of caspase-like activities between 4 and 12 DAF in the endosperm fraction may be at least in part correspond to PCD in ESR as well.

PCD of ESR in maize and wheat has been described at histological level many years ago [48,49]. Here, we document nuclei degradation in barley ESR shortly after cellularization starting from the cells facing the embryo (Fig. 1C, F). The ESR can be subdivided into three different regions distinguished by vacuole size and degree of cellular vacuolization [47]. The highly vacuolated cells facing the embryo degrade firstly followed by the deeper cell layers. This pattern is reminiscent to that of the nucellar projection [14] where the degrading cells at the margins contribute to nutrient transfer to the endosperm [11]. In analogy, we hypothesize here that PCD of ESR cells is important for the nutrient supply to the embryo releasing cell contents and cell remnants into liquid-filled embryonic space. Besides, PCD of ESR provides space for the growing embryo. In the embryoless mutants of maize, the endosperm develops a normal-sized embryo cavity suggesting the existence of an intrinsic program for ESR formation independent from embryo development [51]. PCD of the ESR may be a part of such a program. The nuclei of pericarp cells surrounding the embryo from maternal side are also strongly labeled in TUNEL assay. The degradation of the pericarp mainly occurs in the layer adjacent to the embryo and endosperm (Fig. 1C, F, I). We suppose that growing embryo requires space not only of degenerating ESR but also from the maternal pericarp. It is rather unclear whether degrading pericarp cells also contribute to nutrient delivery to the embryo. The direct nutrient supply from pericarp to the embryo can be anticipated, because nucellar projection and transfer cells are still not developed in the embryo region at this developmental stage (Fig. 1F) [50].

With the establishment of transfer cells and endosperm cellularization, the endosperm serves for accumulation of storage compounds. Highly energetic biosynthesis of starch and storage proteins requires intact and metabolically active cells which have be able to convert large amounts of metabolites into storage compounds. This might be reflected in general decrease of caspase-like activities during main filling phase (10–18 DAF) and absence/low expression of related proteases (Figs. 3, 5). With the decline of storage synthesis, the endosperm cells of maize, wheat and rice undergo PCD [15,16,52]. The numerous TUNEL-positive nuclei in starchy endosperm of barley grains are visible starting from 16 DAF (Fig. 1J–N). The expression of *HvVPE1* [12] and *HvPhS1* (Fig. 5) are increased during seed maturation coinciding with the increase of caspase-1-like and caspase-6-like activities (Fig. 3). It is tempting to speculate these phytaspase and vacuolar processing enzyme are responsible for corresponding activities in maturating endosperm and required for its PCD. A second increase of caspase-4-like and caspase-8-like activities and unique increase in caspase-9-like activity have been also detected during grain maturation (Fig. 3) albeit the corresponding proteases are still unknown.

Some nuclei of the transfer cell layer are also labeled in TUNEL assay at 16 DAF (Fig. 1K). At 18 DAF, almost all nuclei of the transfer cells are positive in the TUNEL assay, indicating massive cellular disintegration. The transfer cells disintegrate after completion of storage product accumulation and thereby interrupt the delivery of nutrients to the starchy endosperm. Such breakdown of metabolite flow may serve as a signal to endosperm cells for switching from storage product accumulation to maturation and grain desiccation.

Detection of PCD in the developing embryo

The zygote developing to the embryo starts to divide later than the fertilized central cell which gives raise to the endosperm. Following cell divisions in the embryo are slower than as in syncytial endosperm. No nuclei degradation in the embryo was

detected during early development (6–10 DAF; Fig. 1C, F, I). However, almost all nuclei in two cell layers of the scutellum and occasional nuclei in other parts of the embryo were TUNEL-positive at 18 DAF (Fig. 1N), indicating massive tissue reorganization during embryo maturation. It is well known that the scutellum is the last grain tissues undergoing PCD in course of germination after accomplishing the supply of nutrients from the starchy endosperm to the growing embryo [53]. Cell disintegration during embryo development in dicots plants is also a well described phenomenon. After the first division of the zygote, the apical daughter cell gives rise to the embryo proper, while the basal cell develops into the suspensor. The latter is a terminally differentiated structure that is removed by PCD [54]. We have detected for the first time the cell disintegrative processes in the late developing embryo and scutellum of grasses. It is possible that such cell disintegration is a result of scutellum reorganization from supporting tissue for developing embryo to feeding tissues for growing embryo during germinating. The molecular mechanisms responsible for PCD in the late embryo as well as its role in embryo development remain to be studied.

To conclude, the spatial and temporal distribution of the TUNEL-positive nuclei suggests that each seed tissue follows individual pattern of development and disintegration, which however harmonizes with growth of the other tissues in order to achieve proper caryopsis development. In analogy to animal system, programmed cell death in the developing barley caryopsis may require a coaction of caspase-like activities. Expression of distinct genes encoding vacuolar processing enzyme and phytaspase largely coincides with caspase-1-like and caspase-6-like activities in the respective tissue and may be responsible for either caspase activity. However, all above assumptions require experimental confirmations. Due to striking similarity of grain development in barley and wheat as well as in other small grain crops, the results and conclusions about PCD in the barley grains may have impact on research of other important cereal crops.

Supporting Information

Figure S1   Negative (A–C) and positive controls (D–F) of TUNEL assay, and standard TUNEL assay performed at 10 DAF (G–H) as well as the localization of nuclear DNA fragmentation detected by the TUNEL assay at 1 (J), 3 (K), 5 DAF (L).
(TIF)

Table S1   Primers used in real-time RT-PCR analyses.
(DOCX)

Acknowledgments

We acknowledge Angela Stegmann, Elsa Fessel, and Uta Siebert for excellent technical assistance. We wish to thank Judith Schmeichel for help with preparation of cDNA libraries from micro-dissected tissues and Winfriede Weschke for critical reading of the manuscript.

Author Contributions

Conceived and designed the experiments: VR. Analyzed the data: VR RR. Contributed to the writing of the manuscript: VR. Performed TUNEL assay: DW. Performed caspase assays: VT RR VR. Performed micro-dissections and cDNA libraries of them: JT. Performed qRT-PCR analyses: RR VR.

References

1. Sabelli PA (2012) Replicate and die for your own good: endoreduplication and cell death in the cereal endosperm. J Cereal Sci 56: 9–20.

2. Visser EJ, Bogemann GM (2006) Aerenchyma formation in the wetland plant Juncus effusus is independent of ethylene. New Phytol 171: 305–314.

3. Han JJ, Lin W, Oda Y, Cui KM, Fukuda H, et al. (2012) The proteasome is responsible for caspase-3-like activity during xylem development. Plant J 72: 129–141.

4. Lucas WJ, Groover A, Lichtenberger L, Furuta K, Yadav SR, et al. (2013) The plant vascular system: evolution, development and functions. J Integr Plant Biol 55: 4.

5. Gunawardena AHLAN (2008) Programmed cell death and tissue remodelling n plants. J Exp Bot 59: 445–451.

6. Della Mea A, Serafini-Fracassini D, Del Duca S (2007) Programmed cell death: similarity and differences in animals and plants. A flower paradigm. Amino Acids 33: 395–404.

7. Lim PO, Hyo JK, Nam HG (2007) Leaf senescence. Annu Rev Plant Biol 58: 115–136.

8. Coll NS, Epple P, Dangl JL (2011) Programmed cell death in the plant immune system. Cell Death Differ 18: 1247–1256.

9. Xiong F, Yu XR, Zhou L, Wang F, Xiong AS (2013) Structural and physiological characterization during wheat pericarp development. Plant Cell Rep 32: 1309–1320.

10. Dominguez F, Moreno J, Javier Cejudo F (2001) The nucellus degenerates by a process of programmed cell death during the early stages of wheat grain development. Planta 213: 352–360.

11. Radchuk V, Borisjuk L, Radchuk R, Steinbiss HH, Rolletschek H, et al. (2006) Jekyll encodes a novel protein involved in the sexual reproduction of barley. Plant Cell 18: 1652–1666.

12. Radchuk V, Weier D, Radchuk R, Weschke W, Weber H (2011) Development of maternal seed tissue in barley is mediated by regulated cell expansion and cell disintegration and coordinated with endosperm growth. J Exp Bot 62: 1217–1227.

13. Hands P, Kourmpetli S, Sharples D, Harris RG, Drea S (2012) Analysis of grain characters in temperate grasses reveals distinctive patterns of endosperm organization associated with grain shape. J Exp Bot 63: 6253–6266.

14. Thiel J, Weier D, Sreenivasulu N, Strickert M, Weichert N, et al. (2008) Different hormonal regulation of cellular differentiation and function in nucellar projection and endosperm transfer cells – a microdissection-based transcriptome study of young barley grains. Plant Physiol 148: 1436–1452.

15. Young TE, Gallie DR (2000) Programmed cell death during endosperm development. Plant Mol Biol 44: 283–301.

16. Nguyen HN, Sabelli PA, Larkins BA (2007) Endoreduplication and programmed cell death in the cereal endosperm. In: Olsen OA (ed) Plant cell monographs 8: Endosperm. Springer, Berlin, Heidelberg, pp. 21–43.

17. Degterev A, Boyce M, Yuan J (2003) A decade of caspases. Oncogene 22: 8543–8567.

18. Ho P, Hawkins CJ (2005) Mammalian initiator apoptotic caspases. FEBS J 272: 5436–5453.

19. Woltering EJ, van der Bent A, Hoeberichts FA (2002) Do plant caspases exist? Plant Physiol 130: 1764–1769.

20. Lombardi L, Casani S, Ceccarelli N, Galleschi L, Picciarelli P, et al. (2007) Programmed cell death of the nucellus during Sechium edule Sw. seed development is associated with activation of caspase-like proteases. J Exp Bot 58: 2949–2958.

21. Borén M, Höglund AS, Bozhkov P, Jansson C (2006) Developmental regulation of a VEIDase caspase-like proteolytic activity in barley caryopsis. J Exp Bot 57: 3747–3753.

22. Coll NS, Vercammen D, Smidler A, Clover C, Van Breusegem F, et al. (2010) Arabidopsis type I metacaspases control cell death. Science 330: 1393–1397.

23. Bonneau L, Ge Y, Drury GE, Gallois P (2008) What happened to plant caspases? J Exp Bot 59: 491–499.

24. Vercammen D, Declercq W, Vandenabeele P, Van Breusegem F (2007) Are metacaspases caspases? J Cell Biol 179: 375–380.

25. Nakaune S, Yamada K, Kondo M, Kato T, Tabata S, et al. (2005) A vacuolar processing enzyme, δVPE, is involved in seed coat formation at the early stage of seed development. Plant Cell 17: 876–887.

26. Hara-Nishimura I, Hatsugai N (2011) The role of vacuole in plant cell death. Cell Death Differ 18: 1298–1304.

27. Julián I, Gandullo J, Santos-Silva LK, Diaz I, Martinez M (2013) Phylogenetically distant barley legumains have a role in both seed and vegetative tissues. J Exp Bot 64: 2929–2941.

28. Hatsugai N, Iwasaki S, Tamura K, Kondo M, Fuji K, et al. (2009) A novel membrane fusion-mediated plant immunity against bacterial pathogens. Genes Dev 23: 2496–2506.

29. Chichkova NV, Shaw J, Galiullina RA, Drury GE, Tuzhikov AI, et al. (2010) Phytaspase, a relocalisable cell death promoting plant protease with caspase specificity. EMBO J 29: 1149–1161.

30. Coffeen WC, Wolpert TJ (2004) Purification and characterization of serine proteases that exhibit caspase-like activity and are associated with programmed cell death in Avena sativa. Plant Cell 16: 857–873.

31. Cai YM, Yu J, Gallois P (2014) Endoplasmic reticulum stress-induced PCD and caspase-like activities involved. Front Plant Sci 5: 41.

32. Matsumoto T, Tanaka T, Sakai H, Amano N, Kanamori H, et al. (2011) Comprehensive sequence analysis of 24,783 barley full-length cDNAs derived from 12 clone libraries. Plant Physiol 156: 20–28.

33. Thiel J, Hollmann J, Rutten T, Weber H, Scholz U, et al. (2012) 454 Transcriptome sequencing suggests a role for two-component signalling in cellularization and differentiation of barley endosperm transfer cells. PLoS One 7: e41867.

34. Radchuk V, Borisjuk L, Sreenivasulu N, Merx K, Mock HP, et al. (2009) Spatio-temporal profiling of starch biosynthesis and degradation in the developing barley grain. Plant Physiol 150: 190–204.

35. Kurepa J, Smalle JA (2008) Structure, function and regulation of plant proteasomes. Biochimie 90: 324–335.

36. Jin Z, El-Deiry WS (2005) Overview of cell death signaling pathways. Cancer Biol Ther 4: 139–163.

37. Slee EA, Harte MT, Kluck RM, Wolf BB, Casiano CA et al. (1999) Ordering the cytochrome c–initiated caspase cascade: hierarchical activation of caspases-2, -3, -6, -7, -8, and -10 in a caspase-9–dependent manner. J Cell Biol 144: 281–292.

38. Vartapetian AB, Tuzhikov AI, Chichkova NV, Taliansky M, Wolpert TJ (2011) A plant alternative to animal caspases: subtilisin-like proteases. Cell Death Differ 18: 1289–1297.

39. Linnestad C, Doan DN, Brown RC, Lemmon BE, Meyer DJ, et al. (1998) Nucellain, a barley homolog of the dicot vacuolar-processing protease, is localized in nucellar cell walls. Plant Physiol 118: 1169–1180.

40. Vierstra RD (2009) The ubiquitin-26S proteasome system at the nexus of plant biology. Nat Rev Mol Cell Biol 10: 385–397.

41. Weschke W, Panitz R, Sauer N, Wang Q, Neubohn B, et al. (2000) Sucrose transport into barley seeds: molecular characterization of two transporters and implications for seed development and starch accumulation. Plant J 21: 455–467.

42. Offler CE, McCurdy DW, Patrick JW, Talbot MJ (2003) Transfer cells: cells specialized for a special purpose. Annu Rev Plant Biol 54: 431–454.

43. Yin LL, Xue HW (2012) The MADS29 transcription factor regulates the degradation of the nucellus and the nucellar projection during rice seed development. Plant Cell 24: 1049–106.

44. Rolletschek H, Weschke W, Weber H, Wobus U, Borisjuk L (2004) Energy state and its control on seed development: starch accumulation is associated with high ATP and steep oxygen gradients within barley grains. J Exp Bot 55: 1351–1359.

45. Borisjuk L, Radchuk V (2014) Seed coat functions with unknown variables. Front Plant Sci (submitted).

46. Olsen OA (2004) Nuclear endosperm development in cereals and Arabidopsis thaliana. Plant Cell 16: S214–S227.

47. Engell K (1989) Embryology of barley: time course and analysis of controlled fertilisation and early embryo formation based on serial sections. Nord J Bot 9: 265–280.

48. Kiesselbach TA, Walker ER (1952) Structure of certain specialized tissues in the kernel of corn. Am J Bot 39: 561–569.

49. Smart MG, O'Brien TP (1983) The development of the wheat embryo in relation to the neighboring tissues. Protoplasma 114: 1–13.

50. Gubatz S, Derksen VJ, Brüß C, Weschke W, Wobus U (2007) Analysis of barley (Hordeum vulgare) grain development using three-dimensional digital models. Plant J. 52: 779–790.

51. Heckel T, Werner K, Sheridan WF, Dumas C, Rogowsky PM (1999) Novel phenotypes and developmental arrest in early embryo specific mutants of maize. Planta 210: 1–8.

52. Kobayashi H, Ikeda TM, Nagata K (2013) Spatial and temporal progress of programmed cell death in the developing starchy endosperm of rice. Planta 237: 1393–1400.

53. Dominguez F, Moreno J, Javier Cejudo F (2012) The scutellum of germinated wheat grains undergoes programmed cell death: identification of an acidic nuclease involved in nucleus dismantling. J Exp Bot 63: 5475–5485.

54. Smertenko A, Bozhkov PV (2014) Somatic embryogenesis: life and death processes during apical-basal patterning. J Exp Bot 65: 1343–1360.

55. Solís MT, Chakrabarti N, Corredor E, Cortés-Eslava J, Rodríguez-Serrano M, et al. (2014) Epigenetic changes accompany developmental programmed cell death in tapetum cells. Plant Cell Physiol 55: 16–29.

56. Rodríguez-Serrano M, Bárány I, Prem D, Coronado MJ, Risueño MC, et al. (2012) NO, ROS, and cell death associated with caspase-like activity increase in stress-induced microspore embryogenesis of barley. J Exp Bot 63: 2007–2024.

57. Kraus D (2014) Daniel's XL Toolbox addin for Excel, version 6.10. http://xltoolbox.sourceforge.net.

10

New Starch Phenotypes Produced by TILLING in Barley

Francesca Sparla[1], Giuseppe Falini[2], Ermelinda Botticella[3], Claudia Pirone[1], Valentina Talamè[4], Riccardo Bovina[4], Silvio Salvi[4], Roberto Tuberosa[4], Francesco Sestili[3], Paolo Trost[1]*

1 Department of Pharmacy and Biotechnology FABIT, University of Bologna, Bologna, Italy, 2 Department of Chemistry "G. Ciamician", University of Bologna, Bologna, Italy, 3 Department of Agriculture, Forestry, Nature & Energy, University of Tuscia, Viterbo, Italy, 4 Department of Agricultural Sciences, University of Bologna, Bologna, Italy

Abstract

Barley grain starch is formed by amylose and amylopectin in a 1:3 ratio, and is packed into granules of different dimensions. The distribution of granule dimension is bimodal, with a majority of small spherical B-granules and a smaller amount of large discoidal A-granules containing the majority of the starch. Starch granules are semi-crystalline structures with characteristic X-ray diffraction patterns. Distinct features of starch granules are controlled by different enzymes and are relevant for nutritional value or industrial applications. Here, the Targeting-Induced Local Lesions IN Genomes (TILLING) approach was applied on the barley TILLMore TILLING population to identify 29 new alleles in five genes related to starch metabolism known to be expressed in the endosperm during grain filling: BMY1 (Beta-amylase 1), GBSSI (Granule Bound Starch Synthase I), LDA1 (Limit Dextrinase 1), SSI (Starch Synthase I), SSIIa (Starch Synthase IIa). Reserve starch of nine M3 mutant lines carrying missense or nonsense mutations was analysed for granule size, crystallinity and amylose/amylopectin content. Seven mutant lines presented starches with different features in respect to the wild-type: (i) a mutant line with a missense mutation in GBSSI showed a 4-fold reduced amylose/amylopectin ratio; (ii) a missense mutations in SSI resulted in 2-fold increase in A:B granule ratio; (iii) a nonsense mutation in SSIIa was associated with shrunken seeds with a 2-fold increased amylose/amylopectin ratio and different type of crystal packing in the granule; (iv) the remaining four missense mutations suggested a role of LDA1 in granule initiation, and of SSIIa in determining the size of A-granules. We demonstrate the feasibility of the TILLING approach to identify new alleles in genes related to starch metabolism in barley. Based on their novel physicochemical properties, some of the identified new mutations may have nutritional and/or industrial applications.

Editor: Joerg Fettke, University of Potsdam, Germany

Funding: Work funded by The University of Bologna, Progetto Strategico STARCHitecture (F. Sparla, GF, VT, RB, SS, RT, and PT). The funders had no role in study design, data collection and analysis, decision to publish, or preparation of the manuscript.

Competing Interests: The authors have declared that no competing interests exist.

* Email: paolo.trost@unibo.it

Introduction

Barley (*Hordeum vulgare* L.) is the fourth most important cereal crop both in terms of cultivated area and tonnage harvested; global production being mostly used as animal feed and for the malting industry (http://faostat.fao.org). Only 5% of the global production of barley is used as ingredients in food preparation, but nevertheless barley grains are a valuable functional food for the high content of soluble dietary fiber [1]. The recent assemblage of the sequence of the 5.1-Gb haploid genome of barley [2] further supports the role of barley as a model species for the Triticeae tribe, which includes very important crops such as wheat (bread and durum) and rye.

Mature barley grains typically contain 50–60% starch on a dry weight basis. Starch is synthesized and stored in granules composed of two types of D-glucose polymers, amylose and amylopectin. Amylose, generally accounting for about 25–30% of starch weight in barley, is essentially a linear polymer of D-glucose units linked by alpha-1,4-glucosidic bonds. The second polymer of starch, amylopectin, is highly branched because of the alpha-1,6-glucosidic bonds that connect short alpha-1,4 linear chains [3], [4], [5], [6], [7], [8], [9].

While most plants contain starch granules of similar size, the Triticeae endosperm presents two classes of starch granules characterized by different sizes and shapes [10], [11]. Most of the starch is stored in large A-granules, but small B-granules prevail in number. In barley, the diameter of A-granules ranges from 10 to 40 μm while B-granules are smaller than 10 μm [10]. Starch granules contain crystalline lamellae in which double helices, composed of parallel linear chains of amylopectin, interact among each other to form different types of crystal packing [3]. Crystalline lamellae are interspersed with amorphous lamellae in which amylopectin branches are concentrated. The exact location of amylose within the semicrystalline architecture of the starch granule is still unknown [9], but certainly amylose influences the global structure of starch granules. For example, starch granules of different composition are characterized by different types of X-ray diffraction patterns [3], [4], [8]. In cereals, the A-type crystal packing is predominant, while the B-type crystallinity, typical of tuber starch, exists in smaller amounts. A third diffraction pattern, named V-type, is associated with lipid-amylose complexes and is little represented in native starches [3].

Both amylose/amylopectin ratios and the architecture of starch granules depend in a complex way on many different enzymes involved in starch metabolism [4], [5], [8], [9]. Biosynthesis of starch polymers in cereal grains strictly depends on the availability of ADP-glucose as a precursor for both amylose and amylopectin polymerization. Starting from ADP-glucose, a single enzyme, the granule-bound starch synthase I (GBSSI), is required for the synthesis of amylose. More complex is the biosynthetic pathway leading to amylopectin production as different classes of soluble starch synthases (SSs) and starch branching and debranching enzymes are required [5], [8], [9].

Starches with different amylose/amylopectin ratios have different properties that influence their possible use for either nutritional purposes or industrial transformations [6], [7], [9], [12], [13], [14]. In barley, mutants with 0–10% amylose (*waxy*) as well as mutants containing up to 70% amylose in the endosperm have been described [15], [16], [17]. Low-amylose starch display higher freeze-thaw stability, an interesting property for food preparation [6]. On the other hand, high amylose starches have interesting nutritional properties due to their positive correlation with resistant starch. This starch fraction is highly resistant to human digestion in the small intestine and reaches the large bowel where it plays a role similar to dietary fiber. Consumption of high-amylose resistant starch is associated with several health benefits, including the prevention of colon cancer, type II diabetes, obesity and cardiovascular diseases [18], [19].

Starch granule size distribution is another important parameter that may affect technological properties and end-use of each particular type of starch [20]. Barley grains are largely used for malting and large A-granules are more readily attached by hydrolytic enzymes than small B-granules [10]. B-granules are apparently protected during malting by a heterogeneous matrix deriving from the grain (proteins and cell walls). As a result, a significant proportion of B-granules escapes degradation and causes several technological problems during beer production [21].

In a previous work using a TILLING strategy, novel allelic variants in genes involved in starch metabolism in barley seeds were identified [22]. Here we describe the starch phenotype of nine mutants carrying either missense or nonsense mutations in five starch-related genes known to be expressed in the endosperm during grain filling: *BMY1* (beta-amylase 1), *GBSSI* (Granule Bound Starch Synthase I), *LDA1* (Limit Dextrinase 1), *SSI* (Starch Synthase I), *SSIIa* (Starch Synthase IIa). Seven mutant lines present starches with potentially interesting features for nutritional uses and/or industrial applications, including an altered amylose/amylopectin ratio or an unusually high percentage of A-granules or A-granules that are larger than in wild type starch.

Materials and Methods

TILLING analysis and plant materials

Details on the TILLING-based molecular screening for the five starch metabolism enzymes were reported in [22] and will only be summarized here. TILLMore is a chemically (sodium-azide, NaN_3) mutagenized barley population including 4,906 $M_{3:4}$ families [23]. TILLMore was screened using a standard TILLING protocol based on LI-COR vertical gel electrophoresis of PCR reactions obtained on 8X bulked genomic DNA samples. Genes tilled were *Beta-amylase 1* (*BMY1*), *Granule-Bound Starch Synthase I* (*GBSSI*), *Limit dextrinase 1* (*LDA1*), *Starch Synthase I* (*SSI*) and *Starch Synthase IIa* (*SSIIa*) (Table 1). For each mutant, plant materials phenotyped in this work were grains (kernels) of M_4 lines (three generations of selfing after mutation

induction), which have been verified to be homozygous for the mutation (not shown). Mutant lines and cv. Morex were grown in open field following standard agronomic practice in 0.5-m long one-row plots (approx. 12 plants per plot) using a randomized design with two replicates. For each line, grains harvested (from all well-grown ears) from two replicates were bulked. The same experiment was carried out in two years. Grains from separated years constituted the biological replicates.

Starch extraction from barley grains

Starch was extracted by grinding the grains to a fine powder in a pepper mill. About 5 g of seeds, corresponding to about 100 seeds, were used for each line and for each biological replicate (except for line 1517-*SSIIa* for which 2.5 g of seeds were used). Starch grains were purified as described in [24]. Briefly, the powder was suspended in 70 ml Extraction Buffer (EB: 55 mM Tris-HCl, pH 6.8, 2.6% SDS, 10% glycerol, 2% ß-mercaptoethanol) and vigorously shaken for 48 h, replacing the EB solution every 24 h. Following the extraction, samples were washed three times in water and filtered through a nylon membrane (cut-off 100 μm) in order to eliminate debris. Filtered samples were spun down for 1 min at 10,000 g. Starch grains were resuspended in 25 ml acetone and spun down again. Once removed the supernatant, starch grains were air-dried under a chemical hood for about 48 h at room temperature.

SDS-PAGE analysis of starch granule proteins

Isolation and electrophoretic separation of starch granule proteins was carried out on mature seeds following the method reported by Zhao and Sharp [25] with some modifications, as reported by Mohammadkhani et al. [26]. Protein bands were visualized by silver staining.

Determination of total starch and amylose content

Total starch content was determined on whole flours using Megazyme Total Starch Assay Kit (Megazyme, Ireland). The relative content of amylose was determined using both the Amylose/Amylopectin assay kit (Megazyme, Ireland) following the manufacturer instructions, and by an iodometric assay as reported in Sestili et al. [27]. Three technical replicas were performed for each mutant and each type of measure. Total starch content and relative amylose content (enzymatic method) were measured on two biological replicas.

Starch morphology

The morphology of starch samples was analyzed using a scanning electron microscopy (SEM) Hitachi S-4000. The samples were glued by a carbon type on an aluminum stub and gold coated (2 nm thick layer) before observations. For each sample two sets of 10 pictures at two magnifications (1000x and 2500x) were randomly collected. These two magnifications allowed a good estimation of the size of large and small granules. The granule size was estimated using the software ImageJ for image processing and analysis. The two main axes for each granule were recorded and 200–500 grains were measured for each sample. Percentage of granules with major axis lower than 8 μm (B-granules) was recorded in 10 pairs of pictures (at different magnification) for each genotype. Statistically significant differences between mutants and wild type mean values were detected by Student's t-test (P<0.01). The frequency analysis was carried out tacking classes of 2 μm.

Table 1. List of TILLING mutant lines carrying either missense or nonsense mutations in five genes related to starch metabolism in barley grains that have been isolated as described in Bovina et al. [22] and phenotypically characterized in this work.

Gene name	Mutant code	Genebank accession	Nucleotide change	Amino acid substitution	Seed and starch phenotype
Beta-amylase 1	2253-*BMY1*	EF175470	G2522A	D277N	Normal
	2682-*BMY1*	EF175470	G2944A	E348K	Normal
Granule-bound starch synthase I	1090-*GBSSI*	AB088761	G2306A	G493E	Low % amylose. Low % V-diffraction pattern
Limit dextrinase 1	905-*LDA1*	AF122050	G1528A	V270I	Low % A- granules
Starch synthase I	1132-*SSI*	AF234163	C5705T	T522I	High % A- granules
	1284-*SSI*	AF234163	G5666A	G509E	Low % A- granules. Large A-granules
	5850-*SSI*	AF234163	G6020A	G576D	High % A- granules. Small A-granules
Starch synthase IIa	1039-*SSIIa*	AY133251	G2453A	G678R	Small A-granules
	1517-*SSIIa*	AY133251	G2449A	W676*	Small and shrunken seeds. Low % total starch. High % amylose. High % V-diffraction pattern. Deformed granules

Starch crystallinity

The X-ray powder diffraction patterns were recorded using a Philips X'Celerator diffractometer with Cu Ka radiation (l = 1.5418 Å) and equipped with a Ni filter. The samples were scanned for 2θ angles between 5° and 30°, with a resolution of 0.02°. The degree of crystallinity of samples was quantitatively estimated following the method of Nara and Komiya [28]. A smooth curve which connected peak baselines was computer-plotted on the diffraction patterns. The area above the smooth curve was taken to correspond to the crystalline portion, and the lower area between the smooth curve and a linear baseline which connected the two points of intensity at 2θ of 30° and 5°. The upper diffraction peak area and total diffraction area over the diffraction angle 5°–30° 2θ were integrated on X'Pert HighScore Plus software (PANalytical B.V. 2008). The ratio of upper area to total diffraction area was taken as the degree of crystallinity.

In the diffraction patterns only peaks associable to A-type and V-type crystallinities were detected. To estimate the relative amount of A-type and V-type crystallinities in the starch samples from the diffraction patterns, only well isolated diffraction peaks were considered. The one at 15.1° is diagnostic of the A-type crystallinity and the one at 19.7° is diagnostic of the V-type one. These diffraction intensities has been normalized on the sum of their intensities.

Results

TILLING molecular analysis

Molecular details about TILLING for five genes involved in starch metabolism were already reported in Bovina et al. [22] and will only be summarized here. TILLING was carried out in TILLMore, a TILLING population in the cultivar (cv.) Morex background which was chemically mutagenized using NaN$_3$ [23]. The analyses identified an allelic series for each of the genes examined with a total number of 29 mutations [22]. Seeds of nine mutant lines carrying either missense or nonsense mutations in the five genes analyzed (*BMY1 GBSSI, LDA1, SSI* and *SSIIa*) were phenotypically characterized in this study (Table 1). Seeds of the mutant lines did not show any macroscopic differences in respect to Morex wild type (wt) with the exception of the line 1517-*SSIIa* (*Starch Synthase IIa*) that showed shrunken kernels with an empty

cavity inside (Figure 1). These seeds were also lighter than wild-type ones (3.6±0.3 vs. 4.9±0.1 g/100-kernels).

Total starch content

Total starch content was measured in whole flours obtained from two biological replicates for each of the nine mutant lines. The water content of the flours was very similar in all samples (≈9%, Table S1 in File S1). Morex wt contained 43% starch on a fresh weight basis, but this value was diminished by one third in mutant 1517-*SSIIa* (27%, P<0.01; Table 2). In no other mutants the total starch content was significantly different to the wild-type value (P<0.05, n = 2).

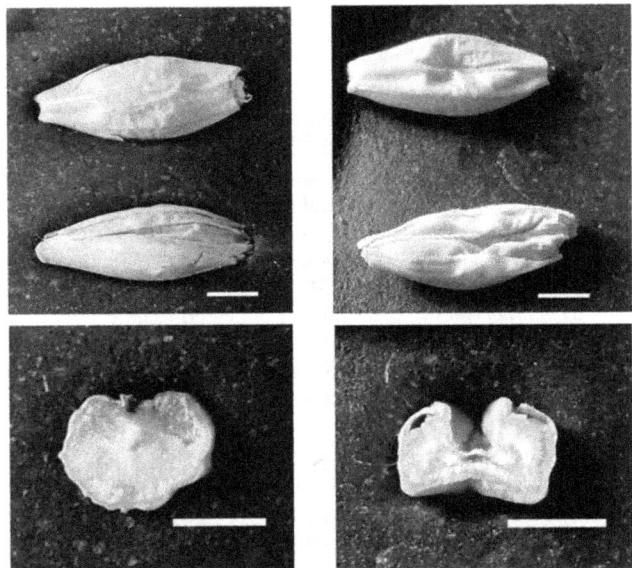

Figure 1. Seed morphology and transverse section of TILLING mutant line 1517-*SSIIa* (*Starch Synthase IIa*) (right) showing a shrunken phenotype, compared with cv. Morex wild type (left). From top to bottom: adaxial and abaxial seed views, and seed cross section. White bars = 2 mm.

Table 2. Content of starch and amylose in seeds of TILLING mutant lines.

	% starch		% amylose (enzymatic)		% amylose (colorimetric)	
Morex	42.5±1.3	(100)	32.3±3.2	(100)	33.4	(100)
2253-BMY1	47.7±2.7	(112)	31.9±1.9	(99)	34.0	(102)
2682-BMY1	46.2±0.6	(109)	34.4±0.9	(107)	29.0	(87)
1090-GBSSI	38.9±0.8	(92)	9.5±3.1 *	(29)	8.8	(26)
905-LDA1	44.2±0.5	(104)	31.0±2.2	(96)	30.8	(92)
1132-SSI	41.8±2.4	(98)	32.6±2.8	(101)	36.1	(108)
1284-SSI	40.6±1.1	(96)	34.5±1.0	(107)	35.9	(107)
5850-SSI	42.1±2.3	(99)	31.0±1.4	(99)	36.7	(110)
1039-SSIIa	43.1±2.1	(101)	34.9±0.7	(108)	36.4	(109)
1517-SSIIa	26.5±0.1 **	(62)	47.5±5.0 *	(147)	48.9	(146)

Total starch content is expressed as percentage of fresh weight (water content in flours was about 9%, with no significant differences among samples, see Table S1 in File S1). Amylose was detected either by enzymatic or colorimetric methods and it is expressed as a percentage of total starch. To facilitate comparisons, all values are also reported in brackets as percentage of the corresponding wild type value. Total starch and relative amylose content was determined on two biological replicates. Significant differences were detected by Student's t-test ($P<0.05$ = *; $P<0.01$ = **). For comparison, colorimetric analysis of amylose was performed on a single biological sample for each genotype, and data shown are means of 3 technical replicates (standard deviations were in all cases below 10% of the mean value).

Amylose content

Whole flours from two biological replicates were also analysed for amylose content by enzymatic assay. For comparison, the relative content of amylose was also determined colorimetrically with similar results (Table 2). Wild type starch contained 32% amylose and similar values were detected in seven over nine mutants (amylose/amylopectin ratio 0.47). However, mutant 1090-*GBSSI*, carrying a missense mutation in *Granule Bound Starch Synthase I*, contained only one third of the normal amylose content in its grain starch (9%, P<0.05; amylose/amylopectin ratio 0.10), and mutant 1517-*SSIIa*, carrying a nonsense mutation in *Starch Synthase IIa*, contained much more amylose than the wild type (47%, P<0.05; amylose/amylopectin ratio 0.88; Table 2).

SDS-PAGE analysis

In order to assess whether the nine mutations identified had an effect on the electrophoretic protein profile typical of the starch granule proteins of barley, SDS-PAGE analysis was performed. With the exception of the lines 1517-*SSIIa* and 1284-*SSI*, all the mutants showed a profile identical to Morex wt, characterized by three major bands corresponding to SSIIa (overlapped with Starch Branching Enzyme II, SBEII), SSI and GBSSI [29] (Figure S1 in File S1). The absence of the SSIIa enzyme was confirmed in the line 1517-*SSIIa*. Notably this mutant appeared to lack also SBEII and SSI isoforms. Moreover, although mutant 1284-*SSI* has no premature stop codon in the *SSI* gene, a drastic reduction of the SSI band was observed in starch granules (Figure 2).

Starch granules morphology

Starch extracts from wild-type and mutant grains were analysed by Scanning Electron Microscopy (SEM). With the exception of 1517-*SSIIa*, starch granules of all remaining samples were quite regularly shaped (Figure 3 and Figure S2 in File S1). A quantitative analysis of granules dimensions was performed by collecting the length of the major and minor axis of 200–500 granules for each biological sample from their SEM digital images. Distribution of granule dimensions was clearly bimodal in all samples (Figure S3 in File S1), with a major sub-population of small spherical granules (major axis <8 μm, B-granules), and a

minor sub-population of larger discoid particles with a major axis varying between 8 and 30 μm (A-granules). Distributions based on minor axis were qualitatively identical to those based on the major axis (not shown). Differently from all other mutants, starch of 1517-*SSIIa* contained irregularly shaped A-granules typically appearing like deflated spheres (Figure 3 and Figure S2 in File S1). B-granules were also irregular in shape and agglomerated. These features prevented a quantitative determination of A and B-type particles in 1517-*SSIIa* samples.

The percentage of B-granules (<8 μm) in wild-type purified starch was 73% (SD). A similar value was found in mutants of *BMY1*, *GBSSI* and *SSIIa* (Figure 4). In the four remaining mutants the percentage of B-granules differed significantly from the wild type Morex (P<0.01). B-granules were less abundant in two soluble starch synthase mutants, 1132-*SSI* (57%) and 5850-

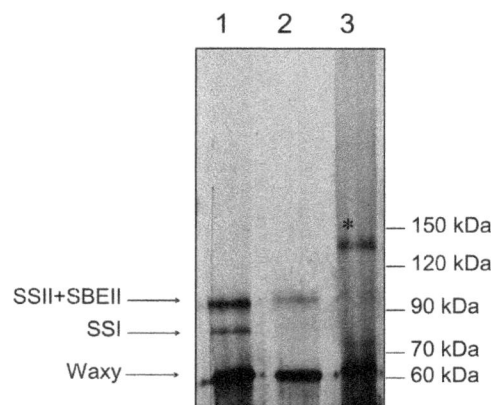

Figure 2. Electrophoretic separation (SDS–PAGE) of starch granule proteins extract from barley wild type cv. Morex (1) and barley mutants 1284-*SSI* (2) and 1517-*SSIIa* (3). The bands corresponding to starch synthase II and starch branching enzyme II (SSII+SBEII), starch synthase I (SSI) and granule-bound starch synthase (GBSSI) are indicated. In lane 3, the high molecular weight band marked with an asterisk is probably due to impurities present in the starch preparation obtained from the shrunken seeds of line 1517-*SSIIa*. Molecular weight standard is schematically reported on the right.

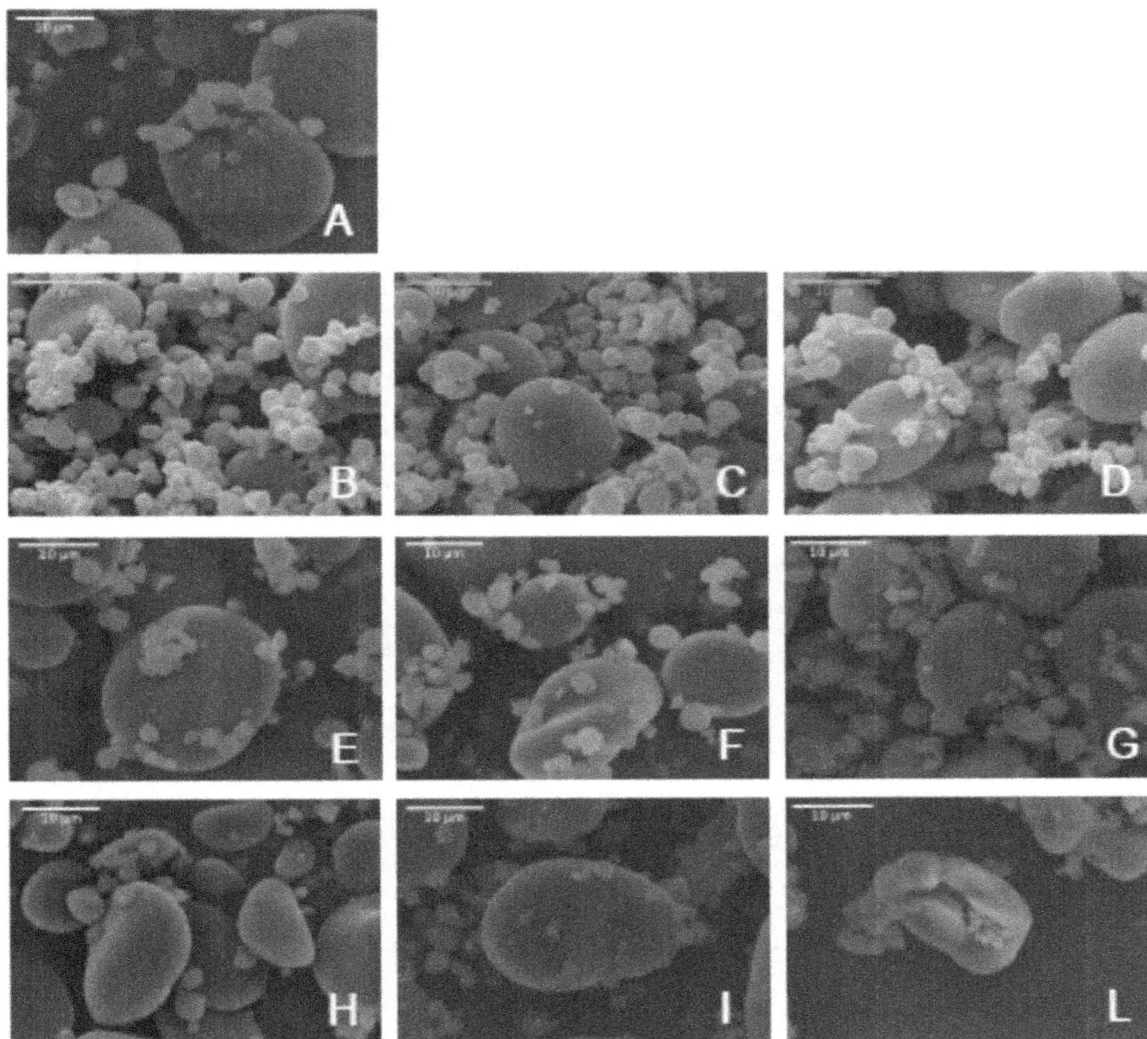

Figure 3. Scanning Electron Microscopy (SEM) analysis of starch granules from barley cv. Morex wild-type (A) and mutants 2253-
BMY1 **(B), 2682-***BMY1* **(C), 1090-***GBSSI* **(D), 905-***LDA1* **(E), 1132-***SSI* **(F), 1284-***SSI* **(G), 5850-***SSI* **(H), 1039-***SSIIa* **(I), 1517-***SSIIa* **(L).** Scale bar: 10 μm.

SSI (62%), but relatively more abundant in mutant 1284-*SSI* of the same gene (85%) and in mutant 905-*LDA1* of limit dextrinase I (85%) (Figure 4). All these four mutants contained missense mutations (Table 1).

The average size of A- and B-granules in each sample was also analysed. In two mutants (5850-*SSI* and 1039-*SSIIa*, both missense mutations), A-type granules were significantly smaller (−25% major axis) than wt ones (17.1 μm) (Table 3; P<0.01). On the other hand, A-granules of 1284-*SSI* mutant were significantly larger (+15%) than wt ones. The average diameter of B-particles varied between 2.3 and 3.5 μm in all samples, with no significant differences between wt and mutants (Table 3). Interestingly, two missense mutants of *SSI* displayed symmetrical properties in A-granules size and frequency: in mutant 1284-*SSI* A-granules were larger but less abundant, while in mutant 5850-*SSI* they were relatively more numerous, but smaller in size (Table 3 and Figure 4).

Crystallinity of starch granules

Crystallinity of starch granules was evaluated by X-ray powder diffraction. The crystallinity of wild-type starch was estimated as 29% and this value ranged between 26 and 33% in all mutants (Table 4), with no clear correlation between the degree of crystallinity and other phenotypic characters previously recorded. On the contrary, the type of crystallinity, as detected from the X-ray diffraction patterns, was more variable. In wild-type starch we estimated a large predominance of the A-type crystal pattern (81%), with a minor contribution of the V-type (Figure 5). No evidence for B-type crystallinity was obtained from diffraction patterns of wild type and mutants. In most of the mutants, the type of crystallinity was similar to that observed in wild type starch, *i.e.* 78–83% A-type and 17–22% V-type. Interestingly, however, in the low-amylose 1090-*GBSSI* mutant, crystallinity was almost exclusively of the A-type (92%) whereas in the high amylose 1517-*SSIIa* mutant crystallinity was prevalently of the V-type (76%) (Figure 5 and Table 4).

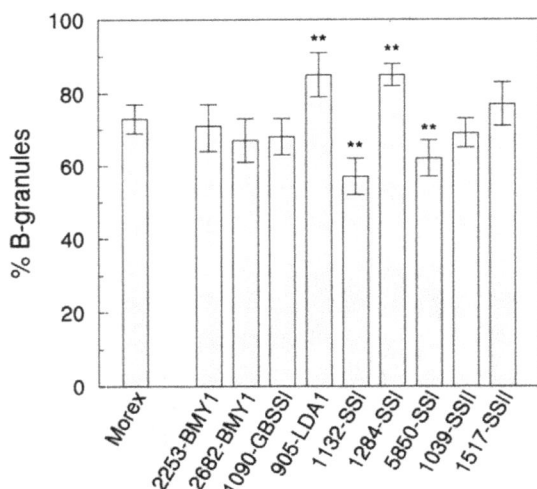

Figure 4. Percentage of B-type granules (diameter <8 μm) in grain starch of barley wild-type cv. Morex and mutant lines. Granules size distribution was determined on 10 couples of SEM images randomly collected for each genotype. Data shown are means ±SD (n = 10). Statistically significant differences between mutants and wild type mean values were estimated by Student's t-test (P<0.01) and are highlighted by a double asterisk (**).

Discussion

Starch structure and chemical composition are genetically determined by a large set of genes [5], [6], [8], [30] and the potential for obtaining different types of starch by screening natural or induced genetic variability is huge. TILLING provides a non-transgenic approach to explore this potential [7], [12], [22], [31], [32], [33], [34], [35], [36]. We exploited TILLING to identify and phenotypically characterize new alleles of five genes involved in grain starch metabolism of barley.

Granule bound starch synthase (GBSSI)

Granule-bound starch synthase I (GBSSI) [4] is specifically expressed in the endosperm of barley [30] where it is known to exert a tight control on the biosynthesis of amylose [37]. GBSSI is coded by the waxy locus and barley cultivars with altered GBSSI

activity contain altered levels of amylose in grains, ranging between 0 and 41% of total starch [29]. Besides amylose, GBSSI is also involved in the synthesis of extra long glucan chains of amylopectin, such that also amylopectin may be modified in waxy mutants [38], [39], [40]. Low-amylose varieties can be used for food applications because of their peculiar starch features (low gelatinization temperature and retrogradation), that confer high freeze-thaw stability and anti-stailing properties to processed food [41], [42].

Here we report a new allele of GBSSI with a G493E point mutation. Grain starch of this mutant (1090-GBSSI) contains less than 10% amylose (vs. 30% of wild-type) and is thus defined low-amylose or near-waxy. Crystallinity of 1090-GBSSI starch was found to be largely A-type, with a minor contribution of the V-type pattern. A similar profile has been previously reported in low amylose barley [43], [44].

Plant starch synthases (both granule bound and soluble isoforms) are proteins of about 60–120 kDa that belong to the glycosyltransferase family GT-5 [4]. They typically contain a catalytic domain formed by two Rossman fold domains delimiting a deep cleft where the catalytic site is located (Figure 6). In plant starch synthases, the catalytic domain is often preceded by an N-terminal sequence of variable length and no clear function. The crystal structure of the catalytic domain of rice GBSSI [45] was used as a template to model barley GBSSI (the two proteins are 84% identical in amino acid sequence). According to the model, glycine-493 belongs to an alpha-helix of the second, C-terminal Rossman fold domain at approximately 10 Å from the ADP binding pocket [45] and 6 Å from the conserved STGGL motif suspected to be involved in catalysis and/or substrate binding [29].

Several near-waxy cultivars are known in barley, all carrying a large deletion in the promoter region of the GBSSI gene that results in strongly diminished expression of the enzyme [29], [46]. Waxy cultivars with no detectable amylose are also known (e.g. cv. CDC Alamo and CDC Fibar) but they contain point mutations in the coding sequence that likely cause complete inactivation of the enzyme without drastically affecting the protein abundance in starch granules [29]. Interestingly, mutation G493E in 1090-GBSSI seems to modulate, rather than inactivate, enzyme activity as demonstrated by the residual content of amylose (10%) detected in its starch. Moreover, this effect is obtained without altering protein expression, as suggested by the SDS-PAGE pattern identical to the wild-type. Indeed, mutation G493E may prove

Table 3. Length of major axis in A-type and B-type starch granules.

	B-granules, major axis [μm]	A-granules, major axis [μm]
Morex	3.01±1.23	17.08±4.88
2253-BMY1	2.50±0.77	17.97±5.17
2682-BMY1	2.80±1.19	18.18±5.47
1090-GBSSI	2.26±0.86	16.49±7.24
905-LDA1	2.83±1.04	16.68±5.10
1132-SSI	3.45±1.09	17.70±4.38
1284-SSI	3.02±1.03	19.69±4.84 **
5850-SSI	3.38±1.86	13.15±3.41 **
1039-SSIIa	3.27±1.74	12.92±3.39 **
1517-SSIIa	Nd	Nd

Data are means ±SD of 200–500 granules measured for each genotype. Statistically significant differences as determined by Student's t-test are indicated (P<0.01 = **). Nd, not determined (starch granules in 1517-SSIIa mutant were irregular in shape).

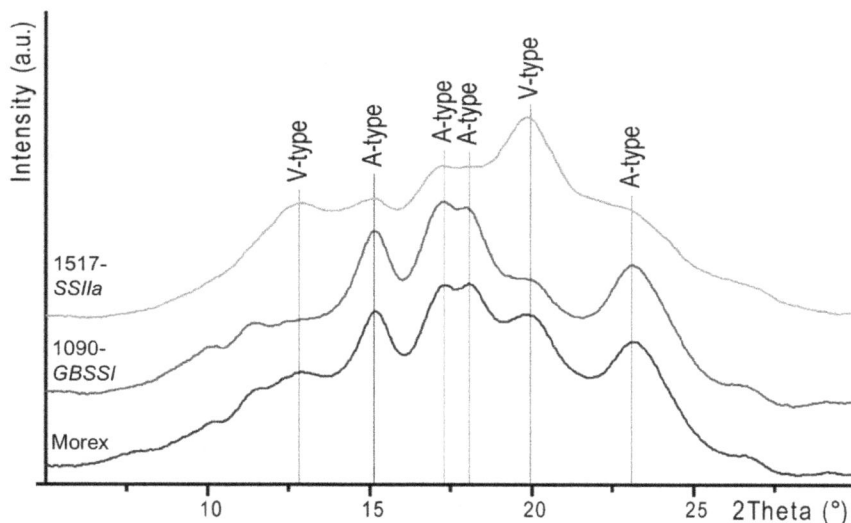

Figure 5. X-ray diffraction patterns of native starch extracts from barley wild type cv. Morex (black) and mutants 1090-*GBSSI* (grey) and 1517-*SSIIa* (light grey). The characteristic peaks of the A-type and V-type polymorphs are indicated.

useful for the understanding the little known catalytic mechanism of GBSSI.

Limit dextrinase (*LDA1*)

Together with isoamylases (ISAs), limit dextrinases (LDAs, also known as pullulanases) constitute the set of starch debranching enzymes. The role of ISAs in removing excess branches of amylopectin formed by branching enzymes is well known [8]. In the absence of isoamylases, starch is synthesized in a highly branched form known as phytoglycogen [47]. Barley limit dextrinase is coded by a single gene (*LDA1*) and is apparently involved both in starch biosynthesis and degradation [48]. *In vivo*, the activity of LDA1 is regulated by a proteinaceous inhibitor LDI [49], and antisense transgenic barley with lower expression of LDI showed higher LDA1 activity and a lower percentage of B-granules (*i.e.* inhibition of granule initiation) and lower amylose/amylopectin ratio [50]. It was suggested that LDA, which is expressed when B-granules are formed, may play a role in reducing the amount of primers that allows the nucleation of small

B-granules. In our study, we have found a mutant of *LDA1* (905-*LDA1*) with a higher percentage of small B-granules that further supports the role of this enzyme in starch granule initiation.

The tridimensional structure of barley *LDA1* has been solved [51]. The protein is made of four domains: the N-terminal domain, the CBM48 domain, the catalytic domain and the C-terminal domain. Valine-270 of *LDA1*, that in mutant 905-*LDA1* is substituted by a more bulky isoleucine, belongs to the carbohydrate binding module CBM48 and is located close to the interface with the catalytic domain (Figure 6). It is plausible that the substitution of valine in isoleucine (V270I) in the CBM48 domain may reduce the capability of the protein to bind glucans and in turn inhibit, albeit indirectly, its scavenging activity toward primers of granules nucleation. Consistently, starch of mutant 905-*LDA1* is formed by a larger number of granules (predominantly small B-granules) and the role of LDA1 in controlling granule nucleation is further supported.

Table 4. Percentage crystallinity and relative intensity of diffraction peaks at 15.1° and 19.7°.

	% crystallinity	I_{A-type}	I_{V-type}
Morex	0.29	0.81	0.19
2253-*BMY1*	0.30	0.82	0.18
2682-*BMY1*	0.33	0.80	0.20
1090-*GBSSI*	0.26	0.92	0.08
905-*LDA1*	0.27	0.81	0.19
1132-*SSI*	0.28	0.79	0.21
1284-*SSI*	0.27	0.79	0.21
5850-*SSI*	0.27	0.80	0.20
1039-*SSIIa*	0.26	0.78	0.22
1517-*SSIIa*	0.29	0.24	0.76

These latter data represent a relative estimation of A-type and V-type crystallinity, respectively.

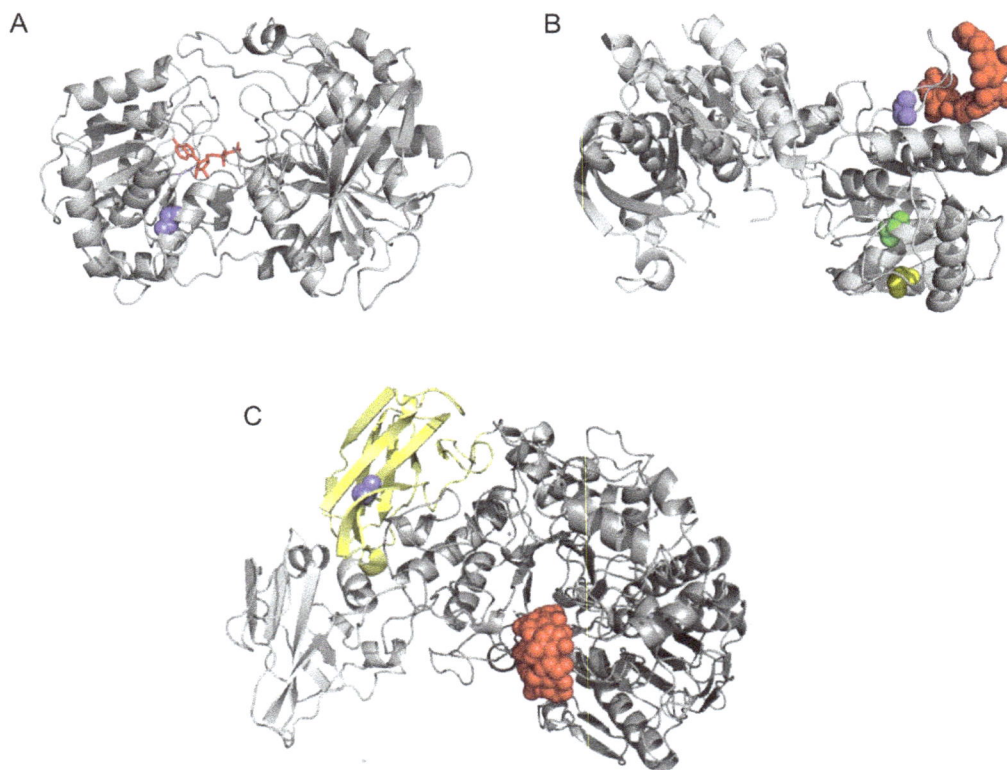

Figure 6. Localization of point mutations in the 3D structures of barley GBSSI, SSI and LDA1. A) Barley GBSSI was modelled by Swissmodel using the catalytic domain of wild-type rice GBSSI complexed with ADP as a template (pdb 3VUF). The two mature proteins are 84% identical in sequence. The main chain of glycine-493 is represented by blue spheres. In mutant 1090-*GBSSI*, glycine-493 is substituted by a glutamate (G493E). Co-crystallized ADP of the rice GBSSI structure (3VUF) is superimposed to highlight the adenine nucleotide binding site. **B)** Crystal structure of barley SSI, co-crystallized with a molecule of maltopentaose (red spheres) (pdb 4HLN). Main chain atoms of mutated residues are represented by coloured spheres: blue (G576D in mutant 5850-*SSI*), green (T522I in mutant 1132-*SSI*) and yellow (G509E in mutant 1284-*SSI*). **C)** Crystal structure of barley LDA1 (pdb 2X4B). The carbohydrate binding module CBM48 is coloured yellow. Residue no. 270 (blue spheres corresponding to main chain atoms) is part of the CBM48 domain. Mutant 905-*LDA1* carries a V270I mutation. A molecule of betacyclodextrine (red spheres) co-crystallized with the protein highlights the putative active site.

Soluble starch synthase I (*SSI*)

In plants, soluble starch synthases are divided into four classes with different specificity (from *SSI* to *SSIV*), and some of them are represented by more than one isoform [4], [8], [9]. While *SSIV* is probably involved in starch granule initiation [52], *SSI* preferentially synthesize short glucan chains using short amylopectin chains as substrate [5], [8]. These short amylopectin chains may then be prolonged by *SSII* and *SSIII* [5], [8], but individual roles and cooperation between different starch synthases in building the starch granule are still undefined, in barley at least. Mutants in rice and wheat clearly suggest that the activity of the different starch synthases is not necessarily sequential and the lack of SSI may be partially compensated *in vivo* [8], [53], [54].

In cereals *SSI* is represented by a single isoform. We have analysed three missense mutations in *SSI* and all of them showed starch phenotypes consisting in modifications in either size or frequency of A- and B-granules. However, our results may appear contradictory: in fact two mutants showed a higher % of A-type granules (1132-*SSI* and 5850-*SSI*), while the third mutant had more B-type granules (1284-*SSI*). Morever, the A-granules of mutant 5850-*SSI* were more abundant but smaller in size and, symmetrically, the A-granules of mutant 1284-*SSI* were larger but less frequent. Only in mutant 1132-*SSI* the higher percentage of A-granules was not compensated by a reduction in size. The unexpected phenotypic difference between the three mutant lines

may be due to currently unknown additional background mutations present in the genome of TILLMore mutant lines [23].

However, some hints could be obtained from the recently solved crystallographic structure of barley *SSI* [55] (Figure 6). The G576D substitution of mutant 5850-*SSI* is localized at the base of a loop involved in the formation of a high affinity binding site for maltooligosaccharides. This site is 30 Å away from the putative catalytic site, but is believed essential for colocalizing branched glucans and *SSI*, thereby favoring catalysis. Moreover, the starch phenotype associated to the G576D mutation (smaller A-granules) may suggest a role of this site in controlling the final size of large starch granules. The other two point mutations here described (Figure 6) are localized in regions of the protein not yet characterized, providing no suggestions to understand their role. Nevertheless, the high percentage of large A-granules in mutant 1132-*SSI* is interesting because this is a desirable trait for malting [10].

Initiation of A and B-granules are separated events, although little is known of the genetic control of this trait in Triticeae. In barley, A-granules are nucleated at 4–14 days post-anthesis, during endosperm cell division, while small B-granules are nucleated later, during endosperm cell growth [56]. A QTL controlling B-granules initiation was recently described in wild wheat Aegilops [10] and in Arabidopsis, SSIV is believed to positively regulate granule initiation [52]. Recently the suppression of *SSI* expression in wheat grains using RNAi technology led to the production of lines with a reduced proportion of B-granules [54]. All three *SSI* barley mutants analysed in this work showed an abnormal distribution between large and small granules. However, because of the complexity of our results, any conclusion about a possible role of SSI in granule initiation in barley is premature.

Soluble starch synthase IIa (*SSIIa*)

SSIIa is the major SS isoform of barley endosperm during grain filling [4]. Mutant *sex6* of barley cv. Himalaya has no *SSIIa* activity and produces shrunken kernels containing starch made of up to 70% amylose [15], [16]. *SSIIa* knock-out mutants are particularly interesting for industrial applications because of their higher level of amylose and resistant starch in the endosperm. Resistant Starch is associated with several human health benefits, including the prevention of the colon cancer, type-II diabetes and obesity [13], [19]. In this work, we identified a mutant line with A-granule of smaller size carrying a missense mutation (1039-*SSIIa*), and a *SSIIa* null mutant (1517-*SSIIa*) characterized by small/shrunken seeds and containing less starch with more amylose (48% of grain starch is made of amylose in this mutant). The SDS-PAGE analysis of starch extracted from 1517-*SSIIa* confirmed the absence of the protein *SSIIa*, together with *SBEII* and *SSI* isoforms. The simultaneous absence of *SSIIa*, *SBEII* and *SSI* in the starch granule was already observed in *SSIIa* mutants of barley, bread and durum wheat [14], [57], [58].

Starch crystallinity of 1517-*SSIIa* was largely characterized by a V-type diffraction pattern suggesting the formation of lipid-amylose complexes, similarly to those observed in the *sex6* mutant [15]. In the missense mutant 1039-*SSIIa*, in spite of the smaller size of A-granules, no other starch parameters including crystallinity were significantly affected. In barley endosperm, *SSIIa* was shown to extend short amylopectin glucan chains of 3–8 glucose units to chains of up to 35 units [15]. Consistently, the lack of *SSIIa* negatively affects amylopectin synthesis, more than amylose synthesis [16], and mutant 1517-*SSIIa* is fully consistent with these results.

Conclusions

TILLING of five genes encoding enzymes involved in starch metabolism enabled us to identify seven new alleles that are associated with new starch phenotypes in terms of amylose/amylopectin ratio, or crystal packing, or distribution of A- and B-granules, or size of A-granules (Table 1). Our results confirmed the role played by granule-bound starch synthase (*GBSSI*) in controlling amylose biosynthesis and, conversely, the role played by soluble starch synthase IIa (*SSIIa*) in controlling amylopectin synthesis. Starch granule initiation appeared to be controlled by limit dextrinase (*LDA1*), and size of A-granules by starch synthases IIa. Thanks to their physical-chemical properties, these new alleles deserve further attention in order to investigate their possible interest in nutritional uses or industrial applications.

Supporting Information

File S1 Figure S1, SDS–PAGE separation of starch protein extract from cv. Morex and barley mutants. Figure S2, SEM analysis of starch granules from cv. Morex (A) and barely mutants 2253-*BMY1* (B), 2682-*BMY1* (C), 1090-*GBSSI* (D), 905-*LDA1* (E), 1132-*SSI* (F), 1284-*SSI* (G), 5850-*SSI* (H), 1039-*SSIIa* (I), 1517-*SSIIa* (L). Scale bars: 10, 20, and 30 μm. Table S1, Water content in whole flours of TILLING mutant lines. Dry weight was obtained after incubation of samples at 80°C for 24 h. Data are means ±SD (n = 4).
(PDF)

Author Contributions

Conceived and designed the experiments: F. Sparla GF VT RB SS RT F. Sestili PT. Performed the experiments: F. Sparla GF EB CP VT RB F. Sestili. Analyzed the data: F. Sparla GF VT RB SS RT F. Sestili PT. Contributed reagents/materials/analysis tools: F. Sparla GF VT RB SS RT F. Sestili PT. Wrote the paper: F. Sparla GF VT RB F. Sestili PT.

References

1. Collins HM, Burton RA, Topping DL, Liao M-L, Bacic A, et al. (2010) Variability in fine structures of noncellulosic cell wall polysaccharides from cereal grains: potential importance in human health and nutrition. Cereal Chem 87: 272–282.
2. The International Barley Genome Sequencing Consortium (2012) A physical, genetic and functional sequence assembly of the barley genome. Nature 491: 711–716.
3. Buléon A, Colonna P, Planchot V, Ball S (1998) Starch granules: structure and biosynthesis. Int J Biol Macromol 23: 85–112.
4. Ball SG, Morell MK (2003) From bacterial glycogen to starch: understanding the biogenesis of the plant starch granule. Annu Rev Plant Biol 54: 207–233.
5. James MG, Denyer K, Myers A (2003) Starch synthesis in the cereal endosperm. Curr Opin Plant Biol 6: 215–222.
6. Jobling S (2004) Improving starch for food and industrial applications. Curr Opin Plant Biol 7: 210–218.
7. Morell MK, Myers AM (2005) Towards the rational design of cereal starches. Curr Opin Plant Biol 8: 204–210.
8. Jeon JS, Ryoo N, Hahn TR, Walia H, Nakamura Y (2010) Starch biosynthesis in cereal endosperm. Plant Physiol Bioch 48: 383–392.
9. Zeeman SC, Kossmann J, Smith AM (2010) Starch: its metabolism, evolution, and biotechnological modification in plants. Annu Rev Plant Biol 61: 209–234.
10. Mazanec K, Dycka F, Bobalova J (2011) Monitoring of barley starch amylolysis by gravitational field flow fractionation and MALDI-TOF MS. J Sci Food Agric 91: 2756–2761.
11. Howard T, Rejab NA, Griffiths S, Leigh F, Leverington-Waite M, et al. (2011) Identification of a major QTL controlling the content of B-type starch granules in Aegilops. J Exp Bot 62: 2217–2228.
12. Slade AJ, Fuerstenberg SI, Loeffler D, Steine MN, Facciotti D (2005) A reverse genetic, nontransgenic approach to wheat crop improvement by TILLING. Nat Biotechnol 23: 75–81.
13. Bird AR, Vuaran MS, King RA, Noakes M, Keogh J, et al. (2008) Wholegrain foods made from a novel high-amylose barley variety (Himalaya 292) improve indices of bowel health in human subjects. Brit J Nutr 99: 1032–1040.
14. Damiran D, Yu P (2010) Chemical profile, rumen degradation kinetics, and energy value of four hull-less barley cultivars: comparison of the zero-amylose waxy, waxy, high-amylose, and normal starch cultivars. J Agric Food Chem 58: 10553–10559.
15. Morel MK, Kosar-Hashemi B, Cmiel M, Samuel MS, Chandler P, et al. (2003) Barley sex6 mutants lack starch synthase IIa activity and contain a starch with novel properties. Plant J 34: 173–185.
16. Li Z, Li D, Du X, Wang H, Larroque O, et al. (2011) The barley amo1 locus is tightly linked to the starch synthase IIIa gene and negatively regulates expression of granule-bound starch synthase genes. J Exp Bot 62: 5217–5231.
17. Asare EK, Jaiswal S, Maley J, Bàga M, Sammynaiken R, et al. (2011) Barley grain constituents, starch composition, and structure affect starch in vitro enzymatic hydrolysis. J Agric Food Chem 59: 4743–4754.
18. Nugent AP (2005) Health properties of resistant starch. Nutrition Bulletin 30: 27–54.
19. Topping D (2007) Cereal complex carbohydrates and their contribution to human health. Trends Food Sci Technol 46: 220–229.
20. Dhital S, Shrestha AK, Hasjim J, Gidley MJ (2011) Physicochemical and structural properties of maize and potato starches as a function of granule size. J Agric Food Chem 59: 10151–10161.
21. MacGregor AW (1991) The effect of barley structure and composition on malt duality, in Proceedings of the European Brewery Convention Congress, Lisbon, Portugal, pp.37–42.
22. Bovina R, Talamè V, Salvi S, Sanguineti MC, Trost P, et al. (2011) Starch metabolism mutants in barley: a TILLING approach Plant Genetic Resources: Characterization and Utilization 9: 170–173.
23. Talamè V, Bovina R, Sanguineti MC, Tuberosa R, Lundqvist U, et al. (2008) TILLMore, a resource for the discovery of chemically induced mutants in barley. Plant Biotechnol J 6: 477–485.

24. Kim K, Johnson W, Graybosch RA, Gaines CS (2003) Physicochemical properties and end-use quality of wheat starch as a function of waxy protein alleles. J Cereal Sci 37: 195–204.

25. Zhao XC, Sharp PJ (1996) An improved 1-D SDS-PAGE method for the identification of three bread wheat waxy proteins. J Cereal Sci 23: 191–193.

26. Mohammdkhani A, Stoddard FL, Marshall DR, Uddin MN, Zhao X (1999) Starch extraction and amylose analysis from half seeds. Starch/Stärke 51: 62–66.

27. Sestili F, Janni M, Doherty A, Botticella E, D'Ovidio R, et al. (2010) Increasing the amylose content of durum wheat through silencing of the SBEIIa genes. BMC Plant Biol 10: 144.

28. Nara S, Komiya T (1983) Studies on the relationship between water-saturation state and crystallinity by the diffraction method for moistened potato starch. Starch/Stärke 35: 407–410.

29. Asare EK, Baga M, Rossnagel BG, Chibbar RN (2012) Polymorphism in the barley granule bound starch synthase 1 (gbss1) gene associated with grain starch variant amylose concentration. J Agric Food Chem 60: 10082–10092.

30. Radchuk VV, Borisjuk L, Sreenivasulu N, Merx K, Mock HP, et al. (2009) Spatiotemporal profiling of starch biosynthesis and degradation in the developing barley grain. Plant Physiol 150: 190–204.

31. Sestili F, Botticella E, Bedo Z, Phillips A, Lafiandra D (2010) Production of novel allelic variation for genes involved in starch biosynthesis through mutagenesis. Molecular Breeding 25: 145–154.

32. Vriet C, Welham T, Brachmann A, Pike M, Pike J, et al. (2010) A suite of Lotus japonicus starch mutants reveals both conserved and novel features of starch metabolism. Plant Physiol 154: 643–655.

33. Botticella E, Sestili F, Hernandez-Lopez A, Phillips A, Lafiandra D. High resolution melting analysis for the detection of EMS induced mutations in wheat SBEIIa genes. BMC Plant Biol. 10: 156.

34. Hazard B, Zhang X, Colasuonno P, Uauy C, Beckles DM, et al. (2012) Induced mutations in the *starch branching enzyme II* (*SBEII*) genes increase amylose and resistant starch content in pasta wheat. Crop Sci 52: 1754–1766.

35. Slade AJ, McGuire C, Loeffler D, Mullenberg J, Skinner W, et al. (2012) Development of high amylose wheat through TILLING. BMC Plant Biol 12: 69.

36. Bovina R, Brunazzi A, Gasparini G, Sestili F, Palombieri S, et al. (2014) Development of a TILLING resource in durum wheat for reverse- and forward-genetics analyses. Crop & Pasture Science 65: 112–124.

37. Nelson OE, Rines HW (1962) The enzymatic deficiency in the waxy mutant of maize. Biochem Biophys Res Comm 9: 297–300.

38. Maddelein ML, Libessart N, Bellanger F, Delrue B, D'Hulst C, et al. (1994) Toward an understanding of the biogenesis of the starch granule: determination of granule-bound and soluble starch synthase functions in amylopectin synthesis. J Biol Chem 269: 25150–25157.

39. Denyer K, Waite D, Edwards A, Martin C, Smith AM (1999) Interaction with amylopectin influences the ability of granule-bound starch synthase I to elongate malto-oligosaccharides. Biochem J 342: 647–653.

40. Yoo SH, Jane J-L (2002) Structural and physical characteristics of waxy and other wheat starches. Carbohydr Polym 49: 297–305.

41. Baik BK, Ullrich SE (2008) Barley for food: characteristics, improvement and renewed interest. J Cereal Sci 48: 233–242.

42. Howard TP, Fahy B, Leigh F, Howell P, Powell W, et al. (2014) Use of advanced recombinant lines to study the impact and potential of mutations affecting starch synthesis in barley. J Cereal Sci 59: 196–202.

43. Waduge R, Hoover R, Vasanthan T, Gao J, Li J (2006) Effect of annealing on the structure and physicochemical properties of barley starches of varying amylose content. Food Res Int 39: 59–77.

44. Naguleswaran S, Vasanthan T, Hoover R, Bressler D (2013) The susceptibility of large and small granules of waxy, normal and high-amylose genotypes of barley and corn starches toward amylolysis at sub-gelatinization temperatures. Food Res Int 51: 771–782.

45. Momma M, Fujimoto Z (2012) Interdomain disulfide bridge in the rice granule bound starch synthase I catalytic domain as elucidated by X-ray structure analysis. Biosci Biotech Bioch 76: 1591–1595.

46. Patron NJ, Smith AM, Fahy BF, Hylton CM, Naldrett MJ, et al. (2002) The altered pattern of amylose accumulation in the endosperm of low-amylose barley cultivars is attributable to a single mutant allele of granule-bound starch synthase I with a deletion in the 5'-non-coding region. Plant Physiol 130: 190–198.

47. Burton RA, Jenner H, Carrangis L, Fahy B, Fincher GB, et al. (2002) Starch granule initiation and growth are altered in barley mutants that lack isoamylase activity. Plant J 31: 97–112.

48. Fujita N, Toyosawa Y, Utsumi Y, Higuchi T, Hanashiro I, et al. (2009) Characterization of pullulanase (PUL)-deficient mutants of rice (Oryza sativa L.) and the function of PUL on starch biosynthesis in the developing rice endosperm. J Exp Bot 60: 1009–1023.

49. Huang Y, Cai S, Ye L, Han Y, Wu D, et al. (2014) Genetic architecture of limit dextrinase inhibitor (LDI) activity in Tibetan wild barley. BMC Plant Biol 14: 117.

50. Stahl Y, Coates S, Bryce JH, Morris PC (2004) Antisense downregulation of the barley limit dextrinase inhibitor modulates starch granule size distribution, starch composition and amylopectin structure. Plant J 39: 599–611.

51. Vester-Christensen MB, Abou Hachem M, Svensson B, Henriksen A. (2010) Crystal structure of an essential enzyme in seed starch degradation: barley limit dextrinase in complex with cyclodextrins. J Mol Biol 403: 739–750.

52. Roldán I, Wattebled F, Mercedes Lucas M, Delvallé D, Planchot V, et al. (2007) The phenotype of soluble starch synthase IV defective mutants of *Arabidopsis thaliana* suggests a novel function of elongation enzymes in the control of starch granule formation. Plant J 49: 492–504.

53. Fujita N, Yoshida M, Asakura N, Ohdan T, Miyao A, et al. (2006) Function and characterization of starch synthase I using mutants in rice. Plant Physiol 140: 1070–1084.

54. McMaugh SJ, Thistleton JL, Anschaw E, Luo J, Konik-Rose C, et al. (2014) Suppression of starch synthase I expression affects the granule morphology and granule size and fine structure of starch in wheat endosperm. J Exp Bot 65: 2189–2201.

55. Cuesta-Seijo JA, Nielsen MM, Marri L, Tanaka H, Beeren SR, et al. (2013) Structure of starch synthase I from barley: insight into regulatory mechanisms of starch synthase activity. Acta Crystallogr D Biol Crystallogr 69: 1013–1025.

56. Jane J-L, Kasemsuwa T, Leas S, Ames IA, Zobel H, et al. (1994) Anthology of starch granule morphology by scanning electron microscopy. Starch 46: 121–129.

57. Yamamori M, Fujita S, Hayakawa K, Matsuki J, Yasui T (2000) Genetic elimination of a starch granule protein, SGP-1, of wheat generates an altered starch with apparent high amylose. Theor Appl Genet 101: 21–29.

58. Lafiandra D, Sestili F, D'Ovidio R, Janni M, Botticella E, et al. (2010) Approaches for the modification of starch composition in durum wheat. Cereal chem 87: 28–34.

11

Using Genotyping-By-Sequencing (GBS) for Genomic Discovery in Cultivated Oat

Yung-Fen Huang[1], Jesse A. Poland[2], Charlene P. Wight[1], Eric W. Jackson[3], Nicholas A. Tinker[1]*

1 Eastern Cereal and Oilseed Research Centre, Agriculture and Agri-Food Canada, Ottawa, Ontario, Canada, 2 Department of Plant Pathology, Kansas State University, Manhattan, Kansas, United States of America, 3 General Mills Crop Biosciences, Manhattan, Kansas, United States of America

Abstract

Advances in next-generation sequencing offer high-throughput and cost-effective genotyping alternatives, including genotyping-by-sequencing (GBS). Results have shown that this methodology is efficient for genotyping a variety of species, including those with complex genomes. To assess the utility of GBS in cultivated hexaploid oat (*Avena sativa* L.), seven bi-parental mapping populations and diverse inbred lines from breeding programs around the world were studied. We examined technical factors that influence GBS SNP calls, established a workflow that combines two bioinformatics pipelines for GBS SNP calling, and provided a nomenclature for oat GBS loci. The high-throughput GBS system enabled us to place 45,117 loci on an oat consensus map, thus establishing a positional reference for further genomic studies. Using the diversity lines, we estimated that a minimum density of one marker per 2 to 2.8 cM would be required for genome-wide association studies (GWAS), and GBS markers met this density requirement in most chromosome regions. We also demonstrated the utility of GBS in additional diagnostic applications related to oat breeding. We conclude that GBS is a powerful and useful approach, which will have many additional applications in oat breeding and genomic studies.

Editor: Xinping Cui, University of California, Riverside, United States of America

Funding: This work was supported by the Canadian Crop Genomics Initiative as part of Agriculture and Agri-Food Canada research grant 1885. The funders had no role in study design, data collection and analysis, decision to publish, or preparation of the manuscript.

Competing Interests: The authors have declared that no competing interests exist.

* Email: Nick.Tinker@AGR.GC.CA

Introduction

Cultivated oat (*Avena sativa* L.) is an allohexaploid ($2n = 6x = 42$) crop species that is grown as a source of food and feed. Oat and other crop species require continuous genetic improvement to meet the agronomic and nutritional needs of modern agriculture and food production. Major crop species such as corn, rice, wheat, canola, and soybean are benefiting considerably from advances in genome science and molecular breeding. These advances include the discovery and marker-assisted selection of single genes and quantitative trait loci (QTL), as well as the use of genomic selection (GS; [1,2]) to identify genotypes with superior performance and breeding value. Oat has also benefited from a long history of genomic and nutritional research [3] and from recent advances provided by a SNP platform and consensus map [4]. However, further advancements and applications of genomic technologies that can be integrated into traditional breeding strategies to accelerate and improve the development of superior oat cultivars are needed.

GS can be more efficient than phenotypic selection or marker-assisted selection for improving complex traits [5], and this has been demonstrated in oat [6]. Beyond GS, genomic characterization of breeding material offers many additional opportunities, including: the ability to monitor, maintain and expand germplasm diversity; the ability to diagnose identity or parentage of unknown material; and the ability to discover and deploy specific beneficial alleles [7,8]. Opportunities also exist for gene discovery, since species such as oat have unique biochemical pathways and adaptations not found in model plant species [9]. There are many technologies that can be applied routinely to whole genome characterization. These include parallel assays that target semi-random polymorphisms, such as Amplified Fragment Length Polymorphism (AFLP) and Diversity Array Technology (DArT), as well as parallel assays for specific single nucleotide polymorphisms (SNPs), such as the Golden Gate assays (Illumina, San Diego, CA). All technologies have strengths and weaknesses. Those that identify semi-random polymorphisms may not provide an adequate density of markers throughout the genome, and the technology may not transfer well between laboratories and different germplasm sets. The application of DArT technology in oat has been successful [10], but only a few hundred of the currently developed markers will segregate in a given population (unpublished results). The recently-developed SNP assay for oat [4] has a similar marker density to the DArT assay, but provides more precise and well-characterized gene-based predictions that may be more uniformly distributed throughout the genome and amenable to comparative mapping.

In oat, as in other polyploid species, genotyping is complicated because of the presence of homoeologous sub-genomes. Markers must be filtered to eliminate those that are confounded by multiple loci. This problem is less prevalent in pre-filtered SNP assays than it is in untargeted assays, although there are still SNP markers known to target different loci in different populations [4]. In all technologies, cost remains a critical factor. Currently, costs for DArT analysis and Illumina-based SNP assays range from \$US 50–60 per sample, which can be prohibitive for routine genomics-assisted breeding where a large number of the lines will be discarded following genotyping and selection.

Recently, a robust genotyping method based on the sequencing of partial genome representations has been developed for parallel high-throughput genotyping. This is referred to as genotyping-by-sequencing (GBS) [11]. GBS utilizes one or more restriction enzymes [12] to digest the genome into fragments that are then sequenced by parallel high-throughput methods. Based on the sequencing data, SNP calling can be done using various bioinformatics pipelines [12–22]. Low per sample cost in GBS is achieved by multiplexing samples from many (*e.g.*, 48, 96, or 384) different genetic entities (hereafter 'lines') simultaneously through the use of short specific 'barcodes' ligated to each sample prior to sequencing. Thus, it is possible to reduce the cost per sample by multiplexing more lines for sequencing. For example, if 96 samples are sequenced in a reaction costing $960, the cost per sample would be $10 over and above the costs of sample preparation and bioinformatic analysis. The cost will also be affected by the choice of single-end or paired-end sequencing. GBS has been useful in a variety of applications in crop plants including: saturating an existing genetic map [19], genome characterization in wheat and barley [12], genomic selection in wheat [23], the genetic ordering of a draft genome sequence in barley [24,25], and the characterization of germplasm diversity in maize and switchgrass [8,16]. These results suggest that GBS could be utilized for basic and applied genomic studies in oat.

Here, we report the development and application of GBS in oat mapping populations and a diverse set of oat germplasm. Our objectives were: (1) to evaluate the effects of different factors that influence the quality and quantity of GBS SNP calls and establish a baseline of operating parameters and expectations for GBS in oat; (2) to compare alternate methods of bioinformatics analysis and establish a pragmatic workflow and nomenclature for GBS data analysis in oat; (3) to saturate a consensus linkage map with GBS loci and establish a positional reference for future work; and (4) to investigate the utility of GBS to address a variety of questions that are typical of potential uses, including: *de novo* linkage mapping, characterizing population structure and linkage disequilibrium, and solving diagnostic issues in breeding germplasm. We discuss these results in the context of where GBS is likely to be most useful in crop development.

Materials and Methods

Genetic materials

Sets of germplasm used in this study are listed in Table 1. Additional diverse oat lines not reported in this study were prepared and sequenced in parallel with this work, which led to a total number of 2,664 oat lines being genotyped with GBS. These samples are mentioned because their presence may have had a minor influence on the parallel sequencing results or global allele-calling pipelines. These effects would be marginal, since more stringent filters were applied within sub-populations.

DNA sample preparation

The isolation of DNA was performed using a variety of methods, as some samples were available from previous studies. The preparation of DNA stocks from the CxH, HxZ, OxT, OxP, and PxG populations was described by Oliver *et al.* [4], while stocks from the KxO population were prepared as described by Wight *et al.* [26]. The latter stocks still contained RNA, which was removed using a standard RNase procedure followed by phenol/chlorofrom extraction and ethanol precipitation.

For the diversity population (IOI panel), eight seeds of each line were germinated in cyg growth pouches (Mega International, Minneapolis, MN, USA). Leaf tissue was harvested in bulk as the second leaves emerged and was put into paper envelopes containing a 5:1 mix of non-indicating and indicating silica gel desiccant. The paper envelopes were then placed in sealed containers for drying. For the VxL population, leaf tissue was harvested from plants growing in the field, then dried in the same manner. DNA was extracted from the VxL and IOI leaf samples using DNeasy Plant Maxi kits (Qiagen Inc., Mississauga, ON, Canada).

GBS library preparation and sequencing

The GBS libraries were constructed in 95-plex using the P384A adapter set (Table S2 in [12]). For each plate, a single random blank well was included for quality control to ensure that libraries were not switched during construction, sequencing, and analysis. Genomic DNA was co-digested with the restriction enzymes *Pst*I (CTGCAG) and *Msp*I (CCGG) and barcoded adapters were ligated to individual samples. Samples were pooled by plate into libraries and polymerase chain reaction-amplified. Detailed protocols can be found in [23]. Each 95-plex library was sequenced to 100 bp on a single lane of Illumina HiSeq 2000 or HiSeq 2500 by the DNA Technologies core facility at the National Research Council, Saskatoon, SK, Canada.

UNEAK GBS pipeline

Sequence results were analysed using the UNEAK GBS pipeline [16], which is part of the TASSEL 3.0 bioinformatics analysis package [27]. This method does not require a reference sequence, since SNP discovery is performed directly within pairs of matched sequence tags and filtered through network analysis. In this method, tags (a tag is an unique sequence representing a group of reads) belonging to complex multi-locus families (as determined by network analysis) are ignored. Parameters in the UNEAK pipeline were set for maximum number of expected reads per sequence file (300,000,000), restriction enzymes used for library construction (*Pst*I-*Msp*I), minimum number of tags required for output (10), maximum tag number in the merged tag counts (200,000,000), option to merge multiple samples per line (yes), error tolerance rate (0.02), minimum/maximum minor allele frequencies (MAF, 0.02 and 0.5), and minimum/maximum call rates (0 and 1). Call rate is defined as the proportion of samples that are covered by at least one tag. The MAF and call rate were set at a low value for global analysis because these parameters were filtered within sub-populations in later steps.

GBS pipeline using population-level filter

A second SNP calling pipeline was employed as described by Poland *et al.* [23]. This pipeline is implemented in TASSEL 3 and was functionally identical to UNEAK to the point of developing a binary presence/absence matrix for each tag across multiple lines. To identify putative SNPs, tags were internally aligned allowing up to 3 bp mismatch in a 64 bp tag. From aligned tags, SNP alleles were identified and the number of lines in the population with each respective tag was tallied in a 2×2 table, counting the number of lines with one or the other tag, both, or neither [23]. A Fisher Exact Test was then used to determine if the two alleles were independent, as would be expected for a single locus, bi-allelic SNP in a population of inbred lines. If the null hypothesis of independence for the putative SNP was rejected ($p < 0.001$), we assumed that the tags were allelic in the population (and, therefore, that the putative SNP was a true single locus, bi-allelic SNP). A significance threshold of $p < 0.001$ was selected for the size of

Table 1. Populations and germplasm samples used in this study.

Genetic material	Abbreviation	Number of lines	Reference*	No. of SNP**
Bi-parental mapping populations				
Otana x PI269616 (F$_6$)	OxP	98	[4]	17,137
Provena x GS7 (F$_8$)	PxG	98	[4]	11,755
Ogle x TAMO-301 (F$_{6:7}$)	OxT	53	[48]	30,726
CDC SolFi x HiFi (F$_7$)	CxH	52	[4]	8,324
Hurdal x Z-597 (F$_6$)	HxZ	53	[4]	4,219
Kanota x Ogle (F$_7$)	KxO	52	[49]	2,582
VAO-44 x Leggett (F$_{4:5}$)	VxL	145	This study	280 (373)
Diversity panels				
Oat diversity panel	IOI	340	[31]	2,155

*First publication that refers to the population
**No. of SNP filtered for subsequent analyses. Please refer to the text for filtering criteria. For VxL, two sets of filtering criteria were used. The only difference between the filtering criteria sets is the heterozygosity level: 8% or 13% (SNP number is between brackets). For IOI, only markers passing filtering criteria with a map position were reported in the table.

population, based on previous work testing false discovery rates in duplicate samples.

Filtering and merging GBS SNP calls

Both of the above pipelines were applied globally to all available sequencing data, except where we deliberately tested SNP identification in partial datasets. This global strategy reduced the need to access large sequencing files repeatedly. However, there was then a need to generate genotype data for specific sub-populations, and to apply population-specific filters for allele frequency, heterozygosity, and data completeness (data completeness is defined as 100% - % missing data; e.g., for a marker genotyped on 100 individuals with 10 individuals showing missing data points, the completeness of the marker is 90%). Furthermore, for genotypes of mapping progeny, it was necessary to recognize the parental phase of alleles, and to represent alleles using conventions required by the mapping software. These filters and secondary analyses were applied using in-house software ('CbyT') written in the Pascal programing language (Text S1). This software provided the additional feature of maintaining a cumulative index of unique SNPs with a consistent naming convention, such that data from different pipelines or subsequent assays could be merged to remove redundancy and to index matching SNPs with the same unique name. Each subsequent analysis required specific filtering criteria, which can be found at the beginning of the method section for each type of analysis.

Linkage mapping

For bi-parental mapping populations, parental lines were genotyped together with the progeny. GBS loci called using both pipelines across six bi-parental RIL populations (OxP, PxG, OxT, CxH, HxZ, and KxO) were filtered at ≥50% completeness, MAF ≥35%, and heterozygosity ≤8% inside each population, which gave a total of 45,117 GBS markers. The SNP data from the six mapping populations reported by Oliver et al. [4] were filtered to the same standards as the GBS SNP data, and the two data types concatenated. Marker phases were determined using parental genotypes when the latter were available and not monomorphic. Monomorphic parental genotypes can result from genotyping errors or genetic variation within the lines used to make the cross. Markers for which there were no good parental data were

converted into both parental phases for further analysis. For each mapping population, the phase of parental alleles was re-checked across the concatenated data by enumerating, for each SNP, the number of linked loci in the same phase (recombination fraction, $r < 20\%$) vs. the number in opposite phase ($r > 80\%$). Loci having a greater number of out-of-phase matches than in-phase matches were rescored in the opposite phase, or were eliminated through a recursive process if this did not improve the in-phase/out-of-phase ratio.

An updated version of the oat consensus map developed by Oliver et al. [4] was generated by placing each new candidate locus (GBS or other non-framework SNP) relative to framework SNPs from the existing map. Pair-wise recombination fraction (rf) was first calculated for all marker pairs, including both framework and non-framework markers. Marker placements were then made relative to the two framework loci showing the smallest rf among any of the six populations. The approximate map position of each placed marker was subsequently estimated by interpolating the cM position proportional to the recombination fraction with the closest two framework loci. When the closest framework locus was at the end of a linkage group, and the recombination with the next-closest framework locus was greater than that between the two framework loci, the candidate was placed distal to the end of the linkage group. In addition to this crude approximation of marker position, a detailed report of each placed marker was produced to show the actual recombination frequencies within each population and across populations between a given marker and all other loci that were within 20% recombination in any of the component populations. Marker data used for marker placement on the oat consensus map are in Table S1.

De novo linkage map construction was performed using MSTMap [28] for the VxL population. GBS loci for *de novo* map construction were called using the UNEAK GBS pipeline and filtered at high stringency (MAF ≥35%, completeness ≥90%) at two different levels of heterozygosity (8% and 13%). The resulting data contained 858 (heterozygosity ≥8%) and 1053 (heterozygosity ≥13%) GBS markers. The choices of 8% and 13% corresponded to the expected heterozygosity at F$_5$ and F$_4$, respectively, factoring in sequencing error and out-crossing rate. For MSTMap, a p-value equal to 10^{-11} was used for the marker clustering threshold. Markers were excluded as unlinked if they were 15 cM away from any other locus or if they belonged to a group containing only two loci. A simple recombination count was used for the objective function. Since map distances estimated by MSTMap are inflated (based on simulated data, result not shown), we re-estimated the recombination fractions between pairs of loci based on the marker order from MSTMap and converted them to map distances using the Kosambi mapping function.

Population structure and LD analysis

For population structure and LD analysis, GBS markers called by the UNEAK pipeline were filtered at $\geq 90\%$ completeness, MAF $\geq 5\%$, and heterozygosity $\leq 5\%$. Population structure was investigated using principal component analysis (PCA). PCA was performed with the 'smartpca' function implemented in EIGEN-SOFT [29]. This function takes into account marker dependency (*i.e.*, markers in LD blocks) through the use of multiple-regression on adjacent markers prior to PCA. The maximum interval distance between markers (ldlimit) was set to 0.001. The number of adjacent markers included in LD adjustment (ldregress) was set at 0, 10, or 50 (designated k0, k10, and k50), such that k0 provided no LD correction and k10 and k50 corresponded to the median and maximum LD block sizes in the IOI dataset. Eigenvalues and Eigenvectors were transferred to the R statistical package [30] for scree plot drawing and other analyses.

A model-based approach was used to investigate the clustering pattern among lines in the diversity panel further, because it determines simultaneously the number of clusters and cluster membership and does not have underlying genetic assumptions that are rarely met [31]. Model-based clustering was based on the first ten PC and conducted using the clustCombi function of the R package mclust [32]. The purpose of clustCombi is to represent a non-Gaussian cluster by a mixture of two or more Gaussian distributions [33]. It first uses the Bayesian information criterion (BIC) to identify the number of Gaussian mixture components and then hierarchically combines components according to an entropy criterion. The final decision concerning the number of clusters to use was made based on an entropy plot; *e.g.*, if six components were identified by BIC and successive component combinations showed no large entropy decrease after four clusters, then the data were represented by four clusters.

Linkage disequilibrium (LD) between two loci was estimated as squared allele-frequency correlations (r^2) by an optimized version (Stéphane Nicolas, personal communication) of LD.Measure in the R package LDcorSV [34]. Four r^2 estimates were calculated: conventional r^2 based on raw genotype data, r^2 with population structure included in the calculation (r_s^2), r^2 with relatedness included in the calculation (r_v^2), and r^2 with both population structure and relatedness included (r_{sv}^2). Population structure was represented by the first four PC after scaling the coordinate identifiers across a range of zero to one. A matrix of relatedness was calculated by A.mat, implemented in the rrBLUP package [35].

The relationship between LD and genetic distance was modeled by fitting two alternate non-linear regression models: a drift-recombination equilibrium model [36] or a modified recombination-drift model including low level of mutation and an adjustment for sample size [37]. Both models were summarized in [38].

Other statistical analyses

We wished to examine how GBS technology could be used to solve diagnostic problems that arise occasionally in any plant breeding program. Germplasm diagnostics were performed using DARwin software [39] to generate clusters based on genetic distances among lines, estimated using simple allele-matching for bi-allelic diploid loci:

$$d_{ij} = 1 - \frac{1}{L} \sum_{l=1}^{L} \frac{m_l}{2}$$

where d_{ij} is the dissimilarity between lines i and j, L is the number of informative loci shared by those lines, and m_l is the number of matching alleles for locus l. Cluster analysis was performed using the un-weighted paired group mean analysis (UPGMA) method.

Results

Library construction, sequencing and coverage

For this study, a total of 38 libraries were generated, each multiplexing 95 lines. A single lane of Illumina HiSeq 2000 or HiSeq 2500 was used to sequence each library. The GBS libraries were constructed as previously described for wheat, with the exception that the forward barcode adapter concentration was reduced to 0.06 pmol for 200 ng of genomic DNA (*vs.* 0.1 pmol used for wheat in [12]). This adapter concentration was found to improve the oat libraries, reducing adapter dimers.

A complete set of short read archives for all GBS oat samples analysed to date has been made available for download from the NCBI short read archive (http://www.ncbi.nlm.nih.gov/sra/) under project accession number SRP037730. Details of these archives, including number of reads and number of good barcoded reads at the level of each flow-cell, single lane, and individual taxon are available in Table S4. Table S4 also provides the key file needed to support re-analysis of the raw short read archives by either of the GBS pipelines reported here.

From a total of 6.3×10^9 reads, 84.4% (5.3×10^9) included the barcode sequence and enzyme cut-site, and had no unreadable base ('N') in the sequence. The UNEAK pipeline found an average of 732,396 tags per oat line in the samples reported here, a total merged tag count across all samples of 358,177,647 tags, and a filtered tag count (tags appearing >10 times) of 17,700,128 that were covered by 564,946,411 matching reads.

Each sequencing lane generated approximately 2×10^8 100 bp reads. After discarding reads that did not have an exact match to one of the barcodes, there were approximately 2×10^6 100 bp reads per sequenced DNA sample. We designated the 2×10^6 100 bp-base reads/DNA sample as one unit of 'depth index'. To test the influence of plexity and sequencing depth on GBS data completeness, we used data from 53 OxT mapping progeny sequenced in three separate lanes. We split the raw sequencing data from two lanes in half and added these incrementally to the un-split lane, which contained a lower number of reads. This provided five different sequencing depths with mixed levels of plexity, from which we computed average depth indices of 0.58, 0.95, 1.33, 1.85, and 2.37. For example, the depth index of 0.58 means there were, on average, $0.58 \times 2 \times 10^6 = 1.16 \times 10^6$ reads/sample for that experiment. The exact read depth for individual samples varied because of sample quality and/or variations in barcode efficiencies. The UNEAK GBS pipeline was run on these five data subsets and SNPs were filtered at four levels of completeness (25%, 50%, 75%, and 90%). The results (Figure 1) showed that an increasing number of SNPs were called as the depth index was increased at all four completeness levels. The response to sequencing depth appears to be linear within the range tested. One of the sequencing runs, added at the second and third levels, had a higher variation in read depth among samples, which explains the lower slope at these levels and the fact that almost no SNPs had a completeness of 90%.

Number of GBS loci called

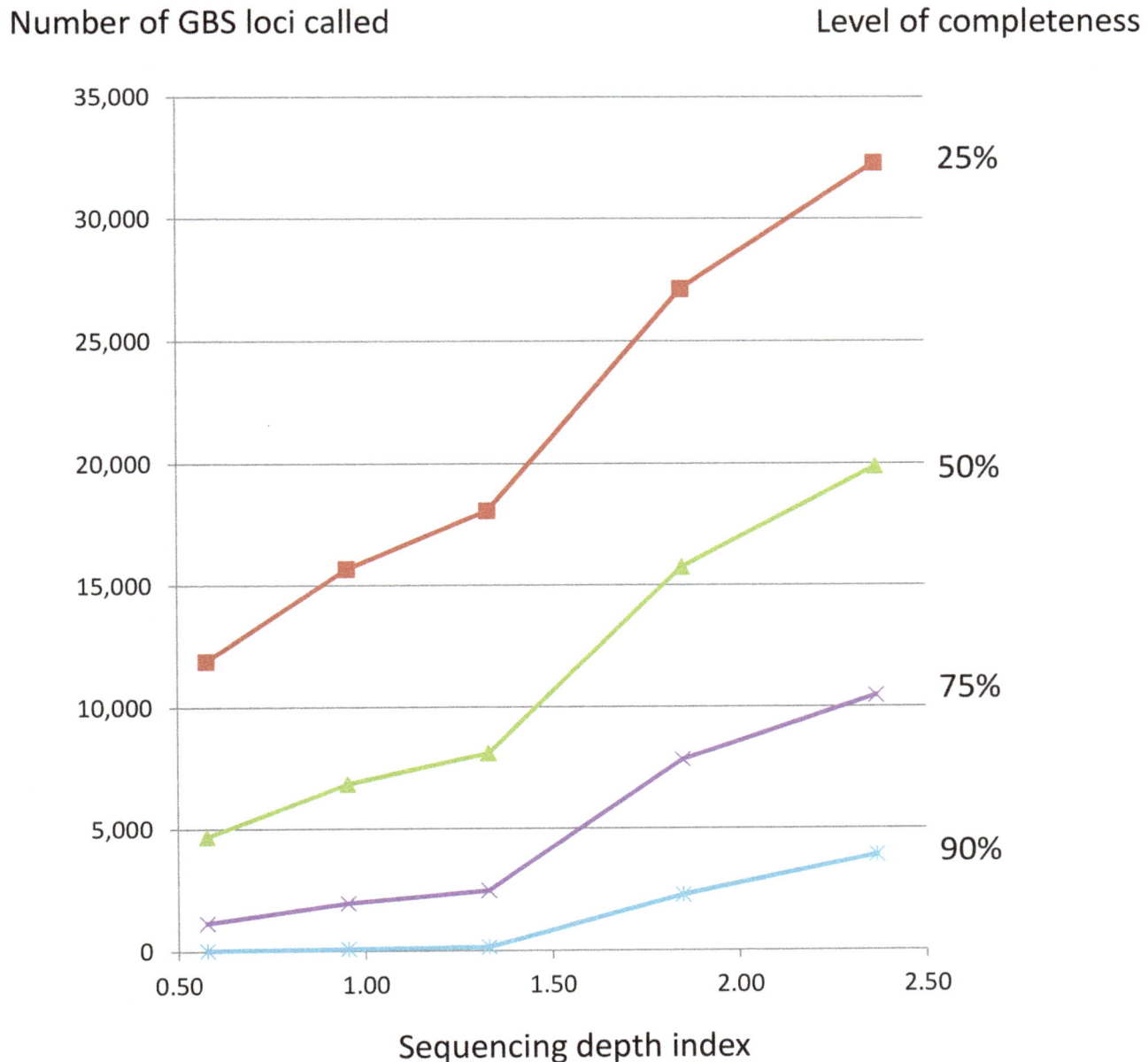

Figure 1. Number of GBS loci *vs.* sequencing depth. Number of GBS SNP loci called in 53 OxT mapping progeny at increasing sequencing depth, filtered at four levels of completeness (25%, 50%, 75%, and 90%). Other filtering parameters were constant, with heterozygosity ≤10% and minor allele frequency ≥30%. A sequencing depth index of 1 represents the average read depth that would be achieved with 95 samples multiplexed in a standard Illumina sequencing run giving approximately 2×10^8 short reads. Thus, an index of 2 would be equivalent to twice this number of reads or half of this plexity.

Population size *vs.* number of GBS SNPs

We examined random subsets of 366 diverse oat varieties, including the IOI set and 26 additional winter oat varieties, to determine how sample size would affect the number of GBS SNPs called at differing levels of completeness. Sample size was varied between 10 and 360 at increments of 10, with two randomly chosen subsets as replicates for each sample size. These data were filtered at a maximum heterozygosity of 10%, MAF of 5%, and minimum completeness of 25%, 50%, 75%, or 90%. At each sample size, the number of SNPs passing these filters was counted. At a low threshold for completeness (25%), the number of SNPs

increased with sample size, plateauing at approximately 50,000 SNPs once 250 of the 360 oat lines had been included (Figure 2). At higher thresholds for completeness (50%, 75%, and 90%), the number of SNPs plateaued at approximately 20,000, 10,000, and 5,000 SNPs, respectively. These plateaus occurred at increasingly smaller sample sizes, and the number of SNPs appeared to decrease slightly as sample size increased beyond the initial plateau.

Number of GBS loci called

Level of completeness

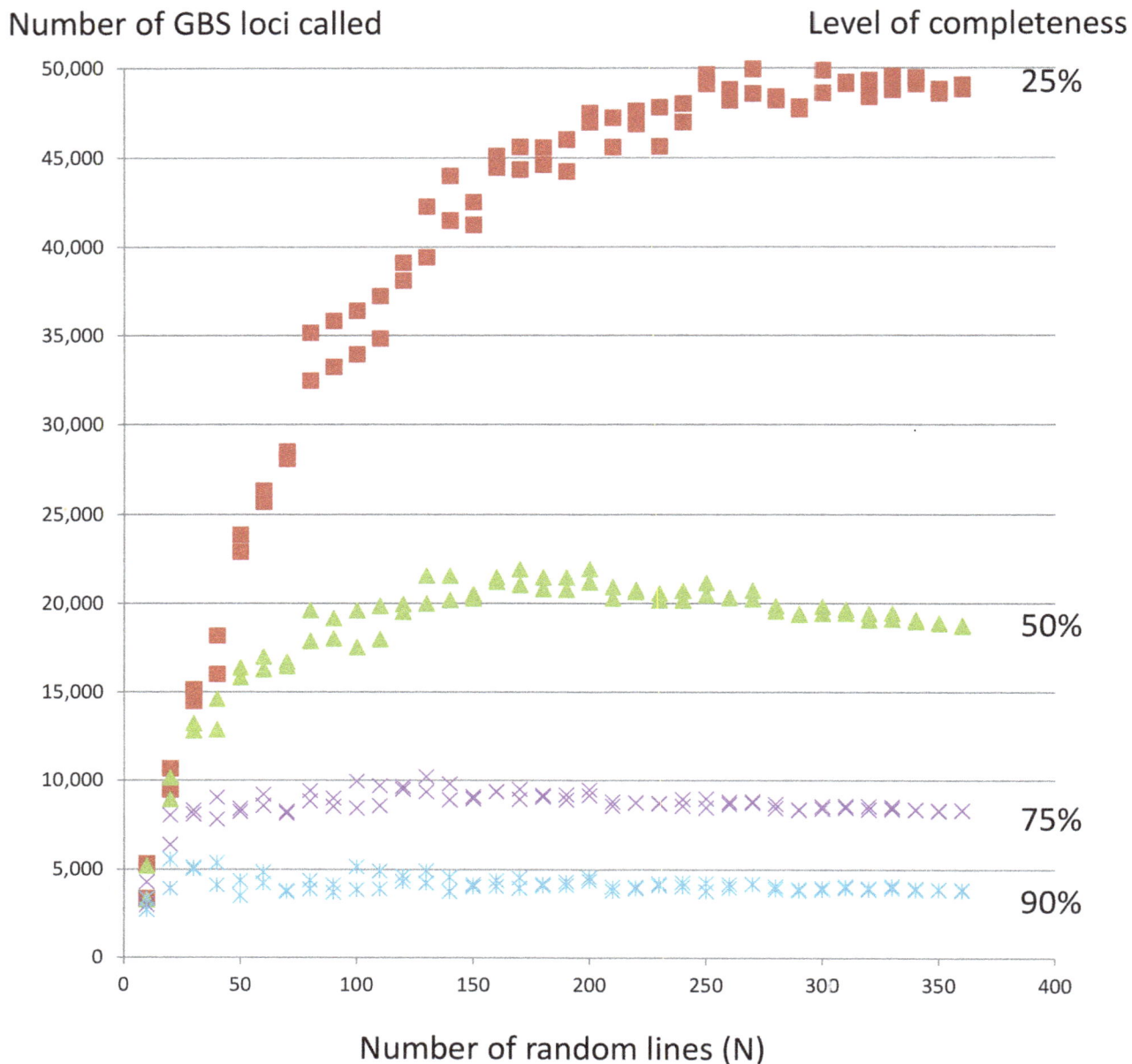

Figure 2. Number of GBS loci *vs.* sample size. Number of GBS loci called in samples of size *N* from a set of 360 diverse oat lines filtered at four levels of completeness (25%, 50%, 75%, and 90%) is shown. Other filtering parameters were constant, with heterozygosity ≥10% and MAF ≥5%.

Multi-allelic SNPs

Although multi-allelic loci were not specifically identified by either of the two GBS pipelines employed in this work, it was apparent that the population-based method would frequently call separate pairs of bi-allelic SNPs at the same site, providing evidence for third and, occasionally, fourth alleles. We used this result to make an approximate estimate of the frequency at which multi-allelic SNPs can be identified from existing pipelines. Of the 355,731 unique bi-allelic SNPs from all oat projects called using both pipelines, there were 343 sets of tri-nucleotide SNPs (*i.e.*, two bi-allelic SNPs having identical context sequence except for a third allele at the SNP position), and 21 sets of tetra-nucleotide SNPs (*i.e.* two bi-allelic SNPs having identical context sequence except for the SNP position). This analysis was not expected to give

comprehensive access to all multi-allelic SNPs, since most of the multi-allelic SNPs would have been filtered out during SNP calling. However, this result suggests that tri- and tetra-nucleotide SNPs are extremely rare, which is expected when SNPs arise primarily as random neutral mutations.

Integration of GBS SNPs with an existing genetic consensus map

45,117 GBS loci, filtered across six bi-parental RIL populations, were placed on the oat consensus map of Oliver *et al.* [4] based on simple counts of recombination fractions. Starting from the initial consensus map, each additional population provided from 2,535 (KxO) to 30,369 (OxT) more loci (Figure S16). As more populations were used for marker placement, fewer new markers were added to the map, but the number of new markers was always proportional to the number of markers available from the source population; *e.g.*, there were always more new loci from OxT (21,894–30,369) than KxO (1,065–2,535) (Figure S16). The complete report for this map is available as a supplementary HTML file in Text S2 (or online at: http://ahoy.aowc.ca/html_link_gbs_text_S2.html). The report format is described in Figure S1. Each individual marker on this HTML map is linked to a separate, detailed report that shows a complete matrix of recombination fractions between the reported marker and the neighbouring loci from the consensus map, as well as other placed markers across all populations.

To investigate the approximate global distribution of GBS loci relative to the SNP consensus map, we divided the map into bins of 5 cM intervals and produced a density histogram which showed the number of placed GBS loci in each bin along the genome (Figure 3). Some bins contained no markers, while some contained a much larger number of markers. In general, GBS loci tended to cluster in the same locations as array-based SNPs. Some clusters probably reflect centromeres, where suppressed recombination causes genetic clustering. However, some chromosomes contained multiple regions of clustering, especially 3C, 4C, 5C, 16A, 19A, 12D, and 21D. This may be caused in part by cytogenetic differences among the parents of the mapping populations, whereby individual maps contain underlying differences in the structure and order of genetic markers. The consensus map would have compressed these differences into a single 'average' map, but the underlying differences among populations remain, and placed markers may appear to cluster at points where the consensus has averaged these differences.

The inclusion of GBS markers appeared to fill gaps within the consensus map. For example, 112 marker intervals larger than 5 cM were present on the original consensus map [4], while only 25 are present on the same map once GBS markers are placed, and the maximum gap size decreased from 26.98 cM to 15.74 cM (Figure S2). These results do need to be interpreted with caution, because the interpolated positions of markers with miss-scored alleles may appear to fill some gaps. An accurate re-interpretation of the consensus map can only be achieved by a complete reanalysis and reinterpretation of the component maps. This work is in progress and will be reported elsewhere together with a complete report on additional SNP loci. Preliminary results of this work (unpublished data) indicate that GBS markers fill some gaps, but that their greatest benefit is to increase the number of loci that are mapped in multiple populations.

Examination of orthology with other crops

The high density of approximately-placed GBS markers provides a new opportunity to examine orthologous relationships between oat and model genomes. The orthology analysis was performed to determine whether matches of short GBS sequence to model genomes would be sufficient to identify major regions of genome co-linearity. Using dot-plots, we explored the locations of sequence similarity between the oat consensus map and pseudo-molecule sequence assemblies from *Brachypodium distachyon* L. (Bd), rice, and barley (Figures S3 to S6). As in a previous analysis using only array-based SNPs [4], we observed that *Brachypodium* had a greater number of matches and better colinearity with oat than did rice. Several stretches of colinearity were observed, such as those on oat 19A (similar to parts of chromosomes Bd1, Bd2, and rice1) and oat 20D (similar to Bd5 and rice4). In the current analysis, it is clear that, in regions of collinear sequence-based matching, both the GBS and other SNPs are contributing similar information. In some cases, the GBS loci appeared to extend the

regions of colinearity (for an example, see Bd1 and Bd4 in Figure S3). It was also clear from the higher density of GBS matches to *Brachypodium* sequences that a greater number of non-coding sequences were similar between *Brachypodium* and oat than between rice and oat, despite the larger genome of rice. This is probably because of the closer ancestral relationship between oat and *Brachypodium*. Barley showed very poor colinearity with oat, based on sequence matching to the newly-available ordered shotgun assembly [25]. We suspect this to be a result of the incomplete nature of the barley assembly. It was also notable that many GBS SNPs showed highly repetitive matches to the barley genome (Figure S5), which were mostly eliminated by removing GBS loci determined to have multiple matches to other GBS loci within oat (Figure S6). This suggests that there are many repetitive elements that are shared between the oat and barley genomes.

GBS SNP annotation

In order to give an approximation of the number and distribution of genic and intergenic GBS SNPs in oat, we compared a complete set of 355,731 context tags of GBS SNPs from all oat projects available at the time of analysis to the chromosome-based genome assembly and accompanying gene predictions from *Brachypodium* (release 2.1; http://www.brachypodium.org) by BLAST. Of these, 19,656 tags (5.5%) showed protein matches (BLASTx) with expectation <0.1, a level that corresponds approximately to a minimum 60% identity over the full tag length or 100% identity over half the tag length. Although there will be oat genes that do not have *Brachypodium* orthologues, it is still likely that fewer than 5% of GBS tags are within transcribed oat genes, because many of the protein signatures matching *Brachypodium* will likely represent vestigial genes in oat. Of the tags with protein matches, 16,712 showed DNA similarity within the transcribed region at the threshold expectation of <0.1. However, we noted that the average BLASTn expectation corresponding with a BLASTx expectation of 0.1 was approximately 0.001; therefore, we conducted further nucleotide matches at this level. This allowed us to identify a total of 46,370 tags (13% of total) with nucleotide matches in *Brachypodium*, among which 30,713 (66% out of matched tags or 8.6% out of total tags) were in intergenic regions, and 15,657 (34% of matched tags or 4.4% of total tags) inside gene regions, of which only 300 did not match a protein. Since gene regions in *Brachypodium* correspond to approximately 43% of the genome, it appears that there is some bias toward GBS nucleotide matches outside of gene regions.

Out of the 46,370 tags matched to *Brachypodium* sequences, 9,684 were positioned somewhere on the consensus map (*cf.* Results/Integration of GBS SNPs with an existing genetic consensus map). On average, 20.5% of mapped loci were positioned inside a gene, ranging from 14.2% to 29.08%, which is a slightly smaller than the proportion in overall tags having BLAST matches to the *Brachypodium* genome. The distribution of *Brachypodium* orthologous SNPs along the oat genome is similar to that of the overall oat GBS SNPs (Figure S17 and Figure 3). Genic and intergenic SNP counts per chromosome and their genome distribution can be found in Figure S17. While this result gives an approximation of the proportion and distribution of genic and intergenic SNPs along the oat genome, care should be taken in interpreting this result to form a general conclusion about oat, not just because of the partial coverage of the present consensus map, but also because of the approximate nature of the annotations made through the use of orthologues.

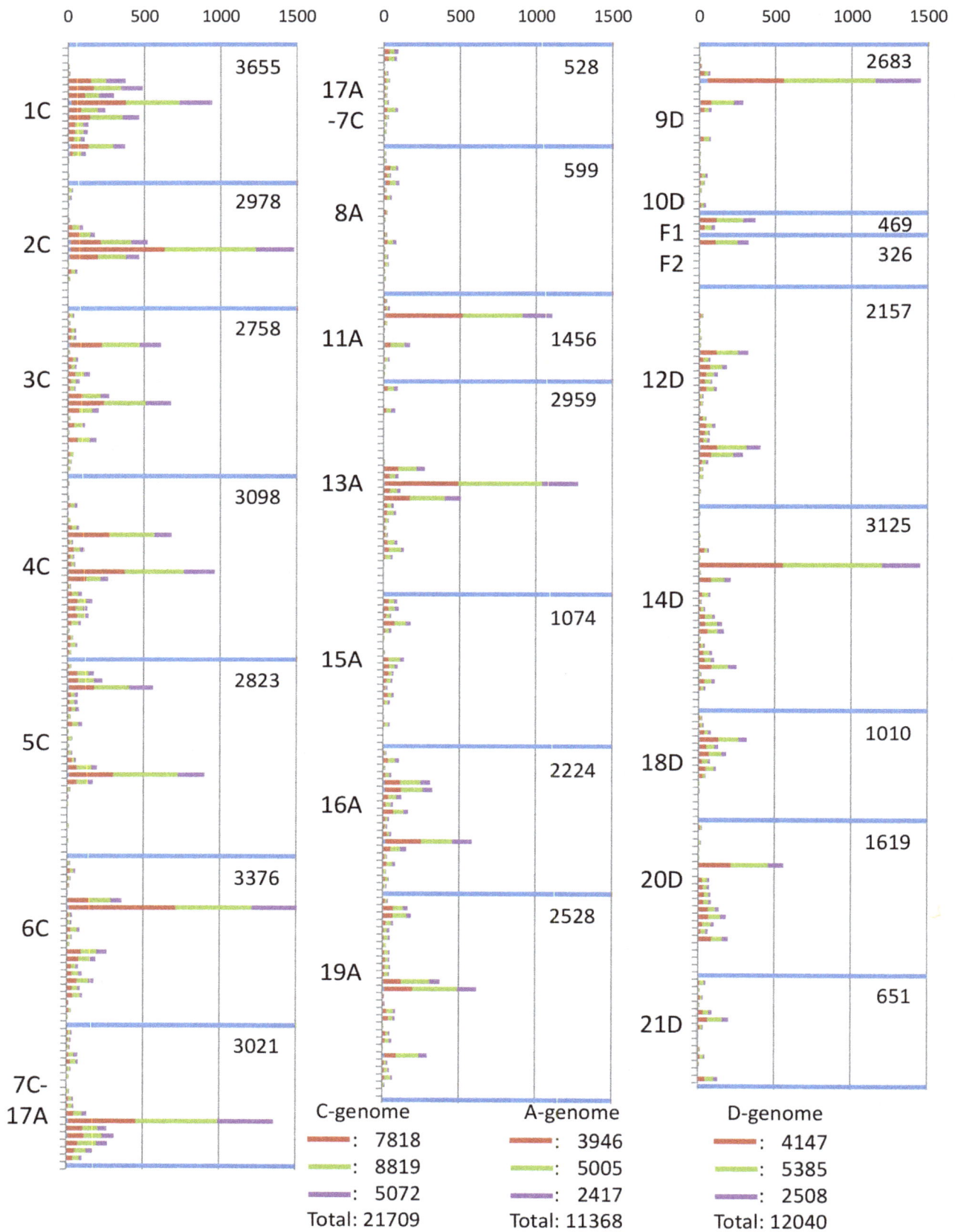

C-genome

    : 7818
    : 8819
    : 5072
Total: 21709

A-genome

    : 3946
    : 5005
    : 2417
Total: 11368

D-genome

    : 4147
    : 5385
    : 2508
Total: 12040

Figure 3. Distribution of GBS loci across the oat genome. Maps of each chromosome (delineated by blue lines and labeled on left) are divided into 5 cM bins with 0 cM starting at the top. Red bars show numbers of loci detected by two pipelines, green shows those detected only by the population-filtering pipeline, and violet shows those detected only by the UNEAK pipeline. Numerals inside boxes show total GBS loci by chromosome. A summary of placed GBS loci by pipeline and by sub-genome is shown.

A *de novo* genetic linkage map using GBS

In the above work, mapping was performed in six populations of reduced size, primarily to approximate the positions of a large inventory of GBS loci relative to an oat consensus map. We also wished to examine the utility of GBS markers for generating a *de novo* linkage map in the absence of other marker types, and to evaluate how well this map could be matched to the current consensus. For this work, we used the VxL population, composed of 145 $F_{4:5}$ RIL families. The GBS loci for map construction were called using the UNEAK GBS pipeline and filtered at high stringency (MAF $\geq 35\%$, completeness $\geq 90\%$) at two different levels of heterozygosity (8% and 13%). The resulting data contained 858 (heterozygosity $\geq 8\%$) and 1053 (heterozygosity $\geq 13\%$) GBS loci. From this, a map with 35 linkage groups having a total length of 1713 cM was constructed. A comparison between this map and the consensus (Figure S7) showed that 280 (heterozygosity $\geq 8\%$) or 373 (heterozygosity $\geq 13\%$) markers were present in VxL but not in the six mapping populations used for consensus map saturation. Most VxL linkage groups corresponded to single consensus chromosomes, and the relative positioning of loci within groups was approximately linear. Several sets of VxL linkage groups (*e.g.*, LG07 and LG20) likely represent single oat chromosomes (in this case, 12D). This suggests that there is good opportunity to perform comparative mapping of traits that are identified in new populations using only GBS technology. The filtering of loci at different levels of heterozygosity provided an opportunity to observe that certain regions of the VxL genetic map are more highly heterozygous than others (red dots in Figure S7, and graphical genotypes in Table S2), as would be expected in an F_4 population. In addition, most of the heterozygous loci were clustered at what are likely centromeres (Figure 3 and Figure S7), explained by the fact that low recombination in centromeric regions has been found to contribute to the retention of residual heterozygosity [40]. This provides good evidence that heterozygous genotype calls in the GBS pipeline are generally accurate and genetically consistent.

Use of GBS to evaluate population structure

PCA and model-based clustering were used to examine the effectiveness of GBS markers to identify population structure in 340 oat lines of global origin. Of these, 41% originated from North American breeding programs (81 lines from Canada and 59 from USA) and the remainder originated elsewhere (Table S3). GBS markers called by the UNEAK pipeline were filtered at $\geq 90\%$ completeness, MAF $\geq 5\%$, and heterozygosity $\leq 5\%$. Of the filtered SNPs, only those that were placed on the consensus map (2155 loci) were considered. Because of genetic clustering, a large number of these SNP loci (1159, or 54%) co-segregated at identical positions, and 1755 (81%) were within 1 cM intervals (Figure S8). Since this dependency is also reflected in LD (see next section), it was likely to distort the Eigenvector/Eigenvalues and to bias the interpretation of population structure [29]. Therefore, we applied two levels of LD correction (k10 and k50), in addition to using an uncorrected analysis (k0) for PCA. At k0, no obvious reflection point was observed in the scree plot, while we could distinguish slight two-stage plateaus in k10 and k50, where the drop of Eigenvalues slowed at approximately PC5 and PC10 (Figure S9). The first ten PC explained 37.6, 32.1, and 31.6

percent of the total variation for k0, k10, and k50, respectively, whereas 25.0, 21.2, and 20.9 percent of the total variation was explained by the first four PC (the approximate point of the first plateau).

Since no obvious groups were separated by the first two PCs, model-based clustering was performed to explore the grouping of oat lines based on the first ten PC. The best solutions for k0, k10 and k50 were four, two, and two clusters, although the entropy plot of k50 could be understood as three clusters (Figure S10). The clusterings from k0, k10, and k50 were generally in agreement and reflected the geographic origins of the oat lines, with European lines tightly clustered together and lines from elsewhere spread out across the plot (Figures 4 and S11). The possible third cluster in k50 was positioned between European and North American lines and was comprised of oat lines from Eastern or Northern Europe, as well as some North American lines. One set of five lines was separated by PC4 and this separation is particularly obvious in the k0 data set (Figure S12). The separation of this cluster seemed to be related to growth habit (three of the five lines are winter oats, Figure S12), but because the number of lines was so small, a definitive conclusion could not be made. Our results showed that this diversity panel does not show substantial structural stratification, and this is in agreement with previous work based on DArT markers [31].

Linkage disequilibrium analysis

From the original 2155 markers, we retained r^2 estimates from 51,850 marker pairs with an average or minimum map distance less than 30 cM. Plotting these r^2 estimates against map distance (Figure 5, Figure S13, and Table 2) showed that LD decays such that, at 0.1, conventional r^2 is equal to an average distance of 21.5 cM (Hill-Weir model) or 13.6 cM (Sved model), while r_v^2 is equal to 2.8 cM (Hill-Weir model) or 2.5 cM (Sved model). The fact that r_v (corrected for relatedness) is much smaller than r^2 (uncorrected) illustrates the necessity of removing the effect of coancestry to reduce the inflation of r^2 estimates, and probably reflects that the IOI panel contained groups of related lines originating from the same breeding programs. Estimates of r_s^2 (accounting for population structure) did not reduce the bias in r^2 as substantially as did the models accounting for coancestry, which is consistent with earlier observations that this population is not highly structured. These results suggest that good genome coverage for GWAS will require a marker spacing of approximately 2.0 cM (r_{sv}^2 k0, min *rf*, Sved model) to 2.8 cM (r_v^2, average *rf*, Hill-Weir model, Table 2). Non-linear model fitting enabled us to estimate the effective population size required for GWAS, which varied from 68 to 110 lines, depending on the choice of r^2 estimates and evolution models (Table 2).

Germplasm diagnostics using GBS

We wished to examine how GBS technology could be used to solve two common diagnostic problems that arise occasionally in any plant breeding program. The first example involved a suspected error in the planting of one replication in an oat variety registration test. The questionable replication could have been discarded, but it had been grown and harvested at a cost that was more than double that required for genotyping the unknown samples, and discarding the replication would jeopardize the

A

B

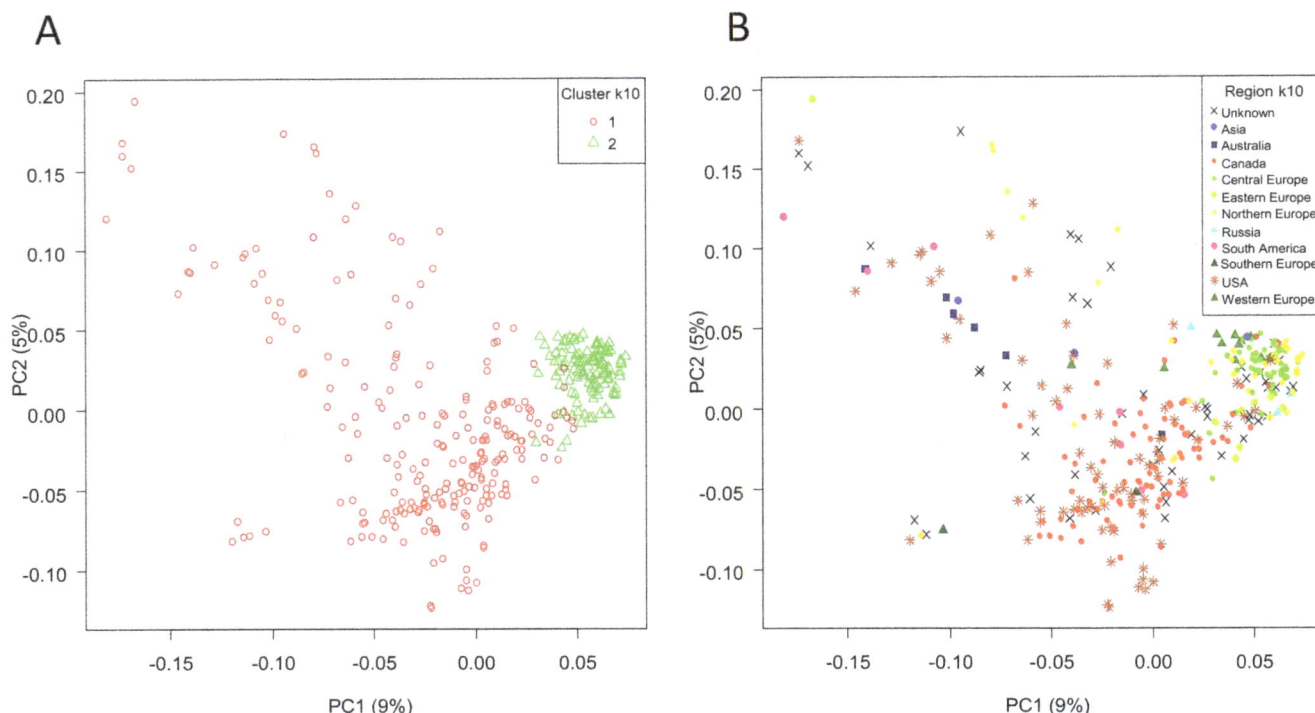

Figure 4. Scatter plots of PC1 *vs.* PC2. The k10 correction is shown: (A) coloured based on clustering from genotypic data, (B) coloured based on geographic origins.

statistical power of the experiment. Problems of this nature can have great economic impact if they delay or improperly influence the release of improved plant varieties. We filtered a set of 1518 diagnostic GBS loci that were polymorphic among known and unknown lines from this test. We then applied UPGMA cluster analysis to the simple allele dissimilarity (*d*) among samples. The results (Figure S14) illustrate that each unknown sample could be paired with one or more known samples. It was then clear that the source of error was a simple reversal of seed envelopes in one of the planting trays. The phenotypic data could then be reassigned to complete the analysis. Although the unknown experimental units could be unambiguously corrected based on the genetic data, a few of the distances between samples known to originate from the same variety (*e.g.*, samples 128 and 232 in Figure S14) were larger than expected. We expect that this is because the DNA samples were prepared from a few seeds sampled randomly from bulks that were harvested by combine from the registration test. If so, this draws attention to the fact that seed harvested from yield trials is often impure and should be used with caution in genetic studies.

The second diagnostic problem was to determine whether an F_2 population originated from a true cross or from selfed seed of a parent. In this case, the progeny appeared very homogeneous and an error was suspected. However, resources had been invested in the cross, and it seemed worthwhile to address this issue before discarding F_3 seed. We filtered GBS loci for ten F_2 progeny and the intended male parent of the cross, together with 340 additional progeny from the IOI diversity panel. The intended female parent was unavailable, but a maternal grandparent (Leggett) was available from the IOI set. Filtering at 10% heterozygosity, 10% MAF, and 90% completeness gave 2205 locus calls. All 2205 loci were completely homogeneous among the suspected progeny and the male parent, with the exception of minor variants that

appeared random and fell within a 1% tolerance for scoring error. A cluster analysis (Figure S15) supported this result. Thus, it was concluded that these seeds were not from a true cross, and that they probably represented a harvesting error in the crossing block.

Discussion

Factors that influence the quality and quantity of GBS SNP calls

The completeness of GBS SNP calls and the number of SNPs filtered at a given completeness are influenced by several factors, including: (1) the actual number of genomic fragments produced by the complexity reduction, (2) the sequencing depth, as determined by the number of reads and the number of samples that are multiplexed, and (3) the underlying density of SNPs, which is related to the diversity and structure of the population [41]. Although other restriction enzymes can be used for complexity reduction (*e.g.*, [11,17]), we limited our present investigation in oat to the *Pst*I-*Msp*I combination which had previously been optimized for the similar-sized genome of wheat [12,23]. This method provided suitable results in oat with minor modifications, and identified tens of thousands of sequence polymorphisms. Although the results in Figure 1 show a somewhat linear response in the number of SNPs identified as sequencing depth increases, the number of SNP calls for all levels of completeness would eventually plateau at a limit (possibly more than 100,000) determined by the complexity reduction and the population. The depth of sequencing required to obtain complete data for all SNP-containing fragments is currently not practical nor cost effective, nor is it required to obtain meaningful genetic data and results. However, we strongly recommend that additional replicate samples be used for the parents of mapping populations or other material that is critical to an experiment to achieve

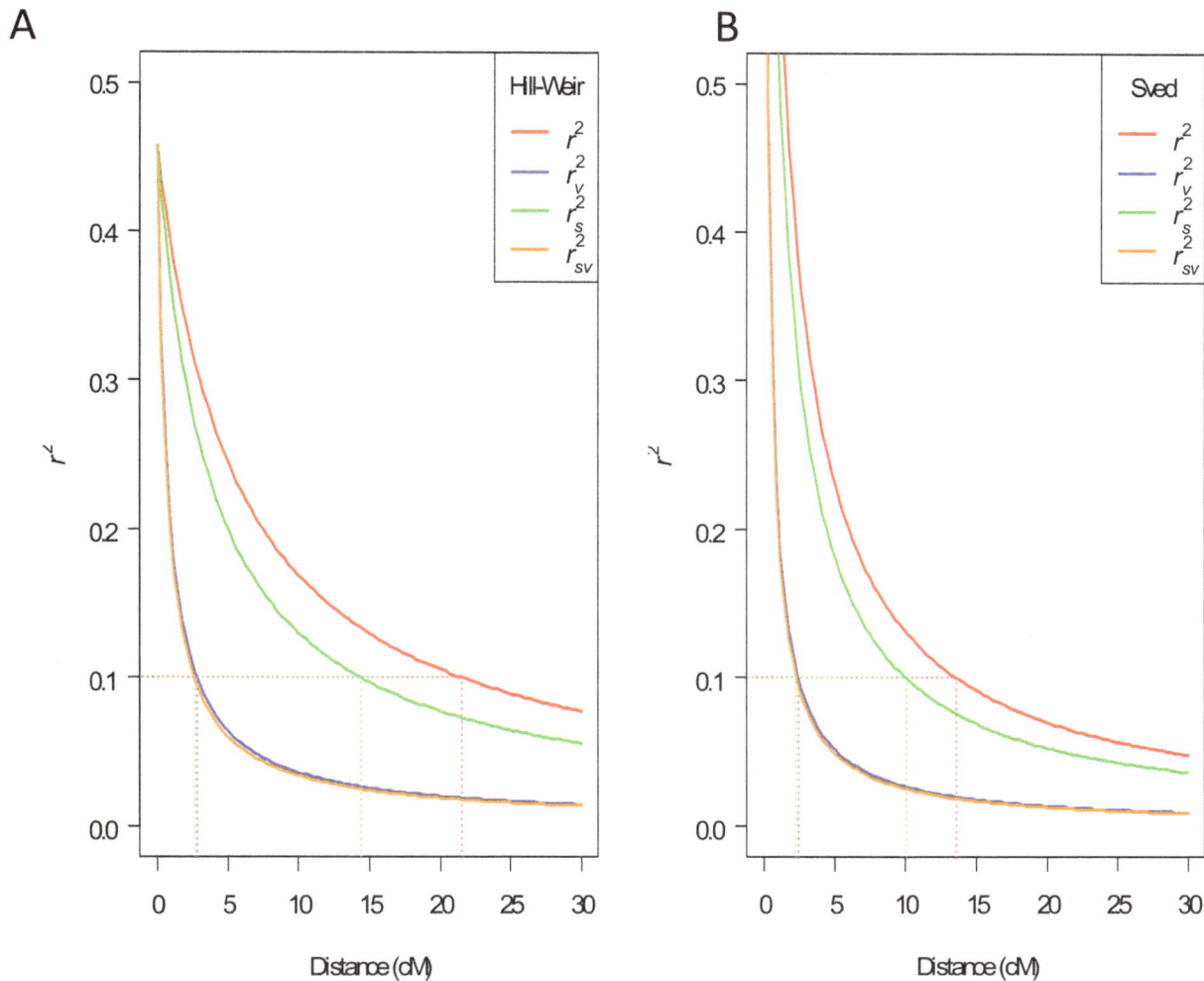

Figure 5. LD decay plot. r^2 estimates were plotted against the average map distance (recombination frequency expressed in cM): (A) relationship fit using the mutation model (Hill-Weir), (B) relationship fit using the recombination-drift model (Sved). Population structure was estimated using the k10 correction.

greater sequencing depth for these lines. In addition, there may be opportunities to produce and analyze reduced complexities generated using selective bases added to primers of existing enzyme systems [18]. Use of selective bases may have value for certain investigations where a more complete and consistent genotyping is required, and would have the advantage of providing data at a subset of loci already characterized in the present work. An example of an application where this might be useful is the routine diagnosis of variety identity or other diagnostic applications that do not rely on a high density of genetic markers. One can also "complete" the missing data points through data imputation. Various imputation methods have been developed which can be highly effective for subsequent analyses (reviewed in [42]). Although most methods require known marker order, often provided by a reference genome sequence, imputation can also be applied to unordered markers and could increase the accuracy of further analyses.

As the number of diverse lines was increased, the number of GBS loci also increased, but this number plateaued at a relatively small number of lines. This result is consistent with the notion that each additional line added an increasing likelihood of identifying rare alleles, such that an increasing number of loci will meet the allele frequency threshold. Meanwhile, larger samples may make it increasingly unlikely that loci will meet filtering criteria and/or that spurious loci will be included. This is possibly why the plateau occurs earlier as data are filtered for higher completeness, and provides a possible explanation for the decrease in SNP number beyond the plateau (Figure 2). In addition, completeness is stochastically variable among samples and among loci. Thus, loci that are included in two diversity subsets are not necessarily those that are the most complete across the entire panel. This phenomenon may make it difficult to obtain consistent loci among different experiments if thresholds for completeness are set too high. For this reason, we recommend that filtering be performed at multiple levels, depending on the purpose of the experiment. For example, the VxL population was filtered at high stringency to develop an initial *de-novo* map, then at low stringency to identify additional GBS loci that mapped across populations.

One of the factors of concern in oat and other polyploid species is that the analysis of genetic loci may be confounded by duplicated homoeologous regions. This factor has made it difficult to discover and apply gene-based SNP loci, and has resulted in some assays where SNPs must be scored with dominant alleles [4,43]. Based on the relative scarcity of BLAST matches in related

Table 2. LD decay estimated in IOI.

r^2	Hill-Weir				Sved			
	Average rf		Min rf		Average rf		Min rf	
	Map distance (cM)	$N_e \pm$ SE	Map distance (cM)	$N_e \pm$ SE	Map distance (cM)	$N_e \pm$ SE	Map distance (cM)	$N_e \pm$ SE
r^2	21.5	9±0.1	19.5	10±0.1	13.6	17±0.1	12.3	18±0.1
r_s^2 k0	14.2	14±0.1	12.7	15±0.2	9.9	23±0.2	8.9	25±0.2
r_s^2 k10	14.4	13±0.1	12.9	15±0.2	10	22±0.2	9	25±0.2
r_s^2 k50	14.5	13±0.1	13	15±0.2	10.1	22±0.2	9.1	25±0.2
r_{sv}^2 k0	2.6	75±0.8	2.2	87±1	2.3	99±1.5	2	111±2
r_{sv}^2 k10	2.6	73±0.8	2.3	85±1	2.3	97±1.5	2.1	109±1.9
r_{sw}^2 k50	2.6	73±0.8	2.3	84±1	2.3	96±1.5	2.1	108±1.9
r_v^2	2.8	68±0.7	2.5	79±0.9	2.5	91±1.3	2.2	102±1.7

The relationship between LD and map distance was modeled by fitting two alternate non-linear regression models: a drift-recombination equilibrium model [36] or a modified recombination-drift model including low level of mutation and an adjustment for sample size [37]. Map distance at $r^2 = 0.1$ was shown. Both average distance across six bi-parental mapping population and minimum distance from available mapping populations were used. N_e effective population size; SE, standard error.

species, the majority of the GBS loci from the *Pst*I-*Msp*I complexity reduction appear to be located in non-genic regions. Furthermore, because the GBS marker calling is based on counts of specific allele variants, all GBS loci are scored with co-dominant alleles. For these reasons, and also because of intense filtering to remove loci with non-diploid inheritance, the GBS method provides good representation of single genetic loci. This would also tend to remove SNPs that fall in conserved genic regions, as these GBS tags would align to all three genomes and the resulting SNPs would not segregate as single loci. Although heterozygote calls are more subject to genotyping errors, our results show the interest of including heterozygous genotypes in certain applications. In particular, we observed that the overall quality of heterozygote determination in the VxL F_4 population was good, and that it enabled meaningful characterization of heterozygous regions in the graphical genotypes (Table S2). However, we have not thoroughly investigated the use of GBS markers in F_2 populations, where it is possible to confuse single loci with duplicated loci having similar genetic ratios. For this reason, GBS should be used with caution in F_2 mapping populations, unless a reference genome or an ordered scaffold is available such that heterozygous regions can be imputed.

Annotation of oat GBS SNPs

We estimated that 5% of oat GBS SNPs were within protein coding sequences. This estimate should be considered preliminary because of the current lack of public oat gene sequences. Using an automated SNP annotation pipeline, Kono *et al.* [44] estimated that only 1.3% of a preliminary subsample of 5000 oat GBS SNPs were genic SNPs. Most of their matches were also based on *Brachypodium*. Their estimate may be lower because of sampling bias and because they used a higher stringency in protein matching ($e < 5 \times 10^{-5}$). Using the same annotation pipeline, Kono *et al.* [44] identified 10.6% of a sample of barley GBS tags as being genic SNPs. The rate in barley may be substantially higher because of the greater availability of barley gene sequences.

Methods of bioinformatics analysis and workflow

We used two bioinformatics pipelines to perform SNP calling. The motivation for using two pipelines was that they were the only two non-reference-genome GBS pipelines of which we were aware. While the UNEAK pipeline gave clear, predictable results, the number of loci passing secondary filtering was low compared to those called by the population-level pipeline. In addition, as observed elsewhere [22], the GBS SNPs identified can vary considerably with different methods. Across all SNPs placed on the oat consensus map, 43% (19,209 out of 45,117) were called only by the population-based method, 22% (9997 out of 45,117) were called only by the UNEAK method, and 36% were called by both methods (Figure 3). Although the population filtering method called more SNPs, these SNPs contained a higher redundancy, and multiple SNPs (linked or unlinked) were sometimes assigned to the same context sequence in the report (data not shown). This is a result of calling SNPs in tags that belong to complex gene families and/or in context sequences of haplotypes that contain multiple SNPs. Such loci are usually ignored by the UNEAK pipeline, which reports only tags with single SNPs. However, in some cases, the direct use of SNPs from the population-filtering report would prevent the correct development of secondary allele assays, if no detailed validation of the assembly was performed to generate unambiguous diagnostic probes for correct SNP interrogation. For these reasons, we performed some of this work using only the UNEAK pipeline. Nevertheless, it was apparent that a much greater number of SNPs were called by the population-filtering method, and these additional SNPs may provide important information which would otherwise be discarded. Until bioinformatics methods can be further refined, we recommend using a combined data set composed of SNPs called by both pipelines. We also recommend the preferential use of SNPs called by the UNEAK pipeline in the development of secondary assays, and the use of appropriate statistical methods to reduce the influence of marker redundancy in subsequent applications such as association mapping.

Saturation of a consensus linkage map with GBS loci

High density maps are required for the precise mapping of important agronomic traits to be targeted in marker-assisted breeding. The high-throughput GBS technology enabled the placement of 45,117 GBS loci identified across six bi-parental mapping populations on the oat consensus map [4]. This high-density map showed marker clustering along chromosomes, similar to a barley map saturated with GBS markers [12]. Since many of these clusters represent centromeres, GBS loci are likely to be more evenly distributed along the physical map. However, gaps of up to 15 cM were still observed in the high-density map. Some gaps may result from lack of polymorphism in the mapping populations, which can be further improved by integrating other mapping populations. Gaps could also be filled in by using GBS libraries produced using different restriction enzymes, as shown in wheat [17]. Gaps and multiple clusters may also be related to the construction of the initial consensus map and it is possible that the consensus map will be improved once GBS markers are fully integrated with additional gene-based SNP loci.

Utility of GBS markers for genetic analysis in oat

We have investigated different GBS applications relevant to oat breeding. In addition to saturating an existing consensus map, GBS markers were suitable for building a *de-novo* linkage map with good genome coverage that revealed colinearity with the consensus. These results were successful because GBS provided a large number of markers that were polymorphic in multiple populations. This will facilitate comparative mapping to validate and refine the location of target alleles in diverse genetic backgrounds, and will increase the options available for molecular breeding.

Analyses of diverse germplasm showed weak population structure in our sample. This weak structure was observed previously when DArT markers were used across a larger oat diversity panel that included the IOI set used in this study [31]. While rice, barley, and maize are known for having strong population structure [45–47], oat, despite having four recognizable types (naked, covered, spring, and winter), has not shown obvious population structure in the samples analysed to date. Although the majority of lines in the IOI set are the spring type (318 out of 340) and covered-seeded (312 out of 340), the remaining naked or winter lines did not form distinct sub-clusters. Instead, the scatterplot tended to reflect the geographic locations of breeding programs and (by inference) the degree of coancestry among lines. A possible explanation for these results could be that, while oat breeders tend to make most crosses among parents that are locally adapted, they have also exchanged elite germplasm with some regularity. The relatively small effective population size compared to the number of IOI lines also supports this interpretation.

The use of GWAS is widely considered to be an attractive alternative to the use of structured (*e.g.*, bi-parental) populations for identifying adaptive alleles for use in molecular breeding. However, effective GWAS requires prior knowledge of LD decay

and an awareness of the population under investigation. Our results highlight that there is a strong gradient of coancestry that needs to be accounted for through the use of an appropriate model, but that other sources of population structure are not important in the population investigated. Although GBS appears to provide a much higher density of markers than required for GWAS, it is possible that target loci are within gaps that do not contain a suitable marker density. For this reason, it may be useful to test additional enzyme combinations for GBS for use in a large association panel when a large investment has been made in phenotyping.

Conclusion

The choice of marker technologies is critical to the success and future application of genetic and genomic research. GBS is attractive because it provides thorough genome coverage and can be applied at low cost with or without a reference genome sequence. However, GBS requires intense bioinformatic analysis, an awareness of the need to filter data, and a tolerance for incomplete data. In this work, we have shown that GBS is an effective method to discover and apply SNPs in the large and complex oat genome, and that GBS integrates and compares favourably with an established SNP technology. The resulting data have provided whole-genome coverage at a density that enables detailed analysis of genetic diversity and high power to detect specific genetic variants.

Our overall conclusion and recommendation is that GBS be used as a cost-effective primary tool in any application similar to those that we have explored in oat. Other applications of GBS in oat, including QTL analysis, genomic selection, and the ordering of genome sequence scaffolds, remain to be fully tested, but are expected to be successful based on indications from this work and from similar use in other species. In future work, we intend to apply GBS routinely to genotype and select among advanced oat breeding lines. As a side benefit to improved selection, we hope to provide new information about the sources and locations of alleles for better adaptation in oat, and to integrate this information with the existing genomic knowledge for oat.

Supporting Information

Figure S1 HTML map format. Instructions for using the HTML-formatted map. The map can be found locally in Text S2 as "HTML_Local_text_S2.html" or online at: http://ahoy.aowc.ca/html_link_gbs_text_s2.html.
(PDF)

Figure S2 Distribution of map gaps in the original and updated oat consensus maps. Empty bars show the distribution of map gaps in the first oat consensus map (Oliver *et al.*, 2013); solid bars show the distribution of map gaps in the consensus map with the GBS markers placed on it (this study). Only gaps larger than 5 cM are shown.
(TIF)

Figure S3 Orthology between oat and *Brachypodium distachyon*. Each dot represents the position of a sequence match (BLASTn, $E < 10^{-12}$) between the oat consensus map (blue dots for GBS loci, red dots for array-based SNPs) and the assembled *Brachypodium distachyon* (Bd) pseudomolecule (release 2.1; http://www.brachypodium.org).
(TIF)

Figure S4 Orthology between oat and rice. Each dot represents the position of a sequence match (BLASTn, $E < 10^{-12}$)

between the oat consensus map (blue dots for GBS loci, red dots for array-based SNPs) and the genome sequence of rice (*Oryza sativa* L., release 6.1 from http://rice.plantbiology.msu.edu). (TIF)

Figure S5 Orthology between oat and barley. Each dot represents the position of a sequence match (BLASTn, $E<10^{-12}$) between the oat consensus map (blue dots for GBS loci, red dots for array-based SNPs) and barley (*Hordeum vulgare* L., cv. Morex, release 2.0 from ftp://ftp.ensemblgenomes.org/pub/plants/release-20/fasta/hordeum_vulgare/dna/ non repeat-masked versions). Barley pseudomolecules are assembled according to chromosome arm (long (2HL to 7HL) or short (2HS to 7HS)), except for chromosome 1H. (TIF)

Figure S6 Orthology between oat and barley (multiple matches removed). Each dot represents the position of a sequence match (BLASTn, $E<10^{-12}$) between the oat consensus map (blue dots for GBS loci, red dots for array-based SNPs) and barley (*Hordeum vulgare* L., cv. Morex, release 2.0 from ftp://ftp.ensemblgenomes.org/pub/plants/release-20/fasta/hordeum_vulgare/dna/ non repeat-masked versions). A subset of matches from Figure S5 is shown: oat sequences that matched other *Hv* sequences more than 6 times at the same BLASTn expectation have been removed. (TIF)

Figure S7 Comparison between the VxL map and the consensus map. Each dot represents a marker shared by the two maps. Red dots highlight markers of higher heterozygosity (between 8 and 13%). (TIF)

Figure S8 Distribution of distances between adjacent markers used for LD analysis. Markers were first sorted according to map position, then the distances were calculated as Position $_m$ minus Position $_{m-1}$. For the first marker of each chromosome, the interval was calculated as the difference between its position and the position of the second marker. (TIFF)

Figure S9 Scree plots of principal components of IOI genotypic data. The first 20 components at three levels of LD correction were used to draw the plots: k0 (without correction), k10 (using 10 adjacent markers for LD adjustment), and k50 (using 50 adjacent markers for LD adjustment). No obvious "elbows" were observed but there was a two-stage decay: at PC5 and at PC10. (TIF)

Figure S10 IOI population structure scatter plot (PC1 vs. PC2) based on genetic clustering. Three levels of LD correction are shown: k0 (A), k10 (B), and k50 (C and D). (TIF)

Figure S11 IOI population structure scatter plot (PC1 vs. PC2) coloured based on the geographical origins of the lines. Three levels of LD correction are shown: k0 (A), k10 (B), and k50 (C). (TIFF)

Figure S12 IOI population structure scatter plot (PC3 vs. PC4) coloured based on genetic clustering (left) or plant habitat (right). Three levels of LD correction are shown: k0 (up), k10 (middle), and k50 (bottom). (TIFF)

Figure S13 LD decay plot. r^2 estimates were plotted against both minimum and average map distance (recombination

frequency expressed in cM): (A) relationship fit using the mutation model (Hill-Weir), (B) relationship fit using the recombination-drift model (Sved). (TIF)

Figure S14 Using GBS markers to resolve an issue in a field experiment. UPGMA cluster analysis of simple allele-matching metric (d) based on 1518 GBS loci with heterozygosity <8%, MAF >20%, and completeness >95%. This evidence was used to correct a planting error in a field experiment. The samples in one replication (red samples) were out of order compared to those in a second, correct replication (green samples), and a set of known controls (blue samples). Analyzing the sub-clusters in the above dendrogram and assigning corrected identities (entry numbers 1–32, above) to the samples in replication 1 made it obvious that the planting order of the first replication had been reversed. (TIF)

Figure S15 Using GBS markers to resolve an issue with breeding materials. UPGMA cluster analysis of simple allele-matching metric (d) based on 2205 GBS loci with >90% completeness. GBS calls were made across samples from 343 diverse oat varieties plus ten putative F$_2$ segregants (green) from a putative cross between SA060123 (red) and a progeny of Leggett (blue). Eight closely related oat cultivars are also shown in this partial cluster dendrogram. Of the 2205 loci, only 131 (6%) showed any variation among the ten progeny plus SA060123, and this variation was within the expectations of heterozygous miscalls. This evidence was used to conclude that the ten progeny were actually from selfed seed of SA060123 rather than true segregants from a hybrid. (TIF)

Figure S16 Effect of adding populations on the number of markers placed on the oat consensus map. Using the consensus map [4] as a framework and starting with a different population each time, markers from the six populations were placed sequentially in all possible combinations (C_k^6, k = 1 to 6). The number of additional markers contributed by the final map at each step is represented by different colours and shapes. The box represents the range between the first and third quartiles and the thick horizontal bar represents the median. (TIFF)

Figure S17 Distribution of annotated GBS markers across the oat consensus map. Maps of each chromosome were divided into 5 cM bins and the number of intergenic/genic markers counted for each bin. Some markers are in the negative range because they are placed off the beginning of the linkage group. (PDF)

Table S1 GBS mapping data for the six bi-parental populations used to update the oat consensus map (Oliver et al., 2013). (ZIP)

Table S2 Graphical genotypes of markers comprising the VxL map. (XLSX)

Table S3 Information about the lines comprising the IOI panel. (XLSX)

Table S4 Raw reads statistics and key file for GBS pipeline. (XLSX)

Text S1 Custom Pascal code for 'CbyT'. (TXT)

Text S2 Updated oat consensus map (HTML_Local_ text_S2.html). See Figure S1 for instructions. (HTML)

Acknowledgments

This work was made possible through excellent technical and professional assistance from the following: Shuangye Wu, for constructing GBS libraries; Rebeca Oliver, Biniam Hizbai, Annick Gauthier, Muriel Jatar, Sophie Ménard, and Paul Gillespie for preparing and handling DNA samples; Andrew Sharpe and Darrin Klassen for performing DNA sequencing; Stéphane Nicolas for sharing R scripts used for statistical analysis; Weikai Yan for sharing breeding material that was used to evaluate genotyping procedures; and Brad de Haan, Steve Thomas, Matthew Hayes, and Kathie Upton for professional assistance with field and greenhouse procedures.

Author Contributions

Conceived and designed the experiments: YFH JAP EWJ NAT. Performed the experiments: YFH JAP CPW NAT. Analyzed the data: YFH JAP CPW NAT. Contributed reagents/materials/analysis tools: JAP EWJ NAT. Wrote the paper: YFH JAP CPW EWJ NAT.

References

1. Meuwissen THE, Hayes BJ, Goddard ME (2001) Prediction of Total Genetic Value Using Genome-Wide Dense Marker Maps. Genetics 157: 1819–1829.
2. Heffner EL, Sorrells ME, Jannink J-L (2009) Genomic Selection for Crop Improvement. Crop Sci 49: 1–12.
3. Molnar SJ, Tinker NA, Kaeppler HF, Rines HW (2011) Molecular Genetics of Oat Quality. In: Webster FH, Wood PJ, editors. Oats: Chemistry and Technology. St. Paul, MN: American Association of Cereal Chemists.
4. Oliver RE, Tinker NA, Lazo GR, Chao S, Jellen EN, et al. (2013) SNP Discovery and Chromosome Anchoring Provide the First Physically-Anchored Hexaploid Oat Map and Reveal Synteny with Model Species. PLoS ONE 8: e58068.
5. Massman JM, Jung H-JG, Bernardo R (2013) Genomewide Selection versus Marker-assisted Recurrent Selection to Improve Grain Yield and Stover-quality Traits for Cellulosic Ethanol in Maize. Crop Sci 53: 58–66.
6. Asoro FG, Newell MA, Beavis WD, Scott MP, Tinker NA, et al. (2013) Genomic, Marker-Assisted, and Pedigree-BLUP Selection Methods for β-Glucan Concentration in Elite Oat. Crop Sci 53: 1894–1906.
7. McCouch SR, McNally KL, Wang W, Sackville Hamilton R (2012) Genomics of gene banks: A case study in rice. American Journal of Botany 99: 407–423.
8. Romay M, Millard M, Glaubitz J, Peiffer J, Swarts K, et al. (2013) Comprehensive genotyping of the USA national maize inbred seed bank. Genome Biology 14: R55.
9. Gutierrez-Gonzalez JJ, Wise ML, Garvin DF (2013) A developmental profile of tocol accumulation in oat seeds. Journal of Cereal Science 57: 79–83.
10. Tinker N, Kilian A, Wight C, Heller-Uszynska K, Wenzl P, et al. (2009) New DArT markers for oat provide enhanced map coverage and global germplasm characterization. BMC genomics 10: 39.
11. Elshire R, Glaubitz J, Sun Q, Poland J, Kawamoto K, et al. (2011) A robust, simple genotyping-by-sequencing (GBS) approach for high diversity species. PLoS ONE 6.
12. Poland J, Brown P, Sorrells M, Jannink J-L (2012) Development of high-density genetic maps for barley and wheat using a novel two-enzyme genotyping-by-sequencing approach. PLoS ONE 7.
13. Beissinger TM, Hirsch CN, Sekhon RS, Foerster JM, Johnson JM, et al. (2013) Marker Density and Read Depth for Genotyping Populations Using Genotyping-by-Sequencing. Genetics 193: 1073–1081.
14. Glaubitz JC, Casstevens TM, Lu F, Harriman J, Elshire RJ, et al. (2014) TASSEL-GBS: A High Capacity Genotyping by Sequencing Analysis Pipeline. PLoS ONE 9: e90346.
15. Liu H, Bayer M, Druka A, Russell J, Hackett C, et al. (2014) An evaluation of genotyping by sequencing (GBS) to map the Breviaristatum-e (ari-e) locus in cultivated barley. BMC Genomics 15: 104.
16. Lu F, Lipka AE, Glaubitz J, Elshire R, Cherney JH, et al. (2013) Switchgrass Genomic Diversity, Ploidy, and Evolution: Novel Insights from a Network-Based SNP Discovery Protocol. PLoS Genet 9: e1003215.
17. Saintenac C, Jiang D, Wang S, Akhunov E (2013) Sequence-Based Mapping of the Polyploid Wheat Genome. G3: Genes|Genomes|Genetics 3: 1105–1114.
18. Sonah H, Bastien M, Iquira E, Tardivel A, Légaré G, et al. (2013) An Improved Genotyping by Sequencing (GBS) Approach Offering Increased Versatility and Efficiency of SNP Discovery and Genotyping. PLoS ONE 8: e54603.
19. Spindel J, Wright M, Chen C, Cobb J, Gage J, et al. (2013) Bridging the genotyping gap: using genotyping by sequencing (GBS) to add high-density SNP markers and new value to traditional bi-parental mapping and breeding populations. Theoretical and Applied Genetics 126: 2699–2716.
20. Uitdewilligen JGAML, Wolters A-MA, D'hoop BB, Borm TJA, Visser RGF, et al. (2013) A Next-Generation Sequencing Method for Genotyping-by-Sequencing of Highly Heterozygous Autotetraploid Potato. PLoS ONE 8: e62355.
21. Ward J, Bhangoo J, Fernandez-Fernandez F, Moore P, Swanson J, et al. (2013) Saturated linkage map construction in Rubus idaeus using genotyping by sequencing and genome-independent imputation. BMC Genomics 14: 2.
22. Mascher M, Wu S, Amand PS, Stein N, Poland J (2013) Application of Genotyping-by-Sequencing on Semiconductor Sequencing Platforms: A Comparison of Genetic and Reference-Based Marker Ordering in Barley. PLoS ONE 8: e76925.
23. Poland J, Endelman J, Dawson J, Rutkoski J, Wu S, et al. (2012) Genomic Selection in Wheat Breeding using Genotyping-by-Sequencing. Plant Gen 5: 103–113.
24. Mascher M, Muehlbauer GJ, Rokhsar DS, Chapman J, Schmutz J, et al. (2013) Anchoring and ordering NGS contig assemblies by population sequencing (POPSEQ). The Plant Journal 76: 718–727.
25. Mayer KFX, Waugh R, Langridge P, Close TJ, Wise RP, et al. (2012) A physical, genetic and functional sequence assembly of the barley genome. Nature 491: 711–716.
26. Wight CP, Tinker NA, Kianian SF, Sorrells ME, O'Donoughue LS, et al. (2003) A molecular marker map in 'Kanota' × 'Ogle' hexaploid oat (Avena spp.) enhanced by additional markers and a robust framework. Genome 46: 28–47.
27. Bradbury PJ, Zhang Z, Kroon DE, Casstevens TM, Ramdoss Y, et al. (2007) TASSEL: software for association mapping of complex traits in diverse samples. Bioinformatics 23: 2633–2635.
28. Wu Y, Bhat PR, Close TJ, Lonardi S (2008) Efficient and Accurate Construction of Genetic Linkage Maps from the Minimum Spanning Tree of a Graph. PLoS Genet 4: e1000212.
29. Patterson N, Price AL, Reich D (2006) Population Structure and Eigenanalysis. PLoS Genet 2: e190.
30. R Core Team (2013) R: A language and environment for statistical computing, R Foundation for Statistical Computing, Vienna, Austria, ISBN 3-900051-07-0, URL http://www.R-project.org/
31. Newell MA, Cook D, Tinker NA, Jannink JL (2011) Population structure and linkage disequilibrium in oat (Avena sativa L.): implications for genome-wide association studies. Theoretical and Applied Genetics 122: 623–632.
32. Fraley C, Raftery AE, Murphy TB, Scrucca L (2012) mclust Version 4 for R: Normal Mixture Modeling for Model-Based Clustering, Classification, and Density Estimation. Department of Statistics, University of Washington.
33. Baudry J-P, Raftery AE, Celeux G, Lo K, Gottardo R (2010) Combining Mixture Components for Clustering. Journal of Computational and Graphical Statistics 19: 332–353.
34. Mangin B, Siberchicot A, Nicolas S, Doligez A, This P, et al. (2012) Novel measures of linkage disequilibrium that correct the bias due to population structure and relatedness. Heredity 108: 285–291.
35. Endelman J (2011) Ridge regression and other kernels for genomic selection with R package rrBLUP. Plant Genome 4: 250–255.
36. Sved JA (1971) Linkage disequilibrium and homozygosity of chromosome segments in finite populations. Theoretical Population Biology 2: 125–141.
37. Hill WG, Weir BS (1988) Variances and covariances of squared linkage disequilibria in finite populations. Theoretical Population Biology 33: 54–78.
38. Remington DL, Thornsberry JM, Matsuoka Y, Wilson LM, Whitt SR, et al. (2001) Structure of linkage disequilibrium and phenotypic associations in the maize genome. Proceedings of the National Academy of Sciences 98: 11479–11484.
39. Perrier X, Jacquemoud-Collet JP (2006) DARwin: Dissimilarity Analysis and Representation for Windows. 5.0.158 (2009) ed. Montpellier: CIRAD.
40. Gore MA, Chia J-M, Elshire RJ, Sun Q, Ersoz ES, et al. (2009) A First-Generation Haplotype Map of Maize. Science 326: 1115–1117.
41. Poland JA, Rife TW (2012) Genotyping-by-Sequencing for Plant Breeding and Genetics. Plant Gen 5: 92–102.
42. Marchini J, Howie B (2010) Genotype imputation for genome-wide association studies. Nat Rev Genet 11: 499–511.
43. Oliver R, Lazo G, Lutz J, Rubenfield M, Tinker N, et al. (2011) Model SNP development for complex genomes based on hexaploid oat using high-throughput 454 sequencing technology. BMC genomics 12: 77.
44. Kono TJY, Seth K, Poland JA, Morrell PL (2014) SNPMeta: SNP annotation and SNP metadata collection without a reference genome. Molecular Ecology Resources 14: 419–425.
45. Garris AJ, Tai TH, Coburn J, Kresovich S, McCouch S (2005) Genetic Structure and Diversity in Oryza sativa L. Genetics 169: 1631–1638.
46. Hamblin MT, Close TJ, Bhat PR, Chao S, Kling JG, et al. (2010) Population Structure and Linkage Disequilibrium in U.S. Barley Germplasm: Implications for Association Mapping. Crop Sci 50: 556–566.

47. Vigouroux Y, Glaubitz JC, Matsuoka Y, Goodman MM, Sánchez GJ, et al. (2008) Population structure and genetic diversity of New World maize races assessed by DNA microsatellites. American Journal of Botany 95: 1240–1253.

48. Portyanko V, Hoffman D, Lee M, Holland J (2001) A linkage map of hexaploid oat based on grass anchor DNA clones and its relationship to other oat maps. Genome 44: 249–265.

49. O'Donoughue LS, Sorrells ME, Tanksley SD, Autrique E, Deynze AV, et al. (1995) A molecular linkage map of cultivated oat. Genome 38: 368–380.

12

Bayesian Inference of Baseline Fertility and Treatment Effects via a Crop Yield-Fertility Model

Hungyen Chen[1], Junko Yamagishi[2], Hirohisa Kishino[1]*

1 Graduate School of Agricultural and Life Sciences, The University of Tokyo, Tokyo, Japan, **2** Institute for Sustainable Agro-ecosystem Services, The University of Tokyo, Tokyo, Japan

Abstract

To effectively manage soil fertility, knowledge is needed of how a crop uses nutrients from fertilizer applied to the soil. Soil quality is a combination of biological, chemical and physical properties and is hard to assess directly because of collective and multiple functional effects. In this paper, we focus on the application of these concepts to agriculture. We define the baseline fertility of soil as the level of fertility that a crop can acquire for growth from the soil. With this strict definition, we propose a new crop yield-fertility model that enables quantification of the process of improving baseline fertility and the effects of treatments solely from the time series of crop yields. The model was modified from Michaelis-Menten kinetics and measured the additional effects of the treatments given the baseline fertility. Using more than 30 years of experimental data, we used the Bayesian framework to estimate the improvements in baseline fertility and the effects of fertilizer and farmyard manure (FYM) on maize (*Zea mays*), barley (*Hordeum vulgare*), and soybean (*Glycine max*) yields. Fertilizer contributed the most to the barley yield and FYM contributed the most to the soybean yield among the three crops. The baseline fertility of the subsurface soil was very low for maize and barley prior to fertilization. In contrast, the baseline fertility in this soil approximated half-saturated fertility for the soybean crop. The long-term soil fertility was increased by adding FYM, but the effect of FYM addition was reduced by the addition of fertilizer. Our results provide evidence that long-term soil fertility under continuous farming was maintained, or increased, by the application of natural nutrients compared with the application of synthetic fertilizer.

Editor: Wenju Liang, Chinese Academy of Sciences, China

Funding: The Grant-in-Aid for Scientific Research (grant number 59440009, PI: Wataru Sunohara) by the Ministry of Education, Science and Culture (MEXT, http://www.mext.go.jp) supported the initial phase (1984–1987) of the long-term fertilizer experiment. The University Farm continued the experiment until 2010 by the regular budget supplied by the University of Tokyo. This work was supported by Grants-in-Aid for Scientific Research (grant number 25280006) by the Japan Society for the Promotion of Science (JSPS, http://www.jsps.go.jp/) to H. K. The funders had no role in study design, data collection and analysis, decision to publish, or preparation of the manuscript.

Competing Interests: The authors have declared that no competing interests exist.

* Email: kishino@lbm.ab.a.u-tokyo.ac.jp

Introduction

Agricultural crop production is directly related to food supply, so agricultural soil productivity must be maintained. Balanced fertilization provides all the essential nutrients for crops to remain healthy and grow productively [1–3]. In a world facing increasing population pressure [4], our highest priority must be to increase crop productivity to ensure food security [2,5–7]. In this context, there has been increasing concern about the long-term productivity of soils on a global scale [8–11].

The relationship between crop yield and soil fertility under different fertilization regimes has been studied for decades. Aref and Wander [12] investigated the long-term trends of corn yield and soil organic matter in different crop sequences and soil fertility treatments. Merick and Németh [13] used results from 60-year field experiments to provide information on the relationship between fertilization and yields of rye and potato. Hallin et al. [14] investigated the relationship between microbial communities and total crop yield and nitrogen content in the crop in a 50-year-old fertilization experiment. Fan et al. [15] studied the trends in grain yields and soil organic carbon (SOC) in a 26-year dryland fertilization trial. By carrying out a 27-year experiment with various fertilization treatments in a rotation cropping system with wheat and maize in a red soil, Zhang et al. [16] investigated trends in SOC, soil nitrogen, and grain yield.

Many researchers have suggested that soil fertility under continuous farming is maintained, or increased, by the application of farmyard manure (FYM) compared with the application of synthetic fertilizer [2,11,17–19]. To ensure gains in crop productivity, it is necessary to maintain and even improve the soil fertility of a field under continuous farming practice [20–23]. In some cases, soils do not contain sufficient amounts of the essential nutrients required for rapid crop growth and high productivity [2,7,24]. As a result, supplemental nutrients, applied as fertilizers, manure or compost, are needed [25–27]. Many studies have analyzed the effects of manure [28–30], chemical fertilizer [31,32], and both [33,34] on soil fertility and crop yield.

Previous models have predicted the crop-yield response to fertilizer application [19] and climate change [35] and have greatly contributed to the software used in agricultural systems research [36,37]. Myers [38] used a static model to estimate the nitrogen fertilizer requirements of cereal crops. Deng et al. [39] proposed a theoretical framework that predicts the optimum planting density and maximal yield for an annual crop plant. Cong et al. [40] indicated that the CENTURY model [41] can simulate fertilization effects on SOC dynamics under different climate and soil conditions.

Soil quality is a complex combination of biological, chemical, and physical properties. In this paper, we focus on the application of these concepts to agriculture and define the baseline fertility as the level of soil fertility that crops utilize for their growth. Therefore, our soil quality is an interaction of the above properties and the crop activities. With this strict definition of baseline fertility, we propose a new crop yield-fertility model that enables quantification of the process of improving baseline fertility and the effects of treatments solely from the time series of crop yields. The model was modified from Michaelis-Menten kinetics and measured the additional effects of the treatments given the baseline fertility. Using more than 30 years of experimental data of maize (*Zea mays*), barley (*Hordeum vulgare*), and soybean (*Glycine max*) yield, we estimated in the Bayesian framework the improvements in baseline fertility and the effects of fertilizer and farmyard manure (FYM). We compared the efficiency of separate applications of fertilizer or FYM to the three crops (i.e., maize, barley and soybean) using estimated model parameters. The temporal variations in the baseline fertility of each crop were estimated and compared for six treatments and two soil types.

Materials and Methods

Field experiments

The long-term fertilizer experiment was conducted between 1980 and 2010 at the University Farm at the Institute for Sustainable Agro-ecosystem Services at the University of Tokyo, Nishitokyo, Tokyo, Japan (35°43′ N latitude and 139°32′ E longitude and an altitude of 53 m above mean sea level). The field site was located in the Kantō Plain, where the soil is covered with pyroclastic material from volcanoes that surrounded the western Kantō region 126,000 years ago. The soil parent material is Tachikawa loam. The surface soil is a black-colored fertile andosol containing a high percentage of humus and the subsurface soil (25 cm below the surface) is red-colored barren clay. Andosols are soils found in volcanic areas formed in volcanic tephra. Andosols have a different composition from chernozems and are not commonly found outside the Pacific "ring of fire". At the beginning of the experiment, the percent nitrogen was 2.23 g/kg in the surface soil and 0.97 g/kg in the subsurface soil. A 2-year crop rotation of maize-barley-soybean-barley was maintained throughout the experimental period and the crop yields were measured. The maize crop was sown at the beginning of July and harvested at the end of September every second year from 1980 onwards. The barley crop was sown in the first half of November and harvested either at the end of May or at the beginning of June of the next year each year from 1980 onwards. The soybean crop was sown at the end of June and harvested in the second half of October every second year from 1981 onwards. The seeds of the three crops were ridge sown using a seeding machine. For maize and soybean, one seed was sown in each hole, and the widths between the ridges and stocks were 71 cm and 23 cm, and 71 cm and 17.5 cm, respectively. For barley, row seeding was applied using a width of 17.8 cm between the ridges.

Six treatment plots containing the NPK fertilizer applications and two levels of farmyard manure (FYM) combined with compost were established in both of the fields with the surface soil (64 m^2 per plot, 8×8 m) and the subsurface soil (56 m^2 per plot, 7×8 m)

and were replicated four times. The percent nitrogen, phosphorus, and potassium in the fertilizer treatments were N:P:K = 12:7.9:13.3 for maize and barley and 3:4.4:8.3 for soybean. Fertilizer was added to the soil at a rate of 1 t ha^{-1} for all crops. The FYM comprised wheat straw and cow dung. The percent nitrogen in the FYM ranged from 0.2–0.4. FYM was added to the plots at two levels, i.e., 20 and 60 t ha^{-1}. The FYM treatments were added to the soil twice a year from 1980 to 1990 before the crops were sown and once a year (before the barley was sown) from 1991 onwards. After the barley harvest in 2008, fertilizer and FYM applications were discontinued for all treatments. Phosphate fertilizer was added simultaneously with the NPK fertilizer and FYM at a rate of 2 t ha^{-1} after the soybean harvest from 1980 until 2001. All aboveground components were removed from the field during harvesting. The dead foliage, below-ground components, and remnant stem sections (i.e., 15 cm for maize and 10 cm for barley and soybean) remained in the field. The harvested material from the six treatments for both the surface and subsurface soils were weighed after drying at 80°C. The dry weight (g m^{-2}) of the aboveground maize components, the barley spike, and the soybean seed were measured.

Testing significance of yield differences in different soils

The crop yields of the surface and subsurface soils without the addition of fertilizer or farmyard manure (FYM) reflect the yield differences that occurred purely as a result of the soil. The crop yields observed for the subsurface soil were 39±25%, 49±20%, and 19±31% (±SD) lower than the yields observed for the surface soil for maize (Fig. S1A), barley (Fig. S1B), and soybean (Fig. S1C), respectively. There was a difference in yield between the two types of soils but it was only significant for barley (t = −3.69, P<0.001; for maize, t = −1.58, P = 0.13 and for soybean, t = −0.94, P = 0.35). Although the crops were planted in the same fields with the same soil, they obtained and used nutrients from the soil in a manner unique to each crop.

The maize and soybean yields did not differ significantly (P> 0.2) between the surface and subsurface soils when fertilizer or FYM or a combination of fertilizer and FYM were applied (Fig. S1A and C for maize and soybean, respectively). The barley yields differed significantly between the surface and subsurface soils when both fertilizer and FYM (t = 2.27, P<0.05) or FYM alone (t = 2.18, P<0.05) were added, but did not differ significantly when only fertilizer was added (t = −0.63, P = 0.52) (Fig. S1B). The difference in yields between the surface and subsurface soils was reduced when either fertilizer or FYM was added to the soil. The application of fertilizer and FYM improved the yields for both the surface and subsurface soils and reduced the yield differences between the soil types. This finding indicates that productivity can be increased to an average level by fertilization even in a field with barren soil. While this may have been prior knowledge, the manner in which fertilizer and FYM contribute nutrients to the soil and how the crops make use of the nutrients obtained from these sources remains unclear.

Testing significance of yearly variance

The maize and soybean yields increased significantly (P<0.05) for both soil types over the 30-year period. The barley yield increased significantly in the subsurface soil (P<0.05) but not in the surface soil (P = 0.15). The soils were fertilized by crop residues (i.e., roots, fallen leaves, and stem sections (15 cm above ground for maize and 10 cm for barley and soybean)) that remained in and on the soil after harvesting. Organic matter accumulated in the soil over time, providing a long-term, slow-release source of nitrogen, phosphorus, sulfur, and other important nutrients for crop growth, which resulted in increased soil productivity.

Testing significance of the treatment effects

We used a nested analysis of variance to compare the yields of the three crops in the surface and subsurface soils in response to the different treatments. The surface soil produced significantly greater barley crop yields compared with the subsurface soil (Fig. S1B, F = 14.68, $P<0.001$). Maize and soybean yields (Fig. S1A, F = 2.91, $P = 0.09$ and Fig. S1C, F = 2.19, $P = 0.14$, respectively) did not differ significantly between the two soil types. The interaction effects of fertilizer and FYM were significant for maize (F = 3.32, $P<0.05$) and barley (F = 10.71, $P<0.001$), but not for soybean (F = 0.69, $P = 0.60$). The soybean yield did not differ significantly among the treatments ($P>0.1$ for both fertilizer and FYM).

The derivation of the crop yield-fertility model

Our model is based on the general pattern of the observed crop yield. The temporal variation in maize yield in response to four treatments applied to the surface soil over 30 years is shown in Fig. 1A. There were clear differences in the yield between the four treatments prior to 1982 (the first two data points). The response of the crop yield descended in the following order: the addition of NPK fertilizer and FYM (F:1, M:1), the addition of NPK fertilizer only (F:1, M:0), the addition of FYM only (F:0, M:1), and no nutrient application (F:0, M:0). The yield differences between the treatments receiving NPK fertilizer and FYM (F:1, M:1) and NPK fertilizer only (F:1, M:0) could not be detected from 1984 onwards. The effect of FYM could no longer be detected once the fertilizer application was continued. From 1994, similar crop yields were observed in the treatments with fertilizer, FYM or a combined application of fertilizer and FYM. In the last year of the study, there were negligible differences between treatments because fertilizer and FYM applications were discontinued in 2008. The soybean data are presented in Fig. 1C. The pattern of the soybean yield differed from maize. There were almost no differences in crop yield among the three treatments that received fertilizer, FYM or a combined application of fertilizer and FYM in the first and later years. The differences in crop yield between the treatments with and without fertilizer were negligible from 1995 onwards. The inherent soil fertility may be sufficient to grow soybean without the addition of fertilizer. The pattern of barley yield (Fig. 1B) was similar to maize but was not as well defined.

Our model can predict crop yields given the total fertility level of the soil. The model can be used to describe the variations in crop yield and the effect of different fertilizer treatments over different years for each crop using the estimated soil fertility. The crop yield-fertility curves for maize, barley, and soybean are presented in Fig. 2A, B, and C, respectively. Soil fertility was increased by the application of fertilizer over a 20-year period. Furthermore, the effect of fertilizer decreased over time for all three crops from 1980–2000.

The crop yield-fertility model

We developed a simple mathematical model to estimate the soil fertility level and to quantify the contributions of fertilizer and FYM to the improvements in crop yields. The model was modified from Michaelis-Menten kinetics [42]:

$$Y_{i,t} \cong \frac{V}{1 + \dfrac{K}{BF_{i,t} + a \times F + b \times M}}. \tag{1}$$

We did not apply the model to describe the kinetics. Rather, we used a form of the model to interpret the effects of baseline fertility, fertilizer, and FYM on the crop yield and the extent of saturation of additional inputs. The model related crop yields ($Y_{i,t}$) to the total fertility ($BF_{i,t} + a \times F + b \times M$) of the soil for treatment i at time t (Fig. 2). The value of the baseline fertility (BF) for the subsurface soil without the addition of fertilizer or FYM in the first year was normalized by setting this value to 1. V is the maximum yield in response to the maximum fertilizer input; K is the fertility level before fertilization that is required to produce half of the maximum yield, V. A large K indicates that more nutrients need to be added to the soil for the crop to grow, so K can infer the sterility of the soil prior to fertilization. $BF_{i,t}$ is the baseline fertility of the crop for treatment i at time t and is assumed to vary gradually over time; F is the level of fertilizer application (0 and 1), M is the level of FYM application (0, 1/3, and 1), and a and b represent the contributions of F and M relative to the baseline fertility, respectively. Hereafter, we refer to V, K, a, and b as the maximum yield, half-saturated fertility, fertilizer contribution, and FYM, respectively. A Bayesian framework was adopted for parameter estimation assuming a gradual change in the baseline fertility.

The likelihood and priors

The likelihood of the yield for treatment i at time t followed a normal distribution with the mean $\dfrac{V}{1 + \dfrac{K}{BF_{i,t} + a \times F + b \times M}}$ and the variance δ:

$$\text{Yield}_{i,t} \sim N\left(\frac{V}{1 + \dfrac{K}{BF_{i,t} + a \times F + b \times M}}, \delta\right) \tag{2}$$

This value was normalized by setting the BF of the subsurface soil without the addition of fertilizer or FYM in the first year to 1. The priors of the BF for the other treatments in the first year followed a gamma distribution with a shape parameter of 1 and a scale parameter of 1. The smoothness priors of the BF from the second year followed a normal distribution with the mean equal to the value of the BF in the preceding year and the variance τ:

$$BF_{i,t} \sim N(BF_{i,t-1}, \tau), \quad t>1. \tag{3}$$

The inverse of δ followed a gamma distribution with a shape parameter of 0.1 and a scale parameter of 10. The inverse of τ followed a gamma distribution with a shape parameter of 0.1 and a scale parameter of 10. The prior of V followed a normal distribution with a mean of 0 and a standard deviation of 1000. The prior of K followed a gamma distribution with a shape parameter of 0.1 and a scale parameter of 10. The priors of a and b followed a gamma distribution with a shape parameter of 1 and a scale parameter of 1. The priors of the estimates were designed to be as non-informative as possible within a realistic range of the

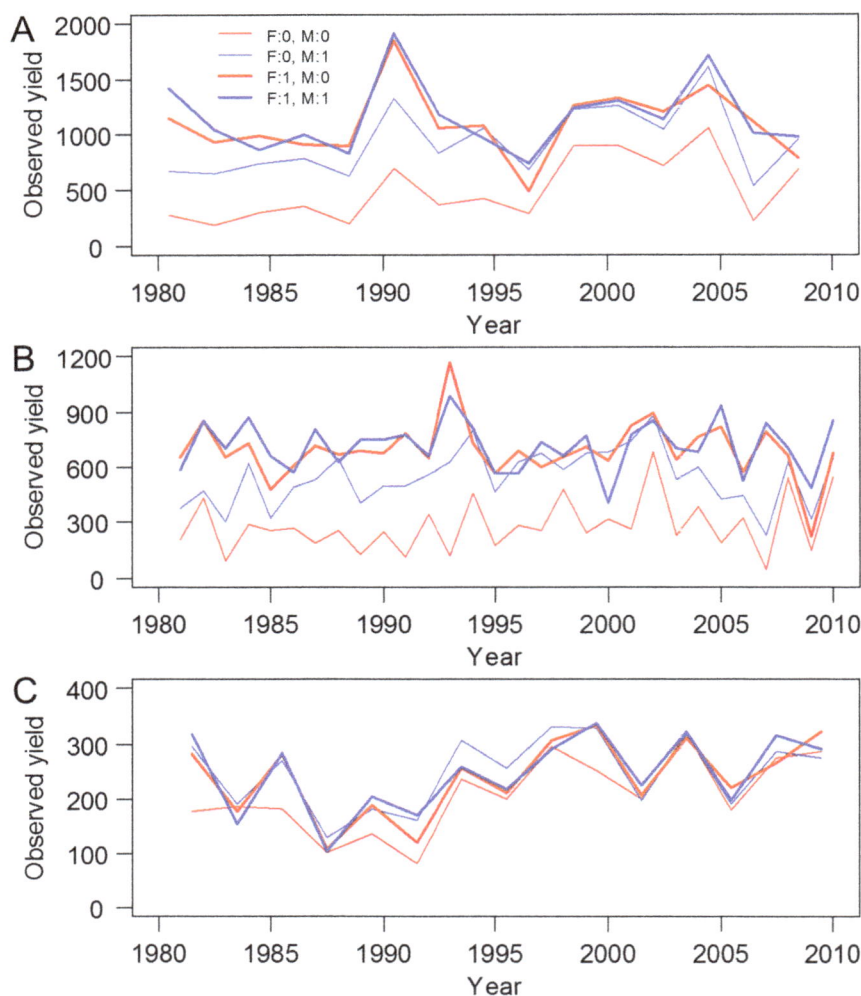

Figure 1. Temporal variations in observed crop yield (g m^{-2}) for four treatments in the surface soil. (A) Maize; (B) Barley; (C) Soybean; F, level of fertilizer; M, level of farmyard manure. See Fig. S1 for all treatments in surface and subsurface soils.

parameter values. All calculations and data analyses were performed using R v2.13.2 [43]. The raw data is available as (Data S1).

Results

Increasing trend of soil fertility

The baseline fertility of the surface and subsurface soils in all treatments increased over the 30-year period for all three crops (Figs. 3, S5–6). The baseline fertility of the subsurface soil was very low for maize and barley but was close to half-saturated for soybean when fertilization was initiated.

The treatment differences in the baseline fertility (i.e., maximum minus minimum) were increased by continuous cropping over the experimental period. The difference in the baseline fertility between the treatments was 0.89 in 1980 and 3.24 in 2008 for maize; for barley, it was 1.20 in 1980 and 2.54 in 2010, and for soybean, it was 0.96 in 1980 and 3.19 in 2009.

We used a 2×2 factorial analysis of variance to compare the baseline fertility of the three crops in response to the different treatments. The interaction effects of fertilizer and FYM were significant for maize (F = 15.09, $P<0.001$) and barley (F = 50.67, $P<0.001$), but not for soybean (F = 1.77, $P = 0.19$). The baseline

fertility of soybean differed significantly for fertilizer (F = 10.26, $P<0.01$) but not for FYM (F = 4.00, $P = 0.05$).

The effects of fertilizer and FYM

The Bayesian estimates of the model for the three crops are shown in Table 1. The traces of the MCMC (Markov Chain Monte Carlo) samples show the well-mixing and convergence to the posterior distributions (Fig. S7). We compared the mean values of the parameters for the three crops. The maize crop had the highest half-saturated fertility level and maximum yield. The soybean crop had the lowest half-saturated fertility level and maximum yield. For maize, the size of the fertilizer contribution was almost equal to the size of the half-saturated fertility, and FYM contribution was 81% smaller than the half-saturated fertility. For barley, the fertilizer contribution was 40% higher than the half-saturated fertility and the FYM contribution was 65% lower than the half-saturated fertility. This indicates that the amount of fertilizer applied in the experiment was 40% more than the fertility level at which the barley yield was half of the maximum yield, but the amount of FYM applied was 65% less than required. For soybean, the values of the fertilizer and FYM

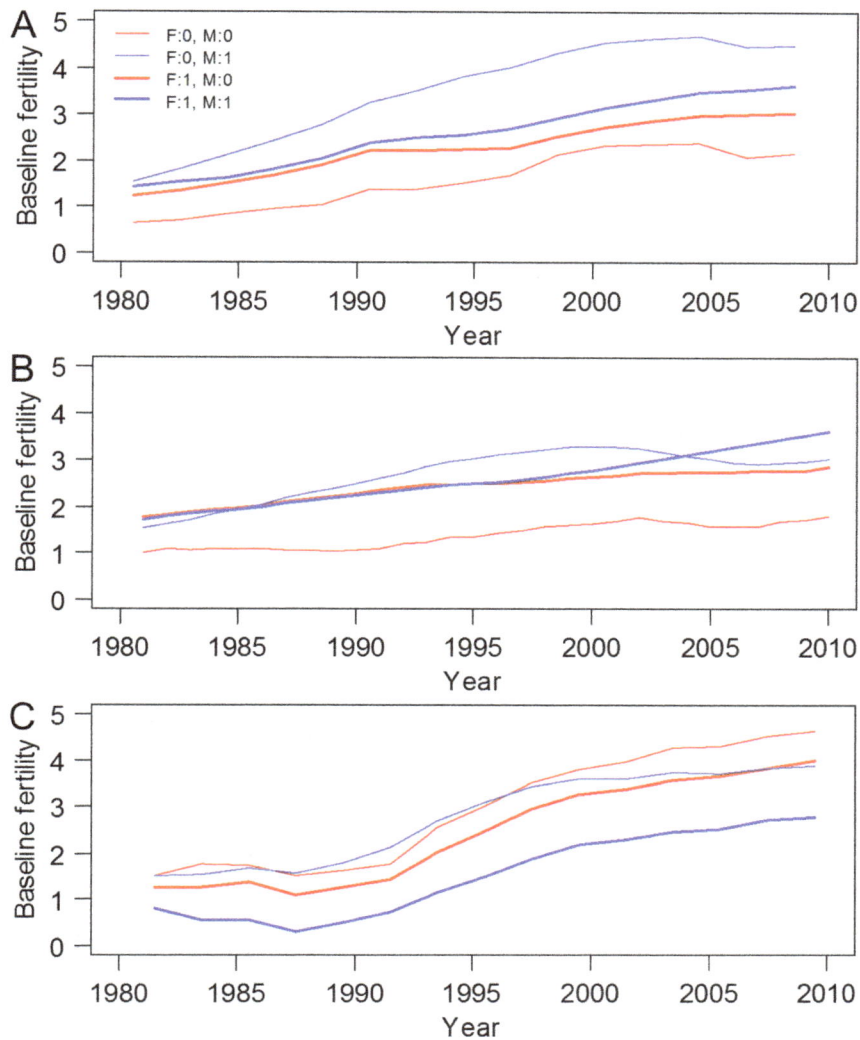

Figure 2. Crop yield-fertility curves for (A) Maize, (B) Barley, and (C) Soybean. The value of 1 for fertility is the fertility without the addition of fertilizer or FYM in the initial year for the subsurface soil. F, level of fertilizer. See Figs. S2A, S3A, and S4A for the band that corresponds to the standard deviation of the curves.

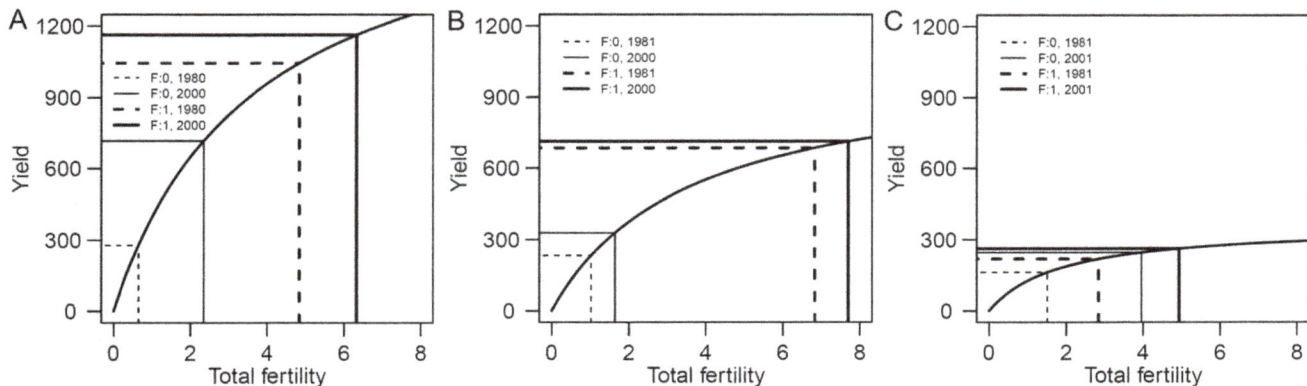

Figure 3. Temporal variations in the baseline fertility for four treatments in the surface soil. (A) Maize; (B) Barley; (C) Soybean; F, level of fertilizer; M, level of farmyard manure. See Fig. S5 for all treatments in surface and subsurface soils.

Table 1. The posterior mean and standard deviation (SD) of the Bayesian estimates.

Crop	V (g m^{-2})		K		A		B	
	Mean	SD	Mean	SD	Mean	SD	Mean	SD
Maize	1833.01	316.56	3.65	1.68	3.61	1.27	0.68	0.53
Barley	1042.87	134.97	3.56	1.18	5.07	1.44	1.25	0.63
Soybean	360.17	65.54	1.87	1.01	1.58	0.58	1.62	0.94

V, maximum yield; K, half-saturated fertility; a, fertilizer contribution; b, FYM contribution.

contributions were similar, i.e., 16 and 13%, respectively, and were lower than the half-saturated fertility.

In contrast with maize and barley, the value of the FYM contribution was higher than the fertilizer contribution for the soybean crop. The application rate of fertilizer and FYM approximated the fertility level at which the soybean yield was half of the maximum yield. The half-saturated fertility level for maize was 3.65, and the half-saturated fertility levels for barley and soybean were 2 and 49% lower, respectively. The fertilizer contribution for barley was 5.07, and the fertilizer contributions for maize and soybean were 29 and 69% lower, respectively. The FYM contribution for soybean was 1.62, and the FYM contributions for maize and barley were 58 and 23% lower, respectively. The model estimates revealed how the different crops used the nutrients from fertilizer and FYM applications to the soil. Maize needed the highest level of fertility to reach half of the maximum yield (i.e., the largest half-saturated fertility) compared with barley and soybean. Fertilizer contributed the maximum amount to the barley yield, and the FYM contributed the maximum amount to the soybean yield.

The total fertility levels among the treatments for both surface and subsurface soils during the 30-year period ranged from 0.54 to 7.22 for maize (Fig. S2A), 0.41 to 9.78 for barley (Fig. S3A), and 0.90 to 5.56 for soybean (Fig. S4A). Barley may be the most efficient crop in terms of fertilizer use among the three crops. The maximum total fertility obtained for soybean was 2.97× the half-saturated fertility, which is larger than the values obtained for maize (1.98) and barley (2.75).

Using the total fertility range, we estimated the predicted range in crop yield using the crop yield-fertility model for each crop. The predicted crop yields ranged from 236.24 to 1217.51 g m^{-2} for maize, 107.70 to 764.56 g m^{-2} for barley, and 117.02 to 269.52 g m^{-2} for soybean. Because of fluctuations in crop yield that could not be described by the model, the predicted yields covered 49, 57, and 42% of the observed yield ranges for maize (35 and 2023 g m^{-2}), barley (14 and 1168 g m^{-2}), and soybean (16 and 381 g m^{-2}), respectively.

The decreasing effect of fertilization over time

In the sterile subsurface soil in the initial year, the application of fertilizer resulted in 153, 189, and 62% increases (the FYM resulted in 42, 65, and 61% increases) in crop yield for maize, barley, and soybean, respectively. In the fertile surface soil in the initial year, the application of fertilizer resulted in 275, 196, and 35% increases (FYM resulted in 149, 98, and 40% increases) in crop yield for maize, barley, and soybean, respectively. After 20 years of farming, the application of fertilizer resulted in 89, 219, and 11% increases (FYM resulted in 74, 150, and 13% increases) in the subsurface soil, and 62, 113, and 7% increases (FYM resulted in 50, 75, and 8% increases) in the surface soil for maize, barley, and soybean, respectively, because of the increased baseline fertility. The greatest yield increases were observed in the initial year in the fertile surface soil with the application of both fertilizer and FYM (compared with the treatment without fertilizer or FYM), and then the yield increment decreased each year (Fig. 4). In the sterile subsurface soil, it took 5–10 years for the yield increase to reach the maximum value.

Predicted crop yields reflected the trend of the observed yields

The estimated fertility levels were used to predict the crop yield using the yield-fertility curves for each crop (Figs. S2A, S3A, and S4A). The trends of the predicted yield were similar to the trends of the observed yield for all three crops. The correlation

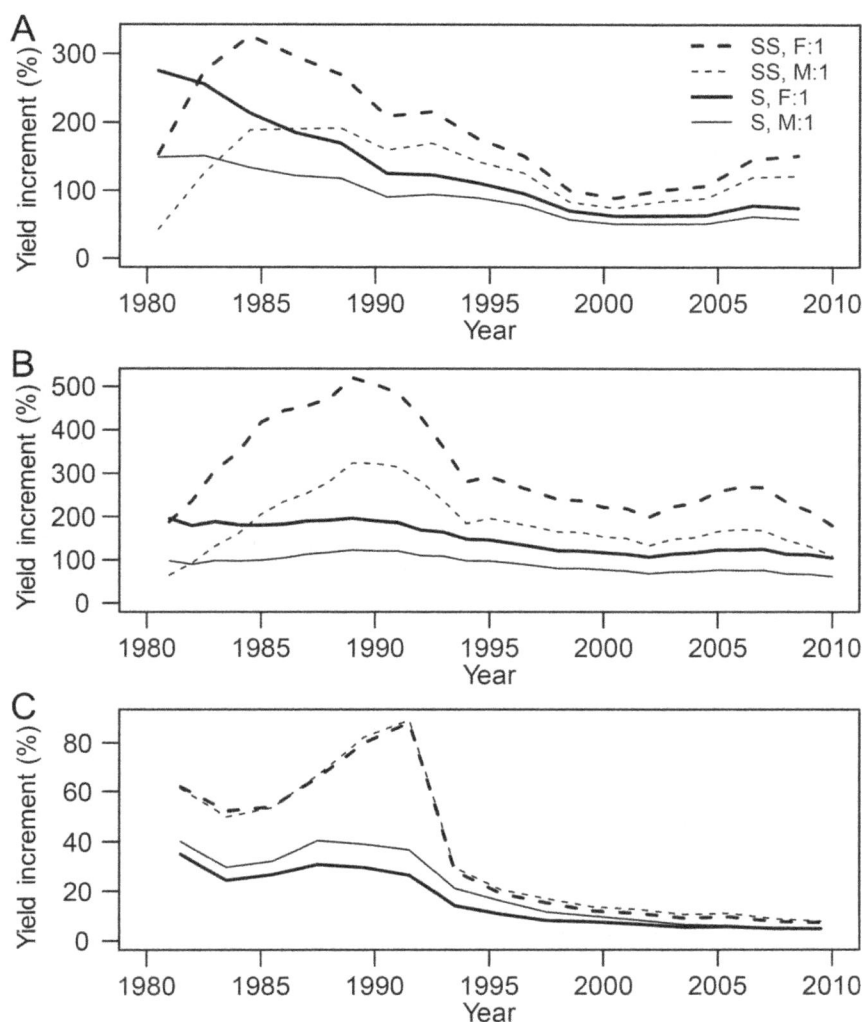

Figure 4. Temporal variations in the yield increment resulting from the application of fertilizer and farmyard manure. (A) Maize; (B) Barley; (C) Soybean; SS, subsurface soil; S, surface soil; F, level of fertilizer; M, level of farmyard manure.

coefficients were 0.758 (Fig. S2B, $P<0.001$) for maize, 0.801 (Fig. S3B, $P<0.001$) for barley, and 0.622 (Fig. S4B, $P<0.001$) for soybean.

Yield fluctuations were observed among the treatments but these fluctuations were not reflected by the predicted crop yields. For maize, the peak yields occurred in 1990 and 2004, and a reduced yield occurred in 1996. For barley, the peak yield for the fertilizer-treated plots occurred in 1993, but the peak yield occurred 1 year later in the plots without fertilizer treatment. The FYM treatments were discontinued in 1991 for the summer harvested crops (maize and soybean). Clear reductions in maize yield in 1991 and soybean yield in 1992 are shown in Fig. 1A and C, respectively, after which the yields for both crops increased until 2008. The estimated fertility levels and predicted yields represented the temporal variations that coincided with treatment changes during the experimental period (Fig. 1). Termination of the phosphate fertilizer application in 2001 was well described by the temporal variations in the baseline fertility of maize and barley. The baseline fertility levels declined, especially in the soil treated with FYM only (Fig. 4). Similar patterns were not observed for the fertilizer-treated soils. A decline in fertility was not observed for

soybean. As previously mentioned, soybean yields were not significantly different among treatments.

The effect of climatic factors

Crop yield can be affected by climatic factors [35] such as precipitation [44,45], temperature [46,47] and recent warming [48]. The impact of climate change may damage crop yields on a global scale [48,49] and lead to decreases in crop production [50]. However, weather variation may play a small role in crop yield on a regional scale [45] and the effect may depend on fertilization and soil type [45,50]. Monthly precipitation and average monthly temperature recorded at the experimental site are shown in Fig. S8A and B. We performed a correlation analysis between the residual of the crop yield (i.e., predicted yield minus observed yield) and the climatic variables for the three crops (Table S1 in File S1). The simple residual analysis indicated that temperature had some effect on maize and soybean yields. We then conducted additional analyses that included the effect of temperature in the crop yield-fertility model (see File S1 and Table S2 in File S1). Based on the results of the minor contribution of temperature, we concluded that the annual fluctuations in crop yields were mainly caused by unknown factors other than the climatic factors.

Discussion

The baseline fertility

The baseline fertility of the subsurface soil was very low for maize and barley. This suggests that the fertility in the subsurface soil prior to the application of fertilizer or FYM was not sufficient for maize or barley.

The increasing baseline fertility trend was also observed in soil without additional fertilizer or FYM. The increased fertility may have resulted from the addition of crop residues to the soil and nitrogen fixation by the soybean crop. These fertility increases may indicate soil maturation (i.e., the accumulation of nutrients) in the experimental fields.

The baseline fertility of maize indicated that when FYM was applied alone, soil fertility increased more than when both FYM and fertilizer were applied, for both soil types (Fig. 3A). The application of FYM increased the organic matter content of the soil, which resulted in increased soil fertility. Similar trends were observed for barley and soybean (Fig. 3B and C), supporting the results observed for maize. The results for soybean suggested that the application of fertilizer could decrease the soil fertility when soybean was planted in a soil with a saturated level of fertilization (Fig. 3C).

The effects of fertilizer and FYM

Our result indicates that the amount of fertilizer applied in the experiment reached the fertility level at which the maize yield was half of the maximum yield, but the amount of FYM required to reach this value was much lower. Barley obtained the highest fertility from the soil and fertilization despite the fact that maize produced the greatest crop yield and had the largest half-saturated fertility level. By using nutrients from the soil and fertilizer, soybean was the crop that most often approached the maximum yield.

Our findings indicate that the sterile subsurface soil needed more time to accumulate nutrients than the surface soil. The percentage increments in crop yield were larger in the subsurface soil than in the surface soil, which indicates that fertilizer application effectively improved the crop yield in the barren field. The fertilizer-induced increment in maize and barley crop yield increased after 2001. This is because the phosphate fertilizer application was discontinued in 2001 and the baseline soil fertility subsequently declined. The soybean crop did not show a similar reduction in yield after 2001.

Long-term field crop fertilization experiment and crop models

Long-term experiments examining fertilizer treatments and crop yield have been widely conducted to assess the effects of applying fertilizer [51], manure [52], or a combination of fertilizer and manure [14,53,54]. In addition, experiments have been conducted to determine the impacts of nutrient addition on the fertility status of different soils [55–57]. However, it is important to note that our long-term record of crop yields with various types of treatments has made this the first study to simultaneously examine the temporal variation in baseline fertility and the contribution of nutrient applications.

Conclusions

By allowing the gradual temporal variation in baseline fertility, it was possible to estimate the process of soil fertilization and the short-term effects of different fertilizer treatments. Our results support the proposal that naturally derived nutrients should be used to maintain soil fertility and synthetic fertilizer should be used to maintain productivity. Our results provide a clear description of the relationship between crop yield and soil, which can be easily understood.

Supporting Information

Figure S1 Temporal variations in observed crop yield (g m^{-2}) for six treatments in a field with a fertile surface soil and a field with a barren subsurface soil from 1980 to 2010. (A) Maize; (B) Barley; (C) Soybean; SS, subsurface soil; S, surface soil; F, level of fertilizer; M, level of farmyard manure. (TIF)

Figure S2 Temporal variations in the observed and predicted yields (g m^{-2}) of maize for six treatments in a field with a fertile surface soil and a field with a barren subsurface soil presented every second year from 1980 to 2008. (A) A crop yield-fertility map. The crop yield-fertility model translates total fertility into predicted yield. The gray band represents the band that corresponds to the standard deviation (±SD) of the crop yield-fertility curve. (B) The relationship between the observed and predicted yields. SS, subsurface soil; S, surface soil; F, level of fertilizer; M, level of farmyard manure; BF, baseline fertility. (TIF)

Figure S3 As for Figure S2 but for barley every year from 1980 to 2010. (TIF)

Figure S4 As for Figure S2 but for soybean every 2 years from 1981 to 2009. (TIF)

Figure S5 Temporal variations in the baseline fertility estimated using the crop yield-fertility model for six treatments in a field with a fertile surface soil and a field with a barren subsurface soil from 1980 to 2010. (A) Maize; (B) Barley; (C) Soybean; SS, subsurface soil; S, surface soil; F, level of fertilizer; M, level of farmyard manure. (TIF)

Figure S6 Temporal variations in the posterior means and the bands that correspond to the standard deviation (±SD) of the Bayesian estimates of baseline fertility for six treatments in a field with a fertile surface soil and a field with a barren subsurface soil. (A) Maize; (B) Barley; (C) Soybean; SS, subsurface soil; S, surface soil; F, level of fertilizer; M, level of farmyard manure. (TIF)

Figure S7 Traces of the MCMC samples of V (g m^{-2}), K, a, and b. (A) Maize; (B) Barley; (C) Soybean. The chain length was set to 1,000,000 steps logging every 100th step. (TIF)

Figure S8 Temporal variation of the climatic variables recorded at the experimental site from January 1980 to December 2011. (A) Monthly precipitation. (B) Average monthly temperature. (TIF)

File S1 Contribution of the climatic factor. The analyses by including a temperature effect in the crop yield-fertility model. (DOCX)

Data S1 The raw data of the crop yields of maize, barley, and soybean from the long-term fertilizer experiment from 1980 to 2010. As for the detail of the experiment, see text. (XLSX)

Author Contributions

Conceived and designed the experiments: HC HK. Performed the experiments: JY. Analyzed the data: HC. Wrote the paper: HC HK. Confirmed the manuscript: HC JY HK.

References

1. Mäder P, Fliessbach A, Dubois D, Gunst L, Fried P (2002) Soil fertility and biodiversity in organic farming. Science 296: 1964–1967.
2. Tilman D, Cassman KG, Matson PA, Naylor R, Polasky S (2002) Agricultural sustainability and intensive production practices. Nature 418: 671–677.
3. Baligar VC, Fageria NK, He ZL (2001) Nutrient use efficiency in plants. Commun Soil Sci Plant 32: 921–950.
4. Holdren JP, Ehrlich PR (1974) Human population and the global environment: population growth, rising per capita material consumption, and disruptive technologies have made civilization a global ecological force. Am Sci 62: 282–292.
5. Glover JD, Reganold RP, Cox CM (2012) Plant perennials to save Africa's soils. Nature 489: 359–361.
6. Alexandratos N (1999) World food and agriculture: outlook for the medium and longer term. PNAS 96: 5908–5914.
7. Cassman KG (1999) Ecological intensification of cereal production systems: yield potential, soil quality, and precision agriculture. PNAS 96: 5952–5959.
8. Isbella F, Reichb PB, Tilmana D, Hobbiea SE, Polasky S, et al. (2013) Nutrient enrichment, biodiversity loss, and consequent declines in ecosystem productivity. PNAS 110: 11911–11916.
9. Cordell D, Drangert J-O, White S (2009) The story of phosphorus: global food security and food for thought. Glob Environ Change 19: 292–305.
10. Rasmussen PE, Goulding KWT, Brown JR, Grace PR, Janzen H, et al. (1998) Long-term agroecosystem experiments: assessing agricultural sustainability and global change. Science 282: 893–896.
11. Matson PA, Parton WJ, Power AG, Swift MJ (1997) Agricultural intensification and ecosystem properties. Science 277: 504–509.
12. Aref S, Wander MM (1997) Long-term trends of corn yield and soil organic matter in different crop sequences and soil fertility treatments on the morrow plots. Adv Agron 62: 153–197.
13. Mercik S, Németh K (1985) Effects of 60-year N, P, K and Ca fertilization on EUF-nutrient fractions in the soil and on yields of rye and potato crops. Plant Soil 83: 151–159.
14. Hallin S, Jones CM, Schloter M, Philippot L (2009) Relationship between N-cycling communities and ecosystem functioning in a 50-year-old fertilization experiment. ISME J 3: 597–605.
15. Fan T, Xu M, Song S, Zhou G, Ding L (2008) Trends in grain yields and soil organic C in a long-term fertilization experiment in the China Loess Plateau. J Plant Nutr Soil Sci 171: 448–457.
16. Zhang H, Xu M, Zhang F (2009) Long-term effects of manure application on grain yield under different cropping systems and ecological conditions in China. J Agr Sci 147: 31–42.
17. Power AG (2010) Ecosystem services and agriculture: tradeoffs and synergies. Philos Soc Trans R Soc Lond B 365: 2959–2971.
18. Altieri MA, Nicholls CI (2003) Soil fertility management and insect pests: harmonizing soil and plant health in agroecosystems. Soil Till Res 72: 203–211.
19. Clark MS, Horwath WR, Shennan C, Scow KM (1998) Changes in soil chemical properties resulting from organic and low-input farming practices. Agron J 90: 662–671.
20. Fan R, Zhang X, Liang A, Shi X, Chen X, et al. (2012) Tillage and rotation effects on crop yield and profitability on a Black soil in northeast China. Can J Soil Sci 92: 463–470.
21. Yanni YG, Dazzo FB (2010) Enhancement of rice production using endophytic strains of *Rhizobium leguminosarum* bv. trifolii in extensive field inoculation trials within the Egypt Nile delta. Plant Soil 336: 129–142.
22. Neupane RP, Thapa GB (2001) Impact of agroforestry intervention on soil fertility and farm income under the subsistence farming system of the middle hills, Nepal. Agric Ecosyst Environ 84: 157–167.
23. Peoples MB, Herridge DF, Ladha JK (1995) Biological nitrogen fixation: an efficient source of nitrogen for sustainable agricultural production? Plant Soil 174: 3–28.
24. Chapin HS III (1980) The mineral nutrition of wild plants. Ann Rev Ecol Syst 11: 233–60.
25. Altieri MA (2002) Agroecology: the science of natural resource management for poor farmers in marginal environments. Agric Ecosyst Environ 93: 1–24.
26. Watson CA, Atkinson D, Gosling P, Jackson LR, Rayns FW (2002) Managing soil fertility in organic farming systems. Soil Use Manage 18: 239–247.
27. Watson CA, Bengtsson H, Ebbesvik M, Lües A-K, Myrbeck A (2002) A review of farm-scale nutrient budgets for organic farms as a tool for management of soil fertility. Soil Use Manage 18: 264–273.
28. Khan AUH, Iqbal M, Islam KR (2007) Dairy manure and tillage effects on soil fertility and corn yields. Bioresource Technol 98: 1972–1979.
29. Zhang H, Xu M, Zhang F (2009) Long-term effects of manure application on grain yield under different cropping systems and ecological conditions in China. J Agr Sci 147: 31–42.
30. Yadvinder-Singh Bijay-Singh, Ladha JK, Khind CS, Gupta RK, et al. (2004) Long-term effects of organic inputs on yield and soil fertility in the rice-wheat rotation. Soil Sci Soc Am J 68: 845–853.
31. Li J, Xu M, Qin D, Li D, Yasukazu H, et al. (2005) Effects of chemical fertilizers application combined with manure on ammonia volatilization and rice yield in red paddy soil. Plant Nutrit Fertilizer Sci 11: 51–56.
32. Malhi SS, Lemke R (2007) Tillage, crop residue and N fertilizer effects on crop yield, nutrient uptake, soil quality and nitrous oxide gas emissions in a second 4-yr rotation cycle. Soil Till Res 96: 269–283.
33. Min DH, Islam KR, Vough LR, Weil RR (2003) Dairy manure effects on soil quality properties and carbon sequestration in alfalfa–orchardgrass systems. Commun Soil Sci Plant 34: 781–799.
34. Körschensa M, Albert E, Armbruster M, Barkusky D, Baumecker M, et al. (2013) Effect of mineral and organic fertilization on crop yield, nitrogen uptake, carbon and nitrogen balances, as well as soil organic carbon content and dynamics: results from 20 European long-term field experiments of the twenty-first century. Arch Agron Soil Sci 59: 1017–1040.
35. Lobell DB, Burke MB (2010) On the use of statistical models to predict crop yield responses to climate change. Agric For Meteorol 150: 1443–1452.
36. McCown RL, Hammer GL, Hargreaves JNG, Holzworth DP, Freebairn DM (1996) APSIM: a novel software system for model development, model testing and simulation in agricultural systems research. Agric Syst 50: 255–271.
37. International Benchmark Sites Network for Agrotechnology Transfer (1993) The IBSNAT decade. Honolulu, Hawaii: Department of Agronomy and Soil Science, College of Tropical Agriculture and Human Resources, University of Hawaii.
38. Myers RJK (1984) A simple model for estimating the nitrogen fertilizer requirement of a cereal crop. Fertil Res 5: 95–108.
39. Deng J, Ran J, Wang Z, Fan Z, Wang G (2012) Models and tests of optimal density and maximal yield for crop plants. PNAS 109: 15823–15828.
40. Cong R, Wang X, Xu M, Ogle SM, Parton WJ (2014) Evaluation of the CENTURY model using long-term fertilization trials under corn-wheat cropping systems in the typical croplands of China. PLOS ONE 9: e95142.
41. Denef K, Six J, Merckx R, Paustian K (2004) Carbon sequestration in microaggregates of no-tillage soils with different clay mineralogy. Soil Sci Soc Am J 68: 1935–1944.
42. Michaelis L, Menten ML (1913) Die Kinetik der Invertinwirkung. Biochem Z 49: 333–369.
43. R Development Core Team (2011) R: a language and environment for statistical computing. Vienna, Austria: R Foundation for Statistical Computing.
44. Spiecker H (1995) Growth dynamics in a changing environment – long-term observations. Plant Soil 168–169: 555–561.
45. Drury CF, Tan CS (1994) Long-term (35 years) effects of fertilization, rotation and weather on corn yields. Can J Soil Sci 75: 355–362.
46. Lobell DB, Hammer GL, McLean G, Messina C, Roberts MJ (2013) The critical role of extreme heat for maize production in the United States. Nat Clim Chang 3: 497–501.
47. Welch JR, Vincent JR, Auffhammerc M, Moyae PF, Dobermann A (2010) Rice yields in tropical/subtropical Asia exhibit large but opposing sensitivities to minimum and maximum temperatures. PNAS 107: 14562–14567.
48. Lobell DB, Field CB (2007) Global scale climate–crop yield relationships and the impacts of recent warming. Environ Res Lett 2: 014002.
49. Schlenker W, Roberts MJ (2009) Nonlinear temperature effects indicate severe damages to U.S. crop yields under climate change. PNAS 106: 15594–15598.
50. Rosenzweig C, Parry ML (1994) Potential impact of climate change on world food supply. Nature 367: 133–138.
51. Shen J, Li R, Zhang F, Fan J, Tang C (2004) Crop yields, soil fertility and phosphorus fractions in response to long-term fertilization under the rice monoculture system on a calcareous soil. Field Crops Res 86: 225–238.
52. Canali S, Trinchera A, Intrigliolo F, Pompili L, Nisini L, et al. (2004) Effect of long term addition of composts and poultry manure on soil quality of citrus orchards in Southern Italy. Biol Fertil Soils 40: 206–210.
53. Poulton PR (1996) The Rothamsted long-term experiments: are they still relevant? Can J Soil Sci 76: 559–571.

54. Jenkinson DS (1991) The Rothamsted long-term experiments: are they still of use? Agron J 83: 2–10.

55. Huang S, Peng X, Huang Q, Zhang W (2010) Soil aggregation and organic carbon fractions affected by long-term fertilization in a red soil of subtropical China. Geoderma 154: 364–369.

56. Zhu P, Ren J, Wang L, Zhang X, Yang X (2007) Long-term fertilization impacts on corn yields and soil organic matter on a clay-loam soil in Northeast China. J Plant Nutr Soil Sci 170 219–223.

57. Vanlauwe B, Diels J, Sanginga N, Merckx R (2005) Long-term integrated soil fertility management in South-western Nigeria: crop performance and impact on the soil fertility status. Plant Soil 273: 337–354.

13

Genome-Wide Association Mapping for Kernel and Malting Quality Traits Using Historical European Barley Records

Inge E. Matthies[1]*[9], Marcos Malosetti[2]*[9], Marion S. Röder[1], Fred van Eeuwijk[2]

1 Department of Gene and genome mapping, Leibniz Institute of Plant Genetics and Crop Plant Research (IPK), Gatersleben, Sachsen-Anhalt, Germany, 2 Biometris, Wageningen University and Research Centre, Wageningen, Gelderland, The Netherlands

Abstract

Malting quality is an important trait in breeding barley (*Hordeum vulgare* L.). It requires elaborate, expensive phenotyping, which involves micro-malting experiments. Although there is abundant historical information available for different cultivars in different years and trials, that historical information is not often used in genetic analyses. This study aimed to exploit historical records to assist in identifying genomic regions that affect malting and kernel quality traits in barley. This genome-wide association study utilized information on grain yield and 18 quality traits accumulated over 25 years on 174 European spring and winter barley cultivars combined with diversity array technology markers. Marker-trait associations were tested with a mixed linear model. This model took into account the genetic relatedness between cultivars based on principal components scores obtained from marker information. We detected 140 marker-trait associations. Some of these associations confirmed previously known quantitative trait loci for malting quality (on chromosomes 1H, 2H, and 5H). Other associations were reported for the first time in this study. The genetic correlations between traits are discussed in relation to the chromosomal regions associated with the different traits. This approach is expected to be particularly useful when designing strategies for multiple trait improvements.

Editor: James C. Nelson, Kansas State University, United States of America

Funding: This study was supported by a grant from the Federal Ministry of Education and Research (BMBF) within the GABI program (GENOBAR, project No. 0315066C). The work of Marcos Malosetti and Fred van Eeuwijk was supported by the Generation Challenge Program (GCP), project No. 221. The funders had no role in study design, data collection or analysis, decision to publish, or preparation of the manuscript.

Competing Interests: The authors have the following interest: Triticarte Pty. Ltd (Canberra, Australia) provided the DArT analyses as a service to the authors.

* Email: matthies@ipk-gatersleben.de (IEM); marcos.malosetti@wur.nl (MM)

[9] These authors contributed equally to this work.

Introduction

Barley (*Hordeum vulgare* L.) is a major cereal crop in Europe. It ranks fourth in worldwide production, after wheat, rice, and maize. It is grown for feed, food, and malting. Most of the malt produced is used for brewing beer and, to a lesser extent, for distilling (e.g., whiskey). In Europe, two-rowed spring cultivars are used mainly for malting and brewing; six-rowed winter barleys are predominantly used for food. However, six-rowed barley has been increasingly used for malting in Europe, following the trend started in the US. Therefore, depending on the end-use, there are two primary aims in breeding barley: 1) superior food and feed quality with high protein content, and 2) high malting quality with high starch and low protein contents. Improving the malting quality is a central goal in breeding, in addition to improving the yield of barley. Malting quality is a complex trait, because it consists of several components, and all are polygenic. Moreover, the definition of high malting quality is not straightforward; it depends on the processing and brewing methods. In general, the main breeding goals for malting barley are high malting extract, low protein content, good solubility properties, good kernel formation, and low glume content.

For the past 80 years, to optimize the malting traits in barley, breeders mainly focused on a narrow gene pool of spring barley types [1]; the most important quality parameters to optimize were the amounts of soluble protein, extract, raw protein, and friability. Further improvements in malting quality must rely on new combinations of genes and germplasms. Molecular marker-assisted selection (MAS) schemes have been applied to developing barley varieties with improved malting quality traits. Those studies have identified many quantitative trait loci (QTL) in barley [2–4]. MAS strategies have facilitated gene pyramiding techniques to acquire advantageous alleles from different loci. With MAS, the breeding efficiency can be improved by eliminating undesired genotypes at early stages, which can reduce time and costs [4–7]. The genome-wide association approach provides a good basis for selection strategies in any breeding program.

The identification of barley genomic regions that influence yield and malting properties will increase our understanding of the genetics and promote the development of cultivars with improved kernel and malting quality. The genetic and biochemical bases of malting quality in barley have been addressed previously [2,8,9]. However, quantification of malting quality parameters requires elaborate, expensive phenotypic analyses.

Typically, the high cost of assessing malting quality in barley lines is due to expensive equipment, laboratory facilities, and experienced personnel. Moreover, assessing malting and brewing quality requires substantial amounts of grain (100–1,000 g), which is often not feasible in the early generations of a breeding cycle. In addition, some malting quality parameters can only be determined in time-consuming, wet lab analyses. These limitations may be overcome with the use of historical phenotypic data recorded in statistical year books, like those from the Deutsche Braugersten-gemeinschaft or the European Brewery Association. These resources may provide a cost-effective approach. The complex dataset considered in the present study may serve as a valid resource for breeding barley varieties with high malting quality.

In addition, the identification of marker-trait associations (MTAs) may represent a cost effective strategy for selecting traits that are typically expensive to identify in MAS schemes [2,3,10]. Molecular markers and QTLs have been described for numerous traits in barley, and major genes have been detected in segregating populations derived from biparental crosses [3,10–14]. The use of genome-wide association mapping for QTL detection has attracted interest in agricultural settings, due to the recent availability of high-throughput genotyping technology and the development of new statistics methodologies [15–17].

Association mapping, also known as linkage disequilibrium (LD) mapping, represents an interesting alternative to traditional linkage analysis. It provides the advantages of (i) wider genomic diversity than provided by biparental segregating populations, (ii) high mapping resolution, by exploiting historical recombinations in the population, and (iii) rapid results, because it is not necessary to create a segregating population [18,19]. In combination with high-density genotyping, association mapping can resolve complex trait variation down to the sequence level by incorporating historical recombination events that occurred at the population level [20,21].

Two association mapping methodologies are widely used in plants. The first is a candidate gene approach, which relates polymorphisms in candidate genes to phenotypic variations in traits. The second approach is a genome-wide association study (GWAS), which relates polymorphisms of anonymous markers to trait variations [16,22]. Candidate gene studies are widely conducted in crop plant species, including barley and maize. Those studies aim to detect functional markers that directly impact the trait of interest [23–27]. The GWAS approach has recently benefitted from the advent of cost-effective high throughput marker technology, like Diversity Array Technology (DArT) [28] and Illumina Bead Chips or Bead arrays [29,30]. High marker coverage is required for conducting a GWAS, but the potential of this approach has been demonstrated in barley [15,22,30–35].

DArT markers are bi-allelic, dominant markers. A single DArT assay can genotype thousands of SNPs and insertions/deletions across the genome simultaneously. Barley was one of the first plant species for which DArT markers became available [36–38]. The integrated barley consensus map now contains 3,542 markers, including DArT markers. This map has been used to locate meaningful associations [39]. The first examples of applying DArT marker technology to *Hordeum* included a GWAS conducted to detect yield-associated genes [40] and a QTL mapping study conducted to identify net blotch resistance in a segregating population [41]. Other examples included the study of linkage disequilibrium (LD) and population structures in association studies that aimed to identify powdery mildew and yield components in barley [42–45]. Another study associated DArT markers with malting quality characteristics from two row Canadian barley lines [46]. In another GWAS, 138 wild barley accessions were genotyped with DArT markers and SNP markers from the Illumina Golden Gate Assay to detect genomic regions associated with spot blotch resistance [47].

An important issue in GWAS is that the population structure which arises from heterogeneous genetic relatedness between entries in the association panel can cause high LDs between unlinked loci [48]. When LDs between markers and traits occur as a consequence of the population structure, they are called false positives or spurious associations. Therefore, a statistical model must account for genetic relatedness, typically by choosing an appropriate mixed linear model that accommodates genetic covariance between observations [44,49,50]. A wide range of models have been proposed that account in one way or another for relationships between genotypes [18,19,49–54]. Population structure is particularly prominent in self-pollinating barley [17,33]; it causes clear spurious associations between spike morphologies (two- versus six-rowed types) and between growth habits and vernalization requirements (winter and spring genotypes) [22,55–59]. The barley collection used in the present study exhibited a combination of those characteristics. Therefore, proper consideration of population structure was required to assure meaningful MTAs.

This study aimed to identify chromosomal regions that influenced kernel and malting quality parameters in barley, based on a diverse set of cultivars and historical phenotypic data. The approach included (i) genotyping the germplasm with DArT markerDArT markers, (ii) investigating the degree of intrachromosomal LD decay within this barley collection, and (iii) performing a GWAS with a mixed linear model approach.

Results

Phenotypic data analysis

Based on the available historical data, we obtained best linear unbiased estimators (BLUEs) for grain yield, eight kernel traits, and ten malting quality traits for each cultivar (Table S1). Inspection of residual plots showed no deviations from model assumptions (Figure S1). Traits expressed in percentages were log-transformed before analysis to stabilize the variance. Summary statistics of the adjusted means are shown in Table 1. A large range of variation was observed for most traits, including soluble nitrogen (solN), grain yield (GY), thousand grain weight (TGW), soluble protein (SolP), and saccharification number (VZ45). In general, broad sense heritabilities were above 0.4, with few exceptions, which indicated that a relatively large genetic component was involved in the determination of the observed trait variation (Table 1).

The correlations among all considered traits are shown in Figure 1. In general, correlations were moderate, with absolute values ranging between 0.30 and 0.70. Strong positive correlations were found between malt extract and the malting quality index (MQI) (r = 0.88), malt extract and friability (r = 0.81); and between MQI and friability (r = 0.82). Furthermore, a high correlation (r = 0.78) was observed between soluble nitrogen (SolN) and soluble protein (SolP). SolN and SolP were also related to color and to the saccharification number (VZ45), as reflected in their relatively high correlations with VZ45 (r = 0.73, and r = 0.76, respectively). Four highly negative correlations were observed between friability and viscosity (r = −0.91); between a larger grain fraction (>2.8 mm) and two smaller grain fractions (r = −0.85 and r = −0.71); and between friability and kernel raw protein (K_RP; r = −0.71). Strongly negative correlations were also observed between malt extract and sieve fractions (SF) <2.2 mm, the K_RP, and viscosity (r = −0.67, −0.67, and r = −0.66, respec-

Table 1. Summary statistics of the nine agronomic and 10 malting quality traits considered in this study.

Trait	Units	N	Min	Max	Mean	Std. dev.	h^2
Agronomic traits							
GY	[dt/ha]	103	52.5	83.1	71.0	7.5	0.60
TGW	[g]	174	37.8	59.1	46.2	3.7	0.65
HLW	[kg]	165	65.3	72.8	68.7	1.4	0.47
KF	[1–9]	131	2.3	6.4	4.6	0.9	0.78
GF	[1–9]	131	1.6	6.7	4.4	1.1	0.78
SF <2.2 mm	[%]	159	0.2	7.5	1.9	1.5	0.39
SF 2.2–2.5 mm	[%]	157	1.0	30.3	8.0	5.9	0.24
SF >2.8 mm	[%]	174	20.0	89.6	60.6	17.0	0.57
K_RP	[%]	168	9.1	13.5	11.2	0.9	0.55
Malting quality traits							
M_RP	[%]	126	8.9	11.8	10.1	0.6	0.22
solN	[mg/100 g DM]	126	559.0	817.9	706.4	57.8	0.45
solP	[%]	126	33.9	52.3	44.0	3.9	0.50
Visc	[mPas]	95	1.4	2.4	1.6	0.1	0.79
Col	[EBC]	112	2.7	6.0	3.6	0.4	0.12
Fria	[%]	125	33.9	95.4	79.9	13.0	0.54
VZ45	[%]	125	31.2	49.6	40.7	4.0	0.39
Extr	[%]	125	77.1	85.1	81.7	1.6	0.64
FiAt	[%]	114	77.7	83.8	81.5	1.2	0.21
MQI	–	114	3.4	10.2	7.5	1.4	0.62

N = number of varieties, h^2 = broad sense heritability. h^2 was obtained by fitting genotypes as random terms in the statistical model. Trait abbreviations: GY = grain yield, TGW = thousand grain weight, HLW = hectoliter weight, KF = kernel formation, GF = glume fineness, SF = sieve fraction (less than 2.2 mm, 2.2–2.5 mm, or more than 2.8 mm), K_RP = raw kernel protein content, M_RP = raw malt protein content, solN = soluble nitrogen, solP = soluble protein, Visc = viscosity, Col = color, Fria = friability, VZ45 = saccharification number, Extr = malt extract, FiAt = final attenuation, MQI = Malting quality index.

Trait correlation	GY	TGW	HLW	KF	GF	SF <2.2	SF 2.2-2.5m	SF >2.8m	K_RP	M_RP	SolN	SolP	Visc	Col	Fria	VZ45	Extr	FiAt	MQI
GY	1.00																		
TGW	0.39	1.00																	
HLW	-0.33	0.10	1.00																
KF	0.47	-0.16	-0.45	1.00															
GF	0.55	0.00	-0.68	0.71	1.00														
SF <2.2 mm	0.31	-0.51	-0.25	0.55	0.39	1.00													
SF 2.2-2.5mm	0.17	-0.44	-0.19	0.50	0.30	0.68	1.00												
SF >2.8mm	0.05	0.65	0.16	-0.57	-0.23	-0.71	-0.85	1.00											
K_RP	0.22	-0.09	-0.06	0.06	0.18	0.37	0.42	-0.41	1.00										
M_RP	-0.25	-0.32	0.22	-0.26	-0.22	0.30	0.13	-0.24	0.73	1.00									
SolN	-0.39	0.00	0.06	-0.44	-0.49	-0.40	-0.34	0.32	-0.12	0.20	1.00								
SolP	-0.23	0.18	-0.09	-0.30	-0.32	-0.57	-0.38	0.44	-0.56	-0.43	0.78	1.00							
Visc	0.48	-0.01	-0.09	0.31	0.45	0.61	0.35	-0.33	0.47	0.13	-0.53	-0.60	1.00						
Col	-0.19	-0.06	-0.01	-0.03	-0.08	-0.01	-0.02	0.06	-0.26	-0.21	0.32	0.44	-0.22	1.00					
Fria	-0.44	0.29	0.02	-0.34	-0.46	-0.66	-0.41	0.49	-0.71	-0.48	0.51	0.77	-0.91	0.29	1.00				
VZ45	-0.34	0.08	0.04	-0.32	-0.37	-0.41	-0.41	0.40	-0.45	-0.13	0.73	0.76	-0.68	0.23	0.62	1.00			
Extr	-0.48	0.34	0.02	-0.54	-0.56	-0.67	-0.52	0.62	-0.67	-0.57	0.37	0.70	-0.66	0.23	0.81	0.57	1.00		
FiAt	0.39	0.10	0.02	-0.15	-0.12	-0.17	-0.07	0.25	-0.26	-0.17	0.31	0.43	-0.37	0.16	0.40	0.46	0.31	1.00	
MQI	-0.45	0.27	-0.02	-0.51	-0.59	-0.54	-0.48	0.59	-0.58	-0.36	0.49	0.69	-0.69	0.24	0.82	0.67	0.88	0.36	1.00

Figure 1. Correlation matrix of yield, kernel quality, and malting quality parameters based on BLUES for each cultivar. The Pearson coefficients of the two sided test are given only for the significant phenotypic trait correlations. BLUES = best linear unbiased estimators, GY = grain yield, MY = marketable yield, TGW = thousand grain weight, HLW = hectoliter weight, KF = kernel formation, GF = glume fineness, SF = sieve fraction, K_RP = raw kernel protein content, M_RP = raw malt protein content, solN = soluble nitrogen, solP = soluble protein, Visc = viscosity, Col = color, Fria = friability, VZ45 = saccharification number VZ45°C, Extr = malt extract, FiAt = final attenuation, MQI = malting quality index.

tively); and between viscosity and VZ45 (r = −0.68). The GY and TGW showed only moderate or low correlations with agronomic and malting quality traits. Both, hectoliter weight (HLW) and color showed low correlations with all other traits. Overall, the correlations among malting quality traits were higher than the correlations among agronomic quality traits. In general, the correlations among agronomic and malting quality traits were moderate to low, which hinted that the genetic determinants of these two types of quality parameters were relatively independent.

Genotyping with DArT markers

The original set of 1088 DArT markers was reduced to 839, because we discarded monomorphic markers, markers with rare alleles (minor allele frequency <0.05), and markers missing more than 10% of the values. The marker map showed a high genomic coverage, with a density of about one marker in every 5 cM for most genomic regions, except for chromosome 4H, which showed some inter-marker distances larger than 15 cM (Table 2; Figure S2).

Population structure and intrachromosomal linkage disequilibrium (LD)

The first three principal components were found to be significant factors by the eigen analysis, and they cumulatively explained 28% of the total variation (20.2% and 5.3% with the first and second axes, respectively). The plot of the first two principal components showed a clear division of the germplasm into three subpools, which largely coincided with the row number and seasonal habit (2-rowed spring, 2-rowed winter, and 6-rowed winter; Figure 2). The only exceptions were five cultivars, which clustered differently, according to what their *a priori* classifications would suggest. These included the 2-row spring varieties "Fergie" and "Phantom", which were located in the 2- and 6-row winter pools, respectively; the 2-row winter variety "Cordoba", which

was grouped with the 6-row winter types; the 6-row winter variety "Tilia", which was located in the 2-row spring pool; and the 2-row spring variety "Stella", which appeared isolated between the 2-row spring and the 6-row winter pools. These results were consistent with results observed in other barley studies [45]. Based on this principal component analysis (PCA), we concluded that the first three principal components represented the major structure in the population. Therefore, we decided to use principal component scores as covariates in other models as an effective strategy to correct for population stratification (i.e., when assessing LD between markers, and when testing for associations between markers and traits).

After correcting for population structure, the intrachromosomal LD was studied in all seven barley chromosomes by inspecting the plot of the associations between linked markers (r^2 values) and their map distances, in cM (Figure 3; Figure S3). Taking a value of $r^2 = 0.20$ as a strict threshold, based on the upper 0.95 quantile of the observed r^2 values between unlinked markers, we found that the markers were, on average, in LD up to a distance of 5 cM. When we imposed the more liberal threshold of $r^2 = 0.10$ (upper 0.80 quantile), the markers were, on average, in LD up to a distance of 10 cM.

In addition to assessing the marker density, the LD-decay information was used to define a Bonferroni-like multiple testing correction factor that we applied in the GWAS. The correction factor was defined as the total number of genome-wide independent tests, which was calculated as the number of chromosome blocks which were in LD, summed over all chromosomes. The corresponding correction factors were used for evaluating the significance (expressed as $-\log_{10}P$) of MTAs. We used significance thresholds of $-\log_{10}(P) > 3.65$ and $-\log_{10}(P) > 3.35$, based on the strict ($r^2 = 0.20$) and more liberal ($r^2 = 0.10$) thresholds for LD, respectively. Cumulative p-values obtained by

Table 2. Summary statistics of the 839 DArT markers used in this study.

Chromosome	Length [cM]	No. of markers	Median inter-marker distance	95th percentile of the inter-marker distance
1H	148.1	98	0.2	5.4
2H	163.2	172	0.2	4.9
3H	177.6	138	<0.1	6.1
4H	147.4	57	0.3	16.9
5H	185.3	112	0.5	5.5
6H	141.4	119	<0.1	6.9
7H	160.2	143	<0.1	5.2
Genome	**1123.1**	**839**	**0.1**	**5.8**

the naïve model and the MLM with correction for population structure and kinship by PCA were compared (Figure S4).

Genome-wide association study (GWAS)

The inflation factors for all traits and models are shown in Table 3. As expected, a large inflation factor was observed for nearly all traits when the model did not account for genetic relatedness (naïve model). However, we observed that, even with the naïve model, the inflation factor was low for final attenuation (FiAt), TGW, raw protein in malt (M_RP), and beer color, which indicated that few strong associations were expected between the markers and those traits (as confirmed with the other models). The inflation factor fell substantially in all five models that accounted for genetic relatedness. The kinship model showed the steepest fall in inflation factor, with values very close to 1 (values below 1 indicated that the correction was too conservative). In models that used groups (based on population structure) and principal component scores to correct for genetic relatedness, the inflation factors decreased substantially, but not as much as the drop observed with the kinship model. Little difference was observed when the correction was considered a fixed or random term in a given model. However, on average, across all traits, the model that used principal component scores as fixed covariables performed slightly better (lower inflation factor) than the other models. Therefore, the following discussion is focused on the MTAs found with the model that used PCA scores as fixed terms.

With a threshold of −log₁₀(P) >3.35, we identified 140 MTAs. With a more strict threshold of −log₁₀(P) >3.65, we found 101 significant MTAs. These numbers are remarkably large, considering the relatively low sample size of this study (Table 4, Figure 4). We also observed an association between the heritability (h²) of the traits and the number of detected MTAs. More MTAs were found for high-heritability traits than for low-heritability traits. For example, kernel formation (KF) and glume fineness (GF) had nearly the highest h² values (both 0.78; Table 1) and were associated with high numbers of markers (10 and 13, respectively; Table 4). In contrast, final attenuation had one of the lowest h² values (Table 1), and it was not associated with any markers (no MTAs; Table 4).

The complete list of MTAs is shown as supporting information in Table S2. Most of the 140 MTAs were observed on barley chromosomes 1H and 5H (Table 4 and Figure 4). Chromosome 5H clearly stood out, with about one third of the detected MTAs (41 out of 140). This was followed by chromosomes 1H and 2H with 30 and 28 MTAs, respectively. Only one MTA was detected on chromosome 4H (Table 4). The locations of markers and their associated traits are displayed in Figure 4 and summarized in Table 4 and Table 5. Some markers were associated with multiple traits. Most of the MTAs that were associated with multiple traits were located on chromosomes 1H, 2H, and 5H, and to some extent on chromosomes 3H and 7H. Some MTAs for different traits were co-localized within a small region of a chromosome (cluster). Most MTA clusters were located on 1H, 2H, 3H, and 5H (Figure 4). The region around 94.5 cM on chromosome 1H was tagged with many MTAs that were associated with multiple traits (SolN, SolP, and VZ45) consistent with the high correlations observed among these traits (Figure 1). The region between 110 and 165 cM on 2H was tagged with MTAs for several different traits, which indicated another hot spot relevant to malting and brewing quality. Furthermore, MTAs for GY, TGW, friability, and K_RP were located on chromosome 5H in the region between 13.8 and 18.0 cM. On the same chromosome, another dense concentration of MTAs was found in the region between 159 and 180 cM (Figure 4, Figure 5, and Figure S5). A summary of the common MTAs is provided as supporting information (Table S2). In addition, Table S2 shows the marker-allele substitution effects for all 140 MTAs.

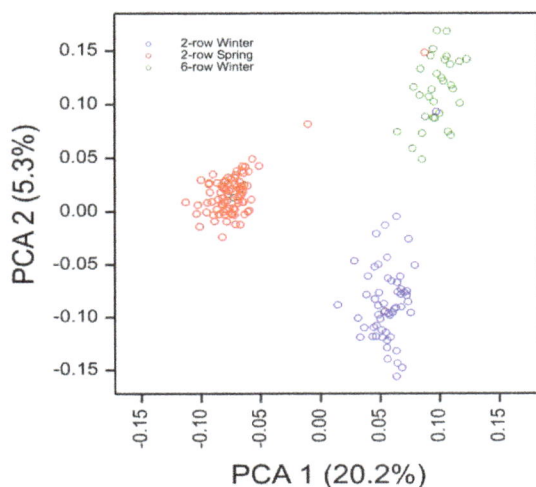

Figure 2. Scatter plot of the first two principal components show the distribution among the cultivars. Cultivars are classified by type: 2-row spring (blue), 2-row winter (red), and 6-row winter (green). The variance explained by each principal component is given in the axis heading.

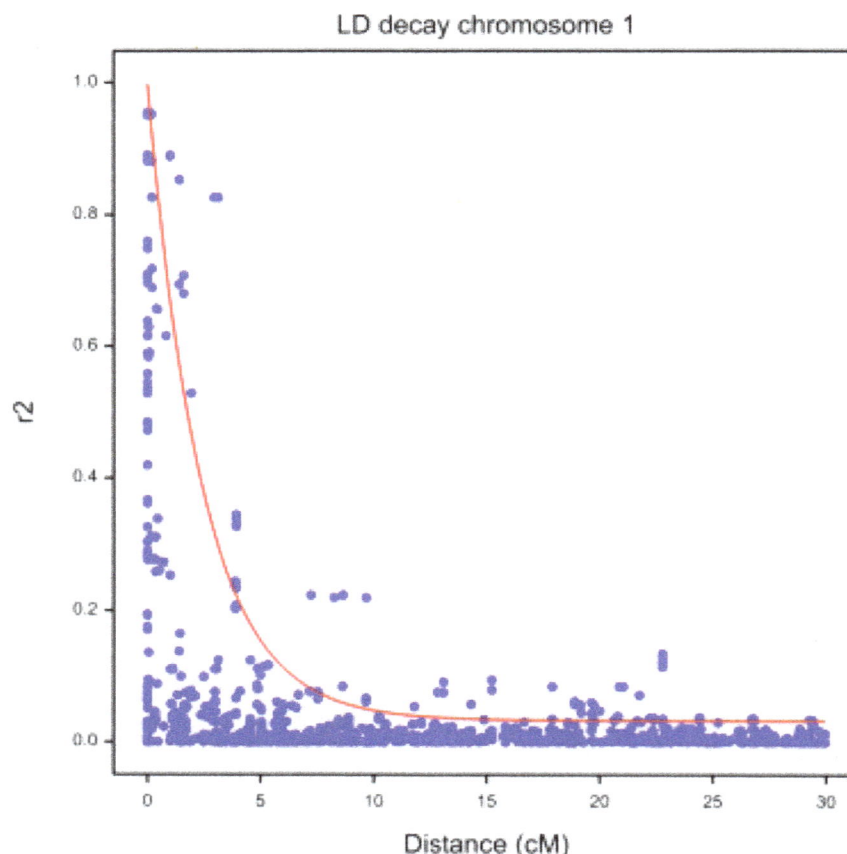

Figure 3. Intrachromosomal Linkage disequilibrium (LD)-decay between all pairs of DArT markers for chromosome 1H. LD between markers (r^2) is a function of marker distances (cM).

Discussion

This study employed an association mapping approach to reveal the genetic basis of several kernel and malting quality parameters in barley. The MTAs identified in this study suggested that some genetic regions are highly important for breeding barleys with enhanced kernel and malting qualities. Some of the identified MTAs confirmed previously known QTLs [[12,46,60–80], but others were identified for the first time in this study (Table 5). Here, we discovered MTAs on all seven barley chromosomes. We found high MTA concentrations on chromosomes 1H, 2H, and 5H (Figure 4), which mostly represented former identified QTL-hotspots [62].

We discovered some novel genomic regions associated with malting quality on chromosomes 2H, 5H, 6H, and 7H. For example, we found associations for M_RP, SF (2.2–2.5 mm), and color on chromosomal regions of 6H that had not been identified before as QTLs. Furthermore, new MTAs were detected for K_RP, extract, KF, and GF, on chromosome 2H; for TGW, friability, and MQI on chromosome 5H; and for viscosity and MQI on chromosome 7H. Overall, the DArT markers located in previously described QTLs and the MTAs that we found were generally associated with the same or similar characteristics (Table 5). For example, the MTA found at 58.7–59.4 cM on chromosome 1H was associated with malt extract, viscosity, and friability, consistent with QTLs reported elsewhere [60–65]. Moreover, two strong QTLs for grain yield on chromosomes 1H and 7H coincided with those described previously [12,60,61] and

[62,71,80], respectively. MTAs for GY and TGW on 5H and 6H were comparable to those detected in other studies [12,78,79]. Furthermore, the MTA found on 5H at 184.4 cM for SolP matched a QTL in this genomic region [66] and two other QTLs for yield on 7H [12]. Some of the MTAs that were associated with phenotypic characteristics also mapped to genomic regions close to QTLs associated with related traits. For example, the QTL QYld.StMo-3H.1 for GY [3,63,72–74] was located at 48.3 cM on chromosome 3H, where we found seven DArT markers that were strongly associated with kernel formation and glume fineness. On chromosome 2H, we detected some markers that were associated with these two traits in genomic regions that were previously reported to have an impact on yield parameters [62,66–68]. We also discovered two DArT markers, bPb-0994 and bPb-6822 (located on chromosome 2H at 113.2 and 114.4 cM, respectively) associated with grain protein content, which were also found in another study [69]. We also found that the marker bPb-0994, located on chromosome 2H, was highly significant for kernel raw protein content. This marker was previously shown to be significant for grain protein content [69], in addition to three markers on 2H and 3H. Two other DArT markers on chromosome 3H, which were associated with grain protein content [69], were related to other traits in our study, including sieve fraction >2.2 mm (bPb-5298 at 145.5 cM) and friability (bPb-9599 at 149.8 cM), because we used a different germplasm.

It was not always straightforward to make comparisons with other known QTLs reported in the literature, because different

Table 3. Genome-wide inflation factors of the naïve (uncorrected) model and five other models that account for genetic relatedness.

Trait	Naïve	Groups random	Groups fixed	PCA scores random	PCA scores fixed	Kinship
GY	8.27	1.17	1.21	1.33	1.38	1.12
TGW	2.29	1.46	1.46	1.54	1.55	1.10
HLW	3.01	1.55	1.53	1.46	1.50	1.02
KF	4.03	1.44	1.43	1.38	1.51	1.13
GF	6.91	1.85	1.82	1.57	1.73	1.20
SF <2.2 mm	7.05	1.45	1.46	1.42	1.44	1.02
SF_2.2–2.5 mm	5.09	1.31	1.28	1.34	1.33	0.10
SF_>2.8 mm	4.29	1.49	1.45	1.47	1.46	1.00
K_RP	11.98	1.94	1.87	1.93	1.80	1.31
M_RP	2.06	1.72	1.69	1.57	1.65	1.13
SolN	3.25	1.47	1.59	1.46	1.55	1.06
SolP	4.84	1.73	1.79	1.61	1.68	1.03
Visc	6.50	1.31	1.32	1.19	1.23	1.15
Col	2.09	1.44	1.42	1.46	1.42	1.06
Fria	8.89	1.84	1.86	1.69	1.75	1.07
VZ45	5.06	1.59	1.59	1.46	1.46	1.06
Extr	8.03	1.66	1.64	1.50	1.51	1.10
FiAt	1.38	1.38	1.73	1.38	1.36	1.01
MQI	6.29	1.85	1.87	1.65	1.71	1.11

studies used different reference maps, marker types, germplasms, experimental sites, and trait measuring protocols [46,62,81–86]. For example, we did not find MTAs identical to those found by Beattie et al. 2010 [46], because different germplasms were studied (North American *vs.* European material). A similar explanation can account for differences in GWAS that mapped GY in landraces cultivated in the high- and low-yield environments of Spain and Syria, respectively [40]. Only one of their associated DArT markers was also detected in our European elite germplasm (bPb-9163 on 5H). Again, this discrepancy was probably caused by the lack of correspondence between the different genetic backgrounds used by Pswarayi et al. (2008) [40] and our study. An association study with kernel quality parameters in a restricted subset of 101 almost identical winter barley varieties (48 2-row and 61 6-row types) was performed based on Illumina-SNP-markers [83]. Only the MTAs that we primarily associated with grain yield and hectoliter weight on chromosomes 1H and 5H matched the findings in that study [83]. Other barley association panel results that were based on different marker systems and germplasm pools (e.g., the Barley CAP germplasm) [87–88] also showed little congruence with our results.

Genetic correlations among traits typically result from either pleiotropic or tightly-linked QTLs. In the present study, we found many MTAs for different traits that co-localized to a single chromosomal region. This co-localization may drive genetic correlations among barley quality traits. In particular, we found clusters of MTAs for malting quality traits on chromosomes 1H, 2H, and 5H, which pointed to hot spots for barley quality. This was consistent with findings from Szücs et al. 2009 [62] and with another study that reported evidence of multilocus clusters that may regulate or control barley malting characteristics [71].

In general, our findings were consistent with the literature and corresponded well with the observed correlations among traits [75,77,87–94]. Most co-localized MTAs represented traits with high phenotypic correlations. For example, MTAs that correlated with the malting quality parameters, SolP, SolN, and VZ45, were detected in the same region on 1H (and 5H). This information may provide valuable guidance for understanding a multivariate response to a protocol designed to select for these traits.

Some traits, like MQI, friability, and extract are genetically correlated with each other. This correlation was reflected in our results by the finding that these traits were significantly associated with the same markers. Traits such as viscosity and friability or final attenuation, VZ45, and extract interact to define malting properties, which contribute to important phenotypic effects. It is crucial to be aware of these interactions to understand the trade-offs implied in the optimization of cultivars.

For example, breeders should be aware that high grain protein concentration is associated with low levels of malt extract. High grain protein increases the likelihood of a chill haze in beer, and barleys with low grain protein concentrations are more economically efficient in the malting process [90]. High protein reduces efficiency, because the grain takes up water slowly and unevenly during the germination process. In addition to producing low levels of malt extract, the resulting beer has a longer filtration time, develops cloudiness, and possesses a shorter shelf life. On the other hand, insufficient levels of grain protein may limit the growth of yeast during the fermentation process; it also reduces the stability of the beer head because the beer foam cannot cling effectively to the side of the glass. Consequently, maltsters prefer a GPC close to 10.5% [90]. It behooves the breeder to know which parameters are correlated, because these parameters require balancing to achieve the optimal outcome.

Table 4. Number of significant (# sign) MTAs, based on high and low thresholds of significance, identified by applying the GWAS model with eigenvalues as fixed effects (fixed PCA score model).

Trait	# sign. MTAs	# sign. MTAs	# sign. MTAs	Chromosomes
	−log10(P) >3.65	3.35 < −log10(P) <3.65	−log10(P) >3.35	with MTAs
GY	6	2	8	1H, 5H[2], 7H
TGW	3	5	8	1H, 5H[2]
HLW	2	0	2	1H, 5H
KF	9	1	10	2H[2], 3H
GF	13	0	13	1H, 2H[2], 3H, 7H
SF_<2.2 mm	3	1	4	3H
SF_2.2–2.5 mm	1	0	1	6H
SF_>2.8 mm	1	5	6	2H
K_RP	11	4	15	2H[3], 5H[3], 6H, 7H
M_RP	8	0	8	5H, 6H
SolN	8	2	10	1H,7H
SolP	7	3	10	1H, 5H[2]
Visc	4	0	4	7H[4]
Col	2	0	2	1H, 6H
Fria	6	7	13	3H, 4H, 5H[2]
VZ45	5	4	9	1H, 3H
Extr	5	1	6	1H, 2H, 5H
FiAt	0	0	0	–
MQI	7	4	11	2H[3], 5H[2], 7H[2]
Total	101	39	140	

Markers located within 5 cM of each other were considered to be the same MTA. The superscript number shown in parentheses next to a chromosome number indicates the number of MTAs on that chromosome.

Conclusions

The current work contributed to the understanding of the genetic basis of kernel and malting qualities in barley. The use of a broad phenotypic data collection that spanned a long time range and several locations provided a means to de-emphasize environmental effects on barley trait expression. We found that combining this historical phenotypic dataset with high-density, low-cost markers such as DArTs facilitated the discovery of new MTAs for barley. As shown previously, we found that association mapping was a powerful, promising approach for dissecting the complexities of malting and brewing qualities in barley. In addition to confirming various known QTLs, we identified some new MTAs; e.g., markers for MQI and viscosity. The MTAs identified in this study will be useful for selecting favorable genotypes in this germplasm that can be used to develop improved barley varieties. The findings of this study should be validated in future field experiments. Our research demonstrated the advantage of combining more than 20 years of expensive phenotyping information with high-density, low-cost marker technology.

Materials and Methods

Germplasm and phenotypic data

A set of 174 European barley cultivars that included 85 two-rowed spring and 89 winter types (57 two-rowed and 32 six-rowed) were included in this study. Historical phenotypic data were available for all 174 cultivars in the annual statistical reports from the German Brewery Association (http://www.braugerstengemeinschaft.de). The historical data were collected from 1985 to 2007 and stored in a database called "MetaBrew" [95]. The following nine agronomic traits were considered: grain yield (GY), thousand grain weight (TGW), hectoliter weight (HLW), kernel formation (KF), glume fineness (GF), three sieve fractions (SF), and raw kernel protein content (K_RP). Ten malting quality traits were also considered: raw malt protein content (M_RP), soluble nitrogen (solN), soluble protein (solP), viscosity (Visc), color (Col), friability (Fria), saccharification number VZ45°C (VZ45), malt extract (Extr), final attenuation (FiAt), and malting quality index (MQI). Malting quality parameters were assessed with standard procedures recommended by the European Brewery Convention (EBC) and the 'Mitteleuropäische Brautechnische Analysenkommission (MEBAK)'.

Genotyping with DArT markers

Seeds for all cultivars were obtained from the breeders. Seeds were grown into young plantlets. Leaf material was harvested from five to six seedlings that were 10 days old. The material was bulked, and genomic DNA was extracted according to the requirements of Triticarte Pty. Ltd. (Canberra, Australia), as described previously [25,26]. A dense, whole genome scan was performed with Diversity Array Technology (DArT), which generated 1,088 mapped and 774 unmapped biallelic markers for this population, according to the published DArT consensus map [37]. The locus designations used by Triticarte Pty. Ltd. were adopted in this study, and DArT markers were named with the prefix "bPb," followed by a unique numerical identifier. We removed markers with minor allele frequencies less than 0.05.

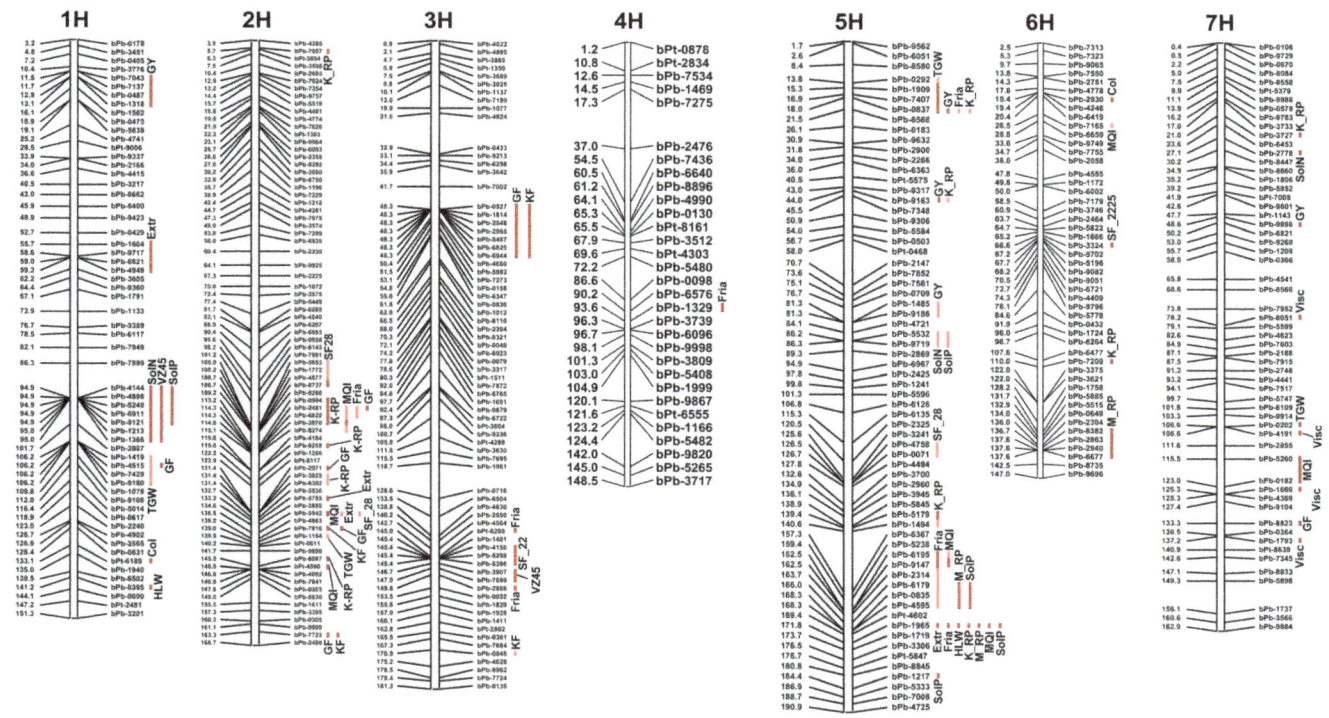

Figure 4. Barley consensus map with DArT markers significantly associated with kernel and malting quality traits. Only markers with the highest effect on a given chromosomal position are depicted. All MTAs that reflect kernel and malting quality parameters are defined either with a strict significance threshold = -log10(P)>3.65 (dark red) or a liberal significance threshold = -log10(P) <3.35 (light red). DArT = Diversity Array Technology, MTA = marker-trait association, GY = grain yield, TGW = thousand grain weight, HLW = hectoliter weight, KF = kernel formation, GF = glume fineness, SF = sieve fraction, K_RP = raw kernel protein content, M_RP = raw malt protein content, solN = soluble nitrogen, solP = soluble protein, Visc = viscosity, Col = color, Fria = friability, VZ45 = saccharification number VZ45°C, Extr = malt extract, FiAt = final attenuation, MQI = malting quality index.

Then, a set of 839 mapped DArT markers was selected for the GWAS to provide coverage that was evenly distributed over the seven barley linkage groups (Figure S2).

Analysis of phenotypic data

Our objective was to concentrate on major, stable MTAs. Therefore, we calculated cultivar-adjusted means over locations and years for each of the traits. Prior to the analysis, traits that were expressed in percentages were log-transformed, such as sieve fraction (SF) and raw protein in kernel or malt (K_RP, M_RP). The following mixed model was used to estimate adjusted cultivar means (random terms underlined):

$$y_{ijk} = \mu + G_i + \underline{Y_j} + \underline{GY_{ij}} + \underline{\varepsilon_{ijk}}$$

$\underline{y_{ijk}}$ is the observation of the ith cultivar, in the jth year, and the kth replicate (location) nested within year j; μ is an intercept; G_i is the fixed cultivar effect; $\underline{Y_j}$ is the random year effect, where; $\underline{Y_j} \sim N(0, \sigma_Y^2)$; $\underline{GY_{ij}}$ is the interaction between the cultivar and the year, where; $\underline{GY_{ij}} \sim N(0, \sigma_{GY}^2)$; and $\underline{\varepsilon_{ijk}}$ is a residual term, where; $\underline{\varepsilon_{ijk}} \sim N(0, \sigma_\varepsilon^2)$. Note that, because locations are considered replications within a year, location effects (and corresponding interactions with cultivars) do not appear explicitly in the statistical model, but are pooled within the residual term, $\underline{\varepsilon_{ijk}}$. Evaluating locations as replicates was justified in this type of trial network,

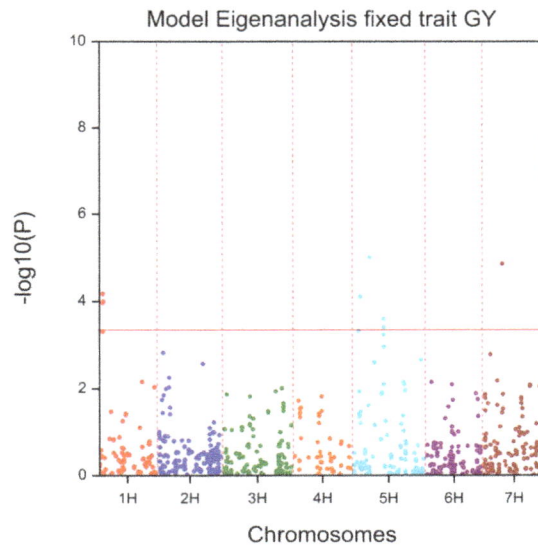

Figure 5. Manhattan plot for GWAS of grain yield (GY), considering eigenvalues (PCA scores) as fixed effects. The significance threshold is -log10(P) >3.65.

Table 5. Genome-wide marker-trait associations (MTAs) detected at a significance threshold of −log$_{10}$(P) >3.35 with the PCA scores fixed model are compared with known reference quantitative trait loci (QTLs) in the same chromosomal regions.

Chr	Chr-Pos [cM]	DArT marker	Traits with MTAs	Traits with Ref QTL	Position of Ref QTL [cM]	Designation of Ref QTL	Literature References
1H	11.5–13.1	bPb-7137	GY	GY	21.2	QYld.pil-1H	[12,60,61]
	58.7–59.2	bPb-6621, bPb-4949, bPb-9717	Extr	Extr, GY	53.5, 60.7, 63.9	QMe.StMo-1H,3, QMe.SIAl-1H	[60–65]
	–	bPb-4144, bPb-4898,	–	Fria, MQI	58.7	QFRI1 1H gP68M59_200-Bmac90 QKp.HaMo-2H,	[65]
	94.9	bPb-5249, bPb-6911,	SolN, SolP,	KF, Extr	92.6 – 100.8	QKp.nab-2H,	[62]
		bPb-9121, bPb-1213, bPb-1366	VZ45			QMe.StMo-1H,4	[66]
	106.2	bPb-1419, bPb-7429, bPb-9180	TGW	Extr TGW,	106.4	QMe.nab-1H.2 QTw.HaMo-1H	[63,66]
	106.2	bPb-4515	GF	HLW	106.4 – 126.7	QTw.HaMo-1H	[62,67]
	133.1	bPt-6189, bPb-0395	Col, HLW	Yield	144.2	QYld.HaMo-1H.2	[30,67]
2H	5.8	bPb-7057	K_RP	–	–	–	–
	108.1 – 108.7	bPb-3653, bPb-1772, bPb-8737	SF>2.8 mm	GY, TGW	92.2–100.8	QYld.StMo-2H.2, QTw.HaMo-2H, QTw.nab-2H, QTw.TyVo-2H.1	[62,63,66–68]
	113.2 – 115.2	(bPb-0994, bPb-2481, bPb-6822, bPb-3870, (bPb-8274)	K_RP, GF, Fria, MQI	GPC (K_RP)	113.2, 114.4	bPb-0994, bPb-6822	[69]
	119.9	bPb-9258	K_RP	–	–	–	–
	131.5	bPb-2971, bPb-3925, bPb-8302	GF, K_RP	Yield	192.2	QYld.pil-2H.1	[12]
	133.3	bPb-5755	Extr	–	–	–	–
	136.6	bPb-5942	SF>2.8 mm, Extr, MQI	–	–	–	–
	139.0 – 139.9	bPb-7816, bPb-1154	GF, KF, TGW	Extr, TGW	138.2, 138.8	QMe.BIKy-2H.2, QTw.BIKy-2H.2	[62,70]
	145.6 – 146.6	bPb-6087, bPt-4590	K_RP, MQI	solP	147.6	bPb-1986	[46]
	1633	bPb-7723	GF, KF	–	–	–	–
3H	48.3	bPb-0527, bPb-1814, bPb-2548, bPb-2965, bPb-5487, bPb-6825, bPb-6944	KF, GF	GY, MQ_RP, solN/solP	44.1–48.3	QYld.StMo-3H.1, QS/T.DiMo-3H	[62,63,71–74]
	145.1 – 145.5	bPt-8299, bPb-4156, bPb-5298, bPb-5396,	Fria, SF<2.2 mm	GPC (K_RP)	145.5	bPb-5298	[69]
	146.8, 147.9	bPb-3907, bPb-7689	VZ45	–	–	–	–
	149.8	bPb-2888, bPb-9599	KF, Fria	GPC (K_RP)	149.8	bPb-9599	[69]
	171.0	bPb-0848	KF	Yield	163.8–181.3	QYld.HaTR-3H	[75]
4H	93.6	bPb-1329	Fria	Yield, GPC (K_RP)	97.7	QYld.HaTR-4H; QGpc.HaTR-4H.2	[62,75–77]

Table 5. Cont.

Chr	Chr-Pos [cM]	DArT marker	Traits with MTAs	Traits with Ref QTL	Position of Ref QTL [cM]	Designation of Ref QTL	Literature References
5H	13.8	bPb-0292	TGW	–	–	–	–
	15.4	bPb-1909	TGW	–	–	–	–
	18.0	bPb-0837, bPb-2872	GY, TGW, K_RP, Fria	–	–	–	–
	44.0	bPb-9163	GY, K_RP	FiAt, Fria, MQI	48.8–51.0	QEV3_5H gP69M61_211-gE32M58_387, QFRI2_5H cP70M48_294-cP68M59_571, QMQI4_5H cP70M48_294-cP68M59_571	[65]
	81.3	bPb-1485, bPb-9186	GY	GY	79.3	QPgw.Blky-4H	[70]
	86.3	bPb-5532, bPb-9179	SolP, SolN	Extr	59.4–93.8	gE33M55_533	[65]
		bPb-0071, bPb-5179	SF>2.8 mm	Visc	132.1	QEv.HaTR-5H	[562]
	140.7	bPb-1494	K_RP	K_RP, GY, MQ_RP*,	139.4–140.7, 144.3*	QGpc.DiMo-5H.2	[12,62]
	159.4	bPb-5238	SF>2.8 mm	GY	150.7	QYld.pil-5H.3, QYdp.S42IL-5H.	[12,78]
	162.6	bPb-6195, bPb-9147	Fria, MQI	Fria, MQI	–	–	–
	166.1 & 168.3	bPb-6179 & bPb-0835, bPb-4595	M_RP, solP, Fria	M_RP, solP, Fria	169.4	QMe.DiMo-5H.3	[62,71]
	171.9	bPb-1965	HLW, K_RP, M_RP, solP, Extr, Fria, MQI	HLW	173.2	QTw.HaTR-5H.2	[62,75]
	184.4	bPb-1217	solP	solP, solN/SolP	182.8	QS/T.HaMo-5H	[66,67]
6H	19.4	bPb-2930	Col	Extr	28.4	QFcd.HaTR-6H	[62,75,76]
	26.5	bPb-7165	MQI	TGW, HLW, fine coarse difference (SF)	28.4	QTw.HaMo-6H.1, QFcd.HaTR-6H	[62] [62,67,75,76,79]
	68.2	bPb-9082	SF2.2–2.5 mm	Yield (GY)	68.5	QYld.StMo-6H, QTgw.S42IL-6H.	[63,71,72,78]
	110.1	bPb-7209	K_RP	GY, TGW	97.9, 105.1	QYld.Blky-6H, QTw.HaTR-6H	[62,70,75]
	136.7–137.7	bPb-8382, bPb-2863, bPb-2940 bPb-6677,	M_RP	Sol_Prot	140.8	QELG8_6H GBM1008-GBM1022	[65]
7H	21.1	bPb-3727	K_RP	TGW	17.0	QTw.Blky-7H.1	[70]
	27.1	bPb-2778	solN	GF	28.4	QFcd.HaTR-6H	[62,76]
	48.6	bPb-9898	GY	GY SF	35.2–48.6, 45.2	GYw1.2, QFcd.DiMo-7H	[62,71,76,80]
	78.2	bPb-8051	Visc	K_RP	73.9	bPb-7952	[46]
	106.6	bPb-0202, bPb-4191	TGW, Visc	GY	106.6	QYld.pil-7H.2, QYdp.S42IL-7H.	[12,78]
	115.6	bPb-5260	MQI	–	–	–	–
	123.1	bPb-0182	MQI	–	–	–	–
	125.4	bPb-1669	Visc	–	–	–	–
	133.4	bPb-8823	GF	TGW, SF		QTw.HaTR-7H.2, QFcd.HaTR-7H	[62,75,76]
	137.2	bPb-1793	Visc	GY	138.8	QYld.pil-7H.3	[12]

Abbreviations: Chr = chromosome; Chr-Pos = position of MTA on the chromosome; DArT: diversity array technology; Ref QTL = reference QTL; GY = grain yield, TGW = thousand grain weight, HLW = hectoliter weight, KF = kernel formation, GF = glume fineness, SF = sieve fraction, K_RP = raw kernel protein content, M_RP = raw malt protein content, solN = soluble nitrogen, solP = soluble protein, Visc = viscosity, Col = color, Fria = friability, VZ45 = saccharification number VZ45°C, Extr = malt extract, FiAt = final attenuation, MQI = malting quality index.

because all testing sites were mainly selected to represent the same target production environment. The best linear unbiased estimates (BLUEs) obtained from this model were used in the subsequent GWAS (Table S1).

Analysis of linkage disequilibrium (LD) between markers

According to previous studies [96,97], the LD between every pair of markers (m, n) in the same linkage group was assessed with the following statistical model:

$$\underline{m_i} = \beta_0 + \sum_p s_{ip}\phi_p + n_i\beta_1 + \underline{\varepsilon_i}$$

where; $\underline{m_i}$ and n_i are the scores of markers, m and n, of genotype i (with values -1 or 1 for either of the two homozygous genotypes); s_{ip} denotes the scores of the first p principal components from an eigenanalysis (singular value decomposition of the molecular marker matrix), as described in [53]. This term represents the effect of population structure. The magnitude of the LD between the markers was assessed by the partial r^2 associated with the $n_i\beta_1$ term. An empirical threshold for LD was determined by randomly sampling 1000 pairs of independent markers (i.e., markers known to map to different linkage groups). Two thresholds were used: one was strict, based on the upper 95% quantile of the distribution of r^2 values and the other was more liberal, based on the upper 80% quantile of the observed r^2 values. To assess how far the LD extended on a particular chromosome, we used the intersection of the threshold r^2 with a 95% quantile non-linear regression line fitted to the observed r^2 values on the particular chromosome. The non-linear quantile regression fitting was based on the method of Koenker & D'Orey [98], which has been implemented in GenStat 16 software [99]. The strict threshold was used to define a lower limit of the LD extension and the liberal threshold used to define an upper limit of the LD extension.

In turn, the LD-decay information for each chromosome was used to define a multiple testing correction threshold for the GWAS, as described previously [96]. This approach was based on a Bonferroni correction, but instead of using all markers as the denominator, it used the number of effective (independent) tests performed genome-wide. We defined the number of independent tests as $n_e = \sum_c \frac{l_c}{d_c}$, where; l_c is the length in cM of chromosome c, and dc is the extension of LD for chromosome c, and calculated the P-value significance threshold, as follows (on a log scale) where; P^* is the genome-wide threshold level (set as 0.05 in our study):

$$-\log(P) = -\log\left(\frac{P^*}{n_e}\right)$$

Genome-wide marker-trait association analysis (GWAS)

GWAS was performed with models that accounted for the genetic relatedness between varieties. Genetic relatedness was expressed in several alternate ways, including the realized kinship (model K), a group factor, based on population structure (fixed or random group models), or a set of individual principal components scores that served as fixed or random covariables in the model (fixed or random PCA score models, respectively). We also used a model that did not include a correction for genetic relatedness (naïve model).

$$\underline{y_i} = \mu + x_i\alpha + \underline{g_i} + \underline{\varepsilon_i} \underline{g_i} \sim N(0, A\sigma^2),$$
$$A = 2K, \underline{\varepsilon_i} \sim N(0,\sigma^2) \quad \text{(model K)}$$

$$\underline{y_i} = \mu + S_k + x_i\alpha + \underline{\varepsilon_i}, \underline{\varepsilon_i} \sim N(0, I\sigma^2) \quad \text{(fixed group model)}$$

$$\underline{y_i} = \mu + \underline{S_k} - x_i\alpha + \underline{\varepsilon_i} \underline{S_k} \sim N(0,\sigma_S^2),$$
$$\underline{\varepsilon_i} \sim N(0, I\sigma^2) \quad \text{(random group model)}$$

$$\underline{y_i} = \mu + \sum_p s_{ip}\phi_p + x_i\alpha + \underline{\varepsilon_i}, \underline{\varepsilon_i} \sim N(0, I\sigma^2)$$
(fixed PCA score model)

$$y_i = \mu + \sum_p s_{ip}\underline{\phi_p} + x_i\alpha + \underline{\varepsilon_i}\phi_p \sim N(0, I\sigma_\phi^2), \underline{\varepsilon_i} \sim N(0, I\sigma^2)$$
(random PCA score model)

$$\underline{y_i} = \mu + x_i\alpha + \underline{\varepsilon_i}, \underline{\varepsilon_i} \sim N(0, I\sigma^2)\text{(naïve model)}$$

In the models above, μ is a constant (intercept); x_i is a marker covariate with values -1 or 1 to denote one of the two homozygous marker genotypes; α is the marker effect; $\underline{g_i}$ is a random polygenic effect; is the fixed group effect (random when underlined); s_{ip} denotes the scores of the first p principal components, and ϕ_p is the associated fixed effect (random when underlined).

The significance of each MTA was assessed with the Wald test, and results are expressed with the associated P-values on a $-\log_{10}$ scale. The performances of the different models were compared by their inflation factors. We focused the discussion of significant MTAs on results from the model that performed best (fixed PCA score model).

All models were fitted with GenStat version 16 [99] with the available features for LD mapping. The mixed linear model (MLM) was fitted with the residual maximum likelihood (REML) method. Graphical mapping of the most significant MTAs was performed with QGene version 4.3.7 [100].

Comparison with known QTLs

For comparisons between significantly-associated DArT markers and known QTLs, the marker and chromosome position information from GrainGenes (http://www.graingenes.org) and from Barley World (http://www.barleyworld.org/) were compared to the reference DArT map created previously [37]. Some marker-associated traits assessed in this study were similar to those identified with known QTLs reported for barley in the GrainGenes database or literature. When the trait designation was missing, but similar, or limited information was available, results were compared between traits with similar features. For example, in some cases, HLW was compared to test weight; KF was compared to kernel length and plumpness; kernel weight was compared to TGW; plan test weight was compared to yield; friability was compared to milling energy and malt tenderness; and SolP was compared to the ratio of soluble/total protein (= Kolbach index).

Supporting Information

Figure S1 Histograms of the phenotypic trait distribution among cultivars. GY = grain yield, MY = marketable yield, TGW = thousand grain weight, HLW = hectoliter weight, KF = kernel formation, GF = glume fineness, SF = sieve fraction, K_RP = raw kernel protein content, M_RP = raw malt protein content, solN = soluble nitrogen, solP = soluble protein, Visc = viscosity, Col = color, Fria = friability, VZ45 = saccharification number VZ45°C, Extr = malt extract, FiAt = final attenuation, MQI = malting quality index.
(ZIP)

Figure S2 Chromosomal distribution of the 839 DArT markerDArT markers used for the genome-wide association analysis. Distances are given in cM.
(TIF)

Figure S3 Intrachromosomal LD-decay between all pairs of DArT markers shown for each barley chromosome, 1H to 7H, after correcting for population structure.
(TIF)

Figure S4 Comparison of P-values obtained by applying the naïve model (blue line) and the mixed linear model (MLM). The MLM incorporates corrections for population structure and kinship, based on PCA scores (red line). The comparison permits a check of the quality of the association results depicted for four traits (a) Grain yield (GY), (b) hectoliter weight (HLW), (c) kernel formation (KF), and (d) thousand grain weight (TGW).
(ZIP)

Figure S5 Manhattan plots show GWAS results from the PCA-corrected model for all kernel and malting parameters. GY = grain yield, MY = marketable yield, TGW = thousand grain weight, HLW = hectoliter weight, KF = kernel forma-

tion, GF = glume fineness, SF = sieve fraction, K_RP = raw kernel protein content, M_RP = raw malt protein content, solN = soluble nitrogen, solP = soluble protein, Visc = viscosity, Col = color, Fria = friability, VZ45 = saccharification number VZ45°C, Extr = malt extract, FiAt = final attenuation, MQI = malting quality index.
(ZIP)

Table S1 Summary of phenotypic parameters for all 174 cultivars. BLUES and means are shown for the breeder, origin, seasonal habit (SH), including spring (S) or winter (W), row number (RN), and phenotypic traits. Abbreviations are: BLUES = best linear unbiased estimators, GY = grain yield, MY = marketable yield, TGW = thousand grain weight, HLW = hectoliter weight, KF = kernel formation, GF = glume fineness, SF = sieve fraction, K_RP = raw kernel protein content, M_RP = raw malt protein content, solN = soluble nitrogen, solP = soluble protein, Visc = viscosity, Col = color, Fria = friability, VZ45 = saccharification number VZ45°C, Extr = malt extract, FiAt = final attenuation, MQI = malting quality index; Sheet 1: The average BLUES (best linear unbiased estimators) for the nine kernel traits and ten malting quality traits considered in this genome-wide association study. Each cultivar represents the average of several accessions of the same variety. Sheet 2: Individual accessions for each cultivar variety; the location of each accession is given with the BLUES (best linear unbiased estimators) for each trait. Sheet 3: Phenotypic means for each cultivar across all accessions. Group assignments indicate the population structures.
(XLSX)

Table S2 Allele frequencies for DArT markers and marker effects on 19 phenotypic parameters based on MTAs identified in the GWAS in barley. Results are shown for MTAs identified with strict (-$\log_{10}$(P) >3.65) or liberal (-$\log_{10}$(P) >3.35) significance thresholds.
(XLSX)

Acknowledgments

The authors gratefully acknowledge the excellent work of Triticarte Pty. Ltd (Canberra, Australia) for the DArT analyses.

Author Contributions

Conceived and designed the experiments: IEM MM MSR FvE. Performed the experiments: IEM. Analyzed the data: IEM MM FvE. Contributed reagents/materials/analysis tools: IEM MSR MM FvE. Wrote the paper: IEM MM MSR FvE.

References

1. Fischbeck G (1992) Barley cultivar development in Europe. Success in the past and possible changes in the future. Barley Genetics VI Munksgaard, Copenhagen. 887–901.
2. Han F, Romagosa I, Ullrich SE, Jones BL, Hayes PM, et al. (1997) Molecular marker-assisted selection for malting quality traits in barley. Mol Breed 3: 427–437.
3. Romagosa I, Han F, Ullrich SE, Hayes PM, Wesenberg DM (1999) Verification of yield QTL through realized molecular marker-assisted selection responses in barley cross Mol Breed 5: 143–152.
4. Dreher K, Khairallah M, Ribaut JM, Morris M (2003) Money matters (I): costs of field and laboratory procedures associated with conventional and marker-assisted maize breeding at CIMMYT. Mol Breed 11: 221–234.
5. Morris M, Dreher K, Ribaut JM, Khairallah M (2003) Money matters (II) costs of maize inbred line conversion schemes at CIMMYT using conventional and marker-assisted selection.
6. Schmierer DA, Kandemir N, Kudrna DA, Jones BL, Ullrich SE, et al. (2004) Molecular marker-assisted selection for enhanced yield in malting barley. Mol Breed 14: 463–473.

7. Xu Y, Lu Y, Xie C, Gao A, Wan J, et al. (2012) Whole-genome strategies for marker-assisted plant breeding. Mol Breed 29: 833–854.
8. Fox GP, Panozzo JF, Li CD, Lance RCM, Inkerman PA, et al. (2003) Molecular basis of barley quality. Australian J Agric Res 54: 1081–1101.
9. Swanston JS, Ellis RP (2002) Genetics and Breeding of Malt Quality Attributes. In: Barley Science – Recent Advances from Molecular Biology to Agronomy of Yield and Quality. Eds: Slafer GA, Molina-Cano JL, Araus JL, Romagosa I. Food Products press, New York, London, Oxford. 85–114.
10. Igartua E, Edney M, Rossnagel BG, Spaner D, Legge WG, et al. (2000) Marker-based selection of QTL affecting grain and malt quality in two-row barley. Crop Sci 40: 1426–1433.
11. Ullrich E, Wesenberg DM (2000) QTL analysis of malting quality in barley based on doubled-haploid progeny of two elite North American varieties representing different germplasm pools. Theor Appl Genet 101: 173–184.
12. Pillen K, Zacharias A, Léon J (2003) Advanced backcross QTL analysis in barley (*Hordeum vulgare* L.). Theor Appl Genet 107: 340–352.

13. Schmalenbach I, Léon J, Pillen K (2009) Identification and verification of QTLs for agronomic traits using wild barley introgression lines. Theor Appl Genet (2009) 118: 483–497.

14. Schmalenbach I, Pillen K (2009) Detection and verification of malting quality QTLs using wild barley introgression lines. Theor Appl Genet (2009) 118: 1411–1427.

15. Waugh R, Jannink JL, Muehlbauer GL, Ramsay L (2009) The emergence of whole genome association scans in barley. Curr Opin Plant Biol 12: 1–5.

16. Rafalski JA (2010) Association genetics in crop improvement. Curr Opin Plant Biol 13: 1–7.

17. Hamblin MT, Close TJ, Bhat PR, Chao S, Kling JG, et al. (2009) Population structure and linkage disequilibrium in U.S. barley germplasm: implications for association mapping. Crop Sci 50: 556–566.

18. Zhu C, Gore M, Buckler ES, Yu Y (2008) Status and prospects of association mapping in plants. Plant Genome 1: 5–20.

19. Yu J, Buckler ES (2006) Genetic association mapping and genome organization of maize. Curr Opin Biotechnol 17: 155–160.

20. Remington DL, Thornsberry JM, Matsuola Y, Wilson LM, Whitt SR (2001) Structure of linkage disequilibrium and phenotypic associations in the maize genome. Proc Natl Acad Sci USA 98: 11479–11484.

21. Thornsberry JM, Goodman MM, Doebley J, Kresovich S, Nielsen D, et al. (2001) Dwarf8 polymorphisms associate with variation in flowering time.Nat Genet 28: 286–289.

22. Rostoks N, Ramsay L, MacKenzie K, Cardle L, Bhat PR, et al. (2006) Recent history of artificial outcrossing facilitates whole-genome association mapping in elite inbred crop varieties. Proc Natl Acad Sci USA 103: 18656–18661.

23. Harjes CE, Rocheford TR, Bai L, Brutnell TP, Kandianis CB, et al. (2008) Natural genetic variation in Lycopene epsilon cyclase tapped for maize biofortification. Science 319: 330–333.

24. Haseneyer G, Stracke S, Piepho HP, Sauer S, Geiger H, et al. (2010) DNA polymorphisms and haplotype patterns of transcription factors involved in barley endosperm development are associated with key agronomic traits. BMC Plant Biol 10: 5.

25. Matthies IE, Weise S, Röder MS (2009) Association of haplotype diversity in the α-amylase gene amy1 with malting quality parameters in barley. Mol Breed 23: 139–152.

26. Matthies IE, Weise S, Förster J, Röder MS (2009) Association mapping and marker development of the candidate genes (1→3),(1→4)-β-D-Glucan-4-glucanohydrolase and (1→4)-β-Xylan-endohydrolase 1 for malting quality in barley. Euphytica 170: 109–122.

27. Matthies IE, Sharma S, Weise S, Röder MS (2012) Sequence variation in the barley genes encoding sucrose synthase I and sucrose phosphate synthase II, and its association with variation in grain traits and malting quality. Euphytica 184: 73–83.

28. Jaccoud D, Peng K, Feinstein D, Kilian A (2001) Diversity arrays: a solid state technology for sequence information independent genotyping. Nucleic Acids Res 29: e25.

29. Oliphant A, Barker DL, Stuelpnagel JR, Chee MS (2002) BeadArrayTechnology: Enabling an accurate,cost-effective approach to high-throughput genotyping. BioTechniques 32: S56–S61.

30. Close TJ, Bhat PR, Lonardi S, Wu Y, Rostoks N, et al. (2009) Development and implementation of high-throughput SNP genotyping in barley. BMC Genomics 10: 582.

31. Rostoks N, Mudie S, Cardle L, Russell J, Ramsay L, et al. (2005) Genome-wide SNP discovery and linkage analysis in barley based on genes responsive to abiotic stress. Mol Genet Genom 274: 515–527.

32. Lapitan NLV, Hess A, Cooper B, Botha AM, Badillo D, et al. (2009) Differentially expressed genes during malting and correlation with malting quality phenotypes in barley (Hordeum vulgare L.). Theor Appl Genet 118: 937–952.

33. Cockram J, White J, Zuluaga DL, Smith D, Comadran J, et al. (2010) Genome-wide association mapping to candidate polymorphism resolution in the unsequenced barley genome. Proc Natl Acad Sci 107: 21611–21616.

34. Ramsay L, Comadran J, Druka A, Marshall DF, Thomas WTB, et al. (2011) INTERMEDIUM-C, a modifier of lateral spikelet fertility in barley, is an ortholog of the maize domestication gene TEOSINTE BRANCHED 1. Nature Genet 43: 169–173.

35. Comadran J, Kilian B, Russell J, Ramsay L, Stein N, et al. (2012) Natural variation in a homolog of Anthirrhinum CENTRORADIALIS contributed to spring growth habit and environmental adaptation in cultivated barley. Nature Genet 44: 1388–1393.

36. Wenzl P, Carling J, Kudrna D, Jaccoud D, Huttner E, et al. (2004) Diversity arrays technology (DArT) for whole-genome profiling of barley. Proc Nat Acad Sci 101: 9915–9920.

37. Wenzl P, Li H, Carling J, Zhou M, Raman H, et al. (2006) A high-density consensus map of barley linking DArT markers toSSR, RFLP and STS loci and agricultural traits. BMC Genomics 7: 206.

38. Varshney RK, Glaszmann JC, Leung H, Ribaut JM (2010) More genomic resources for less-studied crops. Trends Biotechnol 28: 452–460.

39. Alsop BP, Farre A, Wenzl P, Wang JM, Zhou MX, et al. (2010) Development of wild barley-derived DArT markers and their integration into a barley consensus map. Mol Breed 27: 77–92.

40. Pswarayi A, van Eeuwijk FA, Ceccarelli S, Grando S, Comadran J, et al. (2008) Changes in allele frequencies in landraces, old and modern barley cultivars of marker loci close to QTL for grain yield under high and low input conditions. Euphytica 163: 435–44.

41. Grewal TS, Rossnagel BG, Pozniak CJ, Scoles GJ (2008) Mapping quantitative trait loci associated with barley net blotch resistance. Theor Appl Genet 116: 529–539.

42. Zhang LY, Marchand S, Tinker NA, Belzile F (2009) Population structure and linkage disequilibrium in barley assessed by DArT markers. Theor Appl Genet 119: 43–52.

43. Comadran J, Thomas WTB, van Eeuwijk FA, Ceccarelli S, Grando S, et al. (2009) Patterns of genetic diversity and linkage disequilibrium in a highly structured Hordeum vulgare association-mapping population for the Mediterranean basin. Theor Appl Genet 119: 175–187.

44. Comadran J, Russell JR, Booth A, Pswarayi A, Ceccarelli S, et al. (2011) Mixed model association scans of multi-environmental trial data reveal major loci controlling yield and yield related traits in Hordeum vulgare in Mediterranean environments. Theor Appl Genet 122: 1363–1373.

45. Matthies IE, van Hintum T, Weise S, Röder MS (2012) Population structure revealed by different marker types (SSR or DArT) has an impact on the results of genome-wide association mapping in European barley cultivars. Mol Breed 30: 951–966.

46. Beattie AD, Edney MJ, Scoles GJ, Rossnagel BG (2010) Association mapping of malting quality data from western Canadian two-row barley cooperative trails. Crop Sci 50: 1649–1663.

47. Roy JK, Smith KP, Muehlbauer GJ, Chao S, Close TJ, et al. (2010) Association mapping of spot blotch resistance in wild barley. Mol Breed 26: 243–256.

48. Flint-Garcia SA, Thornsberry JM, Buckler ES (2003) Structure of linkage disequilibrium in plants. Ann Rev Plant Biol 54: 357–374.

49. Malosetti M, van der Linden CG, Vosman B, van Eeuwijk FA (2007) A mixed-model approach to association mapping using pedigree information with an illustration of resistance to Phytophthora infestans in potato. Genetics 175: 879–889.

50. Yu J, Pressoir G, Briggs WH, Bi IV, Yamasaki M, et al. (2005) A unified mixed-model method for association mapping that accounts for multiple levels of relatedness. Nature Genetics 38: 203–208.

51. Stich B, Utz FU, Piepho HP, Maurer HP, Melchinger AE (2010) Optimum allocation of resources for QTL detection using a nested association mapping strategy in maize. Theor Appl Genet 120: 553–561.

52. Breseghello F, Sorrells ME (2006) Association mapping of kernel size and milling quality in wheat (Triticum aestivum L.) cultivars. Genetics 172: 1165–1177.

53. Patterson NJ, Price AL, Reich D (2006) Population structure and Eigenanalysis. PLoS Genet 2: e190.

54. Price AL, Patterson NJ, Plenge RM, Weinblatt ME, Shadick NA, et al. (2006) Principal components analysis corrects for stratification in genome-wide association studies. Nat Genet 38: 905–909.

55. Wang M, Jiang N, Jia T, Leach L, Cockram J, et al. (2012) Genome-wide association mapping of agronomic and morphologic traits in highly structured populations of barley cultivars. Theor Appl Genet 124: 233–246.

56. Komatsuda T, Pourkheirandish M, He CF, Azhaguvel P, Kanamori H, et al. (2007) Six-rowed barley originated from a mutation in a homeodomain-leucine zipper I-class homeobox gene. Proc Natl Acad Sci USA 104: 1424–1429.

57. Pourkheirandish M, Komatsuda T (2007) The importance of barley genetics and domestication in a global perspective. Ann Bot 100: 999–1008.

58. Yan L, Fu D, Li C, Blechl A, Tranquilli G, et al. (2006) The wheat and barley vernalization gene VRN3 is an orthologue to FT. Proc Natl Acad Sci USA 103: 19581–19586.

59. von Zitzewitz J, Szücs P, Dubcovsky J, Yan L, Francia E, et al. (2005) Structural and functional characterization of barley vernalization genes. Plant Mol Biol 59: 449–467.

60. Li JZ, Huang XQ, Heinrichs F, Ganal MW, Röder MS (2005) Analysis of QTLs for yield, yield components, and malting quality in a BC3-DH population of spring barley. Theor Appl Genet 110: 356–363.

61. Kalladan R, Worch S, Rolletschek H, Harshavardhan VT, Kuntze L, et al. (2013) Identification of quantitative trait loci contributing to yield and seed quality parameters under terminal drought in barley advanced backcross lines. Mol Breed 32: 71–90.

62. Szücs P, Blake VC, Bhat PR, Chao S, Close TJ, et al. (2009) An integrated resource for barley linkage map and malting quality QTL alignment. Plant Genome 2: 134–140.

63. Hayes PM, Liu BH, Knapp SJ, Chen F, Jones B, et al. (1993) Quantitative trait locus effects and environmental interaction in a sample of North American barley germplasm. Theor Appl Genet 87: 392–401.

64. Barr AR, Jeffries SP, Broughton S, Chalmers KJ, Kretschmer JM, et al. (2003) Mapping and QTL analysis of the barley population Alexis × Sloop. Austr J Agric Res 54: 1117–1123.

65. Krumnacker K (2009) Untersuchung der funktionellen Assoziation von Kandidatengenen in Zusammenhang mit der Malzqualität der Gerste durch Transkriptomkartierung. PhD-Thesis. Technische Universität München (TUM), Lehrstuhl für Pflanzenzüchtung.

66. Marquez-Cedillo LA, Hayes PM, Jones BL, Kleinhofs A, Legge WG, et al. (2000) QTL analysis of malting quality in barley based on the doubled-haploid progeny of two elite North American varieties representing different germplasm groups. Theor Appl Genet 101: 173–184.

67. Marquez-Cedillo LA, Hayes PM, Kleinhofs A, Legge WG, Rossnagel BG, et al. (2001) QTL analysis of agronomic traits in barley based on the doubled haploid

progeny of two elite North American varieties representing different germplasm groups. Theor Appl Genet 103: 625–637.

68. Kjaer B, Jensen J (1996) Quantitative trait loci for grain yield and yield components in a cross between a six-rowed and a two-rowed barley. Euphytica 90: 39–48.

69. Cai S, Yu G, Chen X, Huang Y, Jiang X, et al. (2013) Grain protein content variation and its association analysis in barley. BMC Plant Biology 13: 35.

70. Bezant J, Laurie D, Pratchett N, Chojecki J, Kearsey M (1997) Mapping QTL controlling yield and yield components in a spring barley (Hordeum vulgare L.) cross using marker regression. Mol Breed 3: 29–38.

71. Oziel A, Hayes PM, Chen FQ, Jones B (1996) Application of quantitative trait locus mapping to the development of winter-habit malting barley. Plant Breed 115: 43–51.

72. Romagosa I, Ullrich SE, Han F, Hayes PM (1996) Use of the additive main effects and multiplicative interaction model in QTL mapping for adaptation in barley. Theor Appl Gen 93: 30–37.

73. Zhu H, Gilchrist L, Hayes P, Kleinhofs A, Kudrna D, et al. (1999) Does function follow form? Principal QTLs for Fusarium head blight (FHB) resistance are coincident with QTLs for inflorescence traits and plant height in a doubled-haploid population of barley. Theor Appl Genet 99: 1221–1232.

74. Larson SR, Kadyrzhanova D, McDonald M, Sorrels M, Blake TK (1996) Evaluation of barley chromosome-3 yield QTLs in a backcross F2 population using STS-PCR. Theor Appl Genet 93: 618–625.

75. Tinker NA, Mather DE, Rossnagel BG, Kasha KJ, Kleinhofs A, et al. (1996) Regions of the Genome that Affect Agronomic Performance in Two-Row Barley. Crop Sci 36: 1053–1062.

76. Mather DE, Tinker NA, LaBerge DE, Edney M, Jones BL, et al. (1997) Regions of the genome that affect grain and malt quality in a North American two-row barley cross. Crop Sci 37: 544–554.

77. Bradbury P, Parker T, Hamblin MT, Jannink JL (2011) Assessment of power and false discovery rate in genome-wide association studies using the barley CAP germplasm. Crop Science 51: 52–59.

78. Schnaithmann F, Pillen K (2013) Detection of exotic QTLs controlling nitrogen stress tolerance among wild barley introgression lines. Euphytica 189: 67–88.

79. Varshney RK, Paulo MJ, Grando S, van Eeuwijk FA, Keizer LCP, et al. (2012) Genome-wide association analyses for drought tolerance related traits in barley (Hordeum vulgare L.). Field Crops Research 126 (2012) 171–180.

80. Xue DW, Zhou MX, Zhang XQ, Chen S, Wie K, et al. (2010) Identification of QTLs for yield and yield components of barley under different growth conditions. J Zhejiang Univ-Sci B (Biomed and Biotechnol) 11: 169–176.

81. Kraakman ATW, Niks RE, Van den Berg PMMM, Stam P, Van Eeuwijk FA (2004) Linkage disequilibrium mapping of yield and yield stability in modern spring barley cultivars. Genetics 168: 435–446.

82. Kraakman ATW, Martinez F, Mussiraliev FA, Van Eeuwijk FA, Niks RE (2006) Linkage disequilibrium mapping of morphological, resistance, and other agronomically relevant traits in modern spring barley cultivars. Mol Breed 17: 41–58.

83. Rode J, Jutta Ahlemeyer J, Friedt W, Ordon F (2012) Identification of marker-trait associations in the German winter barley breeding gene pool (Hordeum vulgare L.) Mol Breed 30: 831–843.

84. Heuberger AL, Broeckling CD, Kirkpatrick KR, Prenni JE (2013) Application of nontargeted metabolite profiling to discover novel markers of quality traits in an advanced population of malting barley. Plant Biotech J: 1–14.

85. Bauer AM, Hoti F, von Korff M, Pillen K, Léon J, et al. (2009) Advanced backcross-QTL analysis in spring barley (H. vulgare ssp. spontaneum) comparing a REML versus a Bayesian model in multi-environmental field trials. Theor Appl Genet 119(1): 105–23.

86. Tondelli A, Xu X, Moragues M, Sharma R, Schnaithmann F, et al. (2013) Structural and temporal variation in genetic diversity of European spring two-row barley cultivars and association mapping of quantitative traits. The Plant Genome 6(2): 1–14.

87. Blake VC, Kling JG, Hayes PM, Jannink JL, Jillella SR, et al. (2012) The Hordeum toolbox: the barley coordinated agricultural project genotype and phenotype resource, The Plant Genome 5: 81–91.

88. Iwata H, Jannink JL (2011) Accuracy of genomic selection prediction in barley breeding programs: a simulation study based on the real single nucleotide polymorphism data of barley breeding lines. Crop Science 51: 1915–1927.

89. See D, Kanazin V, Ken Kephart K, Blake T (2002) Mapping genes controlling variation in barley grain protein concentration. Crop Sci. 42: 680–685.

90. Emebiri LC, Moody DB, Horsley R, Panozzo J, Read BJ (2005) The genetic control of grain protein content variation in a doubled haploid population derived from a cross between Australian and North American two-rowed barley lines. J Cer Sci 41: 107–114.

91. Molina-Cano JL, Moralejo M, Elía M, Munoz P, Russell JR, et al. (2007) QTL analysis of a cross between European and North American malting barleys reveals a putative candidate gene for β-glucan content on chromosome 1H. Mol Breed 19: 275–284.

92. Edney MJ, Mather DE (2004) Quantitative trait loci affecting germination traits and malt friability in a two-rowed by six-rowed barley cross. J Cereal Sci 39: 283–290.

93. Moralejo M, Swanston JS, Munoz P, Prada D, Elía M, et al. (2004) Use of new EST markers to elucidate the genetic differences in grain protein content between European and North American two-rowed malting barleys. Theor Appl Genet 110: 116–125.

94. Von Korff M, Wang H, Léon J, Pillen K (2008) AB-QTL analysis in spring barley: III. Identification of exotic alleles for the improvement of malting quality on spring barley (H. vulgare ssp. spontaneum). Mol Breed 21: 81–93.

95. Weise S, Scholz U, Röder MS, Matthies IE (2009) A comprehensive database of malting quality traits in brewing barley. Barley Genet Newsl 39: 1–4.

96. Mangin B, Siberchicot A, Nicolas S, Doligez A, This P, et al. (2012) Novel measures of linkage disequilibrium that correct the bias due to population structure and relatedness. Heredity 108: 285–291.

97. Adetunji I, Willems G, Tschoep H, Bürkholz A, Barnes S, et al. (2014) Genetic diversity and linkage disequilibrium analysis in elite sugar beet breeding lines and wild beet accessions. Theor Appl Genet127: 559–571.

98. Koenker RW, D'Orey V (1987) Algorithm AS229 computing regression quantiles. Applied Statistics 36: 383–393.

99. VSN International (2013) GenStat for Windows 16th Edition. VSN International, Hemel Hempstead, UK. Web page: GenStat.co.uk.

100. Joehanes R, Nelson JC (2008) QGene 4.0, an extensible Java QTL analysis platform. Bioinf 24: 2788–2789.

14

The Kill Date as a Management Tool for Cover Cropping Success

María Alonso-Ayuso, José Luis Gabriel, Miguel Quemada*

School of Agriculture Engineering, Technical University of Madrid, Madrid, Spain

Abstract

Integrating cover crops (CC) in rotations provides multiple ecological services, but it must be ensured that management does not increase pre-emptive competition with the subsequent crop. This experiment was conducted to study the effect of kill date on: (i) CC growth and N content; (ii) the chemical composition of residues; (iii) soil inorganic N and potentially mineralizable N; and (iv) soil water content. Treatments were fallow and a CC mixture of barley (*Hordeum vulgare* L.) and vetch (*Vicia sativa* L.) sown in October and killed on two different dates in spring. Above-ground biomass and chemical composition of CC were determined at harvest, and ground cover was monitored based on digital image analysis. Soil mineral N was determined before sowing and after killing the CC, and potentially mineralizable N was measured by aerobic incubation at the end of the experiment. Soil water content was monitored daily to a depth of 1.1 m using capacitance sensors. Under the present conditions of high N availability, delaying kill date increased barley above-ground biomass and N uptake from deep soil layers; little differences were observed in vetch. Postponing kill date increased the C/N ratio and the fiber content of plant residues. Ground cover reached >80% by the first kill date (~1250°C days). Kill date was a means to control soil inorganic N by balancing the N retained in the residue and soil, and showed promise for mitigating N losses. The early kill date decreased the risk of water and N pre-emptive competition by reducing soil depletion, preserving rain harvested between kill dates and allowing more time for N release in spring. The soil potentially mineralizable N was enhanced by the CC and kill date delay. Therefore kill date is a crucial management variable for maximizing the CC benefits in agricultural systems.

Editor: Anil Shrestha, California State University, Fresno, United States of America

Funding: This work was funded by the Spanish Comisión Interministerial de Ciencia y Tecnología (project AGL 2011-24732) and the Regional Government of Madrid (Project AGRISOST, S2009/AGR-1630). The funders had no role in study design, data collection and analysis, decision to publish, or preparation of the manuscript.

Competing Interests: The authors have declared that no competing interests exist.

* Email: miguel.quemada@upm.es

Introduction

The potential for cover crops (CC) to provide ecological services has been documented in diverse cover cropping systems and environments [1]. Replacing fallow periods with CC may enhance soil aggregate stability [2], water retention capacity [3], nutrient supply [4] and disease suppression [5]. Moreover, plant residues covering the soil following crop kill can improve soil protection over time and help control weeds, preserve soil moisture, ameliorate compacted soils, and reduce soil erosion [6]. However, improperly managed CC may have a detrimental effect on the cash crop either by competing for water and nutrients, building up diseases or retarding seed germination [6,7]. Overall, there are two crucial management factors that determine CC success: species selection and CC termination [8]. In this manuscript, we focus on the termination of CC incorporated into field crop rotations, particularly on kill date. Previous experiments have shown that changing kill date by a few weeks in spring may have a large effect on the water and N pre-emptive competition with the subsequent cash crop, nitrate leaching control, and soil moisture conservation [7,9–12], but there is no agreement on the advantages and limitations of postponing CC termination.

Therefore, a better understanding of the effect of the living CC and its dead mulch on the dynamics of N and water might improve our ability to choose a kill date that maximizes the benefits of CC in the system.

The effect of kill date on soil water availability is a balance between the water extracted by the living CC and the evaporation prevented by the residue mulch [9]. Integrating CC into the cropping system increases above-ground biomass and resistance to decomposition, which enhances the development of a dead mulch covering the soil surface [13]. Some authors have found that, despite the soil water depletion, moisture conservation was improved when the CC was killed later [9,11]. However, other experiments have shown that an earlier CC kill reduced water pre-emptive competition, preserved top soil moisture, and increased water availability for the subsequent crop [7,12]. Similarly, postponing kill date enhanced N pre-emptive competition as soil available N is depleted by the CC uptake and only slowly replenished by the mineralization of a high C/N residue [6]. Legume species may compensate for this effect by fixing more N_2 when killed late [10]. The water and N balance is driven by various factors that are all affected by kill date, such as above-ground biomass accumulation, ground cover (GC), and N uptake

[6]. The chemical composition of crop residues, such as lignin content and the C/N ratio, are important characteristics governing the decomposition process and determining the persistence of the residue layer covering the soil [14]. However, there are few experiments that combine multiple approaches to study these variables, and to our knowledge, none that involves continuous monitoring of soil water dynamics.

We hypothesize that a combined study of CC growth, GC and water and N dynamics would contribute to understanding the effect of kill date and improve management practices. This experiment was conducted over two years to study the effect of kill date on the following factors: (i) the growth and N content of the CC; (ii) the chemical composition and residue quality of the CC; (iii) soil inorganic N dynamics and its potentially mineralizable N; and (iv) the soil water content. A CC mixture of barley and vetch was selected because mixing legumes and non-legumes is known to be an efficient technique to merge N and water management benefits of the individual species [9,15].

Materials and Methods

1. Field experiment

The experiment was conducted from October 2011 to November 2013 at La Chimenea field station (40°03′N, 03°31′W, 550 m a.s.l.), which is located in the central Taxus river basin in Aranjuez (Madrid, Spain). The soil is a silty clay loam (Typic Calcixerept) [16] with a high content of organic matter and carbonates, pH ~8 and low stone content throughout the profile. The most relevant soil characteristics are presented in Table 1. Total organic C was measured by the Walkley-Black method [17]. Calcium carbonate was determined by a volumetric calcimeter method [18]. The climate is semiarid Mediterranean [19] with high interannual variability (Figure 1). Annual rainfall averages ~350 mm with less rain in the summer and more in the autumn and a mean annual temperature of 14.2°C. Weather data were recorded throughout the experimental period by a data logger (CR10X. Campbell Scientific Ltd, Shepshed, UK) located ~100 m from the field site.

The experiment consisted of twelve plots (15 m×12 m) randomly assigned to three treatments (fallow and two different CC kill dates) with four replications. Cover crop treatments were sown with a barley (Hordeum vulgare L., cv. Vanessa) + vetch (Vicia sativa L., cv. Aitana) mixture in early October (October 6, 2011; October 8, 2012). The mixture, 30% (54 kg ha^{-1}) barley seeds and 70% (45 kg ha^{-1}) vetch, was broadcast by hand and buried (~5 cm) with a shallow cultivator. All treatments received

~18 mm irrigation on the sowing date with a sprinkle irrigation system (9.5 mm h^{-1}) to ensure uniform CC establishment. Plots from the first kill treatment (FK) were killed in mid-March (March 13, 2012; March 14, 2013) when barley was at the end of the booting growth stage and vetch at the stem elongation stage. Plots from the second kill treatment (SK) were killed in mid-April (April 9, 2012; April 13, 2013) when barley was at the emerging inflorescence growth stage and vetch was at the stem elongation stage. Cover crops were killed by application of glyphosate (N-phosphonomethyl glycine, 0.7 kg a.e. ha^{-1}) followed by a shredder. The last day of April was chosen as a hypothetical planting date (HPD) for both years, as it is a typical date for planting summer cash crops in the area. The HPD represented the beginning of the mulch period during which the CC residue remained on the ground. To compare the residue mulch effect on water conservation, SWC was homogenized by adding water to the drier plots with a drip irrigation system (30.5 mm to FK treatment and 35.5 mm to SK) in April 2012. As rainfall was abundant in March 2013, no supplemental irrigation was applied. Fallow plots were kept free of weeds by hand weeding during the trial period. The experiment was conducted in a field that was cultivated with triticale during the two previous years and had not received organic amendments or N fertilizer for four years prior to the beginning of the trial.

2. Cover crop analysis

To measure the above-ground biomass and GC in each plot, four 0.5 m×0.5 m squares were marked after sowing the CC. Ground cover was monitored in each square based on digital photos taken from a nadir perspective at 1.2 m height every other week [20]. The images were taken with a Nikon S210 Coolpix camera that had a lens resolution of 5 Mpixels. The images were saved at a resolution of 1200×800 pixels and the GC percentage was determined with the software SigmaScan Pro (Systat software, Chicago, IL, USA) by means of the "Turf analysis" macro developed by Karcher and Richardson [21]. The process consisted of creating a pixel layer by selecting the hue and saturation ranges that identified the surface covered by the crop. The GC was determined to be the number of pixels of the layer divided by the total number of pixels constituting the image of the marked squares [20]. The GC evolution was adjusted to the Gompertz function [22], and the thermal times (°C days) until the GC reached 30% (t_{30}) and 80% (t_{80}) were calculated.

Just before killing the CC, the above-ground biomass of the marked squares was hand harvested by cordless grass shear at the ground level, separated by species, oven dried for 48 h at 65°C,

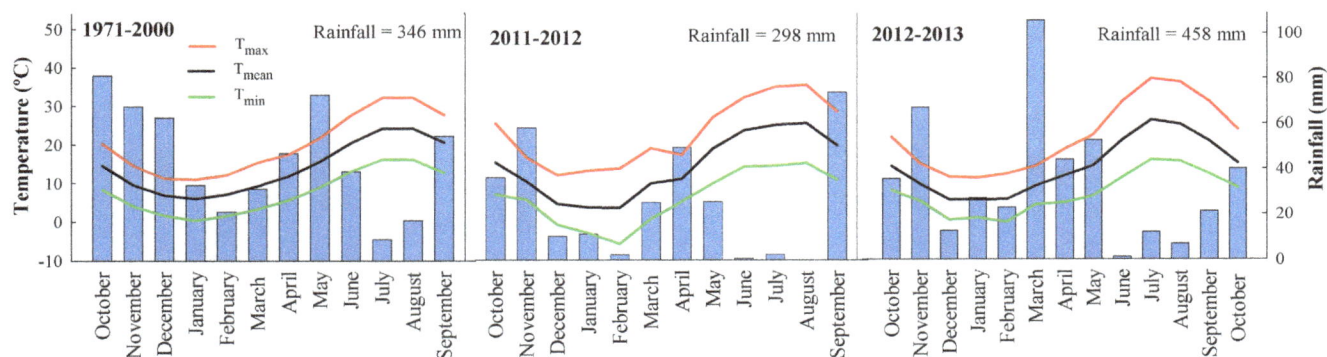

Figure 1. Monthly rainfall (bars) and average maximum, mean and minimum temperatures from 1971 to 2000 and during the two experimental seasons at Aranjuez (Madrid, Spain).

Table 1. Soil properties at the beginning of the experiment.

Depth (cm)	0–23	23–40	40–70	70–120
pH (1:2.5)	8.16	8.06	8.02	7.84
Organic Matter (g kg^{-1})	31.8	29.2	21.9	22.3
CO_3 (g CO_3^{2-} kg^{-1})	198.0	201.3	159.0	181.0
Sand (g kg^{-1})	260	250	250	250
Silt (g kg^{-1})	490	510	520	460
Clay (g kg^{-1})	250	240	230	290

and weighed to determine biomass of barley, vetch, and mixture in kg ha^{-1}. The C and N concentrations in the above-ground biomass were determined for a subsample of each species from each plot using the Dumas combustion method (LECO CHNS-932 Analyzer, St. Joseph, MI, USA). Total N content was calculated for each specie as the product of above-ground biomass times N concentration. The atmospheric N_2 fixation by the legume was estimated by the natural abundance method [23] based on the $\delta^{15}N$ (‰) determination (Europe Scientific 20–20 IRMS Analyzer, Crewe, UK) for subsamples from the vetch and a barley reference cultivated as a sole crop in an adjacent field. Cover crop residue quality was assessed by measuring neutral detergent fiber (NDF), acid detergent fiber (ADF), and lignin (L) with the Goering and Van Soest method [24] in subsamples of each crop species from each plot.

3. Soil inorganic nitrogen content (N$_{min}$)

Four soil cores were taken from each plot to a depth of 1.2 m in 0.2 m intervals with an Eijkelkamp helicoidal auger (Eijkelkamp Agrisearch Equipment, Geisbeek, Netherlands) just before sowing the CC, after the second kill date each year, and at the end of the experiment. Soil cores were combined by depth to provide a composite profile of six samples. For each plot, soil N$_{min}$ was

calculated for each layer. Soil samples were placed in a plastic box and immediately firmly closed then transported, and refrigerated (4–6°C). Within five consecutive days, a soil subsample from each box was extracted with 1 M KCl (~30 g of soil: 150 ml of KCl), centrifuged, decanted, and a subsample of the supernatant volume was stored in a freezer until later analysis. Nitrate concentration in the extracts was determined by the Griess-Ilosvay method [25] after reduction of NO_3^- to NO_2^- with a Cd column. Ammonium in the soil extracts was determined by the salicylate-hypochlorite method [26].

4. Soil nitrogen mineralization potential (N$_0$)

Soil N mineralization potential was estimated by adapting the procedure proposed by Stanford and Smith [27]. A subsample of the top soil layer (0–0.2 m) free from residue contamination was collected from each plot after the second kill date of the second year, air-dried, and sieved (<6 mm). A homogeneous mixture of 30 g of that soil and an equal weight of sand was packed to a depth of 5 cm in plastic syringes (3.5 cm diameter, 10 cm long) to achieve a bulk density of 1.25 g cm^{-3}. Thin glass wool layers were placed between the soil and the bottom of the syringe and over the soil sample to avoid soil loss during leaching and to minimize moisture loss. Initial soil inorganic N was removed by leaching

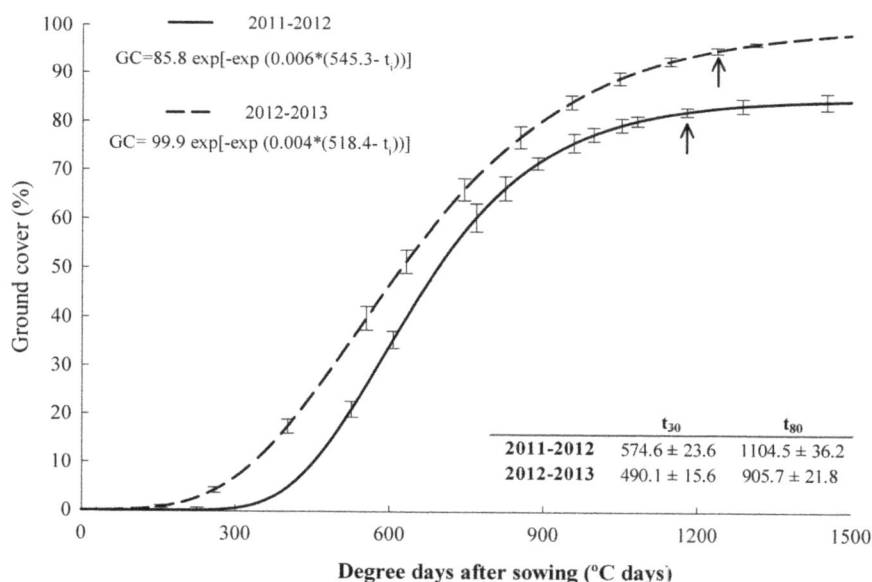

Figure 2. Ground cover (GC) development during the cover crop growth period in both experimental seasons. Arrows show the first kill date. Lines represent the Gompertz model adjusted to the observed values. Fitted models and the thermal time until the ground cover reaches 30 and 80% (t$_{30}$ and t$_{80}$) are shown.

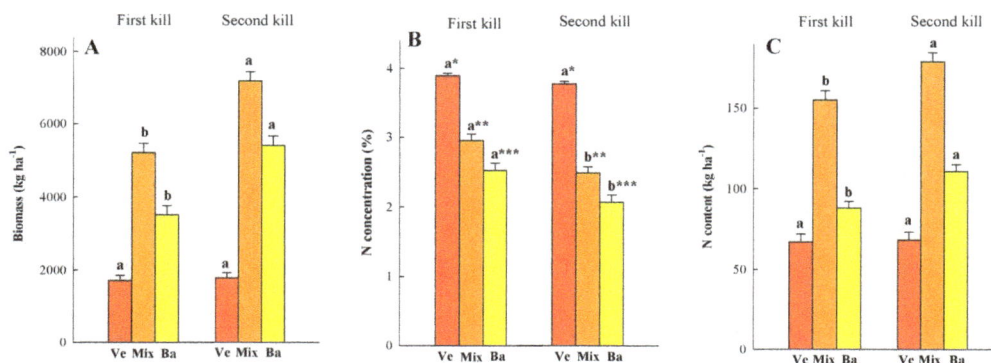

Figure 3. Biomass (A), N concentration (B) and N content (C) for vetch (Ve), barley (Ba) and the mixture (Mix) measured for the first and second kill dates. Values are the mean of two cropping seasons. Letters above bars show significant differences between kill dates for each species, and asterisks between species for each kill date. Small bars represent the standard error.

with 100 ml of 0.01 M $CaCl_2$ followed by 20 ml of a N-free nutrient solution (0.0095 M $CaSO_4$, 0.000047 M KH_2PO_4, 0.00138 M K_2SO_4, 0.0003 M $MgSO_4$). The excess water was removed using a vacuum pump until a weight within 2 g of that measured before leaching was reached. The syringes were covered with a porous parafilm and incubated aerobically at 35°C. Syringes were removed from the incubator at 14, 28, 42, 56, and 60 d after preparation and were leached with 100 ml of 0.01 M $CaCl_2$ solution followed by 20 ml of N-free solution. Leachates were made up to 100 ml with $CaCl_2$, and subsamples were stored in a freezer at −25°C until later analysis. After the leaching procedure, the cores were allowed to drain under vacuum until a weight within 2 g of that measured before leaching was reached. Nitrate concentrations in the leachates were determined by the Griess-Ilosvay method, ammonium by the salicylate-hypochlorite method and total N by the Dumas combustion method, as described above. The N_0 and the mineralization rate constant (k) were estimated after fitting a non-linear regression model ($Nt = N_0 \exp (-k \ t)$) for describing the cumulative N mineralized (Nt) with time (t) in each soil sample.

5. Soil volumetric water content (SWC)

The SWC was monitored daily during the field trial using the EnviroScan capacitance probes (Sentek Pty Ltd, Stepney, Australia) that has been described in detail elsewhere [28]. Seventy-two capacitance sensors were mounted on twelve plastic

extrusions (four repetitions per treatment), introduced into access tubes located in the middle of each plot and connected to three data loggers. Sensor readings were automated and stored in the data loggers and downloaded weekly. To ensure the measurement reliability, a normalization procedure was conducted that obtained reference readings by exposing each sensor to air and water (~20°C). The sensors were centered at 10, 30, 50, 70, 90, and 110 cm below the soil surface in each plastic extrusion, and normalized readings were registered every 6 h. A daily average of the four readings from the 0–20 (10), 20–40 (30), 40–60 (50), 60–80 (70), 80–100 (90), 100–120 (110) cm-deep soil layers was transformed into SWC using a calibration equation that was obtained at the experimental site [29]. The SWC data set was comprised of two CC seasons, which started before sowing in October and lasted until August each year, and used to study the effect of the living CC and the dead mulch on soil moisture.

6. Statistical analysis

Analyses of variance (ANOVA) and t-test were performed in order to determine differences between kill dates for each variable. Year was considered a random effect and, interaction between kill date and year was also tested. Means were separated by Tukey's test at the 0.05 probability level (P≤0.05). Least significant differences (LSD) were calculated for SWC (P≤0.05). Prior to conducting the ANOVA, tests were conducted to verify if the assumptions of ANOVA were met. The Gompertz model was

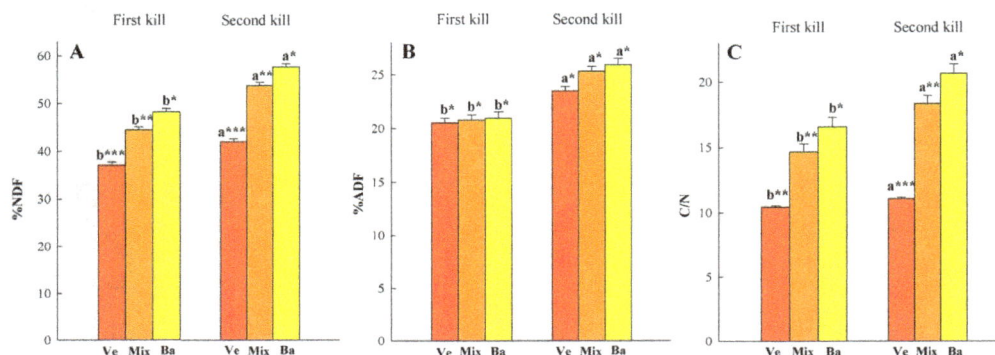

Figure 4. Neutral detergent fiber-NDF- (A) and acid detergent fiber -ADF- (B) fractions, and C/N ratio (C) in vetch (Ve), barley (Ba) and the mixture (Mix), measured for the first and second kill dates. Values are the mean of two cropping seasons. Letters above bars show significant differences between kill dates for each species, and asterisks between species for each kill date. Small bars represent the standard error.

Table 2. Soil inorganic N content (kg N ha^{-1}) in the different soil layers and the entire profile at different sampling dates for the fallow (Fa), first kill (FK) and second kill (SK) treatments.

	Oct 2011	Apr 2012	Oct 2012	Apr 2013	Oct 2013
0–40 cm					
Fa	94 (13.7)	126.8 (7.9) a	135.3 (15.4)	29.7 (3.8) b	71.54 (5.38) c
FK	94 (13.7)	46.9 (7.1) b	121.5 (16.6)	56.9 (3) a	119.37 (28.48) b
SK	94 (13.7)	37 (8.2) b	154 (8)	24.8 (4.5) b	171.72 (2.36) a
40–80 cm					
Fa	24.7 (9.5)	71.6 (10.6) a	86.1 (4.6) a	40.1 (1.8) a	72.22 (7.43)
FK	24.7 (9.5)	30.4 (5.3) b	48.3 (4.4) b	42.6 (3.5) a	65.38 (19.89)
SK	24.7 (9.5)	25.4 (5.1) b	48.1 (7.1) b	27.1 (4.1) b	39.39 (2.68)
80–120 cm					
Fa	26.1 (11.1)	60.6 (3.5)	66.7 (11)	68.8 (3.3) a	100.37 (11.74) a
FK	26.1 (11.1)	56.1 (24.5)	61.5 (29.3)	47 (7.4) b	61.59 (17.91) ab
SK	26.1 (11.1)	34.5 (3.7)	33.5 (6.3)	32.3 (8.3) b	39.24 (5.83) b
Soil profile (0–120 cm)					
Fa	144.8 (22.8)	259 (8.0) a	288.1 (25.4)	138.6 (5.1) a	244.13 (19.05)
FK	144.8 (22.8)	133.32 (27.5) b	231.3 (31.2)	146.6 (10.7) a	246.34 (52.87)
SK	144.8 (22.8)	96.9 (15.4) b	235.6 (14.2)	84.2 (13.3) b	247.34 (7.9)

Means with standard error in parentheses.
Within a date and depth, means with the same letter are not significantly different between kill dates at $P<0.05$.

fitted to the GC and the N mineralization potential model was fitted to the cumulative N mineralized using a non-linear regression procedure. The models were evaluated for their ability to simulate the observed data by comparing the mean of the lack of fit to the mean square due to pure error by using the variance ratio, or F-test. When the lack of fit was significantly smaller than the pure error, the model fitted the data. Further discussion regarding this evaluation procedure can be found elsewhere [30,31]. Statistical analyses were accomplished using the PASW Statistics Software (SPSS, Chicago, IL, USA).

Results

1. Weather conditions

The first experimental year was drier than the second (Figure 1). In the first year, rainfall was substantially lower (298 mm) than the 30-year average although in the second, it was greater (408 mm). Differences mainly occurred in spring; from March to May, the rainfall was 202 mm in 2013 compared with 101 mm in 2012. The rest of the season was rather similar for both years except for intense precipitation in late September 2012. Air temperature followed a typical Mediterranean seasonal distribution over the two years. Winter was cooler in the first year with an average minimum temperature of -3 and $-5°C$ in January and February, respectively.

2. Cover crop: GC, above-ground biomass and N content

The GC followed a classical Gompertz model (Figure 2). The F-test comparing the mean squares due to pure error and lack of fit was not significant at 0.01 level, therefore, the model was adjusted to the GC observations. The first kill occurred between 157 and 159 d after sowing ($\sim1200°C$ days) whereas the second year occurred between 184 and 186 d after sowing ($\sim1500°C$ days). The GC>30%, a crucial threshold for erosion control [32], was attained between 37 and 41 d after sowing (>490°C days). The GC>80%, considered to be full cover for erosion control and direct soil evaporation [32], was reached between 95 and 130 d after sowing (>900°C days). Differences were observed in the maximum GC attained, 85% in 2012 and 100% in 2013. In both

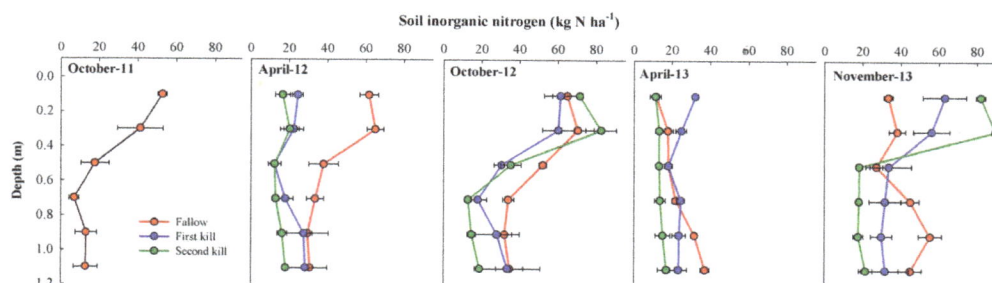

Figure 5. Soil inorganic nitrogen (kg N ha^{-1}) in the upper 1.2 m of the soil profile for the fallow (Fa), first kill (FK) and second kill (SK) treatments at different sampling times.

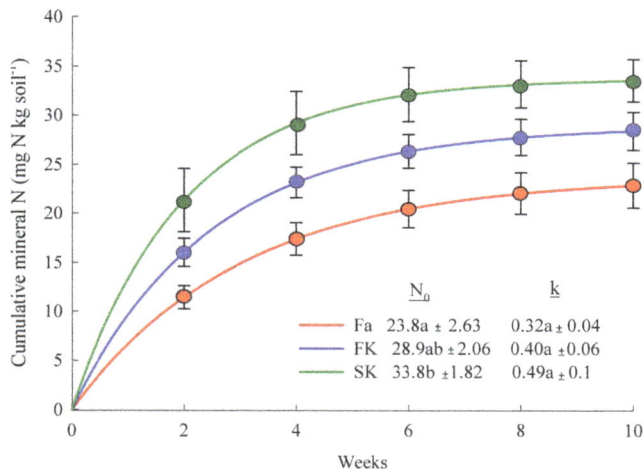

Figure 6. Cumulative N mineralization in soils from the fallow (Fa), first kill (FK) and second kill (SK) treatments during a 10-week aerobic incubation. Soil N mineralization potential (N_0) and N mineralization rate (k) were calculated by fitting a non-linear regression model ($N_t = N_0 \exp(-k\,t)$). Values followed by different letters are significantly different between treatments at ≤0.05 by Tukey's test. Bars represent the standard error.

years, maximum GC was already attained by the first kill date. The ground continued to be covered by the CC residue mulch more than six months after CC killing. Above-ground biomass increased from the FK to the SK by ~2000 kg ha^{-1}, and most of the increase was due to barley (Figure 3). Although barley biomass was greater at the SK than at the FK, no differences were found in vetch biomass between kill dates. Vetch above-ground biomass was double in 2013 than in 2012 whereas barley above-ground biomass did not vary by year. At the end of the mulch period, residues left on the soil surface were greater in the SK (2214 kg ha^{-1} in 2012, and 4014 in 2013) than in the FK treatment (966 kg ha^{-1} in 2012, 2151 in 2013. The ground was fully covered by the residues in both treatments.

Nitrogen concentration in the above-ground biomass was higher in the FK than in the SK. Nitrogen concentration in the mixture decreased from 2.7 to 2.35% in 2012 and from 3.2 to 2.6% in 2013. As expected, vetch had a higher N concentration than barley whereas the N concentration of the mixture was intermediate (Figure 3). Differences in N content between the FK and SK, calculated as the product of above-ground biomass and N concentration, were not significant for the mixture or vetch (Figure 3). However, barley N content in the SK was higher than in the FK; N uptake increased from 79 to 107 kg N ha^{-1} in 2012 and from 97 to 115 kg N ha^{-1} in 2013. The vetch reached high N_2 atmospheric fixation rates during the experiment. In the first season, N_2 atmospheric fixation was >80% of N content and ~100% in the second. No differences in N_2 atmospheric fixation were found between kill dates.

Residue quality fractions varied between kill dates (Figure 4). The NDF and ADF fractions increased when the CC was killed later. An interaction effect between treatment and year occurred for lignin content as differences in the lignin fraction between kill dates were only significant in barley and in the mixture in 2012. Lignin fraction in the mixture increased from 2.2 to 3.6% and in the barley from 1.2 to 2.9% in 2012. The C/N of barley, vetch and mixture residues was always lower for the FK than for the SK.

3. Soil inorganic nitrogen content

At the beginning of the experiment, average N_{min} of the upper layers of the soil profile was 145 kg N ha^{-1} (Figure 5, Table 2). During the first CC growing period, soil N_{min} increased in the fallow treatment, particularly in the upper 0.8 m, while it decreased in the remaining depths. During the mulch period, soil N_{min} in the fallow treatment increased slightly whereas, in the CC treatments, N_{min} increased largely in the upper soil layers, but there was no difference between the treatments by October in 2012.

During the second CC growing season, soil N_{min} was depleted in all treatments. After the CC kill in April 2013, the N_{min} was 84 kg N ha^{-1} in the SK treatment and ~140 kg N ha^{-1} in the others. The FK treatment accumulated more N_{min} in the upper layers, most likely due to the early mineralization that occurred between the CC kill and soil sampling (~4 weeks). At the end of the mulch period in November 2013, a large increase in soil N_{min} occurred in all treatments. Although no difference was observed for average N_{min} (~245 kg N ha^{-1}) over the entire profile, differences between treatments were observed at some soil depths. In the SK, most of the N_{min} (~70%) was in the top layer (0–40 cm) whereas it was ~29% in the fallow treatment. Similarly, while >40% of soil N_{min} was in the bottom layer in the fallow, < 16% was in the bottom layer in the SK. The N_{min} distribution in the FK treatment was intermediate between the others.

4. Soil nitrogen mineralization potential

The one-pool exponential model fit the cumulative mineral N from the aerobic incubation, as the F-test comparing the mean squares due to pure error and lack of fit was not significant at 0.01 level. The N_0 was higher for soils from the SK treatment (34 mg N kg soil^{-1}) than those from the fallow (24 mg N kg soil^{-1}). The N_0 from the FK treatment was intermediate. No differences between treatments were observed in the N mineralization (Figure 6).

5. Soil water content

Soil water content was affected by the presence of CC and by the kill date with differences observed in the upper layers and the whole soil profile (Figure 7, Table 3). During both CC growth periods, SWC followed a similar pattern. At sowing, the three treatments started with low SWC in the entire profile (~220 mm) in both seasons, and only small differences appeared in the top layers. The precipitation during the three months following CC sowing recharged the soil profile, reaching SWC>300 mm in 2013 due to abundant rainfall (Figure 7a). No differences were observed between treatments during this period in any season. However, during the next three months, CC extracted water from the upper layers, and by the first kill date, SWC was higher in the fallow than in the CC treatments. After the first kill date, the SWC in the FK treatment varied compared to the SK depending on the annual weather conditions. During the first season, no differences were observed between the FK and SK treatments, most likely due to low precipitation, and by the second kill date, both had similar SWC and were lower than the fallow (Table 3). During the second season, the high precipitation between both kill dates recharged the soil profile, and the FK treatment SWC was similar to the fallow except in the deeper soil layer (100–120 cm, Figure 7e) where the fallow treatment remained wetter. For the second kill date of 2013, the SK treatment SWC was lower, and differences with the FK were obvious down to 80 cm and down to 120 cm with the fallow (Table 3).

Although initial differences in SWC were observed between the two years of the study during the mulch period, such differences were not evident toward the end of the season. After water

Table 3. Soil water content (mm) in each of the soil layers and the entire profile at different sampling dates for the fallow (Fa), first kill (FK) and second kill (SK) treatments.

		0–20	20–40	40–60	60–80	80–100	100–120	0–120
					— mm —			
2011–2012								
07/10/2011	Fa	23.9 a	40.3 a	38.1 a	40.8 a	37.4 a	40.4 a	221 a
CC sowing	FK	18.7 a	30.8 b	39.4 a	41.7 a	44.2 a	43.4 a	218 a
	SK	18.5 a	35.1 ab	42.6 a	41.0 a	37.0 a	41.0 a	215 a
13/03/2012	Fa	25.2 a	42.7 a	40.5 a	37.5 a	34.1 a	34.9 a	215 a
First kill date	FK	12.4 b	28.7 b	33.3 a	35.5 a	35.5 a	36.5 a	182 b
	SK	11.4 b	31.7 b	35.5 a	32.2 a	33.1 a	35.5 a	179 b
09/04/2012	Fa	33.9 a	42.2 a	40.1 a	38.3 a	35.4 ab	36.3 a	226 a
Second kill date	FK	18.3 b	28.1 b	33.2 a	36.0 a	36.3 a	37.5 a	189 b
	SK	12.6 c	28.6 b	33.8 a	29.8 a	30.8 b	36.4 a	172 c
30/04/2012	Fa	9.4 c	27.6 c	32.6 a	36.2 a	34.1 ab	36.4 a	251 a
HPD	FK	10.6 b	27.4 b	36.9 a	42.7 a	41.5 a	44.0 a	266 a
	SK	12.8 a	32.0 a	38.7 a	37.3 a	37.4 b	43.6 a	269 a
12/09/2012	Fa	9.4 a	27.6 a	32.6 a	36.2 a	34.1 b	36.4 a	176 a
Mulch period end	FK	10.6 a	27.4 a	36.9 a	42.7 a	41.5 a	44.0 a	203 a
	SK	12.8 a	32.0 a	38.7 a	37.3 a	37.4 ab	43.6 a	202 a
2012–2013								
08/10/2012	Fa	43.9 b	55.3 a	31.3 a	33.6 a	32.0 a	34.4 b	231 a
CC sowing	FK	48.5 ab	55.7 a	36.3 a	39.7 a	38.5 a	40.5 a	259 a
	SK	52.7 a	55.2 a	39.0 a	34.7 a	35.3 ab	41.0 a	258 a
15/03/2013	Fa	50.3 a	56.0 a	58.8 a	56.4 a	44.7 a	39.1 a	305 a
First kill date	FK	43.4 b	47.3 b	41.1 b	43.8 b	40.8 b	38.3 a	255 b
	SK	41.2 b	42.2 c	39.3 b	39.3 b	37.4 b	39.2 a	239 b
11/04/2013	Fa	49.4 b	56.1 b	57.7 a	60.0 a	55.9 a	47.7 a	327 a
Second kill date	FK	53.9 a	59.1 a	61.2 a	58.2 a	43.8 b	39.8 b	316 b
	SK	39.1 c	50.4 c	41.0 b	40.1 b	37.8 b	39.3 b	248 b
30/04/2013	Fa	44.2 b	52.6 b	54.6 a	57.6 a	55.6 a	51.8 a	311 a
HPD	FK	52.7 a	55.4 a	57.6 a	59.5 a	50.0 a	42.4 a	307 b
	SK	40.3 c	41.7 c	42.2 b	40.8 b	38.8 b	40.5 b	244 b
01/09/2013	Fa	7.8 a	37.7 a	47.0 a	56.2 a	56.5 a	54.9 a	315 a
Mulch period end	FK	8.5 a	32.5 a	45.3 a	50.9 a	50.1 a	49.2 ab	318 a
	SK	8.8 a	27.5 a	41.4 a	44.0 a	43.3 a	47.2 b	244 b

Within a date and depth, means with the same letter are not significantly different at P<0.05.

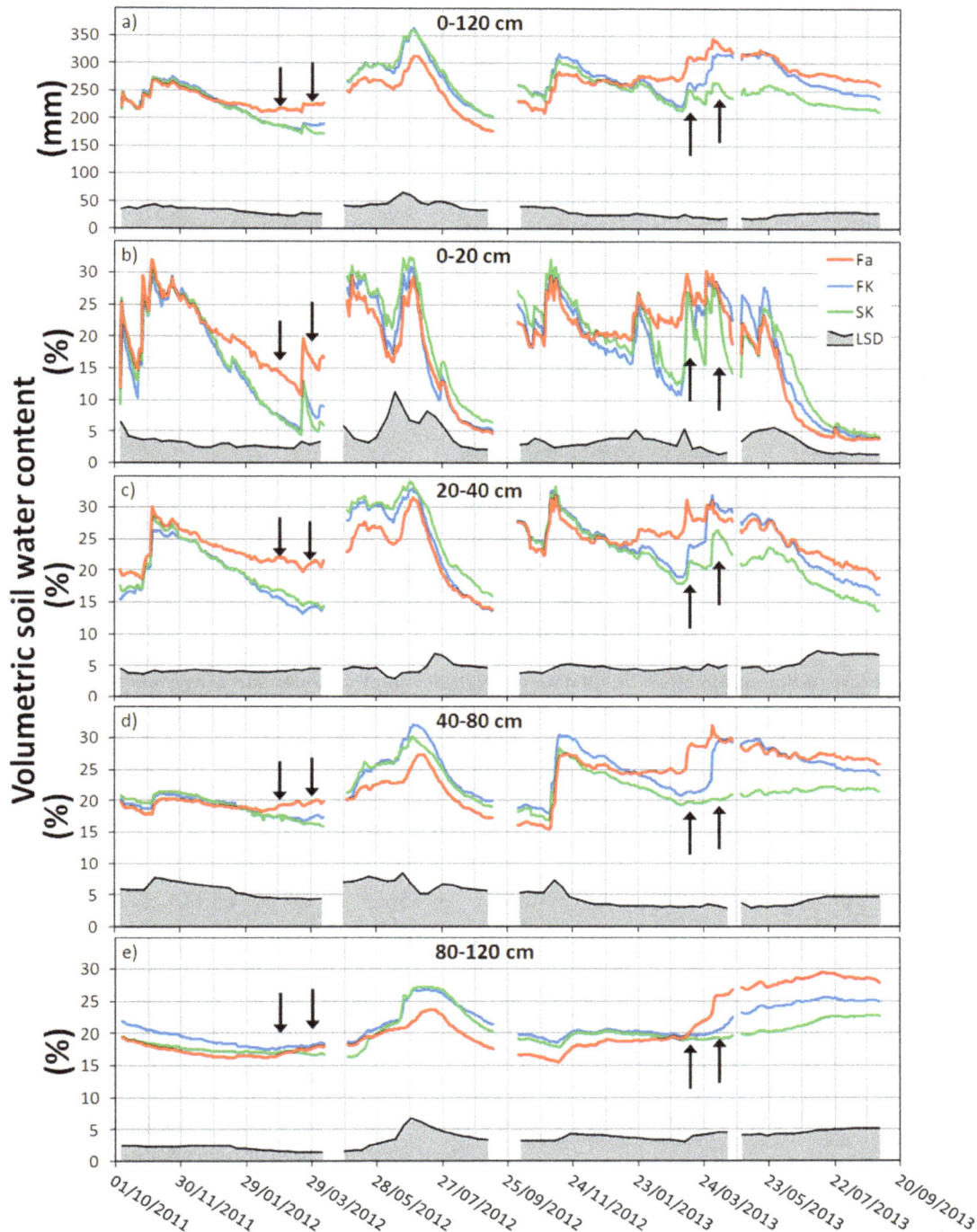

Figure 7. Soil water content monitored with the capacitance sensors at different depths for the fallow (Fa), first kill (FK) and second kill (SK) treatments during the experimental period. Arrows represent the first and second kill dates. The shadowed area in the bottom represents the LSD at 0.05 probability level.

application in April 2012, the SWC in the entire profile at the HPD was similar in all treatments (~262 mm; Table 3). During the wet/dry cycles that occurred in spring and summer, the CC treatments always retained more moisture than the fallow which showed the ability of the residue mulch to reduce soil water evaporation losses (Figure 7). After mid-June, when the soil was left to dry under the summer heat, differences in SWC between treatments disappeared, first in the upper layers and later in the

entire soil profile. Even though slower drying was observed in the CC treatments, no differences were observed between treatments by the end of September (Table 3). In 2013, no water was applied, and the SWC in the entire soil profile was lower for the SK than for the FK and fallow treatments at the HPD (Table 3). At this time, no differences in SWC were observed between the FK and fallow treatments in the entire soil profile. However, the FK upper layer was wetter than the fallow because of the reduction in soil

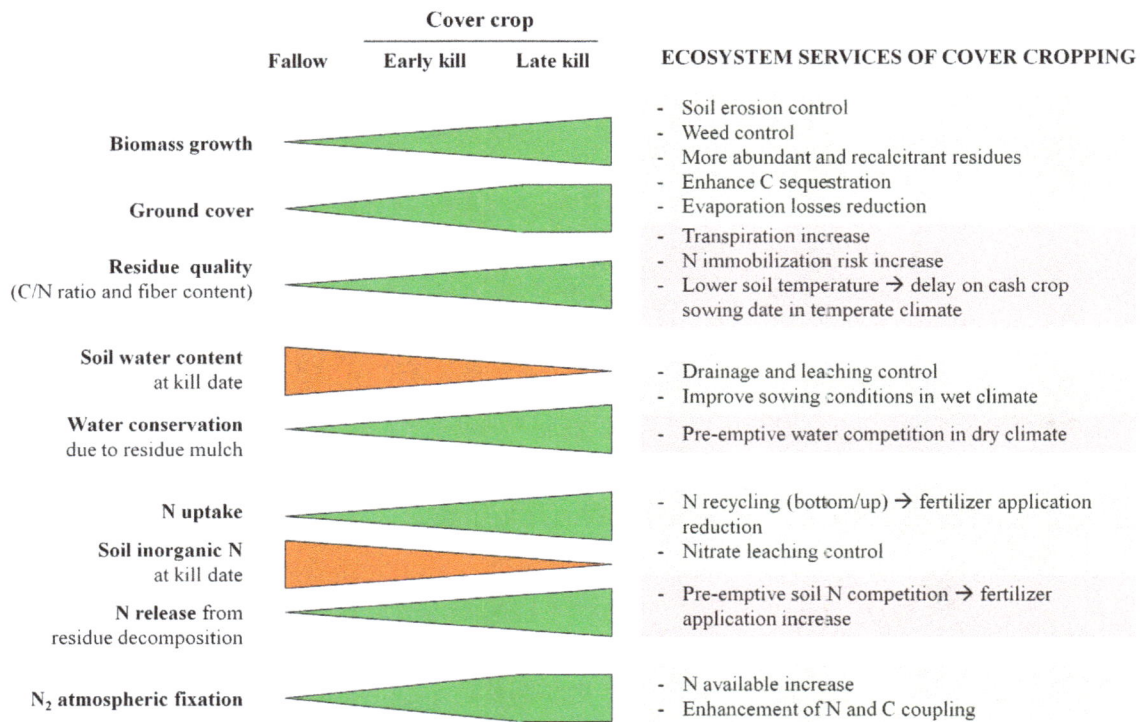

Figure 8. Schematic representation of the effect of kill date on environmental variables and ecosystem services with fallow as a reference condition.

water evaporation caused by the residue mulch (Table 3, Figure 7b). During summer, water losses in the top layer were greater in the fallow than the CC treatments, showing the effect of the mulch. By the end of the experiment, no differences in the SWC of the entire profile were observed between treatments, but some SK subsoil layers remained drier as the result of water depletion caused by the living CC.

Discussion

This study confirmed that CC kill date is an important management strategy as it affected key variables of the soil-plant system, influencing the success of the subsequent cash crop. Previous studies have focused mainly on the effect of N on the succeeding crop [6] and on the CC residue quality [14], but there is a lack of information on the combined effects of CC growth, N dynamics, and water dynamics. This information is necessary to optimize CC benefits and for development of more accurate simulation models.

Delaying kill date did not increase GC whereas the year had an effect on the maximum cover attained. We could not find data to compare the effect of kill date on GC. Nevertheless, our results agreed with those of previous researchers that showed the capacity of the asymptotic Gompertz function to capture the characteristics of the GC evolution of several CC [22,33]. Ground cover is an important variable used in studies related to soil erosion [32], evaporation [34], weed control [35,36] or radiation interception [22]. In addition, the evolution of digital technologies allows for non-destructive monitoring using reliable and efficient techniques [20]. For all of these reasons, we propose that GC should become a common variable for the characterization of CC that facilitates comparisons between species, varieties and management strategies in different regions. In the present study, GC differed with

seasonal weather conditions; the dry 2011–2012 season was less favorable for CC growth than the wet and mild 2012–2013 season. In both seasons, GC>80% was already attained by the first kill date, so no additional benefits in terms of soil erosion or weed control were expected by the second kill date [32,35].

As expected, delaying kill date increased the above-ground biomass of the barley/vetch mixture. This increase was mainly due to barley as there was no difference in vetch biomass between kill dates. The barley/vetch ratio, expressed as seed weight at sowing, was 3:7, but barley was a stronger competitor due to faster establishment and growth rate [20]. The observed increase in biomass is common in many other studies [8,13], but the relationship between the legume and the non-legume differed. In Central Italy, a barley/vetch mixture with a 1:1 sowing rate yielded between 48 and 66% vetch biomass 160 d after planting which increased vetch in the sowing mixture up to 75% and produced between 53 and 76% vetch biomass at harvest [37]. A possible explanation for the large barley dominance in our study is that soil available N was not a limiting factor. Other authors reported similar superior dry matter values for the non-legume compared with the legume in mixtures, and this dominance became stronger with higher levels of N [38]. This hypothesis is corroborated by the plant and soil available N results in our study. Most of the N content in the vetch/barley mixture accumulated before the FK date. Nitrogen content of barley was greater at SK than FK (Figure 3), and in the SK treatment, the inorganic N in the soil was depleted in April, particularly from the deeper layers (Table 2, Figure 5). The high competition with barley for soil N forced the vetch to rely mainly on N_2 fixation for its N requirement, which explains the large degree of N_2 fixation observed in the legume. We were expecting the vetch to fix a significant amount of N_2 between both kill days, as observed in other experiments [10,13], but no increase in vetch N content was observed as in the study of Benincasa et al. [8]. The strong competition with barley for light explains the low vetch growth between kill dates and reinforces the need for sowing mixtures with a large proportion of vetch seeds (>50%) to favor legume activity. Most authors have found that the N content in CC and mixtures increased as kill date was delayed [10,13], but our results show that the main effect of delaying kill date increased above-ground biomass rather than N content.

The effect of postponing kill date on chemical residue composition was in agreement with previous findings in the literature [8,12,39]. Nitrogen concentration decreased from the first to the second kill date, being greater for vetch than for barley (Figure 3). These differences were reinforced in the C/N ratio as a slight increase in C content was observed in the SK samples compared with the FK (Figure 4). Vetch N concentration was similar to that found by other authors, but barley values were greater than those reported in other CC experiments [10,40,41]. In the FK samples, barley N concentration was ~2.9% and ~2.2% in the SK which agrees with data from barley cultivated as a cash crop using synthetic N fertilizer [42]. These data confirm that barley did not suffer from lack of N. The increase in fiber fractions, which is related to a decrease in the labile fraction, from delaying kill date was consistent with other studies and was expected to lead to slower C and N mineralization [43]. Increases in lignin fraction were only significant in barley in the first season, but this is not surprising as there were only three weeks between kill dates, and lignin content is less sensitive to change [39]. In the drier season (2011–2012), barley residue and the mixture had significantly lower fiber and lignin content than in the wetter season for both kill dates. A reduction in NDF, ADF and lignin cereal crop straw in dry years was previously noted in a study conducted over several years and locations in Washington State [44]. Our results confirmed that delaying kill date increased the amount of CC residue and the residues were more recalcitrant to decomposition. This is particularly interesting when the goal is preserving soil moisture [11], controlling weeds [35], or increasing the slow release of the N pool [45]. In the present experiment, a detailed study of CC residue decomposition was not conducted primarily because the lowest part of the residue layer stuck to the soil when sampling, and we decided that it was reliable for determining the amount of the remaining residue but not its quality.

The topsoil N_0 was also affected by the kill date two years after starting the experiment. Soil N mineralization potential assess the capacity of a soil to mineralize N and is a good indicator of land use as it is very sensitive to small changes in labile C and N fractions [46]. The increase of N_0 in the CC treatments proves that CC contribute to the enhancement of labile soil organic pools. The low contribution of CC to total organic C is a known fact as residues contain low quantities of recalcitrant organic matter [2]. However, CC add substrate for soil microorganisms during their growth period (root exudation, root turnover), and residue decomposition increases the organic matter labile fractions [6]. Many studies of N mineralization from CC residues have been published, but there is a lack of studies that focus on the effects on the soil in which they were cropped. In our study, no differences were observed in total soil C, but delaying kill date increased the C supplied by CC and, as a consequence, topsoil N_0. Nitrogen mineralization rate, a parameter that mainly depends on environmental conditions, was not different between treatments and was within the range found in the literature [47].

Consistent with other studies, CC absorbed most of the available soil inorganic N during their growth period, modifying the degree of pre-emptive competition and the risk of N loss with respect to the fallow (Table 2, Figure 5). This study also showed that kill date affected these parameters and could be a means to adapt soil N availability to the requirements of a particular agroecosystem. Delaying kill date increased the risk of pre-emptive competition, a negative effect that was reflected not only in the deeper layers as is common [4,48], but in the upper layers during the second season as well. The early kill date allowed more time for N release from CC residues, and by the HPD, more N was available in the topsoil. This positive effect of an early kill date on reducing pre-emptive competition has not been shown before and was evident when comparing FK with the SK and the fallow in the wet year, in which N_{min} was washed out of the soil during the fall and winter. The well-known CC effect of recycling N in the soil-plant system by uptaking inorganic N from deeper layers and subsequently releasing it to the top layers was clear in this study [6]. In the first season, the soil N_{min} depleted in April by the CC uptake was replenished by October by the N released from the residue mineralization. In the second season, CC prevented most of the available N from being lost from the soil, as happened in the fallow, and released the N by November. The difference in November was not in the amount of N_{min} but its distribution in the soil profile. In the fallow, N_{min} was mostly in the deeper layers and had greater potential for leaching during fall and winter. However, in the SK treatment, N_{min} was in the top layers making it readily available for uptake by the subsequent crop. The N_{min} profile of the FK soil was intermediate. Organic N remained in the residue layer at the end of the experiment, but as mentioned before, we could not obtain a reliable measure of N concentration.

Even if N uptake by CC depleted the soil by April in the first season and seemed to increase the risk of pre-emptive competition with respect to the fallow, we propose that this may be a positive effect in many agricultural systems as it may help control N losses. In a tomato drip irrigation study, it was observed that high soil N_{min} levels at spring planting were linked to large leaching losses as crops are often overwatered to ensure establishment [49]. Excessive irrigation during the crop establishment period may represent up to 80% of total nitrate leaching losses [34,50]. Therefore, keeping soil N_{min} at low levels and correcting fertilizer needs based on crop nutritional status during the cropping season holds promise for controlling N losses and increasing the efficiency of N use [51].

The larger above-ground biomass observed in the SK affected SWC. The CC extracted soil water by transpiration, increasing evapotranspiration losses when compared with the fallow [34]. Spring growth of the vetch/barley mixture was vigorous and soil water depletion was enhanced between the first and the second kill date. The differences in the whole profile SWC at the first kill date between CC treatments and the fallow were >35 mm in the first season and >55 in the second. At the second kill date, the SWC in the FK treatment was similar to the fallow. By the HPD and during the rest of the mulch period, the treatment effect on water availability was variable. While in the first season there was no difference, most likely because the soil was replenished by irrigation, the SWC was 60 mm lower in the SK than in the other treatments during the second season. Similar results were found in California [52,53], where reductions in spring SWC were up to 80 mm due to CC extraction. In Minnesota, the difference in SWC in the upper 0.6 m caused by delaying the rye kill date by three weeks (from April to May) was 27 mm [12]. Our results confirm that late killing of CC increases pre-emptive water competition compared with early killing, a risk that might be mitigated in rainy years or by irrigation if water is available.

A relevant aspect of the increased water uptake by CC is the effect on the upper layers' SWC. Sufficient top layer moisture at planting is crucial to ensure crop establishment and plant survival in many crops [49]. In the present study, the difference in SWC between the fallow and the CC treatments at the first kill date due to water depletion ranged between 13–19 mm in the upper 0.2 m of the soil and was enhanced between the fallow and the SK by the second kill date. However, the effect of CC residue mulching led to a higher SWC in the upper layers of the FK treatment than in the fallow at the HPD. When comparing CC treatments, the results are less clear; more water in the upper layers was retained in the SK than in the FK in the first season, but the SK was drier in the second. The FK treatment always increased water retention in the upper layers with respect to the fallow, but the result varied by year in the SK treatment. The top layer SWC was influenced by CC water uptake during winter, the residue mulch effect, and the amount of rain between the CC kill date and cash crop planting [9]. To clarify the discrepancy between the soil water depletion by CC uptake and conservation by the mulched CC residue, an experiment comparing sheltered and un-sheltered plots in Maryland showed that delaying kill date of a vetch CC produced more biomass and resulted in better soil moisture conservation in the upper 0.3 m [11]. However, recent experiments in Minnesota reported that delaying the rye kill date from the tillering to the booting stage came at the cost of SWC depletion in the top 0.3 m even if the residues continued to cover the soil after termination [12]. In Indiana, early kill of the CC resulted in better water conservation in the topsoil (0.1 m) than late kill. In drought years, early kill date produced an increase in maize yield but had a negative impact in normal precipitation years due to a lowering of soil temperature and workability at the maize planting date [7]. Our results indicate that early kill of the CC enhanced spring topsoil water moisture with respect to late kill, particularly by preserving soil water replenished by precipitation between kill dates.

A limitation found when comparing our study with others that focus on CC killing was that kill dates were always expressed as days after sowing. Given the variety of climatic regions in which cover cropping is practiced, the number of days after sowing is not a valid variable. We propose using degree-days accumulated and growth stage as more suitable variables for comparison between different regions. Soil type also has a major influence on the water balance and N cycling. Simulation models that can take multiple environmental factors into account might allow to overcome these limitations and increase the applicability of this and others studies to different locations and years.

One criticism of this study might be that a crop was not planted after the CC to obtain real data on crop water uptake. We are aware of this situation, but we think that this type of dataset is necessary to advance our understanding of the complex effect of CC kill date on crop resources. Separating the mulching and the effect of transpiration might allow for the calibration of key parameters, proper simulation of both processes, and finally, of the whole system. Replacing fallows by CC is known to provide many benefits, but the agronomic application needs to account for the requirements of specific agroecosystems. Kill date is a management tool to regulate the effects of CC on many environmental variables as presented in Figure 8. In dry environments, residue moisture conservation is the main goal of CC [9] whereas the risk of delaying sowing because of soils that are too wet is a major concern in cooler climates [7]. Depleting soil N enhances pre-emptive competition with the subsequent crop [6] but also controls N losses and creates opportunities for rational fertilizer application [34]. Late termination of CC maximizes their biomass production

and the formation of a thick mulch, enhancing C sequestration, water conservation and weed control [36]. Our work does not close the debate on kill date but contributes to more rational decision-making for the successful use of CC.

Conclusions

Kill date is a major management variable with CC that may lead to increasing water and N use efficiency and controlling N losses and pre-emptive competition with the subsequent crop. The ground cover of CC was >80% by the first kill date so no more benefits in terms of soil erosion or weed control derived from the living CC were expected by the second kill date. However, delaying kill date increased above-ground biomass, providing more CC residues with a higher fiber content and C/N ratio that were more recalcitrant to decomposition and, therefore, more suitable to protect the soil and enhance the slow release of the N pool.

The early kill date decreased the risk of pre-emptive competition by diminishing the N uptake of CC, and allowing more time for N release from CC residues. Cover crops generally uptake inorganic N from deeper layers to be released on the topsoil through residue mineralization. Delaying kill date enhanced this recycling effect that prevented losses of available N by keeping soil inorganic N at low levels. Kill date was a means of controlling this process and showed promise in mitigating N losses and increasing N use efficiency. The N mineralization potential in topsoil was enhanced by the presence of CC and by delaying kill date, proving that CC contribute to the enhancement of labile organic pools.

Delaying kill date increased pre-emptive competition of water, as the water extracted by CC from the first to the second kill date was greater than the water conservation due to the extra residue generated by the SK treatment. The soil water content in the upper layers at the time of planting the subsequent crop increased by the first kill date with respect to the fallow by preserving rain harvested between kill dates, but the soil water content in the late kill could be even lower than in the fallow depending on the climatic conditions of the year.

Supporting Information

Table S1 Biomass, N content and residue left. Above-ground biomass and N content of cover crops at the first kill (FK) and second kill (SK) dates and residue covering the soil at the end of the mulch period (~7 months after cover crop killing). Means with standard error in parentheses. Within a row, means with the same letter are not significantly different between kill dates at P<0.05. (DOCX)

Table S2 Chemical composition of cover crops. Cover crop residue neutral detergent fiber (NDF), acid detergent fiber (ADF) and lignin (L) fractions and C/N ratio for the first kill (FK) and second kill (SK) dates. Means with standard error in parentheses. Within a row, means with the same letter are not significantly different between kill dates at P<0.05. (DOCX)

Image S1 Aerial view of the experimental site. Plots are marked as fallow (FA), first kill (FK) nd second kill (SK) plus the replication number. Image taken between kill dates (March 29, 2012). (TIF)

Image S2 Ground level view of the experimental site. Image taken between kill dates (April 4, 2012). (TIF)

Acknowledgments

We would like to thank the staff of La Chimenea field station (IMIDRA) for their helpful assistance.

References

1. Tonitto C, David MB, Drinkwater LE (2006) Replacing bare fallows with cover crops in fertilizer-intensive cropping systems: A meta-analysis of crop yield and N dynamics. Agriculture, Ecosystems & Environment 112: 58–72.
2. Kuo S, Sainju UM, Jellum EJ (1997) Winter cover crop effects on soil organic carbon and carbohydrate in soil. Soil Science Society of America Journal 61: 145–152.
3. Quemada M, Cabrera ML (2002) Characteristic moisture curves and maximum water content of two crop residues. Plant and Soil 238: 295–299.
4. Gabriel JL, Quemada M (2011) Replacing bare fallow with cover crops in a maize cropping system: Yield, N uptake and fertiliser fate. European Journal of Agronomy 34: 133–143.
5. Abawi GS, Widmer TL (2000) Impact of soil health management practices on soilborne pathogens, nematodes and root diseases of vegetable crops. Applied Soil Ecology 15: 37–47.
6. Thorup-Kristensen K, Magid J, Jensen LS (2003) Catch crops and green manures as biological tools in nitrogen management in temperate zones. Advances in Agronomy 79: 227–302.
7. Stipesevic B, Kladivko EJ (2005) Effects of winter wheat cover crop desiccation times on soil moisture, temperature and early maize growth. Plant Soil and Environment 51: 255–261.
8. Benincasa P, Tosti G, Tei F, Guiducci M (2010) Actual N Availability from Winter Catch Crops Used for Green Manuring in Maize Cultivation. Journal of Sustainable Agriculture 34: 705–723.
9. Clark AJ, Meisinger JJ, Decker AM, Mulford FR (2007) Effects of a grass-selective herbicide in a vetch-rye cover crop system on corn grain yield and soil moisture. Agronomy Journal 99: 43–48.
10. Tosti G, Benincasa P, Farneselli M, Pace R, Tei F, et al. (2012) Green manuring effect of pure and mixed barley-hairy vetch winter cover crops on maize and processing tomato N nutrition. European Journal of Agronomy 43: 136–146.
11. Clark AJ, Decker AM, Meisinger JJ, McIntosh MS (1997) Kill date of vetch, rye, and a vetch-rye mixture: 2. Soil moisture and corn yield. Agronomy Journal 89: 434–441.
12. Krueger ES, Ochsner TE, Porter PM, Baker JM (2011) Winter Rye Cover Crop Management Influences on Soil Water, Soil Nitrate, and Corn Development. Agronomy Journal 103: 316–323.
13. Clark AJ, Decker AM, Meisinger JJ, McIntosh MS (1997) Kill date of vetch, rye, and a vetch-rye mixture: 1. Cover crop and corn nitrogen. Agronomy Journal 89: 427–434.
14. Wagger MG, Cabrera ML, Ranells NN (1998) Nitrogen and carbon cycling in relation to cover crop residue quality. Journal of Soil and Water Conservation 53: 214–218.
15. Tosti G, Benincasa P, Farneselli M, Tei F, Guiducci M (2014) Barley-hairy vetch mixture as cover crop for green manuring and the mitigation of N leaching risk. European Journal of Agronomy 54: 34–39.
16. Soil Survey Staff (2014) Keys to Soil Taxonomy. Washington, DC, USA: USDA, Natural Resources Conservation Service.
17. Nelson DW, Sommers LE (1996) Total carbon, organic carbon and organic matter. In: Sparks D, editor. Methods of soil analysis, part 3: chemical methods. Madison, WI, USA: ASSA and SSSA. 961–1010.
18. Loeppert RD, Suarez DL (1996) Carbonate and Gypsum. In: Sparks D, editor. Methods of soil analysis, part 3: chemical methods. Madison, WI, USA: ASA and SSSA. 437–474.
19. Papadakis J (1966) Climates of the world and their agricultural potentialities. Rome, Italy: DAPCO.
20. Ramirez-Garcia J, Almendros P, Quemada M (2012) Ground cover and leaf area index relationship in a grass, legume and crucifer crop. Plant Soil and Environment 58: 385–390.
21. Richardson MD, Karcher DE, Purcell LC (2001) Quantifying turfgrass cover using digital image analysis. Crop Science 41: 1884–1888.
22. Bodner G, Himmelbauer M, Loiskandl W, Kaul HP (2010) Improved evaluation of cover crop species by growth and root factors. Agronomy for Sustainable Development 30: 455–464.
23. Unkovich M, Herridge D, Peoples M, Cadisch G, Boddey B, et al. (2008) Measuring plant-associated nitrogen fixation in agricultural systems. Camberra, Australia: Australian Centre for International Agricultural Research. 258 p.
24. Goering HK, Van Soest PJ (1970) Forage fiber analysis: apparatus, reagents, procedures, and some applications. Washington, DC: Agricultural Research Service. Handbook no. 379. US Government Printing Office.
25. Keeney DR, Nelson DW (1982) Nitrogen–inorganic forms. In: Page AL, editor. Methods of soil analysis Part 2: Chemical and microbiological properties. Madison, WI, USA: ASA and SSSA. 643–698.
26. Crooke WM, Simpson WE (1971) Determination of ammonium in Kjeldahl digests of crops by an automated procedure. Journal of the Science of Food and Agriculture 22: 9–10.
27. Stanford G, Smith SJ (1972) Nitrogen mineralization potentials of soils. Soil Science Society of America Proceedings 36: 465–472.
28. Paltineanu IC, Starr JL (1997) Real-time soil water dynamics using multisensor capacitance probes: Laboratory calibration. Soil Science Society of America Journal 61: 1576–1585.
29. Gabriel JL, Lizaso JI, Quemada M (2010) Laboratory versus field calibration of capacitance probes. Soil Science Society of America Journal 74: 593–601.
30. Quemada M, Cabrera ML (1995) CERES-N model predictions of nitrogen mineralized from cover crop residues. Soil Science Society of American Journal 59: 1059–1065.
31. Whitmore AP (1991) Method for assessing the goodness of computer simulation of soil processes. Journal of Soil Science 42: 289–299.
32. Quinton JN, Edwards GM, Morgan RPC (1997) The influence of vegetation species and plant properties on runoff and soil erosion: results from a rainfall simulation study in south east Spain. Soil Use and Management 13: 143–148.
33. Ramirez-Garcia J, Gabriel JL, Alonso-Ayuso M, Quemada M (2014) Quantitative characterization of five cover crop species. Journal of Agricultural Science. doi:10.1017/S0021859614000811.
34. Gabriel JL, Muñoz-Carpena R, Quemada M (2012) The role of cover crops in irrigated systems: Water balance, nitrate leaching and soil mineral nitrogen accumulation. Agriculture Ecosystems & Environment 155: 50–61.
35. Teasdale JR, Mohler CL (1993) Light transmittance, soil-temperature, and soil-moisture under residue of hairy vetch and rye. Agronomy Journal 85: 673–680.
36. Saini M, Price AJ, Van Santen E, Arriaga FJ, Balkcom KS, Raper RL (2008) Planting and termination dates affect winter cover crop biomass in a conservation-tillage corn-cotton rotation: Implications for weed control and yield. In: Endale DM editor. Tifton, GA, USA. Southern Conservation Agricultural Systems. 137–141.
37. Tosti G, Benincasa P, Guiducci M (2010) Competition and Facilitation in Hairy Vetch-Barley Intercrops. Italian Journal of Agronomy 5: 239–248.
38. Ofori F, Stern WR (1987) The combined effects of nitrogen-fertilizer and density of the legume component on production efficiency in a maize cowpea intercrop system. Field Crops Research 16: 43–52.
39. Wagger MG (1989) Time of desiccation effects on plant composition and subsequent Nitrogen release from several winter annual cover crops. Agronomy Journal 81: 236–241.
40. Clark AJ, Decker AM, Meisinger JJ (1994) Seeding rate and kill date effects on hairy vetch cereal rye cover crop mixtures for corn production. Agronomy Journal 86: 1065–1070.
41. Benincasa P, Guiducci M, Tei F (2011) The nitrogen use efficiency: Meaning and sources of variation-case studies on three vegetable crops in Central Italy. HortTechnology 21: 266–273.
42. Arregui LM, Quemada M (2008) Strategies to improve nitrogen use efficiency in winter cereal crops under rainfed conditions. Agronomy Journal 100: 277–284.
43. Quemada M, Cabrera ML (1995) Carbon and nitrogen mineralized from leaves and stems of four cover crops. Soil Science Society of America Journal 59: 471–477.
44. Stubbs TL, Kennedy AC, Reisenauer PE, Burns JW (2009) Chemical composition of residue from cereal crops and cultivars in dryland ecosystems. Agronomy Journal 101: 538–545.
45. Quemada M, Cabrera ML, McCracken DV (1997) Nitrogen release from surface-applied cover crop residues: Evaluating the CERES-N submodel. Agronomy Journal 89: 723–729.
46. Cookson WR, Abaye DA, Marschner P, Murphy DV, Stockdale EA, et al. (2005) The contribution of soil organic matter fractions to carbon and nitrogen mineralization and microbial community size and structure. Soil Biology and Biochemistry 37: 1726–1737.
47. Quemada M, Diez JA (2007) Available nitrogen for corn and winter cereal in Spanish soils measured by electro-ultrafiltration, calcium chloride, and incubation methods. Communications in Soil Science and Plant Analysis 38: 2061–2075.
48. Thorup-Kristensen K (1994) The effect of nitrogen catch crop species on the nitrogen nutrition of succeeding crops. Fertilizer Research 37: 227–234.
49. Vázquez N, Pardo A, Suso ML, Quemada M (2005) A methodology for measuring drainage and nitrate leaching in unevenly irrigated vegetable crops. Plant and Soil 269: 297–308.
50. Salmerón M, Isla R, Cavero J (2011) Effect of winter cover crop species and planting methods on maize yield and N availability under irrigated Mediterranean conditions. Field Crops Research 123: 89–99.
51. Vázquez N, Pardo A, Suso ML, Quemada M (2006) Drainage and nitrate leaching under processing tomato growth with drip irrigation and plastic mulching. Agriculture Ecosystems & Environment 112: 313–323.
52. McGuire AM, Bryant DC, Denison RF (1998) Wheat yields, nitrogen uptake, and soil moisture following winter legume cover crop vs. fallow. Agronomy Journal 90: 404–410.
53. Mitchell JP, Peters DW, Shennan C (1999) Changes in soil water storage in winter fallowed and cover cropped soils. Journal of Sustainable Agriculture 15: 19–31.

Author Contributions

Conceived and designed the experiments: MQ. Performed the experiments: MA JLG MQ. Analyzed the data: MA JLG MQ. Contributed to the writing of the manuscript: MA JLG MQ.

15

The Genome of the Generalist Plant Pathogen *Fusarium avenaceum* Is Enriched with Genes Involved in Redox, Signaling and Secondary Metabolism

Erik Lysøe[1]*, Linda J. Harris[2], Sean Walkowiak[2,3], Rajagopal Subramaniam[2,3], Hege H. Divon[4], Even S. Riiser[1], Carlos Llorens[5], Toni Gabaldón[6,7,8], H. Corby Kistler[9], Wilfried Jonkers[9], Anna-Karin Kolseth[10], Kristian F. Nielsen[11], Ulf Thrane[11], Rasmus J. N. Frandsen[11]

1 Department of Plant Health and Plant Protection, Bioforsk - Norwegian Institute of Agricultural and Environmental Research, Ås, Norway, 2 Eastern Cereal and Oilseed Research Centre, Agriculture and Agri-Food Canada, Ottawa, Canada, 3 Department of Biology, Carleton University, Ottawa, Canada, 4 Section of Mycology, Norwegian Veterinary Institute, Oslo, Norway, 5 Biotechvana, València, Spain, 6 Bioinformatics and Genomics Programme, Centre for Genomic Regulation, Barcelona, Spain, 7 Universitat Pompeu Fabra, Barcelona, Spain, 8 Institució Catalana de Recerca i Estudis Avançats, Barcelona, Spain, 9 ARS-USDA, Cereal Disease Laboratory, St. Paul, Minnesota, United States of America, 10 Department of Crop Production Ecology, Swedish University of Agricultural Sciences, Uppsala, Sweden, 11 Department of Systems Biology, Technical University of Denmark, Lyngby, Denmark

Abstract

Fusarium avenaceum is a fungus commonly isolated from soil and associated with a wide range of host plants. We present here three genome sequences of *F. avenaceum*, one isolated from barley in Finland and two from spring and winter wheat in Canada. The sizes of the three genomes range from 41.6–43.1 MB, with 13217–13445 predicted protein-coding genes. Whole-genome analysis showed that the three genomes are highly syntenic, and share >95% gene orthologs. Comparative analysis to other sequenced Fusaria shows that *F. avenaceum* has a very large potential for producing secondary metabolites, with between 75 and 80 key enzymes belonging to the polyketide, non-ribosomal peptide, terpene, alkaloid and indole-diterpene synthase classes. In addition to known metabolites from *F. avenaceum*, fuscofusarin and JM-47 were detected for the first time in this species. Many protein families are expanded in *F. avenaceum*, such as transcription factors, and proteins involved in redox reactions and signal transduction, suggesting evolutionary adaptation to a diverse and cosmopolitan ecology. We found that 20% of all predicted proteins were considered to be secreted, supporting a life in the extracellular space during interaction with plant hosts.

Editor: Yin-Won Lee, Seoul National University, Republic Of Korea

Funding: Nordisk komité for jordbruks- og matforskning (NKJ), Project number: "NKJ 135 - Impact of climate change on the interaction of Fusarium species in oats and barley", and Agriculture and Agri-Food Canada's Genomics Research & Development Initiative funded this work. The funders had no role in study design, data collection and analysis, decision to publish, or preparation of the manuscript.

Competing Interests: One of the co-authors in this manuscript, Carlos Llorens, is employed by the company Biotechvana. The authors have purchased some bioinformatic analysis from Biotchvana, and Biotechvana has no ownership to any results or material.

* Email: erik.lysoe@bioforsk.no

Introduction

Fusarium is a large, ubiquitous genus of ascomycetous fungi that includes many important plant pathogens, as well as saprophytes and endophytes. The genomes of sixteen *Fusarium* spp. have been sequenced during the past decade with a focus on species that either display a narrow host plant range or which have a saprophytic life style. *Fusarium avenaceum* is a cosmopolitan plant pathogen with a wide and diverse host range and is reported to be responsible for disease on >80 genera of plants [1]. It is well-known for causing ear blight and root rot of cereals, blights of plant species within genera as diverse as *Pinus* and *Eustoma* [2], as well as post-harvest storage rot of numerous crops, including potato [3], broccoli [4], apple [5] and rutabaga [6]. *Fusarium*

avenaceum has also been described as an endophyte [7,8] and an opportunistic pathogen of animals [9,10]. The generalist pathogen nature of *F. avenaceum* is supported by several reports on isolates that lack host specificity. One example of this is the report of *F. avenaceum* isolates from *Eustoma* sp. (aka Lisianthus) being phylogenetically similar to isolates from diverse geographical localities or which have been isolated from other hosts [11].

Fusarium avenaceum is often isolated from diseased grains in temperate areas, but an increased prevalence has also been reported in warmer regions throughout the world [12,13]. The greatest economic impact of *F. avenaceum* is associated with crown rot and head blight of wheat and barley, and the contamination of grains with mycotoxins [12]. Co-occurrence of multiple *Fusarium* species in head blight infections is often observed, and several studies covering the boreal and hemiboreal climate zones in the northern hemisphere have revealed that *F. avenaceum* is often among the dominating species [14]. Previously, *F. avenaceum* has been shown to produce several secondary metabolites, including moniliformin, enniatins, fusarin C, antibiotic Y, 2-amino-14,16-dimethyloctadecan-3-ol (2-AOD-3-ol), chlamydosporol, aurofusarin [12,15] and recently also fusaristatin A [16].

The genus *Fusarium* includes both broad-host pathogenic species, utilizing a generalist strategy, and narrow-host pathogenic species, which are specialized to a limited number of plant species. The *F. oxysporum* complex is a well-documented example of the specialist strategy, as each *forma specialis* displays a narrow host range. The genetic basis for this host specialization is dictated by a limited number of transferable genes, encoded on dispensable chromosomes [17]. However, the genetic foundation that allows *F. avenaceum* to infect such a wide range of host plant species and cope with such a diverse set of environmental conditions is currently not well understood. In an effort to shed light on the genetic factors that separates generalists from specialists within *Fusarium*, we sequenced the genomes of three different *F. avenaceum* strains isolated from two geographical locations, Finland and Canada, and from three small grain host plants: barley, spring and winter wheat. Comparison with existing *Fusarium* genomes would further explore pathogenic strategies.

Results and Discussion

Fusarium avenaceum genome sequences

We have sequenced three *F. avenaceum* genomes, one Finnish isolate from barley (Fa05001) and two Canadian isolates from spring (FaLH03) and winter wheat (FaLH27). Assembly of the 454 pyrosequencing based genomic sequence data from Fa05001 resulted in a total genome size of 41.6 Mb, while assembly of the Illumina HiSeq data for FaLH03 and FaLH27 resulted in genome sizes of 42.7 Mb and 43.1 Mb, respectively (Table 1). Gene calling of the three *F. avenaceum* strains resulted in 13217 (Fa05001, gene naming convention *FAVG1_XXXXX*), 13293 (FaLH03, genes named *FAVG2_XXXXX*) and 13445 (FaLH27, genes named *FAVG3_XXXXX*) unique protein coding gene models. Previous comparative genomics studies of filamentous fungi have identified 69 core genes that are found ubiquitously across all fungal clades [18]. All three gene sets included the 69 core genes, suggesting a good assembly and reliable protein-coding gene prediction. Genome sequence data has been deposited at NCBI GenBank in the Whole Genome Shotgun (WGS) database as accession no. JPYM00000000 (Fa05001), JQGD00000000 (FaLH03) and JQGE00000000 (FaLH27), within BioProject PRJNA253730. The versions described in this paper are JPYM01000000, JQGD01000000, and JQGE01000000.

The mitochondrial genome sequence was contained within a single assembled contig for each strain (Fa05001, 49075 bp; FaLH03, 49402 bp; FaLH27, 49396 bp), supporting sufficient coverage and a high quality assembly. Prior to trimming, the FaLH03 and FaLH27 mitochondrial contigs contained 39 and 53 bp, respectively, of sequence duplicated at each end, as expected with the acquisition of a circular sequence. As found in other *Fusarium* mitochondrial genomes [19,20], the *F. avenaceum* mitochondrial genome sequences contain a low G+C content (about 33%) and encode 26 tRNAs and the ribosomal rRNAs *rnl* and *rns*. In addition, the 14 expected core genes (*cob, cox1, cox2, cox3, nad1, nad2, nad3, nad4, nad4L, nad5, nad6, atp6, atp8, atp9*) involved in oxidative phosphorylation and ATP production

are present and in the same order as other *Fusarium* mitochondrial genomes.

Genome structure in *F. avenaceum*

Electrophoretic karyotyping was performed to resolve the number of chromosomes in Fa05001. Previous karyotyping via fluorescence *in situ* hybridization has suggested that *F. avenaceum* isolated from wheat had 8–10 chromosomes [21]. Our attempt to determine the chromosome number in *F. avenaceum* Fa05001 strain by electrophoretic karyotyping was hampered due to the large size of several of the chromosomes. Southern analysis using a telomeric probe did however result in the detection of four distinct bands ranking from 1 to 5 MB, and several diffuse bands above the detection limit of the method (~5 Mb) (Figure S1 in File S1). A high order reordering of the scaffolds from the three sequenced genomes resulted in 11 supercontigs ranging from 0.8 Mb to 6.5 Mb in size, likely corresponding to entire chromosomes or chromosome arms (Figure S2 in File S1). The three genomes display a high level of microcolinearity and only a single putative large genome rearrangement was observed in an internal region of Supercontig 1 between Fa5001 and Canadian isolates (Figure S3 in File S1).

Sequence comparisons between the three genomes revealed a 91–96% nucleotide alignment, with the two Canadian isolates having the fewest unaligned bases. In addition, approximately 1.4–3.2% of the aligned nucleotides exhibited single nucleotide polymorphisms (SNPs), insertions, or deletions between isolates; these were also fewer between the Canadian isolates (Figure 1). These genetic differences were unevenly distributed across the genomes and were largely concentrated at the ends of the supercontigs, while centrally located regions remained relatively conserved. This is similar to what has been previously observed between chromosomes of other *Fusarium* spp. [22]. BLASTn analysis indicated that more than 95% of predicted genes had a significant hit within the two other *F. avenaceum* genomes (Figure S4 in File S1). Together, the results suggest that, despite the large geographical distance between the collection sites, there is a high level of similarity between the three *F. avenaceum* genomes, both in genome structure and gene content. However, some instances of poorly conserved or missing genes were observed in either one or two isolates out of the three (Figure S4 in File S1, File S2, File S3). For example, the three isolates contained some unique polyketide synthases and non-ribosomal peptide synthases. This suggests that there may be some differences in secondary metabolism between the isolates.

Comparison of genome structure to other *Fusarium* species

Phylogenetic analysis of genome-sequenced Fusaria based on *RPB1, RPB2*, rDNA cluster (18S rDNA, ITS1, 5.8S rDNA and 28S rDNA), *EF-1a* and *Lys2* suggest that *F. avenaceum* is more closely related to *F. graminearum*, with greater phylogenetic distance to *F. verticillioides, F. oxysporum* and *F. solani* [23,24]. Phylogeny using *β-tub* alone [24] suggested that *F. avenaceum* is more closely related to *F. verticillioides* than the other three species. The genome data for *F. avenaceum* allowed us to reanalyse the evolutionary history within the *Fusarium* genus based on the 69 conserved proteins, initially identified by Marcet-Houben and Gabaldón [18]. The Maximum Likelihood analysis was based on 25,535 positions distributed on six super-proteins and showed that *F. graminearum* and *F. avenaceum* clustered together in 93% of 500 iterations, with *F. oxysporum* and *F. verticillioides* as sister taxa in 100% of the cases (Figure 2). *Fusarium avenaceum* scaffolds have good alignment with super-

Table 1. Main assembly summary and annotation features of the three *F. avenaceum* genomes.

Strain	Fa05001	FaLH03	FaLH27
Sequencing technology	454	Hiseq	Hiseq
Genome size (Mb)	41.6	42.7	43.2
Sequencing coverage*	21.6x	426.6x	986.2x
Number of contigs	110	180	169
Number of scaffolds	83	104	77
Number of Large Scaffolds (>100 Kb)	40	22	18
Number of Large Scaffolds (>1 Mb)	17	14	11
N50 scaffold length (Mb)	1.43	4.11	4.14
L50 scaffold count	10	5	5
GC content (%)	48%	48%	48%
Number of predicted genes	13217	13293	13445
Average no of genes per Mb	317.7	311.6	311.8
Mean gene length (base pairs)	1554	1557	1552

*Post- removal of mitochondrial genome.

contigs of both *F. graminearum* and *F. verticillioides* with long, similar stretches of syntenic regions (Figure S5 in File S1). The synteny of *F. avenaceum* with *F. graminearum* is visualized in Figure 3, in which long stretches of genes from the same *F. avenaceum* supercontig have orthologs to neighbouring genes on *F. graminearum* chromosomes, indicating a shared genomic architecture. *F. graminearum* genes lacking orthologs in *F. avenaceum* are not distributed uniformly across the supercontig, and are mostly confined to telomeric regions, except for some at interstitial chromosomal sites. Such chromosome regions in *F. graminearum* have been shown to have a higher SNP density [22] and are influencing host specific gene expression patterns [25].

Fusarium avenaceum possesses the genetic hallmarks of a heterothallic sexual life cycle

The observation of two mating-type idiomorphs was another dissimilarity between the *F. avenaceum* isolates [26,27]. The Finnish *F. avenaceum* isolate is of the mating-type *MAT1-1*, possessing the three genes *MAT1-1-1*, *MAT1-1-2*, and *MAT1-1-3* (*FAVG1_07020*, *FAVG1_07021*, and *FAVG1_07022*), while the two Canadian isolates are of mating-type *MAT1-2*, containing the genes *MAT1-2-1* and *MAT1-2-3* (*FAVG2_03853* and *FAVG2_03854* or *FAVG3_03869* and *FAVG3_03870*). Such idiomorphs with different sets of genes has been observed previously in *Fusarium* spp. [26,27], and surveys of *F. avenaceum* populations often find isolates evenly split between mating types [28]. The sexual stage of *F. avenaceum* has been observed [29,30], and both *MAT1-1* and *MAT1-2* transcripts have been detected in this species under conditions favorable for perithecial production in other Fusaria [31], suggesting that *F. avenaceum* is likely capable of heterothallism. This is further supported by our data, in which a single mating-type is present in a given *F. avenaceum* isolate. This is characteristic for heterothallic fungal species, differing from homothallic species such as *F. graminearum*, which contain both mating-types in a single nucleus [32].

Occurrence of few repetitive elements supports the hypothesis that *F. avenaceum* is sexually active

A search for repetitive elements in the Fa05001 genome (using RepeatMasker [33] with CrossMatch) identified 1.0% of the Fa05001 genome as being repetitive or corresponding to transposable elements (Table S1 in File S1). This value is comparable to the 1.12% found for *F. graminearum*, although there were differences in the distributions of the various types of genetic elements. Tad1 and MULE-MuDR transposons were more enriched in *F. avenaceum* while *F. graminearum* had higher proportions of the TcMar-Ant1 transposon and small RNA. The low level of repeats supports the hypothesis that *F. avenaceum* is sexually active in nature, as such low levels are typical for species with an active sexual cycle, such as *F. graminearum*, *F. verticillioides* [1.21% repeats] and *F. solani* [3.8% repeats], while species which rely on asexual reproduction, such as *F. oxysporum*, have higher levels [10.6% repeats]. These results could be somewhat influenced by the fact that the Fusaria genomes are generated with different technologies.

Gene families enriched in *F. avenaceum*

Analysis of the predicted function of the 13217 Fa05001 gene models showed that Fa05001 contains a greater diversity of InterPro families than the other four analyzed Fusaria (Table 2, Figure 4, and File S4). Two highly enriched InterPro categories stand out in Fa05001; "Polyketide synthase, enoylreductase" (IPR020843) and "Tyrosine-protein kinase catalytic domain" (IPR020635), involved in secondary metabolism and signal transduction, respectively. With the exception of *F. solani*, Fa05001 also has the highest number of predicted transcription factors (Table S2 in File S1). Sixty-eight InterPro domains were predicted to be unique to Fa05001, including four tryptophan dimethylallyltransferase (IPR012148) proteins, commonly found in alkaloid biosynthesis. A comparison of Gene Ontology (GO) terms [34] indicated additional differences between the analyzed *Fusarium* species. Functional categories in which Fa05001 had higher numbers of proteins than the other genome-sequenced Fusaria were: "Cellular response to oxidative stress", "Branched-chain amino acid metabolic process", "Toxin biosynthetic process", "Oxidoreductase activity", "rRNA binding" and "Glutathione transferase activity" (Table S3, S4, S5 in File S1).

Reciprocal BLAST revealed that about ¾ of the predicted proteins in Fa05001 have orthologs in *F. graminearum* (76.7%), *F. verticillioides* (76.9%), *F. oxysporum* (78.8%) and *F. solani* (74.1%)

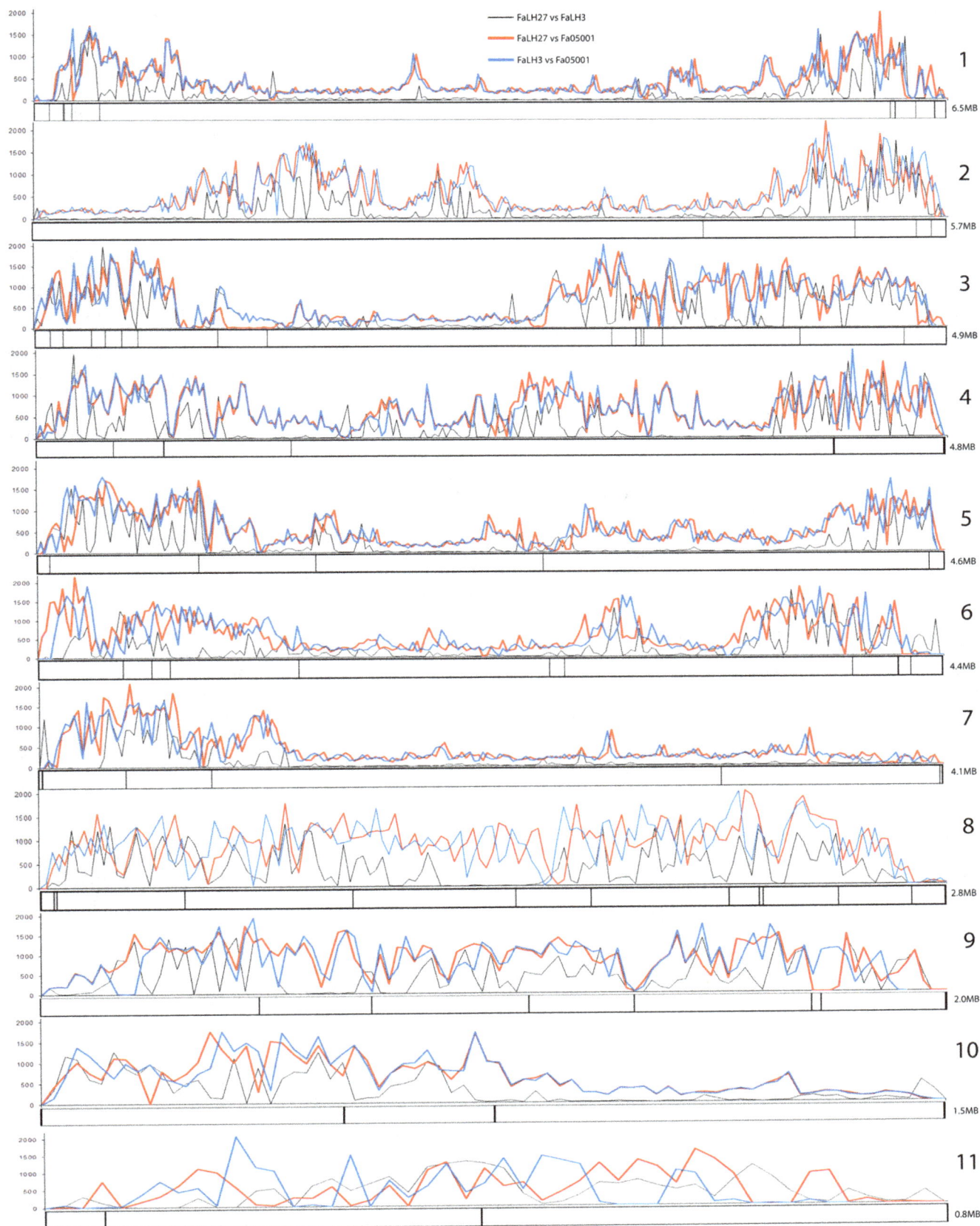

Figure 1. Sliding window map with numbers of SNP's and indels per 20 kb in the three *F. avenaceum* strains Fa05001, FaLH03 and FaLH27 on the 11 supercontigs. Locations of the polyketide synthase and non-ribosomal peptide synthetase genes in the strain FaLH27 are plotted on the supercontigs.

(File S5). The *F. avenaceum* proteins for which no ortholog (no hits or e-value$>10^{-10}$) was found in the other *Fusarium* genomes were especially enriched in GO biological categories "Oxidation-reduction process"; "Toxin biosynthetic process", "Alkaloid metabolic process"; "Cellular polysaccharide catabolic process" and "Transmembrane transport" (Table 3, Table 4, Figure S6 in File S1).

The secretome of *F. avenaceum*

The interplay between the invading fungus and the host plant occurs mainly in the extra-cellular space. The proportion of genes encoding predicted secreted proteins in the Fa05001 genome (File S6, Table S6 in File S1) is remarkably high (~20%; 2,580 proteins) as compared to plant pathogens such as *F. graminearum* (11%) and *Magnaporthe grisea* (13%), saprophytes such as *Neurospora crassa* (9%) and *Aspergillus nidulans* (9%) [22], and the insect pathogen *Cordyceps militaris* (16%) [35]. The secretome appears particularly enriched in proteins involved in redox reactions (Figure S7 in File S1). Inspecting the *F. avenaceum* proteins with no orthologs in other Fusaria (1223 proteins from Figure S6 in File S1), we found that 36% were predicted to be secreted.

Small cysteine-rich proteins (CRPs) can exhibit diverse biological functions, and some have been shown to play a role in virulence, including Avr2 and Avr4 in *Cladosporium fulvum* [36] and the Six effectors in *F. oxysporum* f. sp. *lycopersici* [37]. Other reported functions for CRPs have been adherence [38], anti-microbiosis [39] and carbohydrate binding activity that interferes with host recognition of the pathogen [40]. We found 19 candidates in Fa05001 containing more than four cysteine residues, and an additional 55 containing less than four cysteine residues, but with significant similarity to CRP HMM models (File S7, Table S7 in File S1). Of the predicted *F. avenaceum* CRPs, several are also found in other sequenced *Fusarium* species, but only two were noted with a putative function: CRP5760, with similarity to lectin-B, and CRP5810, a putative chitinase 3.

Metabolic profiling of *F. avenaceum*

A determination of secondary metabolites produced by these three *F. avenaceum* strains was performed on agar media, additionally Fa05001 was also grown *in planta* (barley and oat). Extraction of *F. avenaceum* cultures grown on PDA and YES solid

media revealed the presence of 2-amino-14,16-dimethyloctade-can-3-ol, acuminatopyrone, antibiotic Y, fusaristatin A, aurofu-sarin, butenolide, chlamydosporols, chrysogine, enniatin A, enniatin B, fusarin C, and moniliformin (Table 5). These metabolites have previously been reported from a broad selection of *F. avenaceum* strains [5,16,41,42] whereas the following metabolites previously reported from *F. avenaceum* were not detected in the present study: beauvericin [43], fosfonochlorin [44], diacetoxyscirpenol, T-2 toxin and zearalenone [45–50]. These reports are based on single observations using insufficient specific chemical methods that could yield false positive detection and/or poor fungal identification and no deposition of the strain in a collection for verification. The lack of zearalenone production agrees with the finding that none of the three *F. avenaceum* genomes described here contains the genes required for its production [51,52].

Preliminary genomic analysis had predicted the production of ferricrocin, malonichrome, culmorin, fusarinine and gibberellins [23], but chemical analysis only verified the production of the first two metabolites. The polyketide fuscofusarin (an aurofusarin analogue or intermediate in the biosynthetic pathway) was detected for the first time in *F. avenaceum*. Furthermore, fusaristatin A was found for the first time to be produced *in planta* during climate chamber experiments, and was generally found in *F. avenaceum*-infected barley, but only in trace amounts in oat. In addition, the tetrapeptide JM-47, an HC-toxin analogue reported from an unidentified *Fusarium* strain [53], was detected from all three strains in all cultures, including in barley and oats in climate chamber experiments. Apicidins have been detected in other strains of *F. avenaceum* cultured on YES agar (unpublished results), whereas these compounds were not detected in cultures of the three sequenced strains. Further characterization of the metabolomes on PDA and YES media revealed three major peaks in the ESI+ chromatograms, which were also detected in the barley and oat extracts. that could not be matched to known compounds in Antibase [54]. Since these novel compounds are produced *in planta*, they are candidates for novel virulence factors.

Large potential for secondary metabolite production

The three *F. avenaceum* genomes encode 75 (Fa05001), 77 (FaLH03) and 80 (FaLH27) key enzymes for the production of

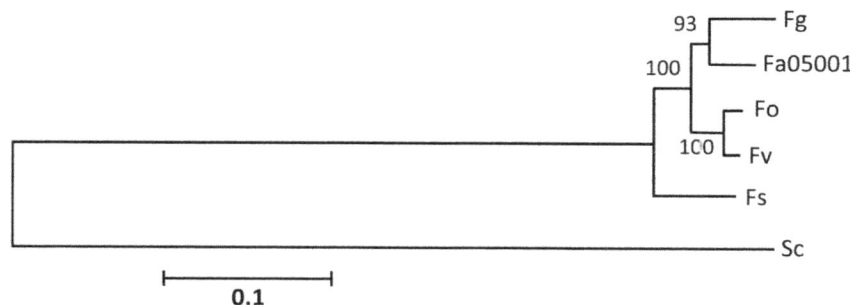

Figure 2. Molecular phylogenetic analysis of Fusarium species based on 69 orthologous proteins. The evolutionary history was inferred by using the Maximum Likelihood method and the tree with the highest log likelihood (−152577,9625) is shown. Bootstrap values, as percentages, are shown next to the individual branches. The tree is drawn to scale, with branch lengths measured in the number of substitutions per site. All positions containing gaps and missing data were eliminated prior to the ML analysis and the final data set contained 25535 positions.

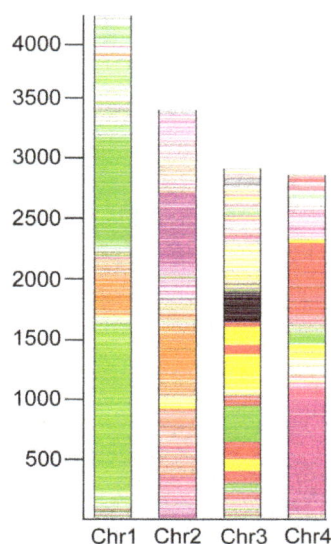

Figure 3. Shared gene homology map between Fa05001 and _F. graminearum_ PH-1 has been created using the four defined _F. graminearum_ chromosomes as templates. Genes in _F. graminearum_ are coloured according to whether genes have corresponding orthologs in Fa05001, with one gene being one strip. Regions of the same color match to the same supercontigs (Figure S2 in File S1) in _F. avenaceum_. White regions represent lack of orthologs in Fa05001.

secondary metabolites, exhibiting a far greater biosynthetic potential than the known secondary metabolites produced by the species. These numbers include genes for 25–27 iterative type I polyketide synthases (PKSs), 2–3 type III PKSs, 25–28 non-ribosomal peptide synthases (NRPSs), four aromatic prenyltransferases (DMATS), 12–13 class I terpene synthases (head-to-tail incl. cyclase activity), two class I terpene synthase (head-to-head) and four class II terpene synthases (cyclases for the class I terpene synthase head-to-head type) (Table 6).

The number of type I PKSs is surprisingly high, considering that the six other public _Fusarium_ genomes (_F. graminearum_, _F. verticillioides_, _F. oxysporum_, _F. solani_, _F. pseudograminearum_ and _F. fujikuroi_) only encode between 13 and 18 type I PKSs. The three _F. avenaceum_ genomes share a core set of 24 type I PKS, and the individual isolates also encode unique type I PKS: oPKS47 (FAVG1_08496) is unique to Fa05001, while the two Canadian isolates share a single oPKS53 (FAVG2_01811, FAVG3_01846) and FaLH27 has two additional PKSs, oPKS54 (FAVG3_02030) and oPKS55 (FAVG3_06174). Of the 55 different type I PKSs described in the seven _Fusarium_ genomes, only two (oPKS3 and oPKS7) are found in all species. Orthologs for 12 of the type I PKSs found in _F. avenaceum_ could be identified in one or several of the publicly available _Fusarium_ genomes, and hence 16 are new to the genus (Figure 5). None of these 16 have obvious characterized orthologs in other fungal genomes or in the GenBank database. The PKSs with characterized orthologs within the _Fusarium_ genus includes the PKSs responsible for formation of fusarubin (oPKS3), fusaristatins (oPKS6), fusarins (oPKS10), aurofusarin (oPKS12) and fusaristatin A (oPKS6) [16,55–57]. Prediction of domain architecture of the 28 PKSs found in _F. avenaceum_ showed that 17 belong to the reducing subclass, four to the non-reducing subclass, two are PKSs with a carboxyl terminal Choline/Carnitine O-acyltransferase domain, four are PKS-NRPS hybrids and one is a NRPS-PKS hybrid. The iterative nature of this enzyme class makes it

impossible to predict the products of these enzymes without further experimental data; this research has been initiated. The _F. avenaceum_ genomes also encode type III PKSs, a class that among fungi was first described in _Aspergillus oryzae_ [58], and recently in _F. fujikuroi_ [59]. The FaLH03 isolate possesses three proteins (oPKSIII-1 to oPKSIII-3), while the two other isolates only have two (oPKSIII-1 and oPKSIII-3).

NRPSs provide an alternative to ribosomal-based polypeptide synthesis and in addition allow for the joining of proteinous amino acids, nonproteinous amino acids, α-hydroxy acids and fatty acids as well as cyclization of the resulting polypeptide [60]. The non-ribosomal peptide group of metabolites includes several well characterized bioactive compounds, such as HC-toxin (pathogenicity factor) and apicidin (histone deacetylase inhibitor) [61,62]. Of the 30 unique NRPSs encoded by the three _F. avenaceum_ isolates, 16 are novel to the _Fusarium_ genus, and include seven mono-modular and nine multi-modular NRPS, with between 2 and 11 modules. Of these, NRPS41 (FAVG1_08623, FAVG2_11354 and FAVG3_11434) is a likely ortholog to gliP2 (similar to gliotoxin synthetase) from _Neosartorya fischeri_ (Genbank accession no. EAW21276), sharing 75% amino acid identity. The other 14 NRPSs are orthologs to previously reported proteins in other _Fusarium_ species [63], and include the three NRPSs responsible for the formation of the siderophores malonichrome (oNRPS1), ferricrocin (sidC, oNRPS2) and fusarinine (sidA, oNRPS6).

All three strains encoded oNRPS31 which shows a significant level of identity to the apicidin NRPS (APS1) described in _F. incarnatum_ and _F. fujikuroi_ [59,64], and the HC-toxin NRPS (HTS1) from _Cochliobolus carbonum_ SB111, _Pyrenophora tritici-repentis_ and _Setosphaeria turcica_ [65,66]. It has not yet been investigated whether NRPS31 is involved in the production of JM-47. Alignment of the genomic regions surrounding oNRPS31 with the APS1 and HTS1 clusters, showed that the three _F. avenaceum_ strains encode nine of the twelve APS proteins found in the other two Fusaria, but lack clear orthologs encoding APS3 (pyrroline reductase), APS6 (O-methyltransferase) and APS12 (unknown function). The _APS_-like gene cluster in _F. avenaceum_ has undergone extensive rearrangements, resulting in the loss of the three _APS_ genes and gain of three new ones (_APS13-15_) (Figure 6 and Table S8 in File S1). Of these, _APS14 (FAVG1_08581, FAVG2_02887_ and _FAVG3_02926_) encodes a fatty acid synthase β subunit, which shares 60% identity with the FAS β subunit (encoded by _FAVG1_03575_, _FAVG2_06952_ and _FAVG3_07032_) involved in primary metabolism. APS14 likely interacts with APS5 (fatty acid synthase α subunit) to form a functional fungal FAS (α6β6) responsible for the formation of the decanoic acid core of (S)-2-amino-8-oxodecanoic acid, proposed by Jin and co-workers [64] to be incorporated into apicidin by APS1. It has previously been hypothesized that APS5 (FAS α) interacts with the FAS α unit from primary metabolism to fulfill this role [64], however, the new model implies that the _F. incarnatum_ genome encodes an unknown APS14 ortholog (no genome sequence available). The _F. fujikuroi_ genome does not contain an APS14 ortholog (TBLASTn against the genome), but apicidin F has been detected in this species [59,67]. Alignment of the apicidin and HC-toxin gene clusters confirmed the observations made by Manning _et al._ [66] and Condon _et al._ [65], regarding the similarities of the two types of gene clusters, which yield very different products (Figure 6).

Dimethylallyltransferase and indole-diterpene biosynthesis proteins are common in the production of bioactive compounds in endophytes [68,69]. The three _F. avenaceum_ genomes encoded four tryptophan dimethylallyltransferase (DMATS), aromatic

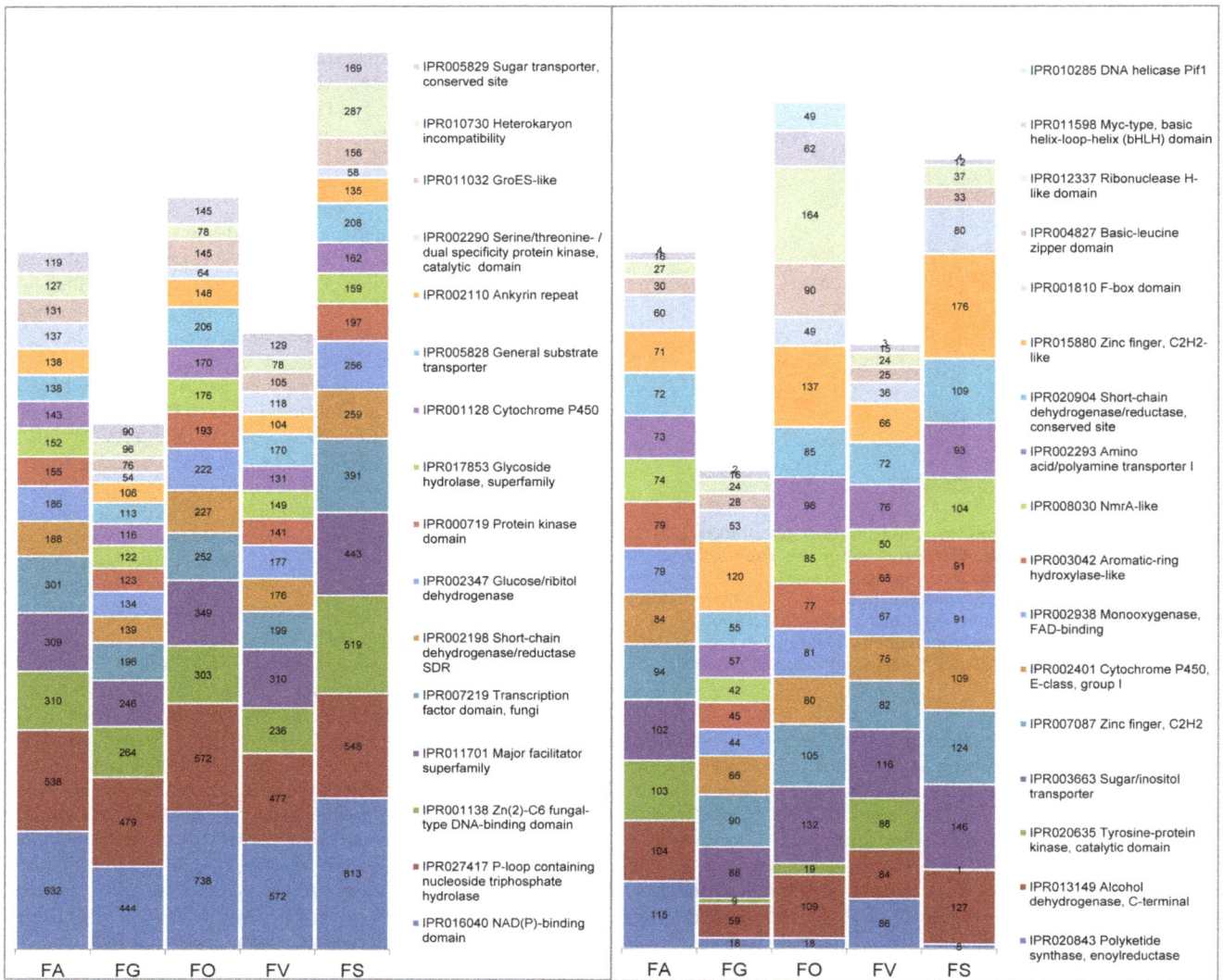

Figure 4. Functional analysis of Fa05001 and other sequenced Fusaria based on InterPro visualizing similarities and differences between the fungi. Categories with most differences between Fa05001 and others are presented with the number of proteins in each category. All details are listed in File S4. FA = *F. avenaceum* Fa05001, FG = *F. graminearum* PH-1, FO = *F. oxysporum* f. sp. *lycopersici* 4287, FV = *F. verticillioides*, FS = *F. solani*.

prenyltransferases, typically involved in alkaloid biosynthesis or modification of other types of secondary metabolites [68]. This is one more than is found in *F. verticillioides* and *F. oxysporum*, and two more than found in *F. fujikuroi*. One putative indole-

diterpene biosynthesis gene was found in all three *F. avenaceum* genomes (*FAVG1_08136*, *FAVG2_00906* and *FAVG3_00934*).

The three *F. avenaceum* genomes are also rich in genes involved in terpene biosynthesis. In the case of the class I terpene synthase,

Table 2. InterproScan analysis and comparison of Fa05001 with other Fusaria.

InterPro analysis	Families	Total domains
Fa05001	5,710	27,474
F. graminearum	5,602	23,560
F. oxysporum	5,609	29,873
F. verticillioides	5,545	25,393
F. solani	5,582	30,538

Table 3. Enriched biological processes in Fa05001 proteins without orthologs in other genome sequenced Fusaria (P<0.05).

F. avenaceum vs F. graminearum

GO-ID	Term	FDR	P-Value
GO: 0055114	Oxidation-reduction process	2.16E-06	9.72E-10
GO: 0009820	Alkaloid metabolic process	0.35	6.33E-04
GO: 0003333	Amino acid transmembrane transport	0.37	0.001
GO: 0009403	Toxin biosynthetic process	0.71	0.004
GO: 0016998	Cell wall macromolecule catabolic process	1	0.02
GO: 0006366	Transcription from RNA polymerase II promoter	1	0.03
GO: 0044247	Cellular polysaccharide catabolic process	1	0.03
GO: 0006573	Valine metabolic process	1	0.03

F. avenaceum vs F. verticillioides

GO-ID	Term	FDR	P-Value
GO: 0055114	Oxidation-reduction process	0.02	3.67E-05
GO: 0009403	Toxin biosynthetic process	0.98	0.003
GO: 0009820	Alkaloid metabolic process	1	0.01
GO: 0044247	Cellular polysaccharide catabolic process	0.02	1
GO: 0018890	Cyanamide metabolic process	1	0.02
GO: 0042218	1-aminocyclopropane-1-carboxylate Biosynthetic process	1	0.03
GO: 0006032	Chitin catabolic process	1	0.03
GO: 0006366	Transcription from RNA polymerase II promoter	1	0.04
GO: 0009636	Response to toxin	1	0.05

F. avenaceum vs F. oxysporum

GO-ID	Term	FDR	P-Value
GO: 0016106	Sesquiterpenoid biosynthetic process	0.44	0.002
GO: 0009403	Toxin biosynthetic process	0.44	0.002
GO: 0055114	Oxidation-reduction process	1	0.02
GO: 0046355	Mannan catabolic process	1	0.04
GO: 0006558	L-phenylalanine metabolic process	1	0.05

F. avenaceum vs F. solani

GO-ID	Term	FDR	P-Value
GO: 0055114	Oxidation-reduction process	0.023	4.28E-05
GO: 0007155	Cell adhesion	0.13	5.08E-04
GO: 0045493	Xylan catabolic process	0.17	8.33E-04
GO: 0009403	Toxin biosynthetic process	0.8	0.005
GO: 0009820	Alkaloid metabolic process	1	0.017
GO: 0000162	Tryptophan biosynthetic process	1	0.033

See also Figure S6 in File S1.

responsible for the head-to-tail joining of isoprenoid and extended isoprenoid units and eventually cyclization, the three genomes encode 12–13 enzymes, of which three were putatively identified as being involved in primary metabolism (ERG20, COQ1, BTS1). In the case of the head-to-head class I systems, *F. avenaceum* encodes two enzymes, similar to the other fully genome-sequenced Fusaria, of which one gene encodes ERG9 and the other is involved in carotenoid biosynthesis. Cyclization of the formed head-to-head type product, if such a reaction occurs, is probably catalyzed by a class II terpene synthase/cyclase, of which *F. avenaceum* encodes four, including an ERG7 ortholog. This is more than the other Fusaria spp. genomes, as *F. graminearum*, *F.*

pseudograminearum and *F. verticillioides* only have ERG7, while *F. solani*, *F. oxysporum* and *F. fujikuroi* have two.

One of the four Type II terpene synthase encoding genes (TS_II_01: *FAVG1_10701*, *FAVG2_04190* and *FAVG3_04223*) shared by all three *F. avenaceum* strains was found to be orthologous to the gibberellic acid (GA) copalyldiphosphate/ent-kaurene synthase (cps/ks) from *F. fujikuroi* (Table S9 in File S1). Previously, the ability to synthesize the GA group of diterpenoid plant growth hormones in fungi has only been found in *F. fujikuroi* mating population A, *Fusarium proliferatum*, *Sphaceloma manihoticola* and *Phaeosphaeria sp.* strain L487 [70–74]. Biosynthesis of GA was first thoroughly characterized in *F.*

Table 4. Genes annotated as reduction-oxidation process in Fa05001 proteins without orthologs in other genome sequenced Fusaria (with expect>1e-10).

Reduction-oxidation function	Genes
4-carboxymuconolactone decarboxylase	FAVG1_07738
ABC multidrug	FAVG1_08208
Acyl dehydrogenase	FAVG1_08563, FAVG1_04763
Alcohol dehydrogenase	FAVG1_04699, FAVG1_12680
Aryl alcohol dehydrogenase	FAVG1_08747
Bifunctional p-450: nadph-p450 reductase	FAVG1_11632
Transcription factor	FAVG1_09453, FAVG1_07648
Choline dehydrogenase	FAVG1_12183
Cytochrome p450 monooxygenase	FAVG1_08576, FAVG1_13151, FAVG1_07923, FAVG1_10161, FAVG1_08627, FAVG1_02807, FAVG1_08721, FAVG1_10699
Delta-1-pyrroline-5-carboxylate dehydrogenase	FAVG1_04749
Dimethylaniline monooxygenase	FAVG1_11196
Ent-kaurene synthase	FAVG1_10701
Glutaryl- dehydrogenase	FAVG1_02842
Homoserine dehydrogenase	FAVG1_09776
l-lactate dehydrogenase a	FAVG1_08604
Mitochondrial 2-oxoglutarate malate carrier protein	FAVG1_08564
Monooxygenase fad-binding protein	FAVG1_10705
Nadp-dependent alcohol dehydrogenase	FAVG1_10174, FAVG1_12103, FAVG1_06950
Nitrilotriacetate monooxygenase component b	FAVG1_07690
Pyoverdine dityrosine biosynthesis	FAVG1_12703
Salicylate 1-monooxygenase	FAVG1_09825
Salicylaldehyde dehydrogenase	FAVG1_09676
Short-chain dehydrogenase	FAVG1_07710, FAVG1_12690, FAVG1_04239, FAVG1_10281, FAVG1_08519

See also Figure S6 in File S1.

fujikuroi and depends on the coordinated activity of seven enzymes, encoded by the GA gene cluster, that convert dimethylallyl diphosphate (DMAPP) to various types of gibberellic acids, with the main end-products being GA_1 and GA_3 [70]. *S. manihoticola's* GA biosynthesis ends at the intermediate GA_4 due to the lack two of genes (DES and P450-3), compared to *F. fujikuroi*, responsible for converting GA_4 to GA_7, GA_3 and GA_1 [73]. Analysis of the genes surrounding the *F. avenaceum* CPS/KS encoding gene showed that six of the seven genes from the *F. fujikuroi* GA gene cluster are also found in all three *F. avenaceum* genomes, with only P450-3 (C13-oxidase) missing (Figure 7). The architecture of the GA cluster in *F. avenaceum* is highly similar to the *F. fujikuroi* cluster, and a single inversion in five of the six genes can explain the relocation of the desaturase (*des*) encoding gene. The inversion could potentially have involved the P450-3 gene and resulted in the disruption of its coding sequence, however the shuffle-LAGAN analysis (Figure 7) and dot plot showed that this has not been the case. The missing gene is not found elsewhere in the genome based on a tblastn search. The presence of six of the seven GA biosynthesis genes suggest that *F. avenaceum* has the potential to produce all the GA's up to G_4 and G_7, but lack the ability to convert these into the GA_1 and GA_3. None of the three *F. avenaceum* isolates have been reported to produce this plant growth hormone.

In summary, *F. avenaceum* has a very large potential for producing secondary metabolites belonging to the PKS, NRPS and terpene classes, with a total of 75–80 key enzymes, see File S8

that summarizes all orthology groups. However, it is expected that multiple enzymes will participate in a single biosynthetic pathway, thereby reducing the potential number of final metabolites. It is possible that some of the metabolites function as virulence factors during infection; however systematic deletion of all PKS encoding genes in *F. graminearum* showed that none of the 15 PKSs in this fungus had significant effect on virulence [75]. It is therefore more likely that at least some of the secondary metabolites function as antibiotics towards competing microorganisms in the diverse set of niches that the species inhabits. When plotting the PKSs and NRPSs on the 11 supercontigs, areas with higher numbers of SNPs, insertions, or deletions between the three strains, were also more populated with secondary metabolite genes (Figure 1), as seen in other Fusaria [22].

Transcriptomics of *F. avenaceum* in barley

To increase our understanding of *F. avenaceum* behaviour *in planta*, we performed RNA-seq on *F. avenaceum*-inoculated barley. Table S10 in File S1 shows a list of *F. avenaceum* genes with the most stable and significant expression (FDR<0.05). Due to putative false positive genes expressed in the host, we applied high stringency in the analysis, and only genes found expressed at 14 days post inoculation (dpi), and which were absent in control, were considered. Genes involved in stress related responses, especially oxidative stress (as defined in the fungal stress response database [76]) were highly represented. This strongly supports our hypothesis formulated from the comparative genomic analysis that

Table 5. Metabolic profiling of the three *F. avenaceum* strains.

	FaLH03		FaLH27		Fa05001				Other strains
	YES	PDA	YES	PDA	YES	PDA	Barley	OAT	*F. avenaceum*
PKS									
2-Amino-14,16-dimethyloctadecan-3-ol	+	+	+	+	+	+	ND	ND	+
Acuminatopyrone	+	+	ND	ND	+	+	ND	ND	+
Antibiotic Y	+	+	+	+	+	+	ND	ND	+
Aurofusarin	+	+	+	+	+	+	+	+	+
Fuscofusarin	+	+	+	+	+	+	ND	ND	NA
Moniliformin	+	+	+	+	+	+	NA	NA	+
Chlamydosporols	+	+	ND	ND	+	+	ND	ND	+
NRPS and mixed NRPS-PKS									
Butenolide	ND	ND	ND	ND	+	+	+	ND	+
Chrysogine	+	+	+	+	+	+	+	+ but 10 × lower than barley	+
Visoltricin	ND	ND	ND	ND	ND	ND	ND	ND	+ (by UV-Vis)
Fusarins C and A	+	ND	+	ND	+	+	ND	ND	+
Enniatins A's and B's	+	+	+	+	+	+	+	+ but 100 × lower than barley	+
Beauvericin	ND	ND	ND	ND	ND	ND	ND	ND	ND
Apicidin	ND	ND	ND	ND	+	ND	ND	ND	+
Fusaristatins	+	+	+	+	+	+	+	Trace	+
Fusariellin A	ND	ND	ND	ND	ND	ND	ND	ND	ND
Ferricrocin	+	ND	+	ND	+	+	ND	ND	NA
Fusarinines	ND	ND	ND	ND	ND	ND	ND	ND	NA
JM-47	+	+	+	+	+	+	+	+	NA
Malonichrome	+	ND	+	ND	+	ND	ND	ND	NA
Other									
Fosfonochlorin	ND	ND	ND	ND	ND	ND	ND	ND	NA
Unknown 26 (NRPS)					+	+	+	+	
Fusarium unknown 31 (NRPS)					+	+	+	+	

ND not detected
NA not analyzed

Table 6. The number of identified signature genes for secondary metabolism in the three *F. avenaceum* strains compared to the other public *Fusarium* genome sequences.

	PKS I	PKS III	NRPS	TS I HT*	TS I HH**	TS II***	DMATS
Fa05001	25 (12)	2 (1)	25 (9)	13 (2)	2(0)	4(0)	4(2)
FaLH03	25 (12)	3 (2)	26 (10)	12 (1)	2(0)	4(0)	4(2)
FaLH27	27 (14)	2 (1)	28 (12)	13 (1)	2(0)	4(0)	4(2)

The number of genes that are unique for *F. avenaceum* is given in the parentheses. Gene classes: type I iterative PKS (PKS I), type III PKS (PKS III), non-ribosomal peptide synthetases (NRPS), terpene synthase class I head-to-tail type (TS I HT), terpene synthase class I head-to-head type (TS I HH), terpene synthases class II (TS II) and aromatic prenyltransferases (DMATS).

*incl. ERG20, COQ1, BTS1,
**incl. ERG9,
***incl. ERG7.

the broad host range of *F. avenaceum* is likely due to a generalized mechanism allowing it to cope with and overcome the innate immune response of plants, such as the generation of reactive oxygen species (ROS) [77].

Genes involved in signal transduction were also overrepresented in the transcriptome, including GO categories for GTP binding, ATP binding, calcium ion binding, and membrane activity (Figure S8 in File S1, File S8). Approximately 33% of all proteins predicted in the genome with Interpro "Tyrosine-protein kinase, catalytic domain" (IPR020635) were found in the transcriptome. This was one of the most highly enriched *F. avenaceum* categories when compared to other *Fusarium* genome sequences, and the *in planta* transcriptome hence supports the comparative results from the genome analysis. During plant infection, fungi need to monitor the nutrient status and presence of host defenses, and respond to or tolerate osmotic or oxidative stress, light and other environmental variables [78]. Stress-signaling/response genes of fungal pathogens are known to play important roles in virulence, pathogenesis and defense against oxidative burst (rapid production of ROS) from the host [79,80]. It is plausible to predict that tyrosine-protein kinases assist in the stress related response. There is a tendency that *F. avenaceum* isolates from one host can be pathogenic on other distantly related plants [11]. This is in contrast to, for example, *F. oxysporum*, a pathogen with a remarkably broad host range at the species level, but where individual isolates often cause disease only on one or a few plant species [81]. Our results support the chameleon nature of *F. avenaceum*, as it is capable of adapting to diverse hosts and environments. This lack of host specialization is likely to be a driving force in *F. avenaceum* evolution. Apart from the general functional categories, the *in planta* transcriptome of *F. avenaceum* also revealed many orthologs of *F. graminearum* pathogenicity and virulence factors, especially those involved in signal transduction and metabolism (Table S11 in File S1, [82]).

Conclusions

In summary, the comparative genomic analyses of *F. avenaceum* to other Fusaria point out several functional categories that are enriched in this fungal genome, and which indicate a great potential for *F. avenaceum* to sense and transduce signals from the surroundings, and to respond to the environment accordingly. *Fusarium avenaceum* has a large potential for redox, signaling and secondary metabolite production, and 20% of all predicted proteins were considered to be secreted. This could suggest that interaction with plant hosts is predominantly in the extracellular space. These genome sequences provide a valuable tool for the discovery of genes and mechanisms for bioactive compounds, and to increase our knowledge of the mechanisms contributing to a fungal lifestyle on diverse plant hosts and in different environments.

Materials and Methods

Sequencing, assembly, gene prediction and annotation

Fusarium avenaceum isolate Fa05001 (ARS culture collection: NRRL 54939, Bioforsk collection: 202103, DTU collection: IBT 41708) was isolated from barley in 2005 in Finland [83]. The strain was grown on complete medium [84] at room temperature for three days on a shaker, before mycelium was vacuum filtered and harvested for storage at $-80\,^{\circ}$C. DNA was isolated using the Qiagen DNeasy Plant Maxi. The sequencing and assembly was performed by Eurofins MGW, using a combination of shotgun (1.5 plate) and paired-end (¼ plate of 3 kb, and ¼ plate of 20 kb) 454 pyrosequencing. Newbler 2.6 (www.roche.com) was used for

Type I PKS

	Fg	Fv	Fo	Fs	Fp	Ff	Fa05001	FaLH03	FaLH27
Fg	1	5	4	5	12	5	10	10	10
Fv		2	9	5	4	12	5	5	5
Fo			0	2	10	5	5	5	5
Fs				7	4	6	4	4	4
Fp					1	4	8	8	8
Ff						4	5	5	5
Fa05001	Shared						0	24	24
FaLH03	Unique							0	25
FaLH27									2
Total	15	16 (+ 1)	14 (+2)	13	13	18 (+1)	25	25	27

Type III PKS

	Fg	Fv	Fo	Fs	Fp	Ff	Fa05001	FaLH03	FaLH27
Fg	0	1	1	1	1	1	1	1	1
Fv		0	1	1	1	1	1	1	1
Fo			0	1	1	1	1	1	1
Fs				1	1	1	1	1	1
Fp					0	1	1	1	1
Ff						0	1	1	1
Fa05001	Shared						0	2	2
FaLH03	Unique							1	2
FaLH27									0
Total	1	1	1	2	1	1	2	3	2

	Fg	Fv	Fo	Fs	Fp	Fa05001	FaLH03	FaLH27	Fl	FFUJ
PKSIII_1	FGSG_08378	FVEG_02316	FOXG_03444	FS123115	FPSE_08994	FAVG1_04458	FAVG2_03323	FAVG3_03340	Yes	FFUJ_05866
PKSIII_2						FAVG1_07768	FAVG2_04248	FAVG3_04292		
PKSIII_3							FAVG2_13006			
PKSIII_4				FS37360						

NRPSs

	Fg	Fv	Fo	Fs	Fp	Ff	Fa05001	FaLH03	FaLH27
Fg	1	10	9	10	16	10	14	14	14
Fv		3	13	9	9	14	10	10	10
Fo			3	9	9	12	10	10	10
Fs				4	10	9	10	10	10
Fp					1	9	13	13	13
Ff						1	10	10	10
Fa05001	Shared						2	22	23
FaLH03	Unique							1	25
FaLH27									2
Total	20	18	16	14	17	16	25	26	28

PKSs in *F. avenaceum*

NR-PKS	4
R-PKS	15 (+1)
PKS-Carn	2
PKS-NRPS	4 (+1)
NRPS-PKS	1
	26 (+2)

Terpene biosynthesis

	Fg	Fv	Fo	Fs	Fa05001	FaLH03	FaLH27	Fp	FFUJ
TS Class I (head-to-tail incl. cyclase) (incl. ERG20, COQ1, BTS1)	13	10 (+2)	10	5	13	12	13	15	13 (+1)
TS Class I (head-to-head) (incl. ERG9)	2	2	2	2	2	2	2	2	2
TS Class II (cyclases for TS class I HH) (incl. ERG7)	1	1	2	2	4	4	4	1	2
Aromatic prenyltransferases (DMATS)	0	3	3	(+1)	4	4	4	0	2
Protein prenyltransferases (Ftas)	1	1	1	1	1	1	1	1	1
Protein prenyltransferases (GGTase-I)	1	1	1	1	1	1	1	1	1

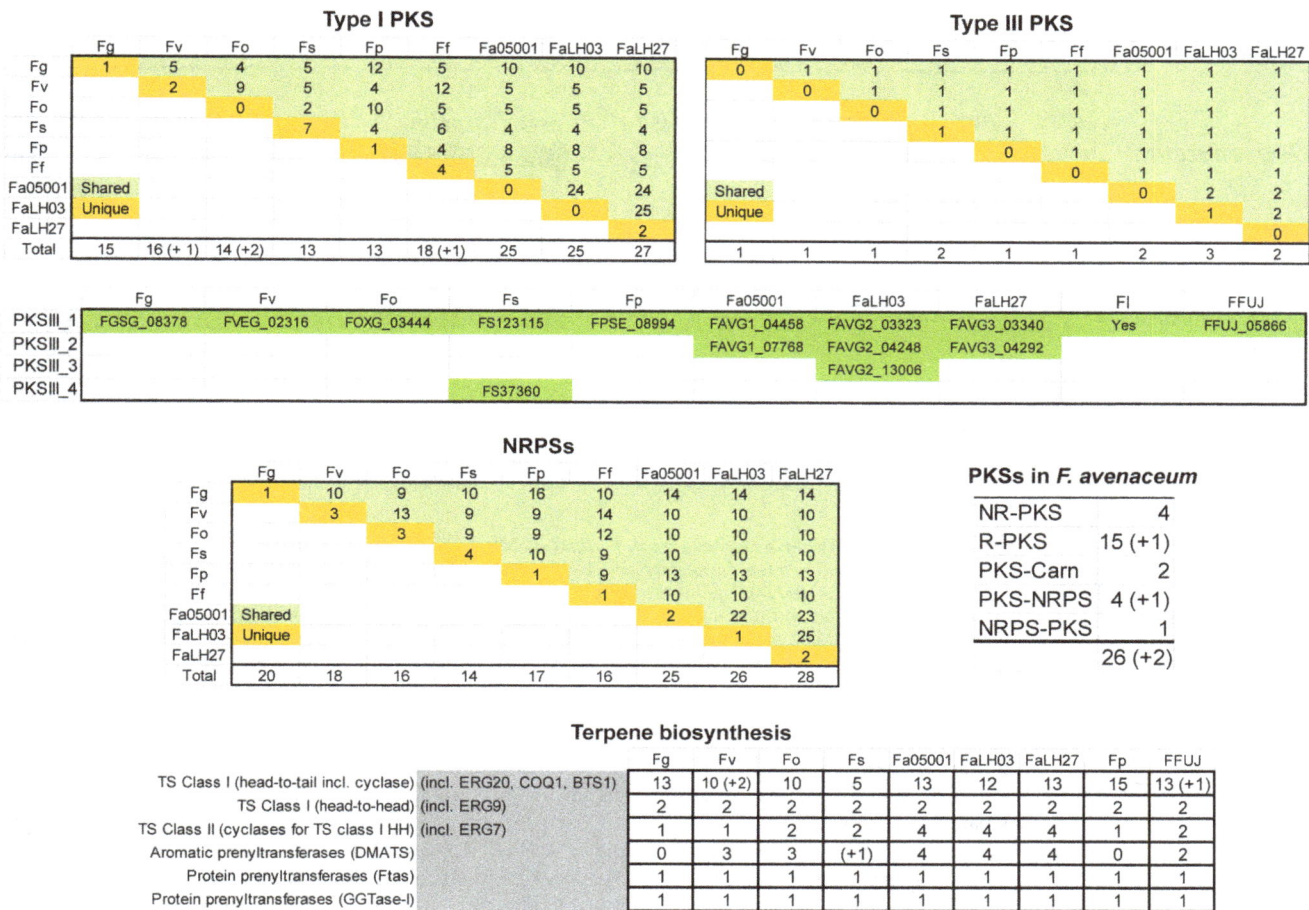

Figure 5. Shared and unique polyketide synthase (PKS), non-ribosomal peptide synthetases (NRPS) and terpene cyclase (TC) encoding genes in public available Fusaria genomes. Green and yellow boxes are the number of shared and unique genes, respectively. *Fg* = *F. graminearum*, *Fv* = *F. verticillioides*, *Fo* = *F. oxysporum*, *Fs* = *F. solani*, *Fp* = *F. pseudograminearum*, *Ff* = *F. fujikuroi* and *Fa05001*, *FaLH03* and *FaLH27* = *F. avenaceum*.

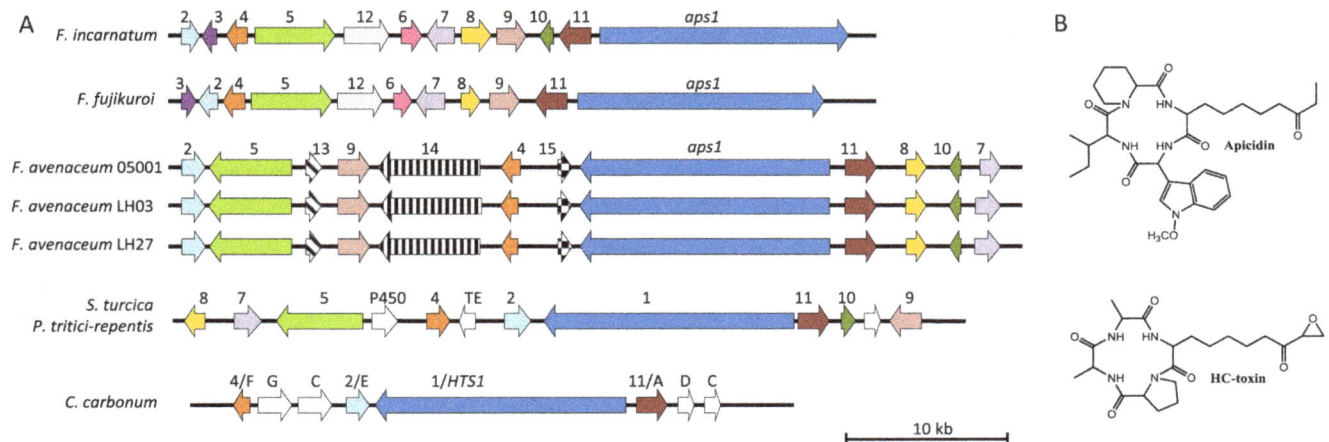

Figure 6. Apicidin-like gene cluster (oNRPS31) in the three *F. avenaceum* strains (Fa05001_Scaffold14, FaLH03_contig11, FaLH27_contig13) compared to the characterized apicidin gene cluster from *F. incarnatum* and the HC-toxin gene clusters from *Cochliobolus carbonum*, *Pyrenophora tritici-repentis* and *Setosphaeria turcica* (A). The genes are colored based on homology across the species. Chemical structure of apicidin and HC-toxin (B). See Table S8 in File S1 for further details.

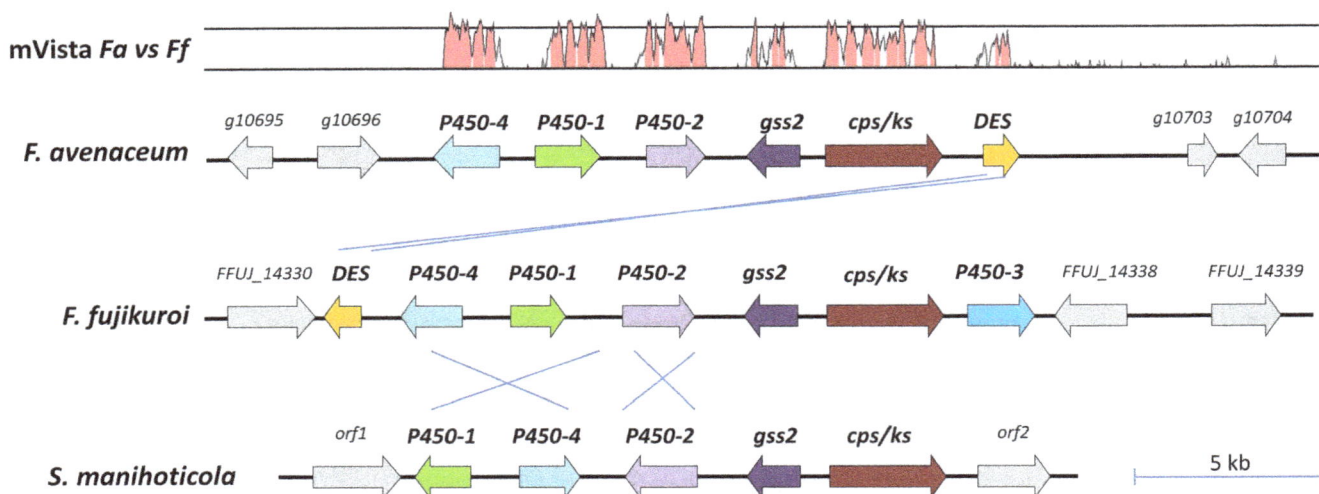

Figure 7. Architecture of the gibberellic acid (GA) gene clusters from *F. fujikuroi* MP-A, *F. avenaceum* and *Sphaceloma manihoticola*. The gene cluster and surrounding genes are identical in the three *F. avenaceum* strains and only Fa05001 is shown (FaLH03 cluster: *FAVG2_04186 - FAVG2_04192* and FaLH27 cluster: *FAVG3_04219 - FAVG3_04224*). The mVista trace shows the similarity over a 100 bp sliding windows (Shuffle-LAGAN plot) between the *F. avenaceum* and *F. fujikuroi* clusters, bottom line =50% and second line =75% identity. Genes: *gss2* = geranylgeranyldiphosphate synthase, *cps/ks* = copalyldiphosphate/ent-kaurene synthase, *P450-4* = ent-kaurene oxidase, *P450-1* = GA_{14} synthase, *P450-2* = C20-oxidase, *P450-3* = 13-hydroxylase and *DES* = desaturase. Note that the intergenic regions are unknown for the *S. manihoticola* GA cluster, while the size of these regions is not to drawn to size.

automatic assembly, and gap closure and further assembly was performed using GAP4 (Version 4.4; 2011, http://www.gap-system.org), a total of 50 primer pairs, and manual editing. PCR products from gaps were Sanger sequenced in both directions, and 38 PCR products were successfully integrated into the assembly. These approaches significantly improved the results, starting from 502 contigs and 89 scaffolds after automatic annotation to 110 contigs and 83 scaffolds (Table 1).

Two Canadian *F. avenaceum* strains, FaLH03 (spring wheat, New Brunswick) and FaLH27 (winter wheat, Nova Scotia), were isolated from wheat samples harvested in 2011 (Canadian Grain Commission, Winnipeg, MB) and deposited in the Canadian Collection of Fungal Cultures (AAFC, Ottawa, ON) with the strain designations DAOM242076 and DAOM242378, respectively. Species identification was confirmed by sequencing the *tef1-α* gene [85]. Strains were single-spored prior to any analysis and confirmed to retain virulence towards durum wheat. After growth for 3 days in glucose-yeast extract-peptone liquid culture, mycelia was collected and freeze-dried. Genomic DNA was extracted using the Nucleon Phytopure genomic DNA extraction kit (GE Healthcare, Baie d'Urfe, Québec) and then used to prepare an Illumina TruSeq library. Each library was sequenced in a single lane on an Illumina HiSeq platform (100 bp paired-end) at the Génome Québec Innovation Centre (Montréal, Québec), yielding 99,386,445 and 233,211,138 reads for FaLH03 and FaLH27, respectively. Reads were assembled in CLC Genomics Workbench 6.0.1. The higher order of the obtained scaffolds in the three isolates was resolved through comparison between the strains. The contigs from the Canadian isolates were ordered to each other by ABACAS [86]. A reiterative reordering approach was used to generate a stable order of the contigs. The successful reordering is illustrated in the alignment of the ordered contigs by MUMMER [87] (Figure S2 in File S1). The scaffolds from the Fa05001 were then ordered according to the Canadian isolates by ABACAS, and aligned to the Canadian isolates by 'MUMMER' (Figure S2 in File

S1). Contig/scaffold overlaps and boundaries in the 'MUMMER' alignments were used to determine the higher order assembly.

Gene prediction was performed with Augustus v2.5.5 (Fa05001) and v2.6 (FaLH03, FaLH27) [88], using default settings and *F. graminearum* as a training set. Protein sequences were annotated and enrichment analysis of gene ontology categories were compared using Blast2GO [89]. Table 1 shows the general statistics of the three genome sequences.

The species phylogenies were constructed based on 69 orthologous proteins from the included species. The 69 protein sets were first aligned individually using MUSCLE with default parameters [18], then manually inspected and concatenated. These super-protein alignments were then analyzed using MEGA6.0 by first identifying the best substitution model and then inferring the evolutionary history of the species using the Maximum Likelihood method, Nearest-Neighbour-Interchange, [90] using the LG+(F) substitution model, uniform substitution rate, removal of all positions with gaps and missing data and bootstrapping with 500 iterations.

Functional analysis and orthology prediction

For comparative genomics analysis, the previously sequenced genomes of *F. graminearum*, *F. verticillioides*, *F. oxysporum* f. sp. *lycopersici* 4287 (*Fusarium* Comparative Sequencing Project, Broad Institute of Harvard and MIT, http://www.broadinstitute.org/) and *F. solani* [91] were re-annotated using Blast2GO [89] concurrently with Fa05001. A functional comparison was performed using the Gene Ontology (GO) categories Biological Process, Molecular Function, and Cellular Compartments at level six, and Interpro. To compare transcription factors, we used the Interpro list from the Fungal Transcription Factor Database [92]. To identify orthologs, protein sets were compared in both directions with BLASTp to identify the best reciprocal hit for each individual protein with expectation values less than 1e-10. RepeatMasker [11] was used on all species to find repetitive sequences in the *Fusarium* genomes, using CrossMatch (http://

www.phrap.org/phredphrap/general.html). BLASTn was used to compare gene sequences between the three *F. avenaceum* genomes.

Electrophoretic karyotyping

Plugs containing 4×10^8/ml protoplasts were loaded on a CHEF gel (1% FastLane agarose [FMC BioProducts, Rockland, Maine] in 0.5×TBE) and ran for 255 hours, using switch times between 1200–4800 s at 1.8 V/cm. Chromosomes of *Schizosaccharomyces pombe* and *Hansenula wingei* were used as a molecular size marker (BioRad). Chromosomes separated in the gel were blotted to Hybond-N+. Southern hybridization was done overnight at 65°C using the CDP star method (GE Healthcare). A 350 bp *Hind*III-*Eco*RI fragment from plasmid pNLA17 was used as a probe, containing the *F. oxysporum* telomeric repeat TTAGGG 18 times.

Prediction of putative secretome

To determine the putative secretome, we employed a pipeline consisting of the following: A combination of WolfPsort (http://wolfpsort.org/), IPsort (http://ipsort.hgc.jp/) and SignalP4.1 [93] to identify subcellular localization and/or signalP motifs of all *F. avenaceum* Fa05001 proteins. Then, we used TMHMM [94] to predict all transmembrane domains. The secretome fasta file with the sequences was created with GPRO [95], with proteins predicted by either the PSORT tools or SignalP that does not contain transmembrane domains. Of the whole set of 13217 predicted proteins in Fa05001, a subset of 2580 sequences has been determined as the putative secretome of *F. avenaceum* (File S6).

Prediction of cysteine-rich proteins

Prediction of putative cysteine-rich proteins CRPs was based on a previously described approach [96]. In brief, this method is based on their expected sequence characteristics, with predicted small open reading frames (ORFs) (20 to 150 amino acids), containing at least four cysteine residues and a predicted signal peptide from the secretory pathway. The 83 Fa05001 scaffolds were used as input against GETORF available in the EMBOSS package [97]. We obtained 565.652 ORFs, which were translated to proteins using the tool Transeq [97]. All proteins were separated into two files (more/less than four cysteine residues). We downloaded a collection of 513 HMM profiles based on CRP models [98], and then created a single HMM database file that was formatted with HMMER3 [99] and used as subject against a HMMER comparison with the two files obtained. Nineteen candidates with more than four cysteine residues were predicted as CRP, and 55 additional sequences with less than four cysteine residues but with significant similarity to particular CRP HMM models were also included (Table S7 in File S1, File S7). We compared these to the supercontigs and scaffolds of the genome sequenced *Fusarium* spp. using BLASTn and TBLASTx, bit-score>50, and BLASTp in NCBI.

Identification of secondary metabolites

The three *F. avenaceum* strains were grown at 25°C in darkness for 14 days as triple point inoculations on Potato Dextrose Agar (PDA, [100]) and Yeast Extract Sucrose agar (YES, [100]). The metabolites of the strains were extracted using a modified version of the micro-scale extraction procedure for fungal metabolites [101]. Six 5-mm plugs from each plate taken across the colonies were transferred to 2 mL HPLC vials and extracted with 1.2 mL methanol:dichloromethane:ethyl acetate (1:2:3 v/v/v) containing 1% (v/v) formic acid. After 1 hr in an ultrasonication bath, extracts were evaporated with nitrogen, the residue dissolved in 150 µL acetonitrile:water (3/2 v/v) and filtered through a standard 0.45-µm PTFE filter.

Barley and oat samples (all in biological triplicates), including none-inoculated samples from climate chamber experiments described below were ground in liquid nitrogen and 50 mg extracted with 1 mL of 50% (vol) acetonitrile in water in a 2 mL Eppendorf tube. Samples were placed in an ultrasonication bath for 30 min, centrifuged at 15,000 g, and the supernatant was transferred to a clean 2 mL vial that was loaded onto the auto sampler prior to analysis. UHPLC-TOFMS analysis of 0.3–2 µL extracts were conducted on an Agilent 1290 UHPLC equipped with a photo diode array detector scanning 200–640 nm, and coupled to an Agilent 6550 qTOF (Santa Clara, CA, USA) equipped with a dual electrospray (ESI) source [102]. Separation was performed at 60°C at a flow rate of 0.35 mL/min on a 2.1 mm ID, 250 mm, 2.7 µm Agilent Poroshell phenyl hexyl column using a water-acetonitrile gradient solvent system, with both water and acetonitrile containing 20 mM formic acid. The gradient started at 10% acetonitrile and was increased to 100% acetonitrile within 15 min, maintained for 4 min, returned to 10% acetonitrile in 1 min. Samples were analyzed in both ESI$^+$ and ESI$^-$ scanning *m/z* 50 to 1700, and for automated data-dependent MS/MS on all major peaks, collision energies of 10, 20 and 40 eV for each MS/MS experiment were used. An MS/MS exclusion time of 0.04 min was used to get MS/MS spectra of less abounded ions.

Data files were analyzed in Masshunter 6.0 (Agilent Technologies) in three different ways: i) *Aggressive dereplication* [103] using lists of elemental composition and the *Search by Formula* (10 ppm mass accuracy) of all described *Fusarium* metabolites as well as restricted lists of only *F. avenaceum* and closely related species; ii) Searching the acquired MS/MS spectra in an in-house database of approx. 1200 MS/MS spectra of fungal secondary metabolites acquired at 10, 20 and 40 eV [102]; iii) all major UV/Vis and peaks in the base peak ion chromatograms not assigned to compounds (and not present in the media blank samples) were also registered. For absolute verification, authentic reference standards were available from 130 *Fusarium* compounds and additional 100 compounds that have been tentatively identified based on original producing strains using UV/Vis, LogD and MS/HRMS [102–104].

Identification of secondary metabolite genes

Type I iterative polyketide synthase (PKS), type III PKS, non-ribosomal peptide synthase (NRPS), aromatic prenyltranferase (DMATS) and class I & II terpene synthase encoding genes were identified by BLASTp using archetype representatives for the six types of genes [105]. Identification of orthologous genes was further supported by comparison to the genomic DNA, using the shuffle LAGAN algorithm with default settings [106,107]. Functional protein domains were identified using the NCBI CDD and pfam databases [108]. Domains specific to non-reducing PKSs, e.g. 'Product template' (PT) and 'Starter Acyl-Transferase' (SAT), were inferred via multiple sequence alignment with the bikaverin PKS (PKS16), which was one of the founding members of the domain group [109]. The nomenclature for PKS and NPS follows that which was introduced by Hansen et al. [63], as indicated by the use of the oPKSx and oNRPSx name, where the prefix 'o' signals that it refers to orthology-groups rather than the original overlapping names schemes used previously in each species. Following the idea regarding transparency in the names, introduced by Hansen and co-works, we applied a similar nomenclature scheme to the type III PKSs (oPKSIII_x) and the various enzyme classes involved in terpene biosynthesis: Terpene Synthase class I head-to-tail (oTS-I-HT_x), Terpene Synthase class I head-to-head (oTS-II-HH_x) and Terpene Synthase class II (oTS-II-x).

Climate chamber infection experiment

Fa05001 was grown on mung bean agar [110] for three weeks at room temperature under a combination of white and black (UVA) light with a 12 h photoperiod. Macroconidia were collected by washing the agar plate with 5 mL sterile distilled water, and diluted with 1.5% carboxymethylcellulose solution to a concentration of 5×10^4 conidia/mL for inoculation. Conidial concentration was determined using a Bürker hemacytometer.

Barley (Hordeum vulgare), cultivar Iron, and oat (Avena sativa) cultivar Belinda were grown in a climate chamber under the following conditions: Two weeks at 10°C/8°C 17 h/7 h 70%RH/60%RH, two weeks 15°C/12°C 18 h/6 h 70%RH/60%RH, three weeks 18°C/15°C 18 h/6 h 70%RH/60%RH and three weeks 20°C/15°C 17 h/6 h 70%RH/60%RH. During anthesis, approximately 1 mL of conidial solution was sprayed on each panicle, a bag was placed over the panicle and removed after 4 days. We used 6 plants per pot, 2 panicles per plant and 3 replicate pots per treatment. At sampling, panicles were immediately stored at −80°C.

Transcriptomics of Fa05001 on barley heads

Panicles from one pot grown in climate chamber experiment were mixed and ground in liquid nitrogen. RNA was extracted from 50 mg subsample from three biological replicates (pots) of untreated control (0 dpi) and F. avenaceum-inoculated tissue (14 dpi), using Spectrum plant total RNA kit (Sigma-Aldrich, Steinheim, Germany), with slight modifications. Due to the high amount of starch in barley heads at 14 dpi, the volume of lysis buffer and binding solution were increased from 500 µL to 750 µL per sample, and samples were incubated for 5 min at room temperature and the lysates were filtered 2 times for 10 minutes. On-column DNase digestion (Sigma-Aldrich, Steinheim, Germany) was used.

PolyA purification and fragmentation, cDNA synthesis, library preparation and 1×100 bp single read module (half a lane Hi-seq 2500) sequencing were done by Eurofins MGW. The resulting fastaq files were trimmed (quality score limit: 0.05, maximum number of ambiguities: 2), and RNA-seq was performed with predicted F. avenaceum genes using CLC Genomics Workbench 6.05, with stringent settings (minimum similarity fraction: 0.95, minimum length fraction: 0.9, maximum number of hits for a read: 10) to subtract host-specific transcripts. Gene expression was calculated using reads per kilobase per million (RPKM) values. A T-test was used to determine significant expression levels in the

biological replicates, comparing F. avenaceum inoculated samples against a control. Transcripts found solely in the F. avenaceum-inoculated plant were used to limit the amount of false positives coming from the host.

Supporting Information

File S1   Supplementary figures and tables.
(DOCX)

File S2   BLASTn results of the three F. avenaceum isolates Fa05001, FaLH03 and FaLH27.
(XLSX)

File S3   Venn diagram of BLASTn results corresponding to Figure S4.
(XLSX)

File S4   Interpro results of F. avenaceum Fa5001, F. graminearum, F. verticillioides, F. oxysporum and F. solani.
(XLSX)

File S5   Reciprocal blast of F. avenaceum Fa5001 vs F. graminearum, F. verticillioides, F. oxysporum and F. solani.
(XLSX)

File S6   Secretome of F. avenaceum Fa5001.
(XLSX)

File S7   Cysteine rich proteins in F. avenaceum Fa5001.
(TXT)

File S8   Summary of secondary metabolite genes.
(XLSX)

File S9   Transcriptome of F. avenaceum Fa5001 on barley heads.
(XLSX)

Acknowledgments

We thank Päivi Parikka, MTT, Finland for the F. avenaceum Fa05001 strain and Tom Gräfenhan, Canadian Grain Commission, for the two Canadian isolates FaLH03 and FaLH27 used for genome sequencing. We thank Danielle Schneiderman and Catherine Brown for technical support and Philippe Couroux for bioinformatics support.

Author Contributions

Conceived and designed the experiments: EL LH SW HD HCK WJ KN UT RF. Performed the experiments: EL LH SW RS HCK WJ AKK KN UT RF. Analyzed the data: EL LH SW RS CL TG KN UT RF. Contributed reagents/materials/analysis tools: EL LH SW RS CL TG AKK KN UT RF. Wrote the paper: EL LH SW HD ER TG HCK AKK KN UT RF.

References

1. Leach MC, Hobbs SLA (2013) Plantwise knowledge bank: delivering plant health information to developing country users. Learned Publ 26: 180–185.

2. Desjardins AE (2003) Gibberella from a (venaceae) to z (eae). Annu Rev Phytopathol 41: 177–198.

3. Satyaprasad K, Bateman GL, Read PJ (1997) Variation in pathogenicity on potato tubers and sensitivity to thiabendazole of the dry rot fungus Fusarium avenaceum. Potato Res 40: 357–365.

4. Mercier J, Makhlouf J, Martin RA (1991) Fusarium avenaceum, a pathogen of stored broccoli. Can Plant Dis Surv 71: 161–162.

5. Sørensen JL, Phipps RK, Nielsen KF, Schroers HJ, Frank J, et al (2009) Analysis of Fusarium avenaceum metabolites produced during wet apple core rot. J Agricult Food Chem 57: 1632–1639.

6. Peters RD, Barasubiye T, Driscoll J (2007) Dry rot of rutabaga caused by Fusarium avenaceum. Hortscience 42: 737–739.

7. Crous PW, Petrini O, Marais GF, Pretorius ZA, Rehder F (1995) Occurrence of fungal endophytes in cultivars of Triticum aestivum in South Africa. Mycoscience 36: 105–111.

8. Varvas T, Kasekamp K, Kullman B (2013) Preliminary study of endophytic fungi in timothy (Phleum pratense) in Estonia. Acta Myc 48: 41–49.

9. Yacoub A (2012) The first report on entomopathogenic effect of Fusarium avenaceum (Fries) Saccardo (Hypocreales, Ascomycota) against rice weevil (Sitophilus oryzae L.: Curculionidae, Coleoptera). J Entomol Acarol R 44: 51–55.

10. Makkonen J, Jussila J, Koistinen L, Paaver T, Hurt M, et al. (2013) Fusarium avenaceum causes burn spot disease syndrome in noble crayfish (Astacus astacus). J Invert Pat 113: 184–190.

11. Nalim FA, Elmer WH, McGovern RJ, Geiser DM (2009) Multilocus phylogenetic diversity of Fusarium avenaceum pathogenic on lisianthus. Phytopathology 99: 462–468.

12. Uhlig S, Jestoi M, Parikka P (2007) *Fusarium avenaceum* - The North European situation. Int J Food Microbiol 119: 17–24.

13. Kulik T, Pszczolkowska A, Lojko M (2011) Multilocus phylogenetics show high intraspecific variability within *Fusarium avenaceum*. Int J Mol Sci 12: 5626–5640.

14. Kohl J, de Haas BH, Kastelein P, Burgers SLGE, Waalwijk C (2007) Population dynamics of *Fusarium* spp. and *Microdochium nivale* in crops and crop residues of winter wheat. Phytopathology 97: 971–978.

15. Sørensen JL, Giese H (2013) Influence of carbohydrates on secondary metabolism in *Fusarium avenaceum*. Toxins 5: 1655–1663.

16. Sørensen L, Lysøe E, Larsen J, Khorsand-Jamal P, Nielsen K, et al. (2014) Genetic transformation of *Fusarium avenaceum* by *Agrobacterium tumefaciens* mediated transformation and the development of a USER-Brick vector construction system. BMC Mol Biol 15: 15. 10.1186/1471-2199-15-15.

17. Ma LJ, van der Does HC, Borkovich KA, Coleman JJ, Daboussi MJ, et al. (2010) Comparative genomics reveals mobile pathogenicity chromosomes in *Fusarium*. Nature 464: 367–373.

18. Marcet-Houben M, Gabaldón T (2009) The tree versus the forest: The fungal tree of life and the topological diversity within the yeast phylome. Plos One 4: e4357.

19. Al-Reedy RM, Malireddy R, Dillman CB, Kennell JC (2012) Comparative analysis of *Fusarium* mitochondrial genomes reveals a highly variable region that encodes an exceptionally large open reading frame. Fung Genet Biol 49: 2–14.

20. Fourie G, van der Merwe NA, Wingfield BD, Bogale M, Tudzynski B, et al. (2013) Evidence for inter-specific recombination among the mitochondrial genomes of *Fusarium* species in the *Gibberella fujikuroi* complex. BMC Genom 14: 1605.

21. Sato T, Taga M, Saitoh H, Nakayama T, Takehara T (1998) Karyotypic analysis of five *Fusarium* spp. causing wheat scab by fluorescence microscopy and fluorescence in situ hybridization. Int Con Plant Pat 2.2.48.

22. Cuomo CA, Guldener U, Xu JR, Trail F, Turgeon BG, et al. (2007) The *Fusarium graminearum* genome reveals a link between localized polymorphism and pathogen specialization. Science 317: 1400–1402.

23. O'Donnell K, Rooney AP, Proctor RH, Brown DW, McCormick SP, et al. (2013) Phylogenetic analyses of RPB1 and RPB2 support a middle Cretaceous origin for a clade comprising all agriculturally and medically important fusaria. Fung Genet Biol 52: 20–31.

24. Watanabe M, Yonezawa T, Lee K, Kumagai S, Sugita-Konishi Y, et al. (2011) Molecular phylogeny of the higher and lower taxonomy of the *Fusarium* genus and differences in the evolutionary histories of multiple genes. Bmc Evolutionary Biology 11: 332.

25. Lysøe E, Seong KY, Kistler HC (2011) The transcriptome of *Fusarium graminearum* during the infection of wheat. Mol Plant Microbe Interact 24: 995–1000.

26. Ma LJ, Geiser DM, Proctor RH, Rooney AP, O'Donnell K, et al. (2013) *Fusarium* pathogenomics. Annu Rev Microbiol 67: 399–416.

27. Martin SH, Wingfield BD, Wingfield MJ, Steenkamp ET (2011) Structure and evolution of the *Fusarium* mating type locus: New insights from the *Gibberella fujikuroi* complex. Fung Genet Biol 48: 731–740.

28. Holtz MD, Chang KF, Hwang SF, Gossen BD, Strelkov SE (2011) Characterization of *Fusarium avenaceum* from lupin in central Alberta: genetic diversity, mating type and aggressiveness. Can J Plant Pathol 33: 61–76.

29. Cook RJ (1967) *Gibberella avenacea* sp. n., perfect stage of *Fusarium roseum* f. sp. *cerealis* 'avenaceum'. Phytopathology 57: 732–736.

30. Booth C, Spooner BM (1984) *Gibberella avenacea*, teleomorph of *Fusarium avenaceum*, from stems of *Pteridium aquilinum*. T Brit Mycol Soc 82: 178–180.

31. Kerényi Z, Moretti A, Waalwijk C, Oláh B, Hornok L (2004) Mating type sequences in asexually reproducing *Fusarium* species. Appl Environ Microbiol 70: 4419–4423.

32. Lee J, Lee T, Lee YW, Yun SH, Turgeon BG (2003) Shifting fungal reproductive mode by manipulation of mating type genes: obligatory heterothallism of *Gibberella zeae*. Mol Microbiol 50: 145–152.

33. Tarailo-Graovac M, Chen N (2009) Using RepeatMasker to identify repetitive elements in genomic sequences. Curr Protoc Bioinformatics 4: 4–10.

34. Ashburner M, Ball CA, Blake JA, Botstein D, Butler H, et al. (2000) Gene Ontology: tool for the unification of biology. Nat Genet 25: 25–29.

35. Zheng P, Xia YL, Xiao GH, Xiong CH, Hu X, et al. (2011) Genome sequence of the insect pathogenic fungus *Cordyceps militaris*, a valued traditional chinese medicine. Genome Biol 12: R116.

36. Thomma BPHJ, Van Esse HP, Crous PW, De Wit PJGM (2005) *Cladosporium fulvum* (syn. *Passalora fulva*), a highly specialized plant pathogen as a model for functional studies on plant pathogenic *Mycosphaerellaceae*. Mol Plant Pathol 6: 379–393.

37. Rep M, van der Does HC, Meijer M, van Wijk R, Houterman PM, et al. (2004) A small, cysteine-rich protein secreted by *Fusarium oxysporum* during colonization of xylem vessels is required for I-3-mediated resistance in tomato. Mol Microbiol 53: 1373–1383.

38. Farman ML, Eto Y, Nakao T, Tosa Y, Nakayashiki H, et al. (2002) Analysis of the structure of the AVR1-CO39 avirulence locus in virulent rice-infecting isolates of *Magnaporthe grisea*. Mol Plant Microbe Interact 15: 6–16.

39. Marx F (2004) Small, basic antifungal proteins secreted from filamentous ascomycetes: a comparative study regarding expression, structure, function and potential application. Appl Microbiol Biotechnol 65: 133–142.

40. de Jonge R, Thomma BPHJ (2009) Fungal LysM effectors: extinguishers of host immunity? Trends Microbiol 17: 151–157.

41. Hershenhorn J, Park SH, Stierle A, Strobel GA (1992) *Fusarium avenaceum* as a novel pathogen of spotted knapweed and its phytotoxins, acetamido-butenolide and enniatin B. Plant Sci 86: 155–160.

42. Thrane U (1988) Screening for Fusarin C production by European isolates of *Fusarium* species. Mycotox Res 4: 2–10.

43. Morrison E, Kosiak B, Ritieni A, Aastveit AH, Uhlig S, et al. (2002) Mycotoxin production by *Fusarium avenaceum* strains isolated from Norwegian grain and the cytotoxicity of rice culture extracts to porcine kidney epithelial cells. J Agricult Food Chem 50: 3070–3075.

44. Takeuchi M, Nakajima M, Ogita T, Inukai M, Kodama K, et al. (1989) Fosfonochlorin, a new antibiotic with spheroplast forming activity. J Antibiot (Tokyo) 42: 198–205.

45. Hussein HM, Baxter M, Andrew IG, Franich RA (1991) Mycotoxin production by *Fusarium* species isolated from New Zealand maize fields. Mycopathologia 113: 35–40.

46. Chelkowski J, Manka M (1983) The ability of Fusaria pathogenic to wheat, barley and corn to produce zearalenone. Phytopathol Z 106: 354–359.

47. Chelkowski J, Visconti A, Manka M (1984) Production of trichothecenes and zearalenone by *Fusarium* species isolated from wheat. Nahrung 28: 493–496.

48. Chelkowski J, Golinski P, Manka M, Wiewiórowska M, Szebiotko K (1983) Mycotoxins in cereal grain. Part IX. Zearalenone and Fusaria in wheat, barley, rye and corn kernels. Die Nahrung 27: 525–531.

49. Ishii K, Sawano M, Ueno Y, Tsunoda H (1974) Distribution of zearalenone-producing *Fusarium* species in Japan. Appl Microbiol 27: 625–628.

50. Marasas W. F O., Nelson P E., Tousson T A. (1984) Toxigenic *Fusarium* species. Identity and mycotoxicology. University Park, Pennsylvania, U.S.A., 328p: Pennsylvania State University Press.

51. Lysøe E, Klemsdal SS, Bone KR, Frandsen RJN, Johansen T, et al. (2006) The *PKS4* gene of *Fusarium graminearum* is essential for zearalenone production. Appl Environ Microbiol 72: 3924–3932.

52. Kim YT, Lee YR, Jin JM, Han KH, Kim H, et al. (2005) Two different polyketide synthase genes are required for synthesis of zearalenone in *Gibberella zeae*. Mol Microbiol 58: 1102–1113.

53. Jiang Z, Barret MO, Boyd KG, Adams DR, Boyd ASF, et al. (2002) JM47, a cyclic tetrapeptide HC-toxin analogue from a marine *Fusarium* species. Phytochemistry 60: 33–38.

54. Laatch H. (2012) Antibase 2012: The natural compound identifier. Wiley-VCH Verlag GmbH.

55. Studt L, Wiemann P, Kleigrewe K, Humpf HU, Tudzynski B (2012) Biosynthesis of fusarubins accounts for pigmentation of *Fusarium fujikuroi* perithecia. Appl Environ Microbiol 78: 4468–4480.

56. Song ZS, Cox RJ, Lazarus CM, Simpson TJ (2004) Fusarin C biosynthesis in *Fusarium moniliforme* and *Fusarium venenatum*. Chembiochem 5: 1196–1203.

57. Malz S, Grell MN, Thrane C, Maier FJ, Rosager P, et al. (2005) Identification of a gene cluster responsible for the biosynthesis of aurofusarin in the *Fusarium graminearum* species complex. Fung Genet Biol 42: 420–433.

58. Seshime Y, Juvvadi PR, Fujii I, Kitamoto K (2005) Discovery of a novel superfamily of type III polyketide synthases in *Aspergillus oryzae*. Biochem Biophys Res Commun 331: 253–260.

59. Wiemann P, Sieber CMK, Von Bargen KW, Studt L, Niehaus EM, et al. (2013) Deciphering the cryptic genome: genome-wide analyses of the rice pathogen *Fusarium fujikuroi* reveal complex regulation of secondary metabolism and novel metabolites. Plos Pathog 9: e1003475.

60. von Döhren H (2004) Biochemistry and general genetics of nonribosomal peptide synthetases in fungi. Adv Biochem Engin/Biotechnol 88: 217–264.

61. Panaccione DG, Scottcraig JS, Pocard JA, Walton JD (1992) A cyclic peptide synthetase gene required for pathogenicity of the fungus *Cochliobolus carbonum* on maize. Proc Natl Acad Sci U S A 89: 6590–6594.

62. Jose B, Oniki Y, Kato T, Nishino N, Sumida Y, et al. (2004) Novel histone deacetylase inhibitors: cyclic tetrapeptide with trifluoromethyl and pentafluoro-oethyl ketones. Bioorg Med Chem Lett 14: 5343–5346.

63. Hansen FT, Sørensen JL, Giese H, Sondergaard TE, Frandsen RJN (2012) Quick guide to polyketide synthase and nonribosomal synthetase genes in *Fusarium*. Int J Food Microbiol 155: 128–136.

64. Jin JM, Lee S, Lee J, Baek SR, Kim JC, et al. (2010) Functional characterization and manipulation of the apicidin biosynthetic pathway in *Fusarium semitectum*. Mol Microbiol 76: 456–466.

65. Condon BJ, Leng YQ, Wu DL, Bushley KE, Ohm RA, et al. (2013) Comparative genome structure, secondary metabolite, and effector coding capacity across *Cochliobolus* pathogens. Plos Genet 9: e1003233.

66. Manning VA, Pandelova I, Dhillon B, Wilhelm LJ, Goodwin SB, et al. (2013) Comparative genomics of a plant-pathogenic fungus, *Pyrenophora tritici-repentis*, reveals transduplication and the impact of repeat elements on pathogenicity and population divergence. G3-Genes Genom Genet 3: 41–63.

67. Niehaus EM, Janevska S, von Bargen KW, Sieber CMK, Harrer H, et al. (2014) Apicidin F: Characterization and genetic manipulation of a new secondary metabolite gene cluster in the rice pathogen *Fusarium fujikuroi*. Plos One 9: e103336. doi: 10.1371/journal.pone.0103336.

68. Lee SL, Floss HG, Heinstein P (1976) Purification and properties of dimethylallylpyrophosphate - tryptophan dimethylallyl transferase, first enzyme of ergot alkaloid biosynthesis in *Claviceps*. sp. SD 58. Arch Biochem Biophys 177: 84–94.

69. Young CA, Tapper BA, May K, Moon CD, Schardl CL, et al. (2009) Indole-diterpene biosynthetic capability of *Epichloe* endophytes as predicted by *ltm* gene analysis. Appl Environ Microbiol 75: 2200–2211.

70. Bomke C, Tudzynski B (2009) Diversity, regulation, and evolution of the gibberellin biosynthetic pathway in fungi compared to plants and bacteria. Phytochemistry 70: 1876–1893.

71. Rim SO, You YH, Yoon H, Kim YE, Lee JH, et al. (2013) Characterization of gibberellin biosynthetic gene cluster from *Fusarium proliferatum*. J Microbiol Biot 23: 623–629.

72. Malonek S, Rojas MC, Hedden P, Gaskin P, Hopkins P, et al. (2005) Functional characterization of two cytochrome P450 monooxygenase genes, P450-1 and P450-4, of the gibberellic acid gene cluster in *Fusarium proliferatum* (*Gibberella fujikuroi* MP-D). Appl Environ Microbiol 71: 1462–1472.

73. Bomke C, Rojas MC, Gong F, Hedden P, Tudzynski B (2008) Isolation and characterization of the gibberellin biosynthetic gene cluster in *Sphaceloma manihoticola*. Appl Environ Microbiol 74: 5325–5339.

74. Kawaide H (2006) Biochemical and molecular analyses of gibberellin biosynthesis in fungi. Biosci Biotechnol Biochem 70: 583–590.

75. Gaffoor I, Brown DW, Plattner R, Proctor RH, Qi WH, et al. (2005) Functional analysis of the polyketide synthase genes in the filamentous fungus *Gibberella zeae* (anamorph *Fusarium graminearum*). Eukaryot Cell 4: 1926–1933.

76. Karányi Z, Holb I, Hornok L, Pócsi I, Miskei M (2013) FSRD: fungal stress response database. Database (Oxford) 2013: bat037.

77. Plancot B, Santaella C, Jaber R, Kiefer-Meyer MC, Follet-Gueye ML, et al. (2013) Deciphering the responses of root border-like cells of *Arabidopsis* and flax to pathogen-derived elicitors. Plant Physiol 163: 1584–1597.

78. Kosti I, Mandel-Gutfreund Y, Glaser F, Horwitz BA (2010) Comparative analysis of fungal protein kinases and associated domains. BMC Genom 11: 1133.

79. Hamilton AJ, Holdom MD (1999) Antioxidant systems in the pathogenic fungi of man and their role in virulence. Med Mycol 37: 375–389.

80. de Dios CH, Roman E, Monge RA, Pla J (2010) The role of MAPK signal transduction pathways in the response to oxidative stress in the fungal pathogen *Candida albicans*: Implications in virulence. Curr Protein Pept Sci 11: 693–703.

81. Dean R, van Kan JAL, Pretorius ZA, Hammond-Kosack KE, Di Pietro A, et al. (2012) The Top 10 fungal pathogens in molecular plant pathology. Mol Plant Pathol 13: 414–430.

82. Urban M, Hammond-Kosack KE (2013) Molecular genetics and genomic approaches to explore *Fusarium* infection of wheat floral tissue. In: Brown DW, Proctor RH, editors.*Fusarium*: Genomics, Molecular and Cellular Biology.-Norfolk, UK: Caister Academic Press. pp.43–79.

83. Kokkonen M, Ojala L, Parikka P, Jestoi M (2010) Mycotoxin production of selected *Fusarium* species at different culture conditions. Int J Food Microbiol 143: 17–25.

84. Harris SD, Morrell JL, Hamer JE (1994) Identification and characterization of *Aspergillus nidulans* mutants defective in cytokinesis. Genetics 136: 517–532.

85. Geiser DM, Jimenez-Gasco MD, Kang SC, Makalowska I, Veeraraghavan N, et al. (2004) FUSARIUM-ID v. 1.0: A DNA sequence database for identifying Fusarium. Eur J Plant Pathol 110: 473–479.

86. Assefa S, Keane TM, Otto TD, Newbold C, Berriman M (2009) ABACAS: algorithm-based automatic contiguation of assembled sequences. Bioinformatics 25: 1968–1969.

87. Kurtz S, Phillippy A, Delcher AL, Smoot M, Shumway M, et al. (2004) Versatile and open software for comparing large genomes. Genome Biol 5.

88. Stanke M, Diekhans M, Baertsch R, Haussler D (2008) Using native and syntenically mapped cDNA alignments to improve de novo gene finding. Bioinformatics 24: 637–644.

89. Conesa A, Gotz S, Garcia-Gomez JM, Terol J, Talon M, Robles M (2005) Blast2GO: a universal tool for annotation, visualization and analysis in functional genomics research. Bioinformatics 21: 3674–3676.

90. Le SQ, Gascuel O (2008) An improved general amino acid replacement matrix. Mol Biol Evol 25: 1307–1320.

91. Coleman JJ, Rounsley SD, Rodriguez-Carres M, Kuo A, Wasmann CC, et al. (2009) The genome of *Nectria haematococca*: contribution of supernumerary chromosomes to gene expansion. Plos Genet 5: e1000618.

92. Park J, Park J, Jang S, Kim S, Kong S, et al. (2008) FTFD: an informatics pipeline supporting phylogenomic analysis of fungal transcription factors. Bioinformatics 24: 1024–1025.

93. Petersen TN, Brunak S, von Heijne G, Nielsen H (2011) SignalP 4.0: discriminating signal peptides from transmembrane regions. Nature Methods 8: 785–786.

94. Krogh A, Larsson B, von Heijne G, Sonnhammer ELL (2001) Predicting transmembrane protein topology with a hidden Markov model: Application to complete genomes. J Mol Biol 305: 567–580.

95. Futami R, Muñoz-Pomer L, Viu JM, Dominguez-Escriba L, Covelli L, et al. (2011) GPRO: The professional tool for annotation, management and functional analysis of omics databases. Biotechvana Bioinformatics 2011-SOFT3 2011.

96. Marcet-Houben M, Ballester AR, de la Fuente B, Harries E, Marcos JF, et al. (2012) Genome sequence of the necrotrophic fungus *Penicillium digitatum*, the main postharvest pathogen of citrus. BMC Genom 13: 646.

97. Rice P, Longden I, Bleasby A (2000) EMBOSS: the European Molecular Biology Open Software Suite. Trends Genet 16: 276–277.

98. Silverstein KAT, Moskal WA, Wu HC, Underwood BA, Graham MA, et al. (2007) Small cysteine-rich peptides resembling antimicrobial peptides have been under-predicted in plants. Plant J 51: 262–280.

99. Finn RD, Clements J, Eddy SR (2011) HMMER web server: interactive sequence similarity searching. Nucleic Acids Res 39: W29–W37.

100. Samson RA, Houbraken J, Thrane U, Frisvad JC, Andersen B (2010) Food and indoor fungi. Utrecht: CBS-KNAW Fungal Biodiversity Centre.

101. Smedsgaard J (1997) Micro-scale extraction procedure for standardized screening of fungal metabolite production in cultures. J Chromatogr A 760: 264–270.

102. Kildgaard S, Månsson M, Dosen I, Klitgaard A, Frisvad JC, et al. (2014) Accurate dereplication of bioactive secondary metabolites from marine-derived fungi by UHPLC-DAD-QTOFMS and a MS/HRMS Library. Mar Drugs 12: 3681–3705.

103. Klitgaard A, Iversen A, Andersen MR, Larsen TO, Frisvad JC, et al. (2014) Aggressive dereplication using UHPLC-DAD-QTOF: screening extracts for up to 3000 fungal secondary metabolites. Anal Bioanal Chem 406: 1933–1943.

104. Nielsen KF, Månsson M, Rank C, Frisvad JC, Larsen TO (2011) Dereplication of microbial natural products by LC-DAD-TOFMS. J Nat Prod 74: 2338–2348.

105. Altschul SF, Madden TL, Schaffer AA, Zhang JH, Zhang Z, et al. (1997) Gapped BLAST and PSI-BLAST: a new generation of protein database search programs. Nucleic Acids Res 25: 3389–3402.

106. Frazer KA, Pachter L, Poliakov A, Rubin EM, Dubchak I (2004) VISTA: computational tools for comparative genomics. Nucleic Acids Res 32: W273–W279.

107. Brudno M, Malde S, Poliakov A, Do CB, Couronne O, et al. (2003) Global alignment: finding rearrangements during alignment. Bioinformatics 19: i54–i62.

108. Marchler-Bauer A, Lu SN, Anderson JB, Chitsaz F, Derbyshire MK, et al. (2011) CDD: a conserved domain database for the functional annotation of proteins. Nucleic Acids Res 39: D225–D229.

109. Crawford JM, Dancy BCR, Hill EA, Udwary DW, Townsend CA (2006) Identification of a starter unit acyl-carrier protein transacylase domain in an iterative type I polyketide synthase. Proc Natl Acad Sci U S A 103: 16728–16733.

110. Dill-Macky R (2003) Inoculation methods and evaluation of *Fusarium* head blight resistance in wheat. In: Leonard KJ, Bushnell WR, editors.Fusarium head blight of wheat and barley. pp.184–210.

16

Biochemical and Molecular Characterization of Barley Plastidial ADP-Glucose Transporter (HvBT1)

Atta Soliman[1,2], Belay T. Ayele[1], Fouad Daayf[1]*

1 Department of Plant Science, Faculty of Agricultural and Food Sciences, University of Manitoba, Winnipeg, Manitoba, Canada, **2** Department of Genetics, Faculty of Agriculture, University of Tanta, Tanta, El-Gharbia, Egypt

Abstract

In cereals, ADP-glucose transporter protein plays an important role in starch biosynthesis. It acts as a main gate for the transport of ADP-glucose, the main precursor for starch biosynthesis during grain filling, from the cytosol into the amyloplasts of endospermic cells. In this study, we have shed some light on the molecular and biochemical characteristics of barley plastidial ADP-glucose transporter, *HvBT1*. Phylogenetic analysis of several BT1 homologues revealed that BT1 homologues are divided into two distinct groups. The HvBT1 is assigned to the group that represents BT homologues from monocotyledonous species. Some members of this group mainly work as nucleotide sugar transporters. Southern blot analysis showed the presence of a single copy of *HvBT1* in barley genome. Gene expression analysis indicated that *HvBT1* is mainly expressed in endospermic cells during grain filling; however, low level of its expression was detected in the autotrophic tissues, suggesting the possible role of HvBT1 in autotrophic tissues. The cellular and subcellular localization of *HvBT1* provided additional evidence that HvBT1 targets the amyloplast membrane of the endospermic cells. Biochemical characterization of *HvBT1* using *E. coli* system revealed that HvBT1 is able to transport ADP-glucose into *E. coli* cells with an affinity of 614.5 µM and in counter exchange of ADP with an affinity of 334.7 µM. The study also showed that AMP is another possible exchange substrate. The effect of non-labeled ADP-glucose and ADP on the uptake rate of [α-^{32}P] ADP-glucose indicated the substrate specificity of HvBT1 for ADP-glucose and ADP.

Editor: Eugene A. Permyakov, Russian Academy of Sciences, Institute for Biological Instrumentation, Russian Federation

Funding: AS was financially supported by a Ph.D. scholarship from the ministry of Higher Education, Egypt. BTA and FD received grant #ARDI-10-1042 from Agri-Food Research and Development Initiative, Manitoba, Canada. The funders had no role in study design, data collection and analysis, decision to publish, or preparation of the manuscript.

Competing Interests: The authors have declared that no competing interests exist.

* E-mail: Fouad.Daayf@umanitoba.ca

Introduction

Starch is the main storage compound in grains of cereals. Its biosynthesis is catalyzed by a number of enzymes, including ADP-glucose Pyrophosphorylase (AGPase) that converts glucose-1-phosphate, using ATP, into ADP-glucose (ADP-Glc). The ADP-Glc acts as the building block of starch. AGPase is located in the endospermic cells of cereal grains and about 85% to 95% of its total activity is found in the cytosol of such cells [1]. The majority of ADP-Glc is synthesized in the cytosol and imported into the amyloplasts by the ADP-glucose transporter located on the membrane of the amyloplasts. In wheat, the ADP-glucose transporter was characterized *in vitro* using reconstituted amyloplasts envelope proteins in proteoliposomes. The results of this study showed that ADP-glucose is transported in counter exchange of AMP and ADP [2]. In maize, the brittle1 (BT1) mutant is deficient in four amyloplasts envelope proteins, including ZmBT1, which is identified as an ADP-glucose transporter [3,4]. This mutant showed a lower rate of ADP-Glc uptake into isolated amyloplasts as compared to the control ones [5]. Using the radioactive (^{14}C) labeled ADP-glucose as a substrate; the ZmBT1 heterologously expressed in *E. coli* was able to transport ADP-glucose with high affinity in counter exchange with ADP [6]. In

vitro starch synthesis in amyloplasts isolated from Risø13, a *lys5* mutant of barley generated by EMS mutagenesis, is suppressed as a result of defect in a major plastidial protein, which was identified as ADP-glucose transporter and designated as HvNST1 [7] and HvBT1 [8]. In non-graminaceous species, two BT1 homologues were identified and characterized, *AtBT1* from *Arabidopsis thaliana*, and *StBT1* from *Solanum tuberosum*. Characterization of *StBT1* using an *E. coli* expression system showed high affinity to AMP, ADP and ATP [9].

Phylogenetic analysis of the BT1 proteins revealed that ZmBT1 and HvNST1 belong to the mitochondrial carrier family (MCF) [7,10]. In cereals, ADP-glucose transporter protein possesses six membrane spanning domains, and the C-terminus and N-terminus of these proteins are located inside the amyloplasts [7]. It was presumed that BT1 homologues are localized in plastids' membranes in autotrophic and heterotrophic tissues and are involved in transporting nucleotides or nucleotide sugars. Recently, subcellular localization analysis of maize and Arabidopsis plants expressing *ZmBT1* and *AtBT1* homologues, respectively, showed dual localization of the two MCF members on the mitochondrial as well as the plastidial membranes. It has also been shown that the transit peptide in the N-terminal targets plastids, while the sequence targeting to the mitochondria is localized within internal domains [11].

The present study characterized barley ADP-glucose transporter, HvBT1, using *E. coli* expression system via monitoring the direct transport of ADP-Glc through the intact *E. coli* cells' plasma membrane. The expression of *HvBT1* in different tissues and the cellular and subcellular localizations of HvBT1 were investigated. In addition the study examined the effect of expressed HvBT1 protein on *E. coli* cell growth and protein expression.

Materials and Methods

Plant material

Barley (*Hordeum vulgare L.*) cv. Harrington plants were grown in a greenhouse (under 8 h dark/ 16 h light photoperiod). Spikes were harvested at 2, 4, 8, and 10 days after anthesis (DAA) and frozen immediately in liquid nitrogen until used for RNA extraction.

Amplification and cloning of *HvBT1*

RNA was extracted using RNeasy Plant Mini Kit (Qiagen, Hilden, Germany). The genomic DNA was digested in columns using RNase-Free DNase I Kit (Qiagen). The purity and integrity of the RNA was determined using agarose gel electrophoresis and NanoDrop spectrophotometer (Thermo Fisher Scientific, Waltham, MA, USA), respectively. The cDNA was synthesized using Revert Aid First Strand cDNA Synthesis Kits (Thermo Scientific). The open reading frame (ORF) of *HvBT1* was amplified using gene specific primers, which were designed based on the reported DNA sequence (GenBank ID: AY560327.2). The amplified ORF was cloned in the pGEM-T-Easy plasmid and introduced into *E. coli* DH5α (Invitrogen, Carlsbad, CA, USA). Following plasmid isolation, the target sequence was verified by sequencing (Macrogen, Rockville, MD, USA) and blast searched against the GenBank database. The resulting ORF was amplified using specific primers flanked by restriction sites for Nde1 (5′-CGTcatatgGCGGCGGCAAT-3′) and BamH1 (5′-TAggatccT-CATGGTCGATCACCG-3′) and cloned into the bacterial expression plasmid pET16b (Novagen, Darmstadt, Germany) in Nde1 and BamH1 restriction sites and further verified by DNA sequencing.

Southern blot analysis

Genomic DNA was extracted using the CTAB method from 2 g frozen leaf tissue and treated with RNase A to eliminate RNA contamination. The genomic DNA was digested using XhoI, BamHI, SalI and KpnI (all do not cut *HvBT1* sequence). The digested products were then separated on 0.8% agarose gel. Denaturation and neutralization processes were performed according to the standard protocol [12]. After transferring the DNA into the nylon membrane (GE Healthcare, Little Chalfont, UK), DNA fragments were fixed into the membrane using UV crosslinker (Stratagene, La Jolla, CA, USA). Gene Images Alkphos Direct Labeling and Detection system (GE Healthcare) was used for preparation of the probe (700 bp). Pre-hybridization, hybridization, washing, and all other procedures were performed according to the manufacturer's instructions. The chemiluminescent signal generated was detected using CDP-Star reagent.

Quantitative real-time RT- PCR (qRT-PCR) analysis of *HvBT1*

Quantitative Real-Time RT-PCR was performed using C1000 Thermal Cycler (Bio-Rad, Herculus, CA, USA) and SSo Fast Eva Green Super mix (Bio-Rad) according to the manufacturer's instructions. *β-Actin* was used as a reference gene. The *HvBT1*

specific sense, (5′-TGTACGACAACCTCCTCCAC-3′); and antisense, (5′-GCAGTGTCTCGTAGGCGTAG-3′) primers and *β-Actin* sense, (5′-CCAAAAGCCAACAGAGAGAA-3′) and antisense, (5′-GCTGACACCATCACCAGAG-3′) primers were used for qRT-PCR. The results were analyzed using $2^{-\Delta\Delta C_T}$ method [13].

Cellular localization of *HvBT1* transcripts

RNA in-situ hybridization was used to investigate the cellular localization of *HvBT1*. Barley caryopsis (8 DAA) were collected and immersed immediately in 4% paraformaldehyde in PBS buffer (pH 7.4) on ice. The caryopsis was cut from both ends to facilitate buffer penetration and exchange. Then the tissue was purged under vacuum for 15 min. The rest of the procedures were performed as described previously [14]. The probe was prepared using DIG-RNA labeling Kit (Roche Applied Science, Mannheim, Germany). Briefly, the full length cDNA was amplified and cloned into pGEM-T-Easy plasmid (Promega, Madison, WI, USA). The ORF was amplified using M13 forward and reverse primers located in the upstream and downstream of T7 and SP6 promoter regions, respectively. The PCR products were used as templates for T7 and SP6 RNA polymerase reactions. Sense and antisense probes were hydrolyzed using bicarbonate buffer (60 mM Na_2Co_3 and 40 mM $NaHCo_3$) to reach 150 bp in length. Pre-hybridization, hybridization, and color development were performed as described previously [15,16].

Subcellular localization of HvBT1 protein

Subcellular localization of HvBT1 was investigated using transient expression of HvBT1 fused with YFP in *Arabidopsis thaliana* protoplast. Plant binary vector pEarleyGate101 containing yellow fluorescent protein (YFP) was prepared using Gateway LR clonase reaction (Invitrogen). The coding sequence of *HvBT1* was cloned into pENTR-1A entry plasmid at EcoRI and XhoI restriction sites. The *HvBT1* specific primers were designed to include the restriction sites of EcoRI and XhoI. The stop codon of *HvBT1* ORF was removed by excluding it from the reverse primer of *HvBT1* to generate continuous ORF that includes both *HvBT1* and *YFP* in the binary plasmid. Phusion High-Fidelity DNA polymerase was used to amplify the coding sequence of *HvBT1* from pGEM-T-Easy plasmid. The PCR product was digested using EcoRI and XhoI. The digested PCR product was cloned in pENTR-1A plasmid using T4 DNA ligase. The LR clonase reaction was performed using pENTR-1A:*HvBT1* and pEarley-Gate101 according to the manufacturer's instruction. The insertion and orientation of *HvBT1* in pEarleyGate101::*HvBT1* were verified by PCR and sequencing (Macrogen). The transient expression was performed by introducing pEarleyGate101::*HvBT1* plasmid into Arabidopsis protoplasts using PEG-mediated transformation approach [17]. The immunolocalization of HvBT1::YFP in the fixed Arabidopsis protoplasts was performed using anti-YFP Tag (mouse monoclonal) primary antibody with dilution of 1:1000. Fluorescein conjugated anti-Mouse IgG (FITC) was used as secondary antibody with dilution of 1:200. The immunolocalization procedures were performed as previously described [18]. The HvBT1-YFP expression was monitored using confocal laser scanning microscope (Zeiss, Oberkochen, Germany). Excitation wavelengths of 488 nm and broad pass 505–530 nm, and 488 nm and long pass 650 nm were used for HvBT1::YFP and chlorophyll, respectively. FITC was detected with excitation of 495 nm and broad pass of 528 nm according to the manufacturer's instructions.

HvBT1 codon optimization

In *E. coli*, the rare codons (RCs) are defined as synonymous codons, which are decoded by low abundant tRNA (rare tRNA). The RCs affect the translation rate of eukaryotic proteins due to the unavailability of the rare tRNA [19]. One of the most limiting steps during expression of eukaryotic genes in *E. coli* is the codon bias. The low usage codons for particular amino acids disrupt the expression of most eukaryotic genes. Gene synthesis methods are able to optimize the DNA sequences by substituting rare codons with other ones that are common in *E. coli*'s translation machinery system. The four problematic amino acids (Proline, Arginine, Leucine and Isoleucine) have more than one codon, thus the challenges are to choose the right codons for these amino acids that can be recognized by the *E. coli* translation machinery and provide the same amino acid sequence up on translation. To this end, the codons of the nucleotide sequence of the target gene were optimized to *E. coli* translation system. The newly synthesized sequence was cloned into pET16b plasmid (Novagen), which has His- tag at the N- terminus to produce fused protein that facilitates protein purification.

Escherichia coli strains and growth conditions

Two *E. coli* strains, Rosetta2 (DE3) plysS and C43 (DE3) (Lucigen, Middleton, WI, USA) were used in this study to investigate the optimal conditions for protein expression and the possible toxic effects of the ADP-glucose transporter protein (HvBT1). Bacterial transformation was performed according to the strain specificity and the manufacturer instructions. Bacterial cells harboring pET16b::*HvBT1*, the original (org) or the optimized sequences (opc), and the control cells harboring the empty pET16b plasmid were grown at 37°C overnight in 5 ml 2xYT medium supplied with 100 μg/ml carbenicillin. This culture was used to inoculate fresh 2xYT culture medium which was then incubated at 37°C for 2 h to reach OD 600 = 0.6. The expression of *HvBT1* was then induced by adding IPTG (final concentration 0.5 mM) and incubation at 20°C overnight. The cells were centrifuged at 5000 *g* and resuspended in the transport buffer (50 mM potassium phosphate, pH 7.2) followed by incubation on ice.

Inhibitory effect of the expressed HvBT1

Escherichia coli cells of C43 and Rosetta2 strains harboring different copies of *HvBT1* (org or opc) were grown in 2xYT medium supplied with 100 μg/mL carbenicillin. After the cell density reaches to OD_{600} ~0.6, the expression of *HvBT1* was induced by adding IPTG (final concentration of 0.5 mM) and incubation at 20°C. The growth rate of cells was monitored over time using spectrophotometer. The same procedures were repeated using 2xYT agar plates. A single colony was spread on agar plates supplied with IPTG (0.5 mM final concentration) and without IPTG as a control, and then incubated at 30°C for 2 days.

Membrane protein extraction and SDS-PAGE

His-tagged HvBT1 protein was extracted and purified with HOOK His Protein Spin purification kit (G-Biosciences, St. Louis, MO, USA) according to the manufacturer's instructions with minor modification. The cell pellets were placed in liquid nitrogen to disrupt the cell wall and then re-suspended in PE LB–Lysozyme buffer provided with the kit with complete protease inhibitor (Roche Applied Science). The cells were incubated at 37°C for 60 min and further disrupted by ultrasonication (250 W, 4×30 s, 4°C). Unbroken cells and cell debris were collected using centrifugation at 4°C, first at 15,000 *g* for 20 min and then at

50,000 *g* for 90 min. The membrane protein fractions (supernatant) were mixed with nickel chelating resin. The rest of the procedures were performed according to the manufacturer's instructions. After elution, His-tagged HvBT1 protein was desalted with Sephadex G50 and the purified membrane protein was subjected to SDS-PAGE analysis as described previously [20].

Transport assay using E. coli C43 (DE3) strain harboring pET16b::HvBT1 (opc)

The transport assay was performed using intact *E. coli* cells. IPTG-induced *E. coli* cells harboring either *HvBT1* recombinant plasmid or the empty one (control) were resuspended in potassium phosphate buffer (pH 7.2). The transport assay was initiated by adding [α-32P] ADP-Glc or [α-32P] ADP as transport substrates to the reaction mixture. [α-32P] ADP-Glc was enzymatically synthesized using a modified protocol [21,22]. The procedures of the assay were as follow: ADP-glucose pyrophosphorylase (AGPase) was extracted from the leaf tissues of two-week-old barley seedlings (0.5 g) with extraction buffer containing 50 mM HEPES (pH 7.4), 2 mM $MgCl_2$, 1 mM EDTA and 1 mM DTT. The mixture was centrifuged at 10,000 *g* for 15 min at 4°C. The supernatant was used directly in the assay. The direct ADP-glucose synthesis was conducted in assay buffer (0.2 mL) containing 100 mM HEPES (pH 7.6), 15 mM $MgCl_2$, 0.025% (w/v) BSA, 0.5 units inorganic pyrophosphatase from baker's yeast (Invitrogen), 0.5 glucose-1-phosphate, 1.5 mM ATP, 15 mM 3-PGA and 4 μL [α-32P] ATP (3000 Ci[111 TBq]/mmol) (Perkin Elmer, Waltham, MA, USA). Supernatant containing AGPase (20 μL) was added to the assay buffer and incubated at 37°C for 20 min and then the reaction stopped by boiling the mixture for 2 min at 95°C.

[α-32P]ADP was enzymatically synthesized from [α-32P] ATP (3000 Ci [111 TBq]/mmol) (PerkinElmer). In brief, 3–4 μL of [α-32P] ATP was added to a 20 μL reaction mixture that contains 20 mL of 50 mM HEPES-KOH (pH 7.2), 5 mM $MgCl_2$, 1 mM glucose, 1 unit of hexokinase, and 10 mM unlabeled ATP. The mixture was incubated for 60 min at room temperature. Hexokinase was deactivated by heating for 2 min at 95°C [23]. The uptake experiment was carried out at 30°C at different time intervals, and terminated by quick transfer of the cells to 0.45 μm filter under vacuum. The filter was washed with 3–4 mL cold potassium phosphate buffer (pH 7.2) to remove unimported radiolabelled substrate. The filter was transferred to scintillation cocktail (5 mL) and the radioactivity of 32P was quantified in a liquid scintillation counter (Beckman, Brea, CA, USA).

Efflux experiment of the putative [α-32P] ADP-Glc substrate was conducted in intact IPTG-induced *E. coli* cells harboring *HvBT1* and empty plasmid as a control. The cells were pre-incubated with 1 μM [α-32P] ADP-Glc in 50 mM potassium phosphate buffer (pH 7.2) for 5 min at room temperature. The uptake buffer was diluted 1000 times using ADP-Glc, ATP, ADP and AMP at final concentration of 1 mM and incubated for different durations. The uptake reaction was stopped by quick filtration with 0.45 μm filter under vacuum. The filter was washed by adding 3–4 mL ice cold potassium phosphate buffer and then placed in scintillation cocktail (5 mL). The radioactivity of 32P was monitored at different time points as described above [6]. The efflux of intracellular [α-32P] ADP was conducted in induced *E. coli* cells harboring *HvBT1*. The cells were incubated with 1 μM [α-32P] ADP for 5 min and the rest of the procedures were performed as described above for [α-32P] ADP-Glc. Non-labeled ADP-Glc and ADP dilution (1000 times) were used to enhance the export of [α-32P] ADP.

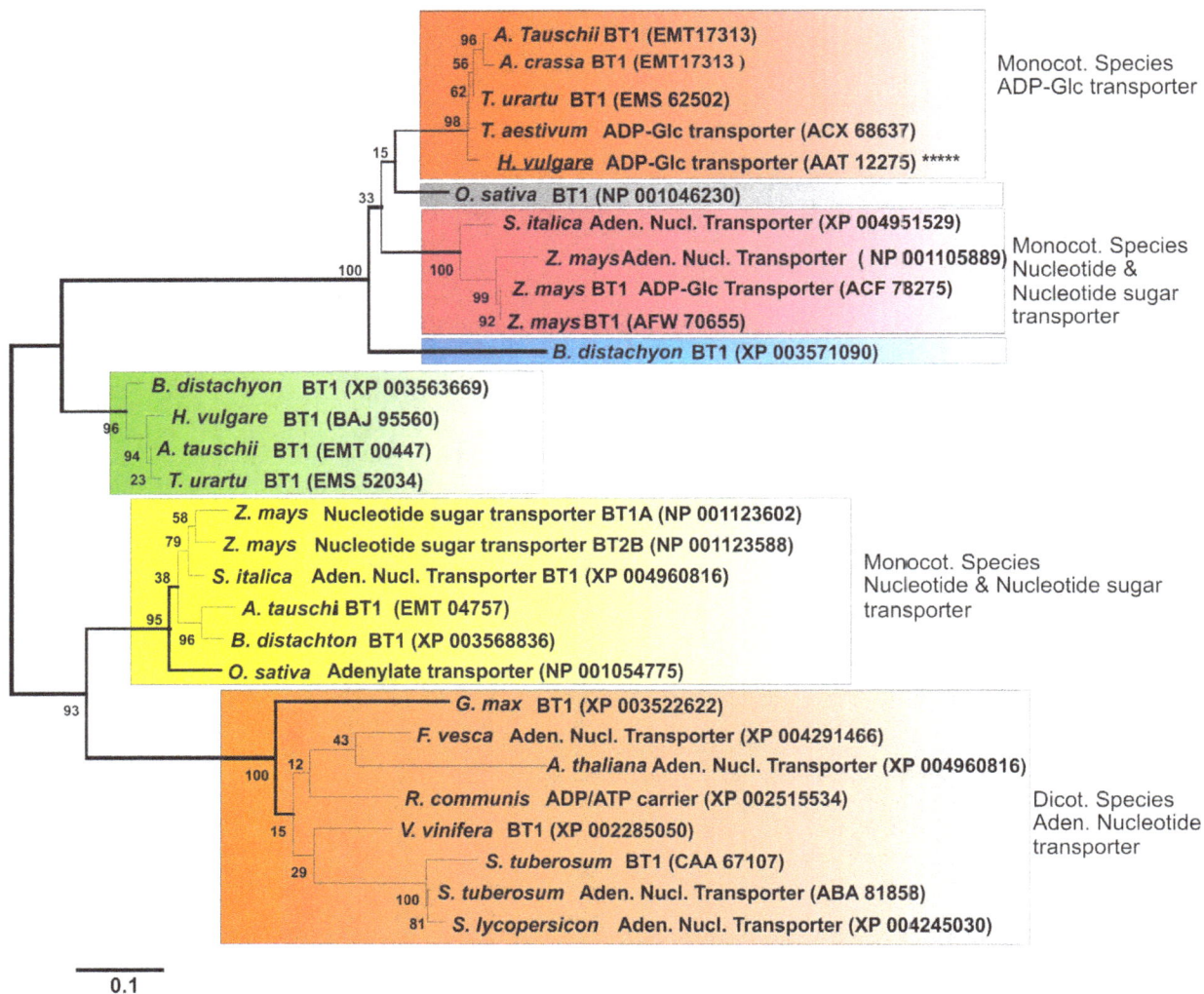

Figure 1. Phylogenetic analysis of BT1 amino acid sequences. The BLASTp [24] program was used to retrieve the amino acid sequences of proteins related to BT1. The retrieved amino acid sequences were aligned with the ClustalX program [25]. Phylogenetic estimates were created by the Molecular Evolutionary Genetic Analysis (MEGA 5.2) program package [26]. The gaps were eliminated from the analysis in MEGA by using complete deletion setting. The phylogenetic tree was generated with the Maximum Parsimony (PARS), Neighbour joining (NJ; setting JTT model), and Maximum likelihood (ML) methods. MEGA 5.2 was also used for determining the best fit substitution model for ML analysis; thus for ML analysis the JTT+G model was applied and for all programs the bootstrap option was selected (1000 replicates) in order to obtain estimates for the confidence levels of the major nodes present within the phylogenetic trees. The phylogenetic tree divided BT1 homologues into two main groups, represent monocotyledonous, and both monocotyledonous and dicotyledonous species. HvBT1 was located in the first group within a distinct subgroup (orange color) with that of wheat (GenBank ID: ACX68637), which was characterized as ADP-glucose transporter [2]. Another subgroup (red color) represent BT proteins from monocot species including that of *maize* (GenBank ID: ACF78275) which is characterized as ADP-glucose transporter. The second group contained BT proteins from both monocotyledonous and dicotyledonous species. Dicotyledonous species were assigned in a distinct subgroup (brown color). They mainly function as nucleotide transporter, for example the potato (GenBank ID: ABA 81858) and *Arabidopsis* (GenBank ID: XP 004960816) BT proteins are characterized as an adenine nucleotide transporter [9,31].

Bioinformatics analysis

BT1 protein sequences were collected using the BLASTp program [24] and were aligned with ClustalX program [25]. The phylogenetic tree was generated by MEGA 5.2 program [26].

Results

Bioinformatics analysis of HvBT1

The phylogenetic analysis divided BT1 proteins into two main groups. The first group represented monocotyledonous species including wheat, barley, maize and rice, and the second group represented both monocotyledonous and dicotyledonous species

(Figure 1). The first group consists of subgroups of BT1 homologues that are classified mainly based on their biochemical functions. The HvBT1 (GenBank ID: AAT12275) is classified in a distinct subgroup (orange color) that includes BT1 homologues from *Triticum aestivum* and *Triticum urartu*. Comparison of the amino acid sequence of HvBT1 with that of other BT homologues showed high similarity with BT homologue from *Triticum aestivum* (GenBank ID: ACX68637; with 97% similarity or 92% identity), *Triticum urartu* (GenBank ID: EMS62502; with 99% similarity or 88% identity), *Aegilopus tauschii* (GenBank ID: EMT17313; with 99% similarity or 89% identity) and *Aegilopus crassa* (GenBank ID: ACX68638; with 97% similarity or 90% identity). The HvBT1

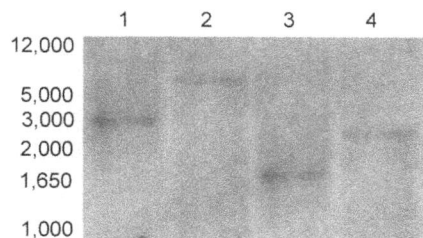

Figure 2. Southern blot analysis of barley _HvBT1_. Barley nuclear DNA was digested with **1**: BamHI, **2**: SalI, **3**: XhoI and **4**: KpnI and subjected to southern blot analysis. DNA probe was prepared with 700 bp of _HvBT1_ cDNA and used for hybridization. Molecular weight of the standard DNA ladder was indicated in the image.

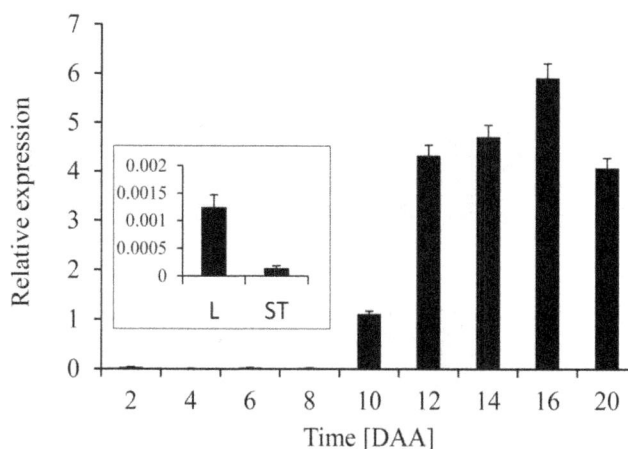

Figure 3. Real-time qPCR analysis of _HvBT1_ in different tissues. Quantitative real-time RT-PCR was used to determine the expression level of _HvBT1_ in different tissues using gene specific primers. _β-actin_, a housekeeping gene from barley, was used as a reference gene Seed samples during grain filling (2 to 20 DAA) and autotrophic tissue samples (stem and leaf) were used for gene expression analysis (see inset).

also showed similarity with homologues from _Zea mays_ (GenBank ID: NP001105889; with 92% similarity or 66% identity) and _Zea mays_ (GenBank ID: ACF78275; with 92% similarity or 71% identity). Other subgroups include BT1 homologues that function as either nucleotide sugar transporter or adenine nucleotide transporter. The second group consists of BT1 homologues from both monocotyledonous and dicotyledonous species. This group is divided into two main distinct subgroups. One of the subgroups includes BT1 from monocotyledonous species that mainly function as either nucleotide sugar transporter or adenine nucleotide transporter. The second subgroup includes BT1 homologues from only dicotyledonous species that mainly function as nucleotide carrier proteins. The BT1 homologues from dicotyledonous species showed low similarity to HvBT1.

Southern blot analysis of _HvBT1_

Restriction analysis of barley genomic DNA with XhoI, BamHI, SalI and kpnI (Figure 2) revealed that only one distinct band can be produced by digestion with different restriction enzymes that do not cut the _HvBT1_ cDNA sequence. This result indicates that there is only one copy of _HvBT1_ in barley genome.

Quantitative real-time RT-PCR analysis of _HvBT1_

The expression profile of _HvBT1_ was investigated in different tissues such as stems, leaves and seeds using quantitative real-time PCR and _β-Actin_ in different tissues. Our showed the presence of high abundance of _HvBT1_ transcripts in the endosperm at different stages of grain filling. The maximum transcript level was detected at 14 to 16 DAA. The transcripts of _HvBT1_ were also detected in the leaf and stem tissues but at lower levels when compared to that found in the endosperm (Figure 3).

Cellular and subcellular localization of _HvBT1_

Cellular localization of _HvBT1_ transcripts was detected by RNA in-situ hybridization. Our results indicated that a strong signal of alkaline phosphatase was detected by the antisense probe in the endosperm, reflecting high accumulation of _HvBT1_ transcripts in the starchy portion of developing caryopsis (Figure 4). Subcellular localization of the HvBT1::YFP fusion protein showed that HvBT1 protein is targeted to the chloroplast membrane (Figure 5A). This result was validated by immunolocalization of HvBT1::YFP; where the fluorescence of FITC was detected in the chloroplast envelop (Figure 5B).

Heterologous expression of HvBT1 in _E. coli_ cells

SDS-PAGE analysis of the purified HvBT1 membrane protein showed that both _E. coli_ strains C43 and Rosetta2 harboring the

optimized ORF of _HvBT1_ are able to express the HvBT1 protein with expected mass of ~45 KDa. Rosetta2, harboring the original ORF was also able to express the HvBT1 at very low level (Figure 6). The Rosetta 2 strain cells harboring either org or opc ORFs of _HvBT1_ showed growth inhibition, while the cell growth appeared to be normal in the case of C43 strain (Figure S1). The use of optimized ORF of _HvBT1_ along with the C43 strain provided an ideal expression system to sustain cell viability and perform the transport assay procedures.

[α-^{32}P]ADP-glucose transport assay

The transport of [α-^{32}P] ADP-Glc was studied using intact _E. coli_ cells harboring the expression plasmid containing _HvBT1_ or the control plasmid (with no _HvBT1_). The results showed that the import of the α-^{32}P labeled substrate follows a non-linear regression trend for Michaelis-Menten kinetics (Figure 7). The affinity of HvBT1 for ADP- glucose was analyzed with different concentrations of [α-^{32}P] ADP-glucose using Wolfram _Mathematica_ 8.0 software (Wolfram, Champaign, IL, USA). Increasing the

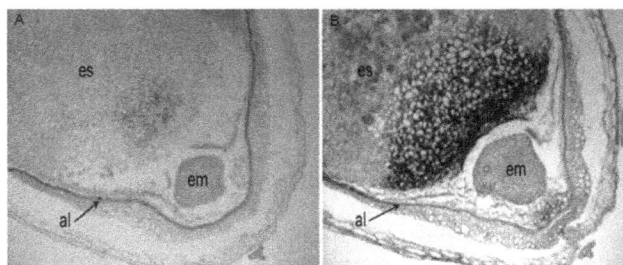

Figure 4. Cellular localization of _HvBT1_. Cellular localization of _HvBT1_ was assayed using RNA-in situ hybridization. **A**: hybridization with the sense probe which produces a very faint signal. **B**: hybridization with antisense probe which produces high signal of alkaline phosphatase. **es**; embryo sac, **al**; aleurone, and **em**; embryo. Strong signal was detected in the embryo sac, which accumulates the endosperm.

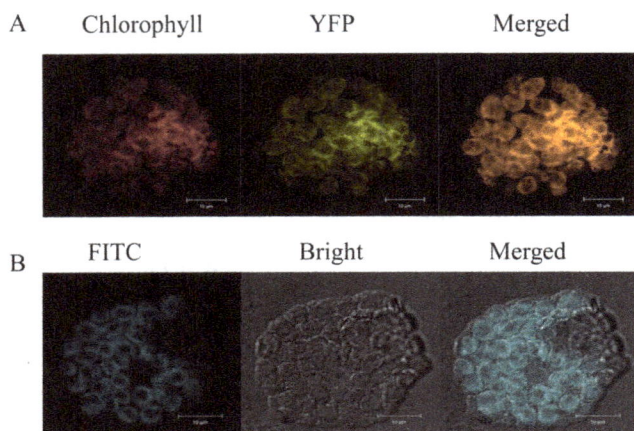

Figure 5. Subcellular localization of HvBT1::YFP. Subcellular localization of HvBT1 was visualized by Zeiss confocal Laser scanning microscope. **A**: transient expression of HvBT1::YFP in living protoplasts; chlorophyll autoflorescence (red color), YFP fluorescence (yellow color) and merged image (orange color). **B**: immunolocalization of HvBT1::YFP was detected using anti-YFP antibody and visualized by the fluorescence of FITC-conjugated antibody. Images represent FITC fluorescence (blue color), bright field (grey image) and the merged image that show the localization of HvBT1::YFP on the chloroplasts membranes.

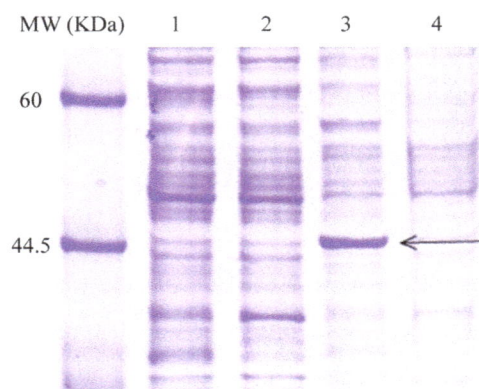

Figure 6. SDS-PAGE analysis of HvBT1 protein. *Escherichia coli* C43 and rosetta2 strains harboring the original (org) or the optimized (opc) ORF of *HvBT1* were grown in 2xYT liquid media supplied with IPTG (0.5 mM final concentration). His-tagged HvBT1 membrane protein was purified and subjected to 12% SDS-PAGE. Lane 1 and 2 represent Rosetta 2 harboring opc and org ORF of *HvBT1*, respectively. Lane 3 and 4 represent C43 harboring opc and org ORF of *HvBT1*, respectively. Black arrows point to a band size of 45 KDa. Protein standard molecular weight is shown.

substrate concentration led to increased radiolabeled ADP-glucose uptake into the intact *E. coli* cells expressing the *HvBT1*. The K_m value of ADP-Glc was calculated to be 614.5 μM and the V_{max} to be 254.1 nmol of ADP-Glc mg of protein^{-1} h^{-1} (Figure 7A). Uptake for [α-^{32}P] ADP was also analyzed as described for [α-^{32}P] ADP-Glc. Likewise; an increase in the import rate of [α-^{32}P] ADP was observed with increased concentrations of [α-^{32}P] ADP. The K_m and V_{max} values for ADP are 334.7 μM and 74.07 nmol of ADP mg protein^{-1} h^{-1}, respectively (Figure 7B).

Efflux of the intracellular [α-^{32}P] ADP-Glc and [α-^{32}P] ADP was monitored using the intact IPTG-induced *E. coli* cells harboring *HvBT1* as described in the materials and methods. Rapid export of [α-^{32}P] ADP-Glc was enhanced by dilution with a high concentration of ADP or AMP, which caused a 75% or 60% reduction from the initial amount, respectively. Meanwhile, dilution of [α-^{32}P] ADP-Glc with a high concentration of non-labeled ADP-Glc or ATP led to 43% or 36% reduction from the initial amount, respectively within 8 min of incubation (Figure 8A). Increased efflux of putative [α-^{32}P] ADP was also observed by dilution of the medium with high concentration of non-labeled ADP-Glc at different incubation periods (Figure 8B). No significant difference in efflux was found between dilution with ADP and the control.

To identify the possible counter-exchange substrates of HvBT1, intact *E. coli* cells were preloaded with ATP, ADP or AMP at a concentration of 1 mM for 10 min at 30°C followed by addition of [α-^{32}P] ADP-Glc (final concentration of 1 mM) and incubation for 10 minutes. Our results indicated that preloading with ADP or AMP has a positive impact on [α-^{32}P] ADP-Glc uptake, while preloading with ATP does not have significant effect (Figure 8C).

To investigate if the [α-^{32}P] ADP-Glc transport activity of HvBT1 is inhibited or activated by other metabolic intermediates, intact *E. coli* cells harboring *HvBT1* were incubated with 100 μM [α-^{32}P] ADP-Glc combined with the different metabolic intermediates at a final concentration of 1 mM. Our results showed that the uptake rate of [α-^{32}P] ADP-Glc was enhanced by ~ 88% or

70% when co-incubated with D-glucose or glucose-1- phosphate, respectively (Table 1). Co-incubation with non-labeled ADP-Glc reduced the uptake of [α-^{32}P] ADP-Glc by ~50% as compared to the control. Similarly, the uptake of [α-^{32}P] ADP-Glc was negatively affected by co-incubation with the nucleotide ADP, whereas the other nucleotides ATP and AMP did not have effect. The reduction of [α-^{32}P] ADP-Glc uptake due to co-incubation of *E. coli* cells harboring *HvBT1* with ADP-Glc or ADP indicated the substrate specificity of HvBT1.

Discussion

In cereal grains, starch is synthesized exclusively in the plastids. ADP-glucose, the precursor of starch, is synthesized mainly in the cytosol of the endospermic cells by the cytosolic form of AGPase. The majority of AGPase activity in cereals is reported to be in the cytosol (extra-plastidial) [27,28]. Various mutants from cereals such as maize shrunken 2 and brittle 2 mutants [29], and the barley Risø16 mutant [30], are characterized by the lack of activity of cytosolic AGPase, which in turn leads to the deposition of less starch as compared to their control. Under normal conditions, ADP-glucose synthesized in the cytosol of the endospermic cells is transported into the amyloplast envelope membranes through a protein carrier. A study with Risø13 mutant of barley showed that the mutant produces low starch yield as compared to the parental control. This is attributed mainly to a decrease in the activity of ADP-glucose transporter caused by a point mutation that led to the substitution of GLU for VAL at position 273 in helx4 [7].

Amino acid sequence-based phylogenetic analysis revealed the clustering of BT1 proteins into two main groups. The first group contained only those isolated from monocotyledonous species including the main cereal crops and other grasses. The HvBT1 (GenBank ID: AAT12275) was assigned to this group, along with proteins from maize, rice, wheat and other grasses. Some of the BT1 proteins in this cluster have been characterized. The ZmBT1 (GenBank ID: ACF78275) is characterized biochemically to function as plastidial ADP-glucose transporter [6]. ZmBT1 is also reported to have dual function of targeting plastids and mitochondrial envelope membranes [11], suggesting that it might

A

B

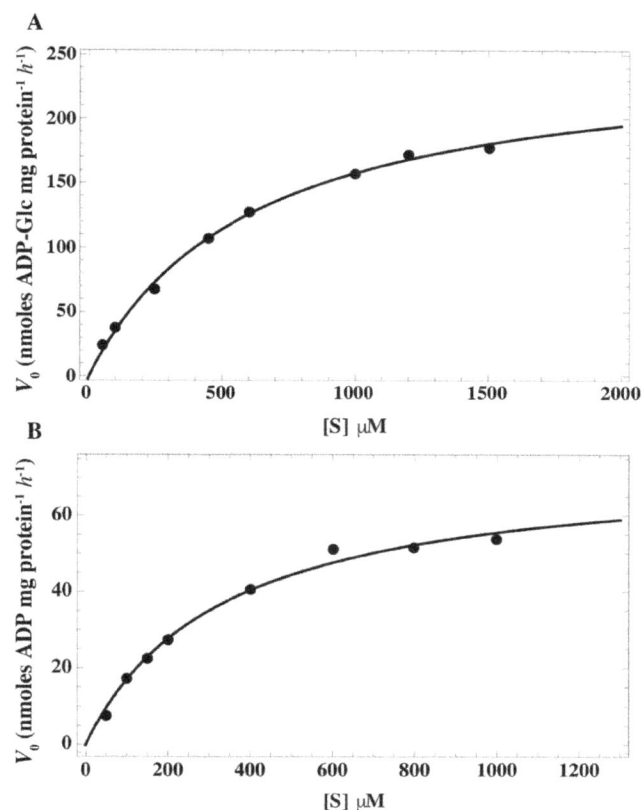

Figure 7. Transport activity of HvBT1 in intact _E. coli_ cells.
Escherichia coli C43 cells harboring the recombinant plasmid and the empty one as a control were incubated with different concentrations of $[\alpha\text{-}^{32}P]$ ADP-Glc. The cells were incubated at 30°C for 10 min. The control values have been subtracted. The data are the mean ± SE of three independent experiments, each with three replicates. **A**: K_m value of ADP-glucose is 614.5±33.24 μM and V_{max} of 254.14 ±19.45 nmol of ADP-Glc mg of protein^{-1} h^{-1}. **B**: K_m and V_{max} values of ADP is 334.7±39.3 μM and of 47.07±3.51 nmol of ADP-Glc mg of protein^{-1} h^{-1}, respectively.

A

B

C

Figure 8. Exchange of the intracellular radiolabeled substrates.
A: _E. coli_ cells harboring the vector containing _HvBT1_ and the control vector were incubated with 1 μM $[\alpha\text{-}^{32}P]$ ADP-Glc at 30°C for 5 min. The assay buffer was diluted with non-labeled ATP, ADP, AMP, and ADP-Glc for indicated time points. The cells were filtered and washed under vacuum, and then measured for radioactivity. The data presented here are the mean ± SE of three independent experiments, each with three replicates. **B**: the procedures for ADP efflux assay was performed as described for ADP-Glc in (A) with ADP and ADP-Glc dilutions. **C**: _E. coli_ C43 cells harboring the vector containing _HvBT1_ and the control vector were preloaded with nucleotides at a final concentration of 1 mM, and then the cells were incubated at 30°C for 5 min. The cells were centrifuged and re-suspended in potassium phosphate buffer (50 mM, pH 7.2) with $[\alpha\text{-}^{32}P]$ ADP-Glc at concentration of 100 μM at 30°C for 8 min. The data presented are the mean ± SE of three independent experiments, each with three replicates.

have a role of transporting some other mitochondrial energy molecules. The barley plastidial ADP-glucose transporter (HvNST1) was identified in the Risø13 mutant that accumulates less starch as compared to the wild type [7], while wheat ADP-glucose transporter is identified by reconstituting amyloplast envelope membrane proteins in liposomes [2]. The second group contained BT1 proteins from both monocotyledonous and dicotyledonous species. Most of these proteins have not been characterized yet but are predicted to act as adenylate transporters. For example, BT1 from potato (GenBank ID: CAA67107) was identified as plastidial adenine nucleotide uniporter [9], and the _Arabidopsis_ (AtBT1) (GenBank ID: NP194966) as a plastidial nucleotide uniporter [31]. The alignment of their deduced amino acid sequence indicates that these BT1 homologues belong to the mitochondrial carrier family (MCF) that possesses membrane spanning domains [9] and conserved motifs designated as mitochondrial energy transfer signature (METS) [32]. A signal peptide of 53 amino acids was also found at the N-terminal of HvBT1.

Southern blot analysis indicated the presence of only one copy of _BT1_ in the barley genome (Figure 2). In contrast, two copies of _BT1_ are detected in maize; _ZmBT1_ and _ZmBT1-2_. The _ZmBT1_ is

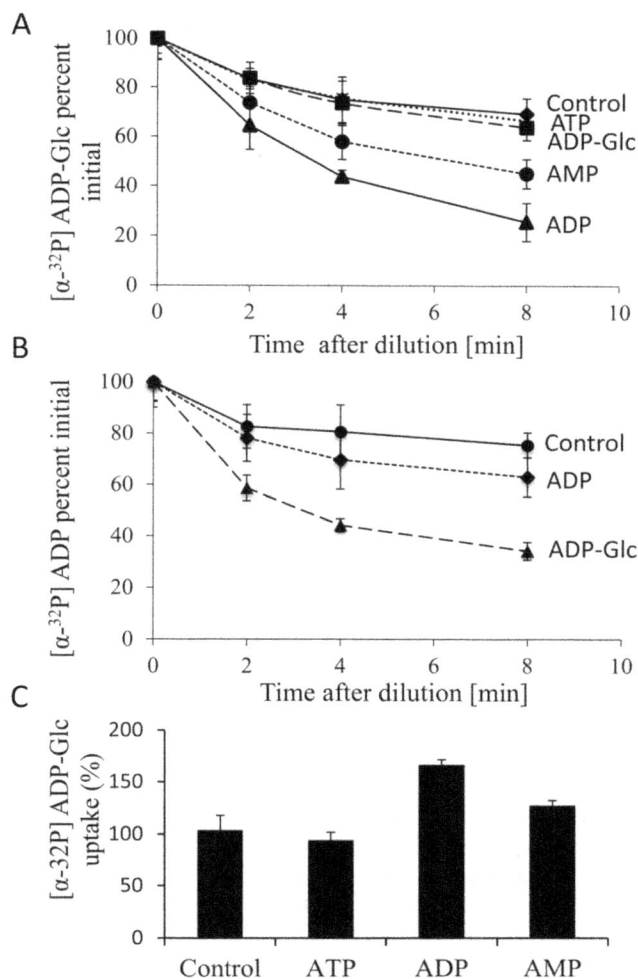

found to be expressed exclusively in the endosperm, while _ZmBT1-2_ in both autotrophic and heterotrophic tissues. The _ZmBT1-2_ might function as an adenine nucleotide transporter that supplies the cells with adenine nucleotides synthesized in the plastids [6]. Gene expression analysis with qRT-PCR and RNA in situ hybridization indicated that _HvBT1_ is exclusively expressed in the endosperm during grain filling (Figure 3 and 4). Detecting the expression of _HvBT1_ at low level in autotrophic leaf and stem tissues might suggest its dual function in the plastidial and mitochondrial envelope membranes like that of the maize plastidial ADP-glucose transporter _ZmBT1_ [11]. However, local-

Table 1. Effects of different metabolites on $[\alpha\text{-}^{32}P]$ ADP-glucose transport activities of HvBT1.

Effectors	$[\alpha\text{-}^{32}P]$ ADP-Glc transport		
	nmol mg protein^{-1}	%	% changes
Control[a]	12.99 ± 0.72[b]	100	0
D-glucose	24.5 ± 0.76	188.54	+88.54
Glucose 1-Phosphate	22.12 ± 0.91	170.17	+70.17
ADP-glucose	6.46 ± 0.35	49.67	−50.33
ATP	11.98 ± 0.98	95.276	−4.73
ADP	8.22 ± 0.85	63.18	−36.82
AMP	11.16 ± 0.78	85.89	−14.11

[a]Control = $[\alpha\text{-}^{32}P]$ ADP-glucose (100 μM).
[b]Data are means ± SE, n = 3.

ization of HvBT1 to the plastid membrane suggested that it acts as an ADP-glucose transporter across the plastid membrane (Figure 5A and B).

The most challenging tasks in understanding the functional and structural properties of membrane proteins are their heterologous expression and protein purification [33]. The most important step is the choice of an ideal *E. coli* strain that can withstand the possible toxic effect of the expressed membrane protein. *Escherichia coli* C43 cells, a mutant of BL21 (DE3), harboring *HvBT1* were able to grow on both liquid and solid media after induction and also express HvBT1 protein at higher level with the optimized sequence of *HvBT1* (Figure 6), indicating the suitability of this strain for studying such membrane protein. The inhibitory effect of HvBT1 on Rosetta2 cells harboring either org or opc ORFs of *HvBT1* (Figure S1) may be due to the rapid expression of the membrane protein that drives the cells to accommodate large amounts of expressed proteins, thereby affecting the natural rhythm of the translation machinery [34]. This ultimately leads to mis-folding of the expressed proteins as well as forming inclusion bodies [33]. The mutations in the *lacUV5* promoter resulted in low expression of the membrane protein, which aided the C43 cells to proliferate normally [35]. Slow expression of HvBT1 in the C43 cells allowed the cells to produce soluble and functional proteins and kept the cells active for downstream applications.

The barley ADP-glucose transporter (HvBT1) transports ADP-Glc in counter-exchange with ADP with apparent affinities of 614 μM and 334 μM, respectively (Figure 7A and B). Similar results have been reported in maize where ZmBT1 was able to transport ADP-Glc in counter-exchange with ADP with apparent affinities of 850 μM and 465 μM, respectively [6]. In wheat, plastidial ADP-glucose transporter was also analyzed by reconstituting amyloplast envelope membrane proteins in proteoliposomes. Its apparent affinities for ADP-Glc, and both ADP and AMP were found to be 430 μM and 200 μM, respectively [2].

Enhancing the efflux of [α-32p] ADP-Glc by dilution of the medium with high concentration of ADP and AMP revealed the potentials of these nucleotides as counter exchange substrates (Figure 8A). This result was supported by the efflux of the putative [α-32p] ADP by dilution with high concentration of non-labeled ADP-Glc (Figure 8B). In agreement with this, efflux study with intact amyloplasts of wheat showed that ADP-glucose transporter protein transports ADP-Glc into the amyloplasts in counter exchange with ADP and AMP [2]. In addition, preloading induced *E. coli* cells harboring *HvBT1* with ADP and AMP resulted in increased ADP-Glc uptake as compared to the control

(Figure 8C). Similar results have been reported for ZmBT1 in maize [6].

Testing several effectors of the $[\alpha\text{-}^{32}P]$ ADP-Glc uptake in the intact *E. coli* cells expressing HvBT1 protein showed that the transport rate of $[\alpha\text{-}^{32}P]$ ADP-Glc increased with co-incubation with glucose or glucose-1-phosphate (Table 1). Interestingly, co-incubation with glucose increased the uptake rate of $[\alpha\text{-}^{32}P]$ ADP-Glc by almost double of the control. Glucose as a sole carbon source plays an important role in the activation of *E. coli* metabolism [36], which subsequently increases the ability of the cells to uptake $[\alpha\text{-}^{32}P]$ ADP-Glc. Glucose-1-phosphate also had a positive effect on the uptake of $[\alpha\text{-}^{32}P]$ ADP-Glc (Table 1). Meanwhile, co-incubation with unlabeled ADP-Glc reduced the uptake of $[\alpha\text{-}^{32}P]$ ADP-Glc by ~50% as compared to the control due to its competition as a substrate with the labeled ADP-Glc. Reduction in the uptake of $[\alpha\text{-}^{32}P]$ ADP-Glc was also observed by co-incubation with ADP, indicating the substrate specificity of HvBT1 for ADP-Glc and ADP.

In summary, the findings of this study show that HvBT1 is able to transport ADP-glucose with high affinity in counter-exchange with ADP and likely with AMP. Its localization to the plastids envelops and exclusively expression in the endospermic cells of the barley grains indicates its significance in determining starch yield in barley.

Supporting Information

Figure S1 Inhibitory effect of HvBT1 on *E. coli* cells. **A**: Growth curve of C43 cells; empty plasmid (C43-E), optimized ORF (C43-OPC), and original ORF (C43-ORG). **B**: Growth curve of Rosetta2 cells; empty plasmid (R-E), optimized ORF (R-OPC), and original ORF (R3-ORG). **C**: Growth of C43 and Rosetta2 cells on plate media. **C** 1 and 2 indicate the un-induced (−) and induced (+) C43 cells harboring original ORF and optimized ORF, respectively. **C** 3 and 4 indicate the un-induced (−) and induced (+) Rosetta2 cells harboring original ORF and optimized ORF, respectively.
(TIF)

Acknowledgments

We would like to thank Dr. Ekkehard Neuhaus (Department of Plant Physiology, University of Kaiserslautern, Germany) for providing us with pET16b::*ZmBT1* plasmid, Dr. Joe O'Neil (Department of Chemistry, University of Manitoba, Canada) for his help in the kinetic analysis of HvBT1, and Dr. Georg Hausner (Department of Microbiology, University of Manitoba, Canada) for his guidance in the phylogenic analysis.

Author Contributions

Conceived and designed the experiments: AS FD. Performed the experiments: AS. Analyzed the data: AS FD BTA. Contributed reagents/materials/analysis tools: FD BTA. Contributed to the writing of the manuscript: AS BTA FD.

References

1. James MG, Denyer K, Myers AM (2003) Starch synthesis in the cereal endosperm. Curr. Opin. Plant Biol. 6: 215–222
2. Bowsher CG, Scrase-Field EFAL, Esposito S, Emes MJ, Tetlow IJ (2007) Characterization of ADP-glucose transport across the cereal endosperm amyloplast envelope. J. Exp. Bot. 58: 1321–1332
3. Cao H, Sullivan TD, Boyer CD, Shannon JC (1995) BT1, a structural gene of the major 39–44 kDa amyloplast membrane polypeptides. Physiol. Plant. 95: 176–186
4. Cao H, Shannon JC (1996) BT1, a protein critical for in vivo starch accumulation in maize endosperm, is not detected in maize endosperm suspension cultures. Physiol. Plant. 97: 665–673
5. Shannon JC, Pien FM, Liu KC (1996) Nucleotides and nucleotide sugars in developing maize endosperms: Synthesis of ADP-glucose in brittle 1. Plant Physiol. 110: 835–843
6. Kirchberger S, Leroch M, Huynen MA, Wahl M, Neuhaus HE, et al. (2007) Molecular and biochemical analysis of the plastidic ADP-glucose transporter (ZmBT1) from *Zea mays*. J. Biol. Chem. 282: 22481–22491
7. Patron NJ, Greber B, Fahy BF, Laurie DA, Parker ML, et al. (2004) The lys5 mutations of barley reveal the nature and importance of plastidial ADP-Glc transporters for starch synthesis in cereal endosperm. Plant Physiol. 135: 2088–2097
8. Bahaji A, Li J, Sánchez-López ÁM, Baroja-Fernández E, Muñoz FJ, et al. (2014) Starch biosynthesis, its regulation and biotechnological approaches to improve crop yields. Biotechnol. Adv. 32: 87–106
9. Leroch M, Kirchberger S, Haferkamp I, Wahl M, Neuhaus HE, et al. (2005) Identification and characterization of a novel plastidic adenine nucleotide uniporter from *Solanum tuberosum*. J. Biol. Chem. 280: 17992–18000
10. Picault N, Hodges M, Palmieri L, Palmieri F (2004) The growing family of mitochondrial carriers in Arabidopsis. Trends Plant Sci. 9: 138–146
11. Bahaji A, Ovecka M, Barany I, Risueno MC, Munoz FJ, et al. (2011) Dual targeting to mitochondria and plastids of AtBT1 and ZmBT1, two members of the mitochondrial carrier family. Plant Cell Physiol. 52: 97–609
12. Sambrook J, Fritsch EF, Maniatis T (1989) Molecular Cloning: A Laboratory Manual, 2 nd ed., Cold Spring Harbor Laboratory, Cold Spring Harbor, NY
13. Livak KJ, Schmittgen TD (2001) Analysis of relative gene expression data using real-time quantitative PCR and the $2^{-\Delta\Delta C_T}$ method. Methods 25: 402–408
14. Belmonte M, Donald G, Reid D, Yeung E, Stasolla C (2006) Alternations of the glutathione redox state improve apical meristem structure and somatic embryo quality on white spruce (*Picea glauca*). J. Exp. Bot. 56: 23–55
15. Canton F, Suarez M, Jose-Estanyol M, Canovas F (1999) Expression analysis of a cytosolic glutamine synthetase gene in cotyledons of Scots pine seedlings: developmental, light regulation and spatial distribution of specific transcripts. Plant Mol. Biol. 40: 623–634
16. Tahir M, Law D, Stasolla C (2006) Molecular characterization of PgAGO, a novel conifer gene of the ARGONAUTE family expre4ssed in apical cells and required for somatic embryo development in spruce. Tree Physiol. 26: 12–57
17. Yoo SD, Cho YH, Sheen J (2007) Arabidopsis mesophyll protoplasts: A versatile cell system for transient gene expression analysis. Nature Protocols 2: 1565–1575
18. Lee DW, Hwang I (2011) Transient expression and analysis of chloroplast proteins in Arabidopsis protoplasts. Methods Mol. Biol. 774: 59–71
19. Chen D, Texada DE (2006) Low-usage codons and rare codons of *Escherichia coli*. Gene Ther. Mol. Biol. 10: 1–12
20. Laemmli UK, MolberE, Showe M, Kelenberger E (1970) Form-determining function of genes required for the assembly of the head of bacteriophage T4. J. Mol. Biol.49: 99–113
21. Ghosh HP, Preiss J (1966) Adenosine diphosphate glucose pyrophosphorylase: a regulatory enzyme in biosynthesis of starch in spinach leaf chloroplasts. J. Biol. Chem. 241: 4491–4504
22. Rösti S, Rudi H, Rudi K, Opsahl-Sorteberg H, Fahy B, et al. (2006) The gene encoding the cytosolic small subunit of ADP-glucose pyrophosphorylase in barley endosperm also encodes the major plastidial small subunit in the leaves. J. Exp. Bot. 57: 3619–3626
23. Tjaden J, Schwöppe C, Möhlmann T, Quick PW, Neuhaus HE (1998) Expression of the plastidic ATP/ADP transporter gene in *Escherichia coli* leads to the presence of a functional adenine nucleotide transport system in the bacterial cytosolic membrane. J. Biol. Chem. 273: 9630–9636
24. Altschul SF, Madden TL, Schäffer AA, Zhang J, Zhang Z, et al. (1997) Gapped BLAST and PSI-BLAST: a new generation of protein database search programs. Nucleic Acids Res. 25: 3389–3402
25. Thompson JD, Gibson TJ, Plewniak F, Jeanmougin F, Higgins DG (1997) The CLUSTAL_X windows interface: flexible strategies for multiple sequence alignment aided by quality analysis tools. Nucleic Acids Res. 25(24):4876–82
26. Tamura K, Peterson D, Peterson N, Stecher G, Nei M, et al. (2011) MEGA5: molecular evolutionary genetics analysis using maximum likelihood, evolutionary distance, and maximum parsimony methods. Mol. Biol. Evol. 28(10):2731–2739
27. Denyer K, Clarke B, Hylton C, Tatge H, Smith A (1996) The elongation of amylose and amylopectin chains in isolated starch granules. Plant J. 10: 1135–1143
28. Tetlow IJ, Davies EJ, Vardy KA, Bowsher CG, Burrell MM, et al. (2003) Subcellular localization of ADP-glucose pyrophosphorylase in developing wheat endosperm and analysis of the properties of a plastidial isoform. J. Exp. Bot. 54: 715–725
29. Giroux MJ, Hannah LC (1994) ADP- glucose pyrophosphorylase in shrunken-2 and brittle-2 mutants of maize. Mol. Gen. Genet. 243: 400–408
30. Johnson PE, Patron NJ, Bottrill AR, Dinges JR, Fahy BF, et al. (2003) A low-starch barley mutant, Riso16, lacks the cytosolic small subunit of ADP-glucose pyrophosphorylase, reveals the importance of the cytosolic isoform and the identity of the plastidial small subunit. Plant Physiol. 131: 684–696
31. Kirchberger S, Tjaden J, Neuhaus HE (2008) Characterization of the Arabidopsis Brittle1 transport protein and impact of reduced activity on plant metabolism. Plant J. 56: 51–63
32. Millar AH, Heazlewood JL (2003) Genomic and proteomic analysis of mitochondrial carrier proteins in Arabidopsis. Plant Physiol. 131: 443–453
33. Bernaudat F, Frelet-Barrand A, Pochon N, Dementin S, Hivin P, et al. (2011) Heterologous Expression of Membrane Proteins: Choosing the Appropriate Host. PLoS ONE 6: 1–17
34. Narayanan A, Ridilla M, Yernool DA (2001) Restrained expression, a method to overproduce toxic membrane proteins by exploiting operator–repressor interactions. Protein Sci. 20: 51–61
35. Wagner S, Klepsch MM, Schlegel S, Appel A, Draheim R, et al. (2008) Tuning Escherichia coli for membrane protein overexpression. Proc. Natl. Acad. Sci. 105: 14371–14376
36. Brown W, Ralston A, Shaw K (2008) Positive transcription control: The glucose effect. Nature Education 1(1):202

17

Lactic Acid and Thermal Treatments Trigger the Hydrolysis of *Myo*-Inositol Hexakisphosphate and Modify the Abundance of Lower *Myo*-Inositol Phosphates in Barley (*Hordeum vulgare* L.)

Barbara U. Metzler-Zebeli[1,2], Kathrin Deckardt[1], Margit Schollenberger[3], Markus Rodehutscord[3], Qendrim Zebeli[1]*

1 Institute of Animal Nutrition and Functional Plant Compounds, Department for Farm Animals and Veterinary Public Health, Vetmeduni Vienna, Vienna, Austria, **2** University Clinic for Swine, Department for Farm Animals and Veterinary Public Health, Vetmeduni Vienna, Vienna, Austria, **3** Institute of Animal Nutrition, University of Hohenheim, Stuttgart, Germany

Abstract

Barley is an important source of dietary minerals, but it also contains *myo*-inositol hexakisphosphate ($InsP_6$) that lowers their absorption. This study evaluated the effects of increasing concentrations (0.5, 1, and 5%, vol/vol) of lactic acid (LA), without or with an additional thermal treatment at 55°C (LA-H), on $InsP_6$ hydrolysis, formation of lower phosphorylated myo-inositol phosphates, and changes in chemical composition of barley grain. Increasing LA concentrations and thermal treatment linearly reduced ($P<0.001$) $InsP_6$-phosphate ($InsP_6$-P) by 0.5 to 1 g compared to the native barley. In particular, treating barley with 5% LA-H was the most efficient treatment to reduce the concentrations of $InsP_6$-P, and stimulate the formation of lower phosphorylated myo-inositol phosphates such as *myo*-inositol tetraphosphate ($InsP_4$) and *myo*-inositol pentaphosphates ($InsP_5$). Also, LA and thermal treatment changed the abundance of $InsP_4$ and $InsP_5$ isomers with $Ins(1,2,5,6)P_4$ and $Ins(1,2,3,4,5)P_5$ as the dominating isomers with 5% LA, 1% LA-H and 5% LA-H treatment of barley, resembling to profiles found when microbial 6-phytase is applied. Treating barley with LA at room temperature (22°C) increased the concentration of resistant starch and dietary fiber but lowered those of total starch and crude ash. Interestingly, total phosphorus (P) was only reduced ($P<0.05$) in barley treated with LA-H but not after processing of barley with LA at room temperature. In conclusion, LA and LA-H treatment may be effective processing techniques to reduce $InsP_6$ in cereals used in animal feeding with the highest degradation of $InsP_6$ at 5% LA-H. Further in vivo studies are warranted to determine the actual intestinal P availability and to assess the impact of changes in nutrient composition of LA treated barley on animal performance.

Editor: Wagner L. Araujo, Universidade Federal de Vicosa, Brazil

Funding: The authors have no support or funding to report.

Competing Interests: The authors have declared that no competing interests exist.

* Email: Qendrim.Zebeli@vetmeduni.ac.at

Introduction

Barley is an important cereal crop used for livestock feeding and human consumption. It contains relatively large amounts of starch, protein, dietary fiber, and minerals which make this cereal a highly valuable ingredient of the diet [1]. It represents an important source of phosphorus (P), with total P content exceeding 4 g per kg dry matter (DM). However, the availability of P for non-ruminants in barley, like in other cereals and legumes, is low because the major part of P is stored in form of *myo*-inositol hexakisphosphate ($InsP_6$) [2], and its salts, also called phytate, serving as a P source for germination [3]. *Myo*-inositol hexakisphosphate is considered an antinutritional factor due to its low digestibility in monogastric animals but also due to its ability to build mineral complexes which inhibit the absorption of cations (e.g., Ca^{2+}, Fe^{2+} and Zn^{2+})

and protein in the gastrointestinal tract [4,5]. Endogenous cereal phytases that catalyse the hydrolysis of $InsP_6$ to inorganic P and lower *myo*-inositol phosphates (InsP), most importantly *myo*-inositol pentaphosphates ($InsP_5$), *myo*-inositol tetraphosphates ($InsP_4$), and *myo*-inositol triphosphates ($InsP_3$) [6], during germination can be activated by luminal conditions (i.e., pH) in the gastrointestinal tract, rendering a certain amount of P available for the host [7]. Compared with other cereals such as rye and wheat, barley grain possesses lower endogenous phytase activity [2], emphasizing the necessity to treat barley grain to improve intestinal P availability.

Up to now intestinal availability of plant P is mostly enhanced by supplementation of microbial phytases in diets for monogastric livestock species [8], thereby relying on optimal gastrointestinal conditions for maximum phytase activity. Because gastrointestinal pH and digesta passage rate may not always support phytase activity, the degradation of InsP$_6$ prior to feeding to animals is of particular interest as lower InsP can be almost completely used by monogastric animals [9]. Traditional processing methods of cereals for human consumption like soaking, malting, germination, and dough fermentation activate endogenous phytase activity thereby promoting the hydrolysis of InsP$_6$ [2,5,10–12]. Similar processing techniques may apply in livestock animal nutrition. However, because these processing methods reduce availability and concentration of other nutrients and thus lower the nutritional value, processing of feed (e.g., soaking and fermentation) prior to feeding is mostly restricted to liquid feeding systems for pigs by far [13–15]. Lowering pH in the grain stimulates endogenous phytase activity [7]. Therefore, treatment of cereal grains with lactic acid (LA), which is naturally produced during soaking and fermentation in cereal grains, may favor InsP$_6$ hydrolysis [5]. Lower concentrations (0.2–0.9%) of LA previously showed to reduce InsP$_6$ in barley [16] and may be a suitable processing method to treat barley grain. Also, hydrothermal treatment can reduce InsP$_6$ in grains and could therefore lead to a further reduction in InsP$_6$ concentration when combined with LA treatment [16–18]. Because LA treatment can have additional benefits on health and performance in livestock animals [19–24], treatment of barley with LA may be of interest in animal feeding. We hypothesized that soaking barley in increasing concentrations of LA in combination with heat may exert an additive effect on InsP$_6$ hydrolyzing properties. The main aim of this study was to evaluate the hydrolyzing capacity of increasing concentrations (0.5, 1 and 5%) of LA alone or in conjunction with heat on InsP$_6$ degradation in barley grain and the appearance of intermediate InsP such as InsP$_3$, InsP$_4$, and InsP$_5$, and their respective isomers. We were also interested in the effects of chemical and thermal processing on changes in the overall chemical composition of barley, which might have consequences for the feeding value of barley grain for livestock animals.

Materials and Methods

Barley Grain and Lactic Acid

Winter 2-row *Eufora* barley (*Hordeum vulgare* L.) grown during the 2011 season in Eastern Austria was used in this experiment [25]. *Eufora* barley represents a common barley variety used in animal feed and human nutrition in Austria and was provided by the Department of Crop Sciences, Division of Plant Breeding, University of Natural Resources and Life Sciences Vienna, Vienna (research group: H. Grausgruber). After harvesting, grains were carefully cleaned and freed of extraneous matter. Food-grade DL-lactic acid solution (85%, wt/wt) used in this study was purchased from Alfa Aesar GmbH & Co KG (Karlsruhe, Germany). LA solutions (0.5, 1 and 5% LA) were prepared using deionized water (vol/vol). The pH of LA solutions was 2.4, 2.2, and 1.8 for 0.5%, 1% and 5% LA, respectively, prior to treatment.

Soaking and Thermal Treatment of Grains

The procedure of LA and thermal treatments was the same as described in our previous study [25]. Triplicate barley subsamples were randomly taken and soaked in increasing concentrations of LA, without or with heat treatment (LA-H; only 1% and 5% LA), resulting in an orthogonally designed experiment (i.e., 0.5% LA, 1% LA, 5% LA, 1% LA-H, and 5% LA-H). Based on our previous study [25], where the impact of heat treatment on changes in nutrient composition of barley was small for 0.5% LA, only effects of heat treatment with 1 and 5% LA were investigated in this study. For treatment, a barley subsample (50 g) was soaked in the respective LA solution (1:1.6 wt/wt) at room temperature (22°C)

or heated at 55°C in an oven for 48 hours. Attention was paid that every grain was sufficiently soaked in the treatment solution. After the 48-hours incubation, treated barley samples were spread on Petri dishes and air-dried at 22°C for 24 hours before being ground prior to chemical analysis. Samples of LA-H treatment were cooled to 22°C prior to air-drying. Triplicate subsamples of the untreated *Eufora* barley were used as control (native barley). Only the barley grains were used for subsequent analyses. Drip losses, and thus potential nutrient losses, of the wet barley samples onto the Petri dish were not recovered after air-drying.

Sample Preparation

Native and dry treated barley samples were ground to pass a 0.5 mm sieve (Type 738, Fritsch, Rudolstadt, Germany). Barley subsamples used for InsP analyses were ground to pass a 0.2 mm sieve, and attention was paid that the ground mass of barley was fine and uniform for analysis. Milled samples were packed in sealed plastic bags and stored at 4°C until further analyses.

Analyses of Inositol Phosphates

For the analysis of InsP$_3$ to InsP$_6$ isomers, the ground material was extracted twice with a solution containing 0.2 M EDTA and 0.1 M sodium fluoride (pH 10) using a rotary shaker. Sample to extractant ratio was 1 g to 15 mL, and the total time of extraction was 1 h. After centrifugation the combined supernatants were ultracentrifuged using a Microcon filter (cut-off 30 kDa) devise (Millipore, Bedford, MA, USA) at 14,000×g for 30 minutes. Throughout the whole extraction procedure the samples were kept below 5°C. Filtrates were analyzed by high-performance ion chromatography (HPIC) and InsP were detected using a UV detector at 290 nm after postcolumn derivatization using an ICS-3000 system (Dionex, Idstein, Germany) equipped with a Carbo Pac PA 200 column and corresponding guard column. Gradient elution was done with increasing amounts of hydrochloric acid (0.05 M to 0.5 M within 33 minutes). Fe(NO$_3$)$_3$ solution (0.1% Fe(NO$_3$)$_3$ × 9 H$_2$O in HClO$_4$) was used as reagent for derivatization according to Philippy and Bland [26].

InsP$_6$ dipotassium salt was obtained from Sigma (Deisenhofen, Germany), InsP$_5$ isomers from Sirius Fine Chemicals (Bremen, Germany), InsP$_3$ and InsP$_4$ isomers, as far as available, from Santa Cruz Biotechnology (Heidelberg, Germany). These standards were used for peak identification. InsP$_6$ was used for calibration. Quantification of lower inositol phosphates was done according to Skoglund et al. [27]. Calibration curves were linear from quantification limit to approximately 10 to 30 μmol/g depending on the InsP isomer.

Quantification limits for InsP-isomers (S/N>10) were 1 μmol/g DM for InsP$_3$ and InsP$_4$ and 0.5 μmol/g DM for InsP$_5$, whereas the detection limits (S/N>5) were 0.5 μmol/g DM for InsP$_3$ to InsP$_4$ and 0.25 μmol/g DM for InsP$_5$. The InsP concentrations were determined as μmol InsP/g DM, and subsequently converted to g P pertaining to each InsP category (i.e., InsP$_3$-P, InsP$_4$-P, InsP$_5$-P, and InsP$_6$-P) based on their molecular weight and the respective content of P in the InsP molecule. Samples were analyzed in duplicate. The abundance of the different InsP$_3$, InsP$_4$ and InsP$_5$ isomers was in untreated barley samples as well as in LA and LA-H treated barley was used to evaluate the nature of the InsP$_6$ degradation caused by LA and heat treatment.

Nutrient Analyses

Dry matter, crude ash (CA), crude protein (CP), starch (total, non-resistant (NRS) and resistant starch (RS)), neutral detergent fiber (NDF), and acid detergent fiber (ADF) of the native and treated barleys were determined. Samples were analyzed for DM by oven-drying at 103°C for 4 h [27]. Crude ash was determined by combustion of samples over night at 580°C [28]. Crude protein was analyzed by the Kjeldahl method [28]. The concentrations of NDF and ADF were determined according to official methods [28,29] using Fiber Therm FT 12 (Gerhardt GmbH & Co. KG, Königswinter, Germany) including heat-stable α-amylase digestion for NDF determination, and were expressed exclusive of residual ash ($aNDF_{OM}$ and ADF_{OM}, respectively). The difference between $aNDF_{OM}$ and ADF_{OM} was considered as the hemicelluloses (HC) fraction. For P determination, samples were analyzed using ICP-OES (Vista Pro, Varian, Darmstadt, Germany) after acid digestion using a combination of sulphuric and nitric acid as described previously [30]. Samples were also analyzed for resistant starch (RS) and non-resistant starch (NRS) using a commercial enzymatic RS assay kit (Megazyme International Ireland Ltd., Bray, Ireland) following manufacturer's protocol, as previously described [25]. Total starch was calculated from RS and NRS fractions. Three subsamples per treatment were analyzed in duplicate.

Statistical Analysis

Data were subjected to two-way ANOVA using the PROC MIXED of SAS (SAS 9.2, SAS Institute Inc., Cary, NC, USA) with polynomial contrasts between control barley and barley treated with 0.5%, 1% and 5% LA as well as orthogonal contrasts between 1% and 5% LA treatments and 1% and 5% LA-H treatments. Linear patterns were analyzed using contrast statement of SAS accounting for unequal spacing among Control and treatments with 0.5, 1, and 5% LA or Control and treatments 1% LA-H and 5% LA-H. Interactions between LA concentration × heat were assessed where applicable. Duplicates per subsample were averaged and used as the experimental unit in the statistical analysis. Processing method served as fixed effect and sample nested within treatment as random effect. Degrees of freedom were approximated by Kenward-Roger method. Differences at $P<0.05$ level were declared significant.

Results

Impact of Lactic Acid and Heat Treatment on the Hydrolysis of Inositol Hexakisphosphate

The concentration of total P was not different between native and LA treated barley. Additional heat treatment reduced total P concentration by 0.3 g in LA-H treated barley compared to the native barley (Table 1). However, total P concentration did not differ between LA and LA-H treated barley. Myo-inositol hexakisphosphate concentration decreased in response to LA treatment and, in particular, when barley was treated with LA and oven-heated at 55°C (Figure 1). Gradual increase in LA concentration from 0 to 5% resulted in a linear decrease in $InsP_6$-P concentration from 2.55 g $InsP_6$-P/kg DM for control barley to 2.24, 2.04, and 1.75 g $InsP_6$-P/kg DM for 0.5, 1, and 5% LA, respectively. The additional heat treatment further lowered the $InsP_6$-P concentration to 1.55 and 1.49 g/kg DM for 1 and 5% LA-H, respectively. $InsP_3$ was present in all treatments in amounts of 0.05–0.11 g $InsP_3$-P/kg DM. $InsP_4$ and $InsP_5$ isomers were only quantifiable for 5% LA as well as 1 and 5% LA-H, ranging from 0.10–0.14 g $InsP_4$-P/kg and 0.16–0.22 g $InsP_5$-P/kg DM.

Proportions of $InsP_6$-P and total InsP-P relative to total P in barley are shown in Figure 2. The $InsP_6$-P proportion decreased ($P<0.01$) in barley when treated with LA (from 62.1 in control to 54.7, 50.8, 45.3% for 0.5%, 1% and 5% LA, respectively), and the extent of reduction was greater ($P<0.001$) when the LA-H treatment was applied (41.4 and 40.5% for 1 and 5% LA-H treatments, respectively; Figure 2), compared to the control. Also, when comparing LA-H with LA treated barley, LA-H treatment reduced the proportion of $InsP_6$-P compared to LA treatment ($P=0.012$). The $InsP_3$-P was only a very small proportion of total InsP-P in the control barley and barley treated with 0.5% and 1% LA; therefore, total InsP-P mainly comprised $InsP_6$-P for these treatments. Due to the increase in $InsP_4$-P and $InsP_5$-P with 5% LA and 1 and 5% LA-H, the proportion of total InsP-P was similar for LA and LA-H treated barley but lower ($P<0.01$) when compared to the control barley (Figure 2).

The $Ins(1,5,6)P_3$ was quantifiable for the control barley and all treatments. In the control group as well as in the treatment with 0.5% LA a peak deriving from one or more of the coeluting isomers $Ins(1,2,6)P_3$, $Ins(1,4,5)P_3$ and $Ins(2,4,5)P_3$ was not detectable, whereas a peak from coeluting isomers was detectable but not quantifiable in barley treated with 5% LA and 1 and 5% LA-H (Figure 3). The $Ins(1,2,5,6)P_4$ was the predominant $InsP_4$ isomer in barley treated with 5% LA and 1 and 5% LA-H (Figure 3). The $InsP_4$ isomer was found in barley treated with 0.5 and 1% LA in amounts below the quantification limit but was not detected in control barley. Furthermore, $Ins(1,2,3,4)P4$ was found in 5% LA and 1 and 5% LA-H treated barley but not in the other treatments. In 5% LA and 1 and 5% LA-H treated barley, $Ins(1,2,3,4,5)P_5$ was the primary $InsP_5$ isomer. $Ins(1,2,3,4,6)P_5$ was detected in the control group and in LA-H treatments, $Ins(1,2,4,5,6)P_5$ were above the detection limit but not quantifiable for all treatments (Figure 4).

Impact of Chemical and Heat Treatment on Barley's Chemical Composition

After soaking barley for 48 h, pH values of barley treated with LA or LA-H raised by 0.3 to 1.2 units (pH 2.5, 2.7, 2.1, 2.7 and 2.1 for 0.5% LA, 1% LA, 5% LA, 1% LA-H and 5% LA-H pre-incubation; and pH 3.7, 3.2, 2.4, 3.9 and 3.0 for 0.5% LA, 1% LA, 5% LA, 1% LA-H and 5% LA-H after 48 h of incubation, respectively).

The greatest change in the chemical composition was observed for the starch content when comparing treated barley samples with the native barley grain (Table 1, Figure 5). In general, LA and in particular LA-H treatment decreased ($P<0.05$) total starch content of barley. Resistant starch, both as g/kg DM and as proportion of total starch, was higher ($P<0.001$) in LA treated barley than in control barley (Figure 5) and peak increase was attained by 5% LA (RS relative to total starch: 0.9 in control vs. 5% in 5% LA). However, when the barley samples that were treated with 5% LA underwent thermal treatment, RS content was comparable to the native barley grain (Figure 5).

The CP content of barley did not change when barley was treated with LA, but additional heat treatment lowered the concentration of CP by 0.4% units compared to the control barley (Table 1). Moreover, LA and LA-H treatment modified the fiber fractions of barley. The contents of $aNDF_{OM}$ and ADF_{OM} increased by 1.7 and 0.8% in response to LA treatment, respectively, whereas the content of HC remained similar for LA treated and control barley. The heat treatment increased $aNDF_{OM}$ and ADF_{OM} concentrations in barley when 1% LA treatment was used compared to the control, whereas heat decreased the $aNDF_{OM}$ concentration by approximately 2% when barley was soaked in 5% LA (Table 1). This finding suggests an interaction ($P<0.01$) between LA and heat treatment for these variables. As a consequence, HC content was reduced by approximately 2.5% with the 5% LA-H treatment compared to the control barley. Barley treated with LA and LA-H also contained less crude ash than the native control barley.

Table 1. Nutrient composition of native barley (CON) or barley steeped in various concentrations of lactic acid at room temperature at 22°C (LA) or oven-heated at 55°C (LA-H).

Item[2]	CON	LA			LA-H		SEM[1]	P-value[3]		
		0.5%	1%	5%	1%	5%		1	2	3
Dry matter (%)	90.7	93.0	92.2	93.1	91.8	92.2	0.09	<0.001	<0.001	<0.001
Starch (% DM)	59.5	55.2	52.5	54.7	55.3	54.5	0.57	<0.001	<0.001	0.038
Crude protein (% DM)	13.4	13.1	13.6	13.1	13.0	12.9	0.07	0.352	0.004	<0.001
NDF (% DM)	14.5	15.0	16.9	16.8	16.1	12.2	0.30	<0.010	0.490	<0.010
ADF (% DM)	5.34	5.97	6.36	6.20	6.36	5.30	0.220	0.010	0.130	0.110
HC (% DM)	9.15	9.04	10.5	10.6	9.77	6.94	0.360	0.090	0.150	<0.010
Ash (% DM)	2.39	1.98	1.97	1.97	1.90	1.83	0.028	<0.001	<0.001	0.002
P (g/kg DM)	4.11	4.07	3.99	3.85	3.79	3.71	0.084	0.220	<0.010	0.100

[1]SEM = standard error of the mean (n = 3).
[2]DM = dry matter, NDF = neutral detergent fiber, ADF = acid detergent fiber, HC = hemicelluloses (NDF – ADF), P = total phosphorus.
[3]Contrasts, 1 = Control vs. LA, 2 = Control vs. LA-H, 3 = LA (1 and 5%) vs. LA-H (1 and 5%+oven-heating).

Discussion

There is an increasing interest in enhancing utilization of minerals from cereal grains used in animal nutrition. This strategy alleviates the dependency on inclusion of large amounts of inorganic P in animal diets with great economical and ecological importance [8,31]. Because of the low availability of P in cereals for monogastric livestock species, a range of feed processing techniques has been applied to reduce their $InsP_6$ concentration. Yet, in the feeding of monogastric livestock species such as swine and poultry, processing techniques used to increase P availability of feeds are often restricted to microbial phytase supplementation [8,32]. Our data indicated that treatment of barley grain with LA and LA-H was able to decrease the $InsP_6$ concentration and thus potentially increase P availability in barley. Most previous studies investigating the effect of LA on phytate degradation focused only on $InsP_6$ disappearance [16,18]. Here, we could show characteristic changes in the accumulation of lower InsP, such as $InsP_3$ to $InsP_5$, related to the LA concentration and heat treatment. These lower InsP may interfere less in intestinal mineral availability than $InsP_6$; however, $InsP_3$ to $InsP_5$ still bind P and can have an inhibitory effect on mineral absorption [33]. Because soaking of cereals in water is current practice in liquid feeding systems for livestock, we abstained from comparing the effects of LA treatment with soaking barley in water in the present study. Also, the present processing of barley grain aimed at being applied in dry feeding systems; therefore, the comparison between the native barley and the LA-treated barley was more relevant for the present study than the comparison between soaking in water and LA.

Overall, the concentration of total P and $InsP_6$ in native barley were in accordance with data from previous studies [34–37] showing comparable $InsP_6$ disappearances when barley was treated with LA and and LA-H [16,18,38]. Accordingly, the $InsP_6$ reducing effect of LA was more pronounced at higher concentrations and potentiated by the heat treatment [16,18,38]. The most effective treatments in the present study, i.e. 5% LA, 1% LA-H and 5% LA-H, converted 17 to 22% of $InsP_6$-P into inorganic P or lower InsP-P in barley grain and the disappearance of $InsP_6$-P was about 10% greater with heat treatment than at room temperature.

Plant phytases and $InsP_6$ are mostly localized in the aleurone layer of cereal grains [39–41]. The two phytases isolated from barley are activated in wet conditions when a slightly acidic pH of 5 and 6 is reached, respectively [39]. Soaking of cereal grains stimulates endogenous LA production causing lower pH with progressing incubation time [16,36,40,42]. Treating barley grains with LA solutions might therefore mimic the endogenous LA production, shortening the time until the critical pH value is reached for phytase activation. However, in this experiment pH values of LA treated barley were much more acidic than the actual pH values for optimum endogenous phytase activity. Possible explanations for $InsP_6$ removal during treatment with LA without or with heat may therefore be that endogenous phytases of barley may have been shortly activated during the soaking process and a certain phytase activity during the drying process cannot be excluded, thereby contributing to the inorganic P release with LA and LA-H treatment. Yet, a reduction in phytase activity was previously found in barley grains soaked in 0.8% LA when compared to barley soaked in water after 48 and 96 h of incubation [16]. Endogenous phytase activity was not determined in the present study. However, it can be assumed that other processes, such as leaching of nutrients and acidic ester hydrolysis, than an enhancement of phytase activity likely contributed to the $InsP_6$ degradation in the present study. Soaking processes are

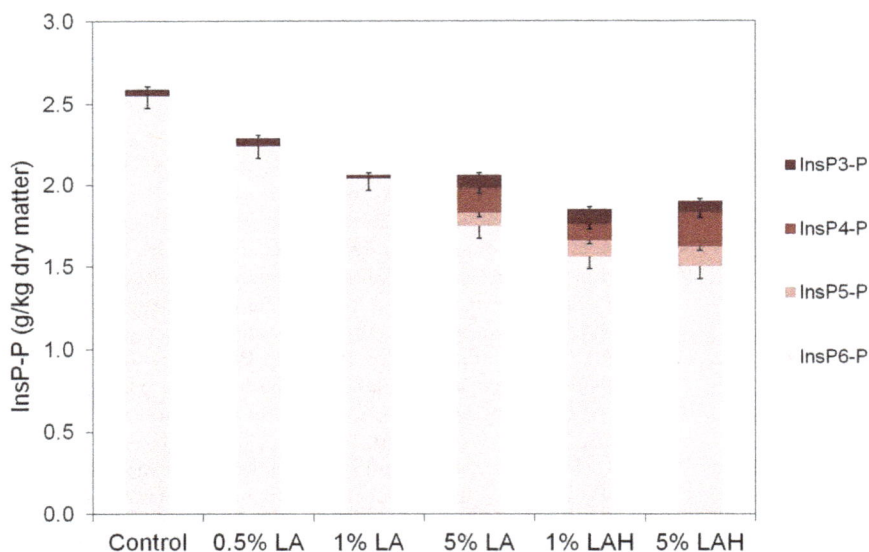

Figure 1. The concentrations of P pertaining to *myo*-inositol tri- to hexakisphosphate (InsP₃-P, InsP₄-P, InsP₅-P, and InsP₆-P) and to the sum of them (total InsP-P) in untreated barley (control) or barley soaked in increasing concentrations of lactic acid at room temperature in 22°C (LA) or oven-heated at 55°C (LA-H). Data are shown as least square means ± standard error of the mean (n = 3). For InsP₃-P: all contrasts $P > 0.10$; for InsP₄-P: control vs. LA $P = 0.50$, control vs. LA-H $P = 0.030$, LA vs. LA-H $P = 0.097$; for InsP₅-P: control vs. LA $P = 0.10$, control vs. LA-H $P < 0.001$, LA vs. LA-H $P = 0.006$; for InsP₆-P: control vs. LA $P < 0.001$, control vs. LA-H $P < 0.001$, LA vs. LA-H $P < 0.001$; for the sum of InsP-P: control vs. LA $P < 0.001$, control vs. LA-H $P < 0.001$, LA vs. LA-H $P = 0.055$.

generally associated with leaching of nutrients including minerals [3,31,43]. Leaching of minerals into the soaking medium may have been indicated by the lower crude ash concentration in treated barley samples and the higher pH of the soaking medium after the 48-hour incubation compared with initial pH values. Heat treatment can potentiate the soaking effect as the heat causes structural changes in the grain leading to a more rapid hydration (i.e. swelling) of the grain [31,40,44]. As we could only observe a reduction in total P of barley when treated with 1 and 5% LA-H, loss of P and with this of InsP₆ by leaching may have been mostly restricted to these treatments. Haraldsson and coworkers [16] estimated that a loss of 5% of InsP₆ during soaking and heat treatment (48°C) of barley with 0.8% LA could be explained by leaching processes in their study.

Another explanation for the reduction in InsP₆ in response to LA and LA-H treatment of barley could be related to the low pH in the soaking medium. Phosphate groups are esterified to the inositol ring of InsP, and can be removed by acidic ester hydrolysis [16,31,45]. Our data suggest an acceleration of acid hydrolysis of InsP₆ in response to additional heat treatment, which is indicated by the lower InsP₆ concentration and the accumulation of InsP₄ and InsP₅ for LA-H treated barley. Because only small amounts of InsP₅ to InsP₃ were detected, it is likely that this treatment might have triggered a complete degradation of lower InsP as soon as the first phosphate group was released from InsP₆ [44]. The accumulation pattern of lower InsP isomers may help to differentiate whether InsP₆ hydrolysis was more related to endogenous phytase activity or pH and heat. In this experiment, the occurrence of Ins(1,2,3,4,5)P₅, Ins(1,2,3,4)P₄, Ins(1,2,5,6)P₄ and Ins(1,2,6)P₃ with 5% LA and 1 and 5% LA-H may indicate the action of cereal phytases because these phytases, like barley phytases P1 and P2, are suggested to be 6-phytases [E.C.3.1.3.26] [35,46,47]. However, an ultimate distinction between endogenous 6-phytase action and pH and heat effects cannot be made using the present experimental design.

In line with previous studies evaluating soaking procedures [43], the LA and LA-H treatment of barley resulted in small losses of other nutrients. Observed changes in nutrient composition may reduce the feed value of LA and LA-H treated barley, with the decrease in total starch as the most critical loss for the feed value as it affects the energy concentration of barley. Despite its indigestibility for the host animal, the greater RS concentration of LA treated barley may increase the functional and thus health-promoting potential of barley for livestock animals, such as pigs [48] and ruminants [19–22,49]. Aside from leaching of nutrients, e.g. minerals, starch, and water-soluble protein into the soaking medium and potentiation of this effect by heat treatment [43], it is thinkable that the low pH in the soaking medium modified the molecule structure of some nutrients; for instance leading to the higher RS content of barley with increasing LA concentration [25]. Interestingly, the combination of the highest LA concentration and heat likely abolished the effect on RS formation, which confirms previous findings [25].

The leaching of certain nutrients into the soaking medium likely caused an increase in concentrations of other nutrients in barley such as fiber fractions. Here, aNDF_{OM} and ADF_{OM} contents increased for all LA and 1% LA-H treated barley samples thereby maintaining a similar HC content among treatments. Yet, low pH combined with heat treatment seemed to catalyze degradation of fibrous components in barley grain as indicated by the lower aNDF_{OM}, ADF_{OM} and HC contents for 5% LA-H treatment compared to all LA and 1% LA-H treatments. Fibrous components can be mostly found in the three aleurone layers of the barley grain and mainly consist of cellulose, arabinoxylan and mixed-linked β-glucan [50]. According to previous studies, the arabinoxylan fraction may be more susceptible to low pH and heat than the cellulose and β-glucan fractions [16,51,52]. The β-glucan fraction in barley may even be stabilized by LA and heat treatment [16].

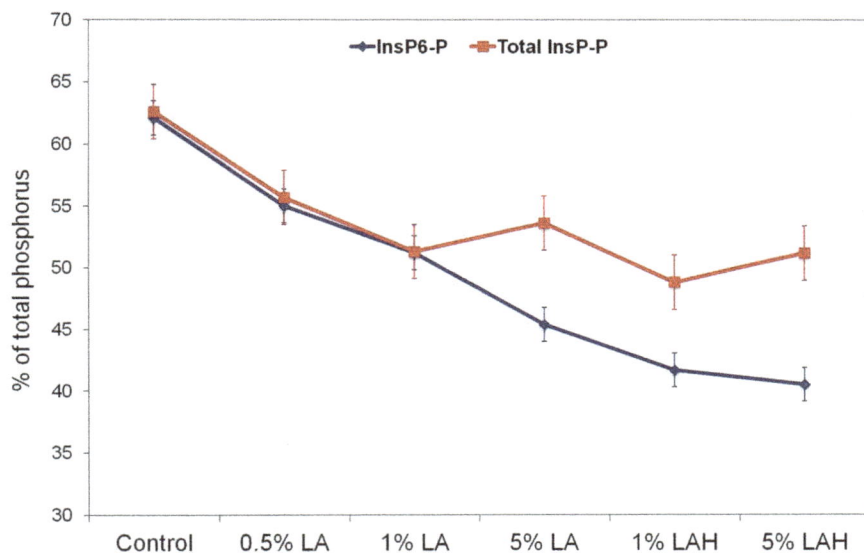

Figure 2. Changes in *myo*-inositol hexakisphosphate (InsP$_6$-P) and the sum of InsP$_3$-P to InsP$_6$-P (total InsP-P) relative to total phosphorus of untreated barley (control) or barley soaked in increasing concentrations of lactic acid at room temperature in 22°C (LA) or oven-heated at 55°C (LA-H). Data are shown as least square means $\pm$ standard error of the mean (n = 3). LA and LA-H effects on InsP$_6$-P: Control vs. LA $P<0.01$, Control vs. LA-H $P<0.001$, LA vs. LA-H $P = 0.012$; LA and LA-H effects on total InsP-P: Control vs. LA $P<0.01$, Control vs. LA-H $P< 0.001$, LA vs. LA-H $P = 0.14$.

Finally, the total InsP$_6$ degradation by LA and LA-H treatment may remain below the degradation extent reported by dietary supplementation of microbial phytase [34]. Yet, the conditions for the pre-treatment of barley grain may be more easily controlled and stabilized than luminal conditions in the gastrointestinal tract which are necessary to guarantee sufficient InsP$_6$ degradation.

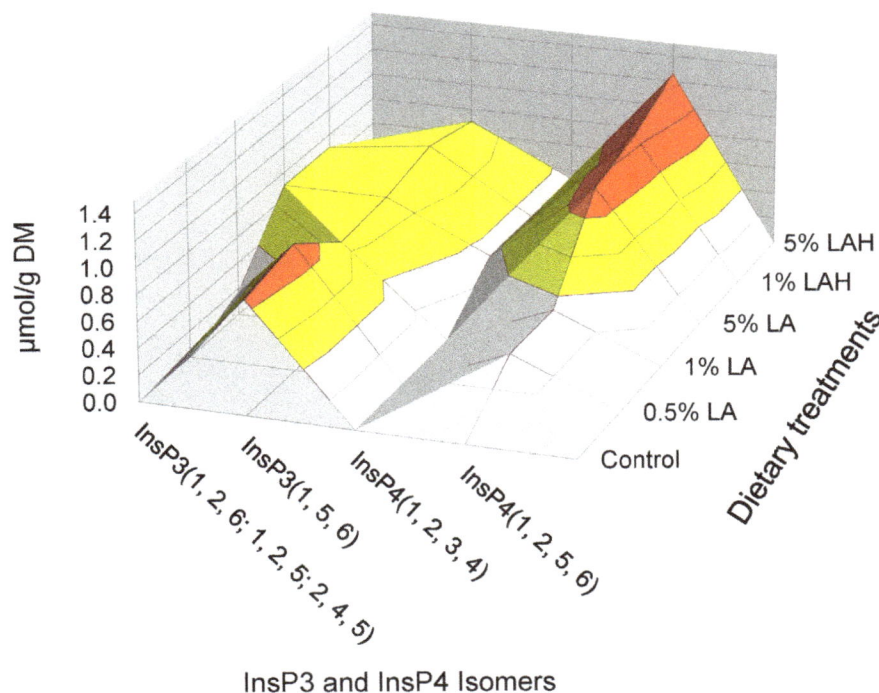

Figure 3. Concentrations of various isomers of *myo*-inositol triphosphate (InsP$_3$) and tetraphosphate (InsP$_4$) in untreated barley grain (control) or barley grain soaked in increasing concentrations of lactic acid at room temperature in 22°C (LA) or oven-heated at 55°C (LA-H). Data are shown as least square means (n = 3). Isomers exceeding a concentration of 1 µmol/g dry matter were quantified (area labeled in red color); isomers having concentrations between 0.5 to 1 µmol/g dry matter (detection limit and measurement threshold, respectively) were detected but could not be quantified (area labeled in yellow color); are below detection limit of these isomers is shown in white color ($<$ 0.5 µmol/g dry matter).

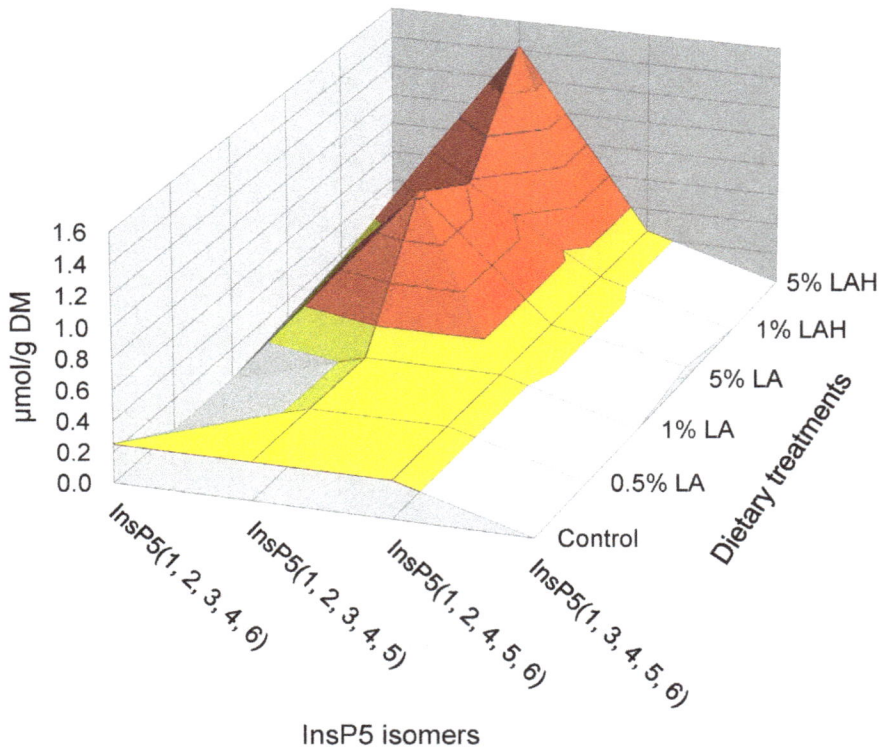

Figure 4. Concentrations of various isomers of *myo*-inositol pentaphosphate (InsP₅) in untreated barley (control) or barley soaked in increasing concentrations of lactic acid at room temperature in 22°C (LA) or oven-heated at 55°C (LAH). Data are shown as least square means (n = 3). Isomers exceeding a concentration of 0.5 μmol/g dry matter were quantified (area labeled in red color); isomers having a concentration between 0.25 to 0.5 μmol/g dry matter (detection limit and measurement threshold, respectively) were detected but could not be quantified (area labeled in yellow); area below detection limit of these isomers is shown in white color (<0.25 μmol/g dry matter).

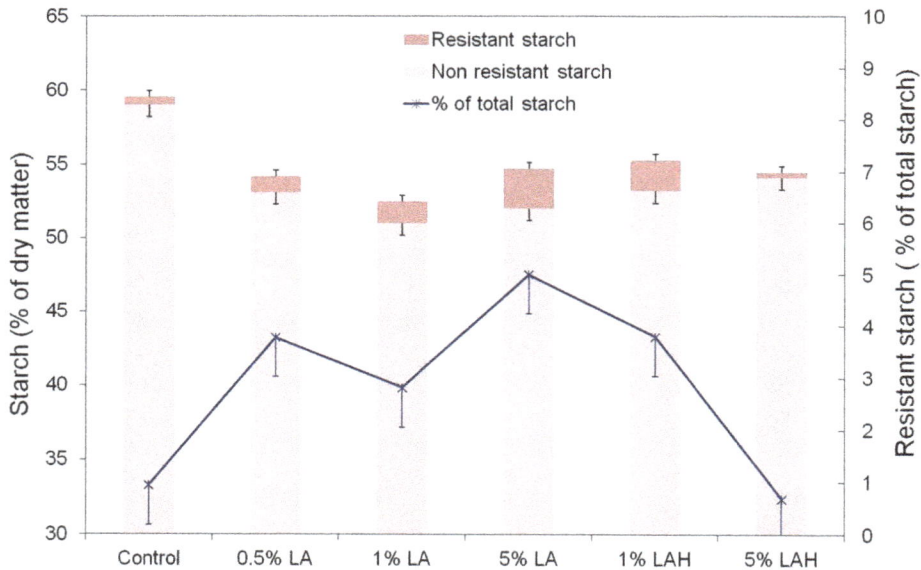

Figure 5. Changes in concentrations of resistant starch (RS) and non-resistant starch (NRS) of untreated barley (Control) or barley soaked in increasing concentrations of lactic acid at room temperature in 22°C (LA) or oven-heated at 55°C (LA-H). Data are shown as least square means ± standard error of the mean (n = 3). LA and LA-H effects on RS: Control vs. LA $P = 0.006$, Control vs. LA-H $P = 0.202$, LA vs. LA-H $P = 0.049$; LA and LA-H effects on RS relative to total starch: Control vs. LA $P = 0.006$, Control vs. LA-H $P = 0.186$, LA vs. LA-H $P = 0.051$; LA and LA-H effects on NRS: Control vs. LA $P < 0.001$, Control vs. LA-H $P < 0.001$, LA vs. LA-H $P = 0.033$.

Optimum microbial or cereal phytase activity depends on gastrointestinal pH and passage rate and may be biased in case luminal conditions are suboptimal in vivo. For instance, in pigs after weaning gastric pH may not reach the necessary acidic pH for microbial phytase activation [53]. Even in cattle nutrition, the inclusion of phytases has been suggested to optimize the utilization of dietary P [54] which indicates that, despite the highly complex rumen microbiota and long ruminal retention times of about 30–48 hours of ingested feed [55], $InsP_6$ hydrolyzing capacity may be limited, in particular when short forage particle size is fed [54]. Additional advantages of LA treatment of barley are improved storage stability by decreasing molding of the grain post-harvest and, when eaten, support of gastric barrier function in monogastric livestock animals [56]. In dairy cows, treating the barley fraction of the diet with 0.5 and 1% LA proved beneficial for rumen fermentation and immune-metabolic health status of the animals [19–22].

In conclusion, treating barley grain with LA or LA-H may be effective processing techniques to reduce the $InsP_6$ concentration of cereals used in animal feeding. The greatest $InsP_6$ hydrolysis was observed with the highest investigated LA concentration of 5% and heat treatment. Lower InsP profiles obtained with 5% LA, 1% LA-H and 5% LA-H treatments are similar to profiles found when microbial 6-phytase is applied. Changes in nutrient composition of barley grain due to LA and LA-H treatment, i.e. lower starch and ash concentrations and accumulation of fiber fractions, may impact the feed value of barley, but might increase its health-enhancing properties, in particular, due to greater concentrations of RS and dietary fiber. To determine the actual intestinal P availability and to assess the effects of changes in nutrient composition and functional abilities of LA and LA-H treated barley, further in vivo studies are needed.

Acknowledgments

The assistance of A. Dockner, S. Leiner and M. Wild (Institute of Animal Nutrition and Functional Plant Compounds, University of Veterinary Medicine Vienna, Austria) for the feed chemical analyses is gratefully acknowledged.

Author Contributions

Conceived and designed the experiments: QZ. Performed the experiments: KD MS. Analyzed the data: BM. Contributed reagents/materials/analysis tools: MS MR QZ. Contributed to the writing of the manuscript: BM QZ MS MR.

References

1. Baik BK, Ullreich E (2008) Barley for food: Characteristics, improvement, and renewed interest. J Cereal Sci 48: 233–242.
2. Egli I, Davidsson L, Juillerat MA, Barclay D, Hurrell RF (2002) The influence of soaking and germination on the phytase activity and phytic acid content of grains and seeds potentially useful for complementary feeding. J Food Sci 67: 3484–3488.
3. Stewart A, Nield H, Lott JNA (1988) An investigation of the mineral content of barley grains and seedlings. Plant Physiol 86: 93–97.
4. Sandberg AS (1991) The effect of food processing on phytate hydrolysis and availability of iron and zinc. Adv Exp Med Biol 289: 499–508.
5. Leenhardt F, Levrat-Verny MA, Chanliaud E, Rémésy C (2005) Moderate decrease of pH by sourdough fermentation is sufficient to reduce phytate content of whole wheat flour through endogenous phytase activity. J Agric Food Chem 53: 98–102.
6. Ariza A, Moroz OV, Blagova EV, Turkenburg JP, Waterman J, et al. (2013) Degradation of phytate by the 6-phytase from hafnia alvei : A combined structural and solution study. PLoS ONE 8: e65062.
7. Pable A, Gujar P, Khire JM (2014) Selection of phytase producing yeast strains for improved mineral mobilization and dephytinization of chickpea flour. J Food Biochem 38: 18–27.
8. Kiarie E, Romero LF, Nyachoti CM (2013) The role of added feed enzymes in promoting gut health in swine and poultry. Nutr Res Rev 26: 71–88.
9. Blaabjerg K, Jørgensen H, Tauson AH, Poulsen HD (2011) The presence of inositol phosphates in gastric pig digesta is affected by time after feeding a nonfermented or fermented liquid wheat- and barley-based diet. J Anim Sci 89: 3153–3162.
10. Proulx AK, Reddy MB (2007) Fermentation and lactic acid addition enhance iron bioavailability of maize. J Agric Food Chem 55: 2749–2754.
11. Afify AE-MMR, El-Beltagi HS, Abd El-Salam SM, Omran AA (2011) Bioavailability of iron, zinc, phytate and phytase activity during soaking and germination of white sorghum varieties. PLoS ONE 6: e25512.
12. Sanz-Penella JM, Frontela C, Ros G, Martinez C, Monedero V, et al. (2012) Application of bifidobacterial phytases in infant cereals: Effect on phytate contents and mineral dialyzability. J Agric Food Chem 60: 11787–11792.
13. Canibe N, Jensen BB (2003) Fermented and nonfermented liquid feed to growing pigs: effect on aspects of gastrointestinal ecology and growth performance. J Anim Sci 81: 2019–2031.
14. Canibe NH, Miettinen H, Jensen BB (2008) Effect of adding Lactobacillus plantarum or a formic acid containing-product to fermented liquid feed on gastrointestinal ecology and growth performance of piglets. Livest Sci 114: 251–262.
15. Plumed-Ferrer C, von Wright A (2009) Fermented pig liquid feed: nutritional, safety and regulatory aspects. J Appl Microbiol 106: 351–368.
16. Haraldsson A-K, Rimsten L, Alminger ML, Andersson R, Andlid T, et al. (2004) Phytate content is reduced and β-glucanase activity suppressed in malted barley steeped with lactic acid at high temperature. J Sci Food Agric 84: 653–662.
17. Tabekhia MM, Luh BS (1980) Effect of germination, cooking, and canning on phosphorus and phytate retention in dry beans. J Food Sci 45: 406–408.
18. Fredlund K, Asp N-G, Larsson M, Marklinder I, Sandberg AS (1997) Phytate reduction in whole grains of wheat, rye, barley and oats after hydrothermal treatment. J Cereal Sci 25: 83–91.
19. Iqbal S, Zebeli Q, Mazzolari A, Bertoni G, Dunn SM, et al. (2009) Feeding barley grain steeped in lactic acid modulates rumen fermentation patterns and increases milk fat content in dairy cows. J Dairy Sci 92: 6023–6032.
20. Iqbal S, Zebeli Q, Mazzolari A, Dunn SM, Ametaj BN (2010) Feeding rolled barley grain steeped in lactic acid modulated energy status and innate immunity in dairy cows. J Dairy Sci 93: 5147–5156.
21. Iqbal S, Terrill SJ, Zebeli Q, Mazzolari A, Dunn SM et al. (2012) Treating barley grain with lactic acid and heat prevented sub-acute ruminal acidosis and increased milk fat content in dairy cows. Anim Feed Sci Technol 172: 141–149.
22. Iqbal S, Zebeli Q, Mazzolari A, Dunn SM, Ametaj BN (2012) Barley grain-based diet treated with lactic acid and heat modulated plasma metabolites and acute phase response in dairy cows. J Anim Sci 90: 3143–3152.
23. Tanaka T, Imai Y, Kumagae N, Sato S (2010) The effect of feeding lactic acid to Salmonella typhimurium experimentally infected swine. J Vet Med Sci 72: 827–831.
24. Willamil J, Creus E, Pérez JF, Mateu E, Martín-Orúe SM (2011) Effect of a microencapsulated feed additive of lactic and formic acid on the prevalence of Salmonella in pigs arriving at the abattoir. Arch Anim Nutr 65: 431–444.
25. Deckardt K, Khiaosa-ard R, Grausgruber H, Zebeli Q (2014) Evaluation of various chemical and thermal feed processing methods for their potential to enhance resistant starch content in barley grain. Starch/Stärke 66: 558–565.
26. Philippy BQ, Bland JM (1988) Gradient ion chromatography of inositol phosphates. Anal Biochem 175: 162–166.
27. Skoglund E, Carlsson NG, Sandberg AS (1997) Determination of isomers of inositol mono- to hexaphosphates in selected foods and intestinal contents using high-performance ion chromatography. J Agric Food Chem 45: 431–436.
28. VDLUFA (Verband Deutscher Landwirtschaftlicher Untersuchungs- und Forschungsanstalten), in: Handbuch der Landwirtschaftlichen Versuchs- und Untersuchungsmethodik, Bd. III Die chemische Untersuchung von Futtermitteln, 4. Erg.-Lfg., VDLUFA-Verlag, Darmstadt, Germany, 1997.
29. Van Soest PJ, Robertson JB, Lewis BA (1991) Methods for dietary fiber, neutral detergent fiber, and nonstarch polysaccharides in relation to animal nutrition. J Dairy Sci 74: 3583–3597.
30. Shastak Y, Witzig M, Hartung K, Rodehutscord M (2012) Comparison of retention and prececal digestibility measurements in evaluating mineral phosphorus sources in broilers. Poultry Sci 91: 2201–2209.
31. Bohn L, Meyer AS, Rasmussen SK (2008) Phytate: impact on environment and human nutrition. A challenge for molecular breeding. J Zhejiang Univ Sci B 9: 165–191.
32. Rutherfurd SM, Chung TK, Moughan PJ (2014) Effect of microbial phytase on phytate P degradation and apparent digestibility of total P and Ca throughout the gastrointestinal tract of the growing pig. J Anim Sci 92: 189–197.
33. Sandberg AS, Brune M, Carlsson NG, Hallberg L, Skoglund E, et al. (1999) Inositol phosphates with different numbers of phosphate groups influence iron absorption in humans. Am J Clin Nutr 70: 240–246.

34. Shen Y, Yin YL, Chavez ER, Fan MZ (2005) Methodological aspects of measuring phytase activity and phytate phosphorus content in selected cereal grains and digesta and feces of pigs. J Agric Food Chem 53: 853–859.

35. Pontoppidan K, Pettersson D, Sandberg AS (2007) The type of thermal feed treatment influences the inositol phosphate composition. Anim Feed Sci Technol 132: 137–147.

36. Blaabjerg K, Nørgaard JV, Poulsen HD (2012) Effect of microbial phytase on phosphorus digestibility in non-heat-treated and heat-treated wheat-barley pig diets. J Anim Sci 90: 206–208.

37. Esmaeilipour O, van Krimpen MM, Jongbloed AW, De Jonge LH, Bikker P (2012) Effects of temperature, pH, incubation time, and pepsin concentration on the in vitro stability of 2 intrinsic phytase of wheat, barley, and rye. Anim Feed Sci Technol 175: 168–174.

38. Bergman EL, Fredlund K, Reinikainen P, Sandberg AS (1999) Hydrothermal processing of barley (cv. Blenheim): optimisation of phytate degradation and increase of free myo-inositol. J Cereal Sci 29: 261–272.

39. Greiner R, Jany KD, Alminger ML (2000) Identification and properties of myo-inositol hexakisphosphate phosphohydrolases (Phytases) from barley (Hordeum vulgare). J Cereal Sci 31: 127–139.

40. Raboy V (2003) Myo-Inositol-1,2,3,4,5,6-hexakisphosphate. Phytochemistry 64: 1033–1043.

41. Dionisio G, Holm PB, Brinch-Pedersen H (2007) Wheat (Triticum aestivum L.) and barley (Hordeum vulgare L.) multiple inositol polyphosphate phosphatases (MINPPs) are phytases expressed during grain filling and germination. Plant Biotech J 5: 325–338.

42. Beal JD, Niven SJ, Brooks PH, Gill BP (2005) Variation in short chain fatty acid and ethanol concentration resulting from the natural fermentation of wheat and barley for inclusion in liquid diets for pigs. J Sci Food Agric 85: 433–440.

43. Hurrell RF (2004) Phytic acid degradation as a means of improving iron absorption. Int J Vitam Nutr Res. 74: 445–452.

44. Blaabjerg K, Jørgensen H, Tauson AH, Poulsen HD (2010) Heat-treatment, phytase and fermented liquid feeding affect the presence of inositol phosphates in ileal digesta and phosphorus digestibility in pigs fed a wheat and barley diet. Animal 4: 876–85.

45. March JG, Simonet BM, Grases F, Salvador A (1998) Indirect determination of phytic acid in urine. Analytica Chimia Acta 1–3: 63–68.

46. Sandberg AS (2001) In vitro and in vivo degradation of phytate. In: Food phytases, Eds. R Reddy and SK Sathe. CRC Press, Boca Raton, FL, USA; 139–156.

47. Greiner R, Farouk AE, Carlsson NG, Konietzny U (2007) myo-inositol phosphate isomers generated by the action of a phytase from a Malaysian waste-water bacterium. Protein J 26: 577–584.

48. Regmi PR, Metzler-Zebeli BU, Gänzle MG, Van Kempen TAG, Zijlstra RT (2011) Starch with high amylose content and low in vitro digestibility increases intestinal nutrient flow and microbial fermentation and selectively promotes bifidobacteria in pigs. J Nutr 141: 1273–1280.

49. Deckardt K, Khol-Parisini A, Zebeli Q (2013) Peculiarities of enhancing resistant starch in ruminants using chemical methods: opportunities and challenges. Nutrients 5: 1970–1988.

50. Selvendran RR, Stevens BJH, DuPont MS, (1987) Dietary fiber: chemistry, analysis and properties. Adv Food Res 31: 117–209.

51. Agger J, Johansen KS, Meyer AS (2011) pH catalyzed pretreatment of corn bran for enhanced enzymatic arabinoxylan degradation. N Biotechnol 28: 125–135.

52. Holopainen-Mantila U, Marjamaa K, Merali Z, Käsper A, de Bot P, et al. (2013) Impact of hydrothermal pre-treatment to chemical composition, enzymatic digestibility and spatial distribution of cell wall polymers. Bioresour Technol. 138: 156–62.

53. de Lange CFM, Pluske J, Gong J, Nyachoti CM (2010) Strategic use of feed ingredients and feed additives to stimulate gut health and development in young pigs. Livest Sci 134: 124–134.

54. Jarrett JP, Wilson JW, Ray PP, Knowlton KF (2014) The effects of forage particle length and exogenous phytase inclusion on phosphorus digestion and absorption in lactating cows. J Dairy Sci. 97: 411–418.

55. Zebeli Q, Tafaj M, Weber I, Dijkstra J, Steingass H, et al. (2007) Effects of varying dietary forage particle size in two concentrate levels on chewing activity, ruminal mat characteristics, and passage in dairy cows. J Dairy Sci. 90: 1929–1942.

56. Heo JM, Opapeju FO, Pluske JR, Kim JC, Hampson DJ, et al. (2013) Gastrointestinal health and function in weaned pigs: a review of feeding strategies to control post-weaning diarrhoea without using in-feed antimicrobial compounds. J Anim Physiol Anim Nutr (Berl) 97: 207–237.

18

De Novo Transcriptome Assembly and Analyses of Gene Expression during Photomorphogenesis in Diploid Wheat *Triticum monococcum*

Samuel E. Fox[1⅃¶¤a], **Matthew Geniza**[1,2¶⅃], **Mamatha Hanumappa**[1⅃], **Sushma Naithani**[1,3], **Chris Sullivan**[1,3], **Justin Preece**[1], **Vijay K. Tiwari**[4¤b], **Justin Elser**[1], **Jeffrey M. Leonard**[4], **Abigail Sage**[1], **Cathy Gresham**[5], **Arnaud Kerhornou**[6], **Dan Bolser**[6], **Fiona McCarthy**[7], **Paul Kersey**[6], **Gerard R. Lazo**[8], **Pankaj Jaiswal**[1,4*]

1 Department of Botany and Plant Pathology, Oregon State University, Corvallis, Oregon, United States of America, 2 Molecular and Cellular Biology Graduate Program, Oregon State University, Corvallis, Oregon, United States of America, 3 Center for Genome Research and Biocomputing, Oregon State University, Corvallis, Oregon, United States of America, 4 Department of Crop and Soil Science, Oregon State University, Corvallis, Oregon, United States of America, 5 Institute for Genomics, Biocomputing and Biotechnology, Mississippi State University, Mississippi State, Mississippi, United States of America, 6 European Bioinformatics Institute, Hinxton, Cambridge, United Kingdom, 7 School of Animal and Comparative Biomedical Sciences, University of Arizona, Tucson, Arizona, United States of America, 8 USDA-ARS, Western Regional Research Center, Albany, California, United States of America

Abstract

Background: *Triticum monococcum* (2n) is a close ancestor of *T. urartu*, the A-genome progenitor of cultivated hexaploid wheat, and is therefore a useful model for the study of components regulating photomorphogenesis in diploid wheat. In order to develop genetic and genomic resources for such a study, we constructed genome-wide transcriptomes of two *Triticum monococcum* subspecies, the wild winter wheat *T. monococcum ssp. aegilopoides* (accession G3116) and the domesticated spring wheat *T. monococcum ssp. monococcum* (accession DV92) by generating *de novo* assemblies of RNA-Seq data derived from both etiolated and green seedlings.

Principal Findings: The *de novo* transcriptome assemblies of DV92 and G3116 represent 120,911 and 117,969 transcripts, respectively. We successfully mapped ~90% of these transcripts from each accession to barley and ~95% of the transcripts to *T. urartu* genomes. However, only ~77% transcripts mapped to the annotated barley genes and ~85% transcripts mapped to the annotated *T. urartu* genes. Differential gene expression analyses revealed 22% more light up-regulated and 35% more light down-regulated transcripts in the G3116 transcriptome compared to DV92. The DV92 and G3116 mRNA sequence reads aligned against the reference barley genome led to the identification of ~500,000 single nucleotide polymorphism (SNP) and ~22,000 simple sequence repeat (SSR) sites.

Conclusions: *De novo* transcriptome assemblies of two accessions of the diploid wheat *T. monococcum* provide new empirical transcriptome references for improving Triticeae genome annotations, and insights into transcriptional programming during photomorphogenesis. The SNP and SSR sites identified in our analysis provide additional resources for the development of molecular markers.

Editor: Girdhar K. Pandey, University of Delhi South Campus, India

Funding: This work was supported by the laboratory startup funds provided by the Department of Botany and Plant Pathology and College of Agricultural Sciences at Oregon State University to PJ and SN. Research funds were also provided by the Department of Crop and Soil Science, Agricultural Sciences at Oregon State University to JL. MG received Anita S. Summers travel award from the Department of Botany and Plant Pathology at Oregon State University and to present this work at the Plant and Animal Genome Conference. Contributions from International collaborators, PK, DB and AK were supported by the 'transPLANT' project funded by the European Commission within its 7th Framework Programme, under the thematic area 'Infrastructures' (contract #283496) and the Triticeae Genomics for Sustainable Agriculture project (BBSRC: #BB/J003743/1). The funders had no role in the study design, data analysis, or preparation of the manuscript.

Competing Interests: The authors have declared that no competing interests exist.

* E-mail: jaiswalp@science.oregonstate.edu

⅃ These authors contributed equally to this work.

¶ These authors are co-first authors on this work.

¤a Current address: Department of Biology, Linfield College, McMinnville, Oregon, United States of America,
¤b Current address: Department of Plant Pathology, Kansas State University, Manhattan, Kansas, United States of America

Introduction

Einkorn wheat is one of three cereal crops domesticated prior to 7000 B.C. that contributed to the Neolithic Revolution [1]. Stands of wild einkorn, subspecies *Triticum monococcum* ssp. *aegilopoides*, are extensive in rocky areas of southeastern Turkey [1]. Domesticated einkorn, subspecies *T. monococcum* L. ssp. *monococcum* L. (2n = 14) originated in the Karacadağ mountains of Turkey [2] and was widely cultivated during the Neolithic period. Domesticated einkorn differs from the wild accessions in possessing plumper seeds and tough rachis phenotypes that prevent seed shattering, a domesticated trait selected for avoiding loss of yield [3].

T. monococcum, carrying the representative diploid wheat A genome (A^mA^m), is closely related to *T. urartu* (A^uA^u), the donor of the A genome of the cultivated hexaploid (AABBDD) wheat (*T. aestivum*) [4]. The genome size of *T. monococcum* is about 5.6 Gb, which is 12 times the size of the rice genome and 40 times the genome of the model dicot plant *Arabidopsis thaliana* [5]. However, in comparison to the ~17 Gb genome size of common hexaploid wheat, the diploid *T. monococcum* offers relative simplicity and has been used extensively as a model [6]. The many existing wild populations of *T. monococcum* growing in their natural habitat have suffered little selection pressure and thus offer opportunities to study its diversity [7]. They also serve as a reservoir of useful alleles and traits, such as salinity tolerance [8] and disease resistance [9,10], and thus have been utilized for generating genetic maps to facilitate comparative mapping [11] and map-based cloning of genes [12,13]. Combining the sequence and positional information of the genes based on recently published barley (*Hordeum vulgare*) [14], *T. urartu* [15] and *Aegilops tauschii* [16,17] genomes with the genetic tools and transcriptome-based resources available for *T. monococcum* reported herein will allow progress in future genetic studies in wheat and other closely-related species.

Light regulates a wide range of plant processes including seed germination, organ, cell and organelle differentiation, flowering [18–21] and metabolism [22]. The germination of a seed in the dark follows skotomorphogenesis (the growth of an etiolated seedling). Upon exposure to light, seedlings go through photomorphogenesis (greening) that is marked by chlorophyll biosynthesis, differentiation of protoplastids into chloroplasts, the initiation of carbon assimilation, elongation and thickening of the hypocotyl, and the activation of the shoot apical meristem leading to the development of the first true leaves [23–25]. Although the transition from skotomorphogenic to photomorphogenic growth has been well-documented in *Arabidopsis* [24,25], the complex gene networks at the genome level controlling this developmental transition in wheat are not well understood.

In order to investigate and identify the complex transcriptional network associated with seedling photomorphogenesis in Einkorn wheat, we conducted Illumina-based transcriptome analyses (RNA-Seq) of two *T. monococcum* subspecies: DV92, a spring Einkorn accession of the cultivated *T. monococcum* ssp. *monococcum* collected in Italy and G3116, a wild winter Einkorn, *T. monococcum* ssp. *aegilopoides*, collected in Lebanon [11]. Computational analysis of the transcriptome data provided functional annotations to the gene models and gene families. We also identified gene loci harboring SSR and SNP sites and predicted their consequences on transcript structure, coding features and expression.

Results

Sequencing and *de novo* assembly of transcriptomes

A total of twelve cDNA libraries were created, six from each of the DV92 and G3116 accessions. These libraries represent three replicates prepared from dark-grown seedlings sampled eight days (8DD) after germination, and three replicates prepared from seedlings grown in the dark for eight days and then exposed to continuous light for 48 hours, sampled eleven days after germination (48LL). The sequencing of cDNA libraries from the 8DD and 48LL samples on the Illumina HiSeq 2000 platform generated 39.56 Gbp of nucleotide sequence from DV92 and 37.65 Gbp from G3116. *De novo* assemblies were performed using Velvet and

Oases [26], resulting in a total number of 120,911 transcripts for DV92 and 117,969 transcripts for G3116 ($\geq$200 bp in length; Table 1). The assemblies of each accession were created in a two-step process: first, two separate assemblies were generated from optimized 31 and 35 K-mer lengths; second, transcript isoforms were clustered to obtain discrete assemblies for DV92 and G3116, representing the total number of unique transcripts after merging. The quality of transcriptome assemblies was assessed with various statistical metrics including the overall number (coverage), average length and diversity of transcripts (the estimated number of discrete loci assembled), and via comparison with published, annotated genomes. The average length for DV92-derived transcripts was 1,847 bp; the average length for G3116-derived transcripts was 1,783 bp (Table 1). The overall frequency distributions of transcript lengths are similar to other *de novo* plant transcriptome assemblies [27–29] and similar to the overall distribution of barley and *T. urartu* gene lengths (Figure S1).

Comparisons with the Triticeae genomes

To annotate, characterize and approximate the coverage of sequenced and assembled transcripts representing common gene loci, we compared the transcripts of DV92 and G3116 to transcripts of other plant species from Poaceae (Table 2) using BLAST [30]. *Triticum* shares a more recent common ancestor with barley than with *Brachypodium* [11], therefore, we chose the barley genome (Gramene 030312 v2.18) as the reference for further comparative analysis. Over 92% of transcripts from both DV92 and G3116 were successfully mapped to the barley genome and show broad coverage of the genome (Table 2; Figure 1). Approximately 77% of DV92 and G3116 transcripts mapped to ~90% of the barley gene models with $\geq$95% percent identity (Figure 1; Tables 2 and 3). In the reciprocal BLAST analysis, we successfully mapped ~91% of the barley gene models to the G3116 transcriptome and ~93% of the barley transcripts to the DV92 transcriptome (Table 3).

Comparison of the DV92 and G3116 transcriptomes with the *T. urartu* (wheat A genome) and the *A. tauschii* (wheat D genome) genomes and gene models [15–17] suggest that ~84% of the *T. monococcum* transcripts from both accessions mapped to the *T. urartu* gene models, while ~86% mapped to the *A. tauschii* gene models (Table 2). 80–85% of the *A. tauschii* and *T. urartu* coding sequences matched DV92 or G3116 transcripts in a reciprocal BLASTn analysis (Table 3).

Functional annotation

InterPro domain annotations were assigned to 54,814 DV92 transcripts and 53,627 G3116 transcripts based on analyses of putative polypeptide encoded by the longest Open Reading Frame (ORF) for a given transcript (Table S1). InterPro domain mappings provided Gene Ontology (GO) annotations for 42,931 DV92 transcripts and 41,983 G3116 transcripts. Blast2GO [31] analysis provided GO annotations for 64,950 DV92 and 61,783 G3116 transcripts (see Data Access section). Using both InterPro and Blast2GO methods, we assigned functional annotation to a total of 71,633 (59.0%) DV92 and 69,437 (58.8%) G3116 transcripts. Overall, 2,897 and 2,867 GO terms were assigned to DV92 and G3116 transcripts respectively, with 2,742 GO terms common to both.

Differential expression of genes during photomorphogenesis

The RNA-Seq short reads from the dark-grown, etiolated (8DD) and light-exposed, green (48LL) samples were mapped

Table 1. Transcriptome assembly statistics.

Transcriptome assemblies	Total number of reads	Number of Transcripts	Largest sequence (bp)	Average length (bp)	Median length (bp)
DV92-31 k-mer	435,806,374	87,972	21,251	1633	1393
DV92-35 k-mer	435,806,374	82,185	13,427	1699	1460
DV92 Merged		120,911	21,331	1847	1600
G3116-31 k-mer	366,215,814	84,491	21,999	1579	1316
G3116-35 k-mer	366,215,814	79,936	13,528	1624	1372
G3116 Merged		117,969	22,045	1783	1525

Transcriptome assembly statistics for *T. monococcum ssp. monococcum* (DV92) and *T. monococcum ssp. aegilopoides* (G3116) generated by Velvet/Oases. The statistics describe the sequence input to the assembler and the number of assembled transcripts and relative transcript length in base pairs (bp). The merged assembly is a feature of Oases that merges transcript isoforms into putative gene loci.

against the respective transcriptomes of DV92 and G3116 to study light-regulated gene expression during photomorphogenesis. 25,742 G3116 and 23,526 DV92 transcripts show ≥2-fold change in expression (p ≤0.05) between 8DD and 48LL samples (Figure 2A and B). G3116 contains more light up-regulated and down-regulated transcripts compared to DV92 (Figure 2A and C). The differentially expressed transcripts from both accessions maps to 7,248 (30%) unique barley homologs. Henceforth, we analyzed differential expression of corresponding putative homologous *T. monococcum* genes in etiolated (8DD) and green (48LL) samples across two accessions DV92 and G3116 in a four-way comparison (Figure 2C). Compared to DV92, more than double the number of unique genes in G3116 are up- and down-regulated by light. Thirty-seven genes (Table S2) show a common profile across all four samples. This set includes homologs of light-harvesting chlorophyll B-binding protein, 3-ketoacyl-COA synthase, pyruvate kinase, tubulin beta chain, red chlorophyll catabolite reductase and cellulose synthase-like protein (Table S2). Interestingly, unique set of fifty-one genes show increased expression in DV92, but

decreased expression in G3116 in response to light (Figure 2C). This set includes homologs of rubisco activase, brassinosteroid-6-oxidase, 3-ketoacyl-CoA-synthase, histone H2A, SEC-C motif-containing protein, ATP-dependent *clp* protease ATP-binding subunit, heat shock protein 90 and cpn60 chaperonin family protein (Table S2). Conversely, a set of forty-one genes shows decreased expression in DV92 but increased expression in G3116 in response to light (Figure 2C). This set includes homologs of germin-like protein 1, plastid transcriptionally active 13, Tetratricopeptide repeat (TPR)-like superfamily protein and CAX interacting protein 1 (Table S2).

For each set of differentially expressed genes (Figure 2C), enrichment of a selected GO molecular function categories is shown in Figure 2D. We found that the 41- and 51-gene sets show enrichment for proteins that are likely to have ion and cation binding, nucleotide binding and transfer activities. The 41-gene set has a greater percentage of hydrolases, whereas, the 51-gene set contains a greater percentage of transporters (Figure 2D). Among the light up-regulated genes common to both DV92 and G3116,

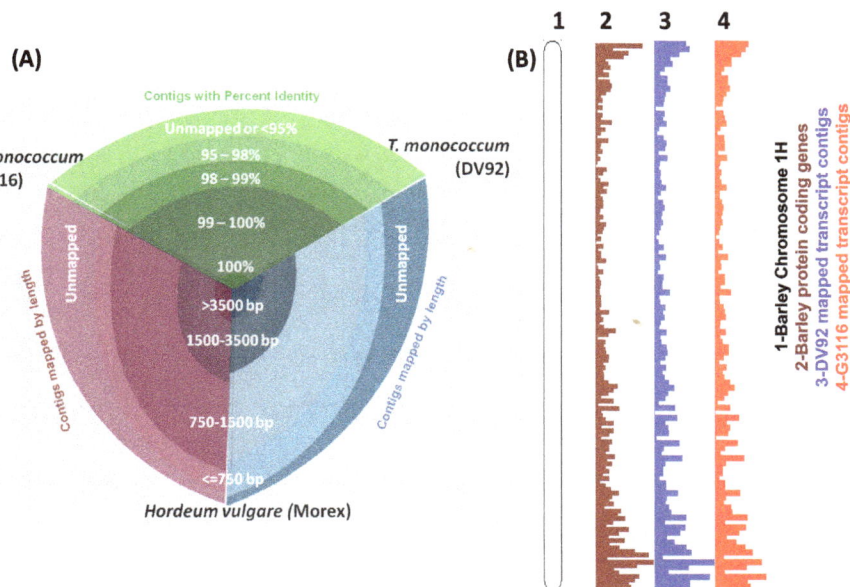

Figure 1. Mappings DV92 and G3116 transcripts to the sequenced *Hordeum vulgare* (barley) genome v1.0 (source: Gramene/Ensembl Plants). (A) A hive plot shows comparison between *Triticum monococcum* accessions G3116 and DV92 vs. the barley genome. (B) A density plot view of the Ensembl Plants genome browser showing barley chromosome-1H karyotype view (track-1) with annotated barley genes (track-2; maroon) and the mapped G3116 transcripts (track-3; blue) and DV92 transcripts (track-4; red).

Table 2. BLAST results.

Target	Query			
	DV92 (120,911)		G3116 (117,969)	
	# hits	% hits	# hits	% hits
DV92	-	-	116,227	98.50%
G3116	117,872	97.50%	-	-
T. urartu (wheat A genome) [*]	118,618	98.10%	115,498	97.90%
T. urartu Transcripts[*]	102,176	84.50%	99,148	84.00%
A. Tauschii (wheat D genome)[*]	120,061	99.30%	117,090	99.25%
A. tauschii Transcripts[*]	104,932	86.70%	101,749	86.25%
T. aestivum Transcripts[§]	115,528	95.50%	113,064	95.80%
T. aestivum Transcripts[^]	115,244	95.30%	112,786	95.60%
H. vulgare genome v2.18[#]	112,442	92.30%	109,816	93.10%
H. vulgare Transcripts v2.16[#]	93,369	77.20%	91,411	77.50%
O. sativa indica ASM465 v1.16[#]	83,775	69.30%	82,176	69.70%
O. sativa japonica MSU6[#]	84,836	70.20%	83,291	70.60%
B. distachyon v1.1[#]	88,655	73.30%	86,990	73.70%

Source: [*]GigaBD; [#]Gramene; [^]Plant GDB GenBank release 175; [§] DFCI release 12.0.
BLASTn (E-value $1e^{-5}$) nucleotide sequence comparisons of *T. monococcum ssp. Monococcum* (DV92) and *T. monococcum ssp. aegilopoides* (G3116) transcripts against gene models and genomes from other sequenced grass species suggesting the coverage represented in the *T. monococcum* transcriptome.

we found enrichment of genes encoding for structural components of cell envelopes, proteins involved in anatomical structure formation and proteins associated with cellular component biogenesis, having cellular component location 'plastid' (GO:0009536) or 'intracellular organelle' (GO:0043229), and enrichment of gene products targeted to 'thylakoid' (GO:0009579). Other categories of genes that show increased expression after exposure to light include components of carbohydrate metabolism, namely, the 'oligosaccharide metabolism' (GO:0009311), cell wall remodeling (GO:0004553; glycosyl hydrolases), and 'post-translational protein modification' (GO:0043687). The light down-regulated genes were associated with the biological process 'phosphate metabolic process (GO:0006796) with enrichment for 'nucleotide diphosphatase activity' (GO:0004551) (Table S3).

In DV92, transcripts encoding red (phytochrome) and blue (cryptochrome) light receptor proteins are down-regulated by 2-fold or more, whereas, orthologous transcripts in G3116 are up-regulated by 2-fold or more during photomorphogenesis (Table S1). A small subset of DV92 and G3116 transcripts mapped to genes with known homologs in plants exhibit differential expression during photomorphogenesis (Table S4). The light-induced genes include *lhcb* coding for chlorophyll a/b binding proteins, *Elongated hypocotyl 5* (*HY5*) coding for a positive regulator of photosynthesis associated nuclear genes, *rbcs* coding for ribulose bisphosphate carboxylase small subunit, homologs of rice *YGL138(t)* gene involved in chloroplast development [32], genes coding for mitochondrial transcription termination factor, late embryogenesis abundant protein LEA, and those coding for Rossmann-like alpha/beta/alpha sandwich fold containing protein (Table S4). Notably, homologs of gene coding for ABA 8'-hydroxylase activity associated with germination are significantly light up-regulated in G3116 but not in DV92. The light down-regulated genes include homologs of wheat *Rht-B1* DELLA protein, a nuclear repressor of gibberellin response, and *TaIAA1*, a primary auxin-response gene [33].

Table 3. The coverage and mapping of *T. urartu*, *A. tauschii* and *H. vulgare* transcripts on DV92 and G3116 transcriptomes using BLASTn (E-value $1e^{-5}$).

Target	Query					
	T. urartu		*A. tauschii*		*H. vulgare*	
	(Transcripts #34,879)[*]		(Transcripts #43,150)[*]		(Transcripts #62,240)[^]	
	# hits	% hits	# hits	% hits	# hits	% hits
DV92	29,784	85.40%	35,618	82.50%	57,781	92.80%
G3116	29,108	83.40%	34,783	80.60%	56,609	90.90%

Source: [*]GigaBD; [^]Gramene.
The number of transcripts and percent of transcripts from each query that hit a transcript from DV92 and G3116 are shown.

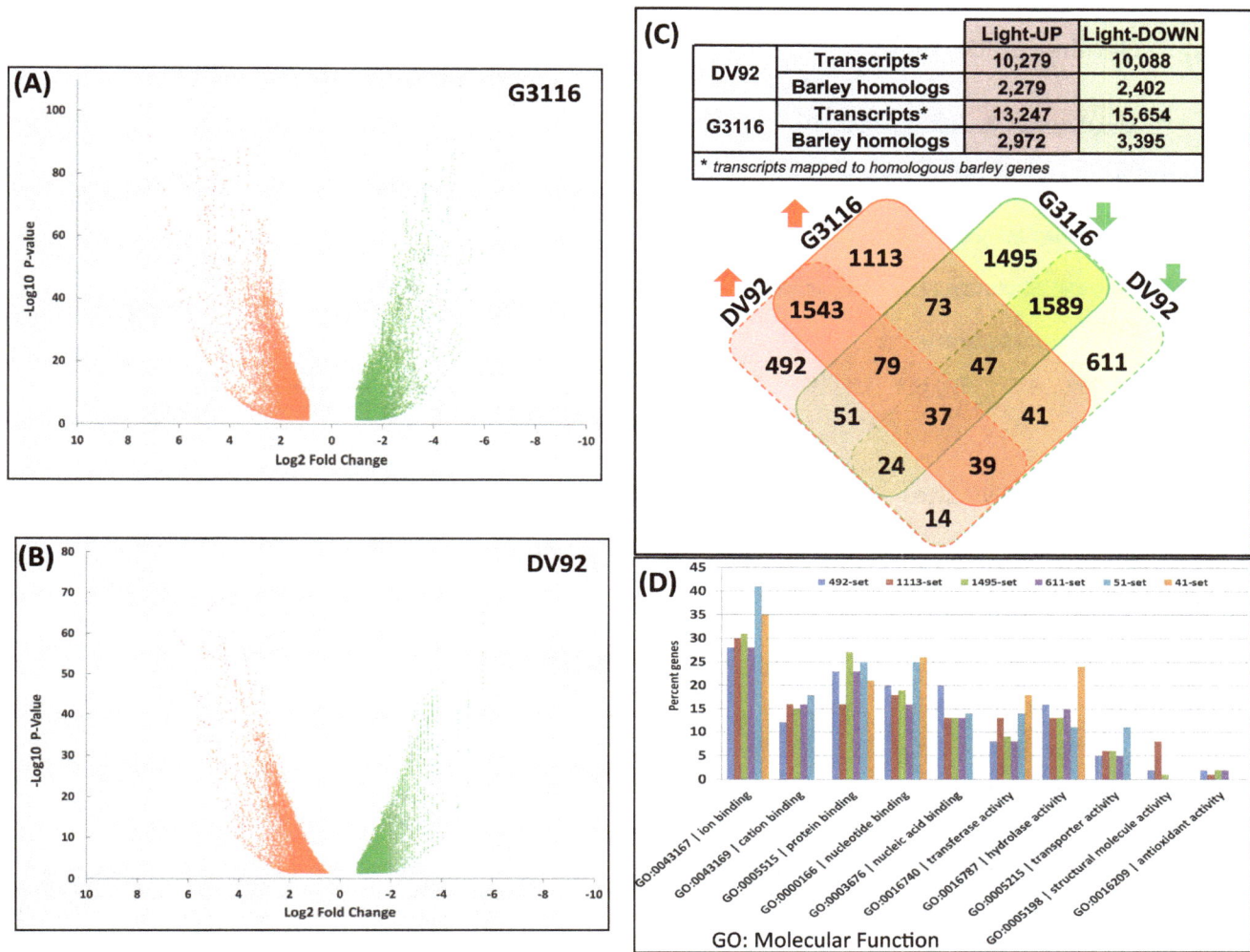

Figure 2. Analyses of the differentially expressed transcripts. A scatter plot of light up- regulated (red colored) and down-regulated (green colored) transcripts from G3116 (A) and DV92 (B) accessions of *T. monococcum*. Each spot represents a single transcript. (C) The table lists counts of differentially expressed transcripts from the DV92 and G3116 accessions shown in the adjacent scatter plots and their barley homologs. The four-way Venn diagram shows the distribution of barley homolog counts with reference to the mapped light up-regulated (red shaded boxes) and light down-regulated (green shaded boxes) transcripts. (D) Barley homologs from various unique sets identified in the Venn diagram (C) and their selected molecular function enrichment.

Developing genetic marker resources from the sequenced transcriptome

Molecular genetic markers are very useful for the analysis of genetic variation and heritable traits. Well established genotyping methods, such as high-throughput genotyping-by-sequencing (GBS) and chip-based methods using genomic DNA facilitate the interrogation of SNP and SSR markers. Similarly, large RNA-Seq data sets can be mined for molecular marker sites [27], which may then be used for genetic trait mapping, diversity analysis and marker-assisted selection in plant breeding experiments. This method permits future systems-level studies to explore the integrated analysis of gene function, expression, and the consequence of sequence variation on gene structure and function.

Identification of SSR marker loci

We mined the DV92 and G3116 transcriptome assemblies for di-, tri-, tetra-, penta-, and hexa-nucleotide SSRs with a minimum of 8, 6, 4, and 3 repeat units, respectively. We identified 29,887

SSR sites in 22,019 unique DV92 transcripts and 28,122 SSR sites in 20,727 unique G3116 transcripts (Figure 3A; Table S5). 3,413 transcripts orthologous between DV92 and G3116 contain identical SSRs, whereas 703 DV92 and G3116 orthologous transcripts contain variable-length SSRs. Some of these 703 sites may represent duplicate SSRs found in transcripts that map to the same or overlapping locus; therefore we aligned our assembled transcripts to the barley genome and identified 148 unique barley gene loci that harbor the variable SSR-containing sequence (Figure 3C). We experimentally verified a small number of SSRs for genotyping the DV92 and G3116 accessions (data not shown), though a majority of the markers will require experimental validation before they can be used.

Identification of SNP marker loci

To identify single nucleotide polymorphism (SNP) sites across the DV92 and G3116 transcriptomes, we used SOAPsnp [34] to align and identify the raw *T. monococcum* sequence reads against the

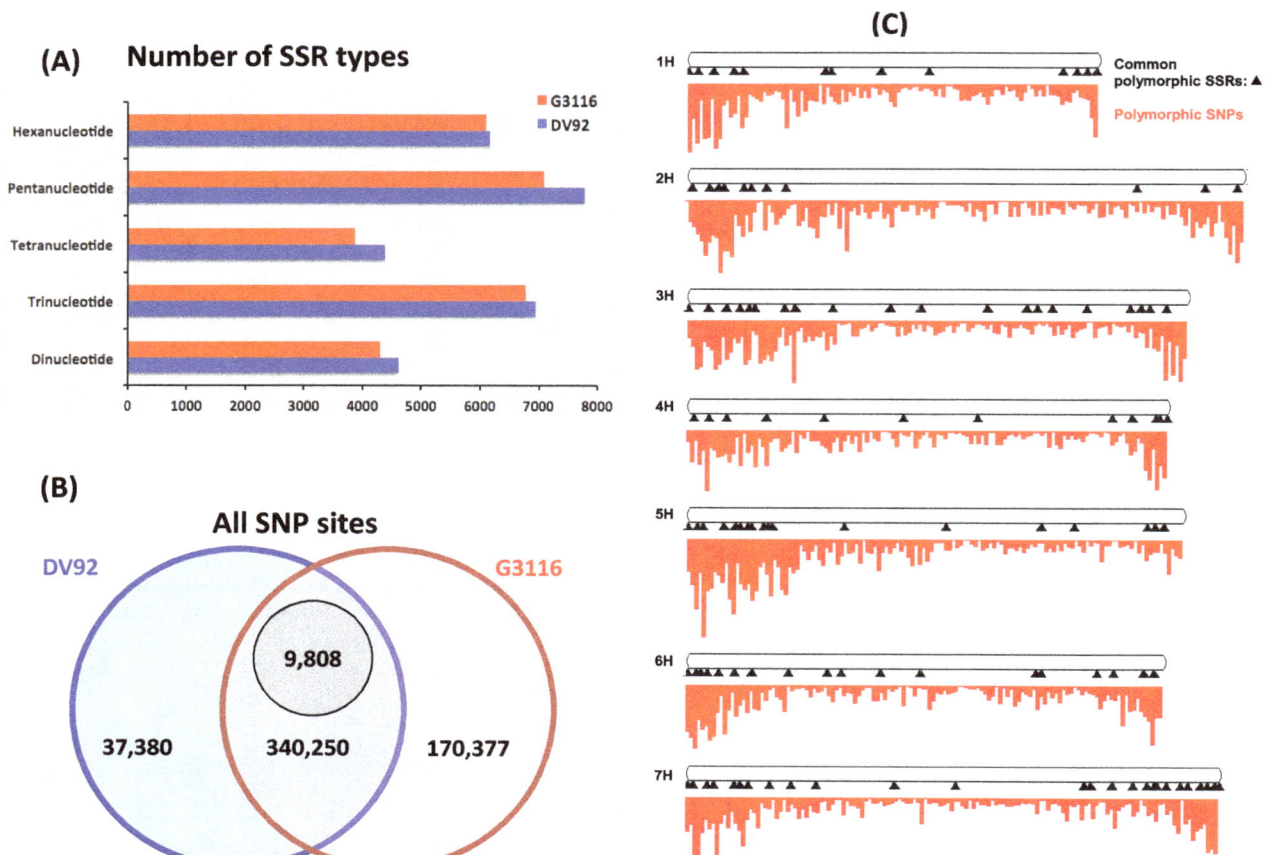

Figure 3. Genetic marker discovery. Polymorphic sites identified in the transcriptome of DV92 (blue) and G3116 (red). (A) Number of SSR identified in the transcriptomes. (B) Number of SNPs identified in the two genotypes by aligning against the sequenced barley reference genome. 9,808 out of 340,250 common SNP sites have polymorphism between DV92 and G3116. (C) Mapping of common, variable 9,808 SNP and 148 SSR sites identified in the DV92 and G3116 transcriptomes on the karyotype view of the reference barley genome hosted by the Ensembl Plants. The SNP sites are shown as red colored density plot and SSR sites are depicted as black triangles along the length of the respective barley chromosomes.

barley genome. We identified 510,627 SNPs with an average of one SNP per 3600 bp of the assembled barley genome. Of these, 170,377 SNP sites were unique to G3116, and 37,380 SNP sites were unique to DV92 (Figure 3B). More than 50% of the SNP sites (330,444) are present in both the DV92 and G3116 accessions. Of these common sites, 9,808 SNP sites were identified with different alleles for DV92 and G3116. These 9,808 SNP sites show a uniform distribution along the barley genome (Figure 3C), thus holding potential utility as genetic markers in wheat breeding programs. These 9,808 SNP sites are present in 5,989 unique protein coding genes, which include a subset of 4,935 GO-annotated genes (Table S6) and 2,543 differentially expressed genes. A greater number of nucleotide transitions were also discovered in DV92 when compared to G3116, which had more transversions (Table S7). In order to address the biological relevance of these SNPs, we predicted the potential effects of the variants and identified a diverse set of consequences on the transcript's structure, splicing and protein coding features with reference to the barley genome and annotated gene models (Table 4). Notably, we identified over 300,000 downstream variants, ~200,000 missense variants, 10,000-18,000 transcript splice site mutations, and more than 400 sites with a gain in stop codons (Table 4). Unique DV92 and G3116 SNPs are distributed

across variance consequence categories in similar proportions to combined SNPs (Table 4).

Discussion

This study provides the *de novo* assembled transcriptomes of two *T. monococcum* sub-species, representing the domesticated accession DV92 and the wild accession G3116. High-throughput RNA-Seq technology, bioinformatics tools and publicly available databases enabled higher quality transcriptome assemblies of these diploid wheat varieties, both of which are closely related to the wheat A-genome progenitor *T. urartu*. However, approximately 15% of the DV92 and G3116 transcriptomes do not map to the *T. urartu* and *A. tauschii* (progenitor of the wheat D genome) gene models (Table 3). We compared these unmapped *T. monococcum* transcripts against the barley genome and found 4,954 DV92 and 5,362 G3116 transcripts bear homology to 2,607 barley genes, suggesting that these genes have not been annotated in the published wheat A and D genomes [15–17]. Furthermore, comparison of the *T. monococcum*, *T. urartu* and barley gene models also revealed other disparities. For example, gene models for the *T. urartu* gene TUIUR3_02586-T1 lack exon-4, 3′ and 5′ UTRs and potentially unspliced introns when compared to the barley homolog MLOC_59496. In our analysis, multiple *T. monococcum* transcript

Table 4. Prediction of SNP variant consequence with reference to the annotated barley genome.

Predicted variant effect	Number of SNP sites with consequences		Unique	
	DV92	G3116	DV92	G3116
3 prime UTR variant	131,758	165,696	6,918	9,022
5 prime UTR variant	86,389	127,854	4,450	9,371
coding sequence variant	21,545	30,920	2,422	3,704
downstream gene variant	328,112	440,765	19,120	26,060
initiator codon variant	364	507	22	49
initiator codon variant, splice region variant	6	8	none	None
intergenic variant	35,753	54,722	2,682	24,136
intron variant	46,901	111,413	4,717	14,217
missense variant	198,794	258,081	9,929	17,763
missense variant, splice region variant	1,145	1,866	89	188
non coding exon variant, nc transcript variant	7	11	1	1
splice acceptor variant	10,094	18,609	572	2,103
splice donor variant	18,433	34,503	1,145	3,985
splice region variant, 3 prime UTR variant	681	962	25	51
splice region variant, 5 prime UTR variant	685	1,137	50	80
splice region variant, coding sequence variant	136	272	15	39
splice region variant, downstream gene variant	2	3	none	None
splice region variant, intron variant	31,692	63,891	2,270	8,074
splice region variant, synonymous variant	3,500	5,083	176	448
stop gained	462	732	40	62
stop gained, splice region variant	4	8	none	2
synonymous variant	451,169	538,115	18,687	29,046

SNP Variant Consequence Prediction based on the *T. monococcum* SNPs identified by aligning the sequenced reads from DV92 and G3116 to the reference barley genome and the barley gene models (v1.0) available from Ensembl Plants database. Listed variant effect types are based on the categories adopted by the Ensembl Plants database.

isoforms aligned with the barley homolog MLOC_59496 support the barley gene model (Figure S2) and thus provide empirical evidence for the missing features in *T. urartu* gene TUIUR3_02586-T1 (Figure S3). Our findings demonstrate the utility of the *T. monococcum* transcriptome data in enriching and improving Triticeae genome annotation, including the recently published A and D genomes.

To our knowledge, this study is the first to provide the relative expression of transcript isoforms (Figure 2, Table S1) in both etiolated seedlings and light-exposed green seedlings of cultivated spring accession DV92 and wild winter accession G3116 of *T. monococcum* (Figure S4). In order to preserve the granularity of the transcript isoform-based expression profile, we avoided projecting a weighted expression profile of the genes. This allowed us to identify a greater number of differentially expressed transcripts in G3116 (Figure 2A). However, for simplicity, the four-way Venn diagram (Figure 2C) was constructed to show comparison between the light up- and down-regulated genes from the two accessions.

In general, the transcriptomes of both DV92 and G3116 suggest up-regulation of the genes involved in chloroplast biogenesis, photosynthesis and carbohydrate metabolism, such as the homologs of *Elongated hypocotyl 5 (HY5)*, *YGL138(t)* [32,35] and photosystem II chlorophyll a/b-binding protein *lhcb* (Table S4). In addition, differentially expressed transcripts encoding for mitochondrial transcription termination factor-like protein (mTERF), late embryogenesis abundant protein (LEA) and Rossmann-like alpha/beta/alpha sandwich fold containing pro-

tein family members were found to be light up-regulated (Table S4). In humans, the mitochondrial transcription termination factor attenuates transcription from the mitochondrial genome, up-regulates the expression of 16S ribosomal RNA, and has high affinity for the tRNA$^{Leu(UUR)}$ gene [36–38]. The Arabidopsis mTERF gene family members are known to play roles in organelles; for example, *SUPPRESSOR OF HOT1-4 1 (SHOT1)*, a mitochondrial protein, is involved in heat tolerance and regulation of oxidative stress [39], *SINGLET OXYGEN-LINKED DEATH ACTIVATOR10 (OLDAT10)*, a plastid protein, activates retrograde signaling and oxidative stress, and *BELAYA SMERT (BSM)* regulates plastid gene expression [40]. The mTERF domain containing proteins from both the DV92 and G3116 accessions showing light up-regulation are predicted to be chloroplast proteins (TargetP value ~0.9) (Figure S5). To our knowledge, this is the first report of light up-regulation of wheat gene family members encoding mTERF, LEA and Rossmann-like alpha/beta/alpha sandwich fold containing proteins.

Other proteins that show light-induced differential regulation are involved in phytohormone metabolism and signaling. Transcripts homologous to *T. aestivum Rht-B1* that code for a DELLA protein were down-regulated by light [41]. DELLA proteins are repressors of gibberellin (GA) signaling and act immediately downstream of GA receptor. When GA synthesis is induced by light, the binding of GA to its receptor causes degradation of DELLAs via the ubiquitin-proteasome pathway [42]. GA is a hormone that is well known to promote seed germination in addition to participating in other parts of the plant life cycle. DELLAs have also been suggested to mediate interaction between GA and abscisic acid (ABA) pathways, as one of its targets, *XERICO*, is known to regulate ABA metabolism [42]. The levels of transcripts homologous to ABA 8′-hydroxylase were significantly higher in G3116 relative to DV92. ABA 8′-hydroxylase degrades ABA, a hormone involved in dormancy [43]. Degradation of ABA results in a decreased ABA-to-GA ratio resulting in the breaking of dormancy [44]. ABA 8′-hydroxylase activity may be one of the difference between winter and spring varieties. Conversely, increased levels of transcripts homologous to gene encoding for brassinosteroid-6-oxidase were found in DV92 in response to light, but not in G3116. Transcripts homologous to *TaIAA1*, an early auxin-response gene from wheat [33], were down-regulated by light in both DV92 and G3116, which is consistent with the previous report [33]. In addition to auxin, the *TaIAA1* gene is also induced by brassinosteroids [33]. Several genes showed accession-specific expression profile, such as the 51 and 41 gene sets (Figure 2C, Table S2), which may reflect differences in anatomical features and the plant's response to its immediate environment. For instance, the levels of transcripts homologous to rice *germin-like protein 1* show decrease in DV92 but increase in G3116 in light-exposed seedlings. The *germin-like protein-1* in rice has been shown to play a role in the regulation of plant height and disease resistance [45]. Transcripts homologous to genes coding for heat shock protein 90 and cpn60 chaperonin family protein increase in DV92, but decrease in G3116 in response to light (Table S2). Changes in the expression levels of transcripts encoding components of hormone biosynthesis, signaling and protein targets suggest that photomorphogenesis is a carefully orchestrated interplay of both developmental signals (often genotype-specific) and light response.

We identified over 500,000 SNP sites and approximately 22,000 SSR/microsatellite sites in the transcriptome assemblies of *T. monococcum*. Of these, 9,808 SNP and 148 SSR sites are common polymorphic sites in both accessions. The 9,808 SNPs overlap 2,543 barley genes that show light mediated up- and down-regulation of homologous transcripts in *T. monococcum*. A few notable genes in this differentially expressed set include (Figure S6 and Table S8) the light down-regulated protein coding genes for CASP-like membrane protein, Xyloglucan endo-transglycosylase activity, Auxin-responsive family protein and a novel protein carrying the DUF1644 domain. Whereas, the light up-regulated protein coding genes includes, photosystem-I subunit PSAK, PSAH, Ribulose-1,5-bisphosphate carboxylase (RUBISCO) small subunit RBCS, Chlorophyll a/b binding protein LHCB, Mitochondrial transcription termination family member and novel uncharacterized proteins (Figure S6 and Table S8). Our data suggest that 170,377 SNPs is unique to G3116 and 37,380 SNPs is unique to DV92 (Figure 3B); this provides an opportunity to study the wild winter and cultivated spring habits of the two accessions in greater detail. The SNP and SSR genetic sites identified in our dataset, along with those identified in other genetic populations [46] and wheat projects [47], will provide useful marker resources for fine mapping experiments and marker-assisted wheat breeding programs.

Along with the *T. monococcum* transcriptomes from two accessions, we have provided additional genomic and genetic resources including their functional annotations, differential gene expression analyses and potential SNPs and SSRs, which can be used to explore Triticeae genome diversity, co-expression networks involved in photomorphogenesis and to develop stochastic and metabolic networks [22,48,49]. In addition, these resources can be used to identify novel genes, transcript models and eQTLs, and to study plant's adaptation to diverse climatic conditions, impacts of domestication on crop plants and evolution of novel genes.

Methods

Plant material and growth conditions

Seeds of the *Triticum monococcum* ssp. monococcum accession DV92, a cultivated spring wheat, and *Triticum monococcum* ssp. aegilopoides accession G3116, a wild winter wheat, were sown into sunshine mix (Sun Gro Horticulture, Agawam, MA, USA). The trays were watered thoroughly and were shifted (in the evening hours) to a dark growth chamber set to cycle temperature between 20°C for 12 hours (8am–8pm) and 18°C for the next 12 hours (8pm–8am). The seedlings were grown in the dark for next 8 days and the soil was kept moist by gently spraying with water every 72 hours. Seeds were not vernalized prior to sowing. Germination was observed within two days for both accessions. The first set of dark-grown seedlings shoot samples (8DD), consisting of three replicate from each accession, were collected at the end of day-8 under green light. (8DD). On day-9 at 10 am, continuous light (120 µmol/m^2/sec at soil surface) was started for 48 hours (48LL) and a second set of seedling shoot samples (48LL), consisting of three replicates from each accession, were collected at the end of 48 hours of treatment on day 11. Each replicate contained shoots of three seedlings of similar height (Figure S4). Harvested samples were immediately frozen in liquid nitrogen and stored at −80°C.

Sample preparation for Illumina sequencing

Total RNA from frozen seedling shoot sample was extracted using RNA Plant reagent (Invitrogen Inc., USA), RNeasy kits (Qiagen Inc., USA), and treated with RNase-free DNase (Life Technologies Inc., USA) as previously described [27,50]. The mRNA concentration, quality were determined using ND-1000 spectrophotometer (Thermo Fisher Scientific Inc., USA) and Bioanalyzer 2100 (Agilent Technologies Inc., USA). Samples were prepared using the TruSeq RNA Sample Preparation Kits (v2) and sequenced on the Illumina HiSeq 2000 instrument (Illumina Inc., USA) at the Center for Genomic Research and Biocomputing, Oregon State University.

De novo transcriptome assembly and annotation

Illumina sequences were processed for low quality at an error rate of 0.00001, parsed for index sequences and pairs, and filtered and trimmed using customized Perl scripts. FASTQ file generation and removal of low quality reads were performed by CASAVA software v1.8.2 (Illumina Inc.). The high-quality sequences used in the assembly process included 435,806,374 and 366,215,814 paired-end 101 bp reads for DV92 and G3116 respectively (Table 1). The samples were assembled with Velvet (Velvet v1.2.08), which uses De Bruijn graphs to assemble short reads [51]. An assembly of 31 and 35 k-mer length was performed separately for both the DV92 and G3116 reads. The assemblies generated by Velvet were analyzed using Oases (Oases v0.2.08), which was developed for the *de novo* assembly of transcriptomes [26], and uses the read sequence and pairing information to produce transcript isoforms.

Similarity searches were conducted with BLASTn [30] (E-value $<= 1e^{-5}$) using assembled transcripts as a query against gene model sequence databases of other species of grasses with sequenced genomes, namely, hexaploid wheat (*T. aestivum*) transcripts (DFCI release 12.0), *T. aestivum* (Plant GDB GenBank release 175), barley (*Hordeum vulgare*) transcripts (Gramene v.2.16), barley genome (Gramene v.2.16), *Oryza sativa* spp. indica (Gramene ASM465v1.16), *Oryza sativa* spp. japonica (Gramene

MSU6.16), *Brachypodium distachyon* transcripts (Gramene v.0.16), and the *Brachypodium distachyon* genome (NCBI). *T. monococcum* transcripts were functionally annotated using a combined approach based upon functional motif analysis and sequence homology. Transcripts were translated into the longest predicted open reading frame (ORF) peptide sequences using the ORFPredictor web application [52] and resulting proteins assigned InterPro identifiers using InterProScan v4.8 [53,54]. These InterPro assignments were also mapped to Gene Ontology (GO) terms. Additionally, we did Blast2GO analysis [31] of *T. monococcum* transcripts to transfer GO annotations from functionally annotated genes in non-wheat genomes. A BLASTx search (E-value $\leq 1e^{-2}$ and percent identity $\geq 90\%$) was performed to identify highly homologous sequences against the NCBI GenBank non-redundant protein database. The resulting best hits with GO annotations were used to project similar GO assignments [55,56] to *T. monococcum* transcripts. GO annotations from both methods were combined and duplicated annotations were removed to produce non-redundant gene ontology annotation files for *T. monococcum* DV92 and G3116. The AgriGO Analysis Toolkit [57] was used to identify statistically-enriched functional groups. This method includes a Fisher's exact test with a Yekutieli correction for false discovery rate calculation. Significance cutoffs included a P-value of 0.05 and a minimum of 5 mapping entries per GO term.

Genetic marker development

The assemblies of DV92 and G3116 were mined for SSRs using Perl code from the Simple Sequence Repeat Identification Tool (SSRIT; [58]; http://archive.gramene.org/db/markers/ssrtool). We identified di-, tri-, tetra-, penta-, and hexa-nucleotide SSRs with a minimum of 8, 6, 4, 3, and 3 repeat units, respectively. We then used custom Perl scripts to identify orthologous DV92 and G3116 transcripts containing common SSRs.

An alignment database was generated using SOAP's 2bwt-builder with the barley genome (version 030312v2). Illumina sequences (FASTQ formatted) of length 51 bp were processed and aligned through SOAP (Version: 2.20) [59] with default options. Alignment data was then separated into different text files based on the chromosome of the hit sequence and each chromosome alignment file was sorted based on hit start position. After separation and sorting, data was processed through SOAPsnp (version 1.02) [34] to identify single nucleotide polymorphisms (SNPs). SOAPsnp was run using standard options for a diploid genome as stated in the documentation. SOAPsnp output files were then reformatted to VCF output, a community standard format developed by the 1000 Genomes project (http://www.1000genomes.org/wiki/Analysis/Variant%20Call%20Format/vcf-variant-call-format-version-41) to make them more accessible for analysis by other downstream programs. To call a SNP, values for novel homozygous prior probability and novel heterozygous prior probability were set at 0.0005 and 0.0001, respectively. The transition/transversion ratio was set to 2:1 in prior probability. The rank sum test was enabled to give heterozygous prior probability further penalty if reads did not have the same sequencing quality for better SNP calling. A maximum read length of 51 bp was used. We used the Ensembl Plants API Effect Predictor tool [60] to infer potential consequences of the SNP variants.

Gene expression analysis

We used CASHX v2.3 to align the DV92 and G3116 reads to their respective transcriptome assembly [61]. Indexed reads were used for each replicate for both dark and light comparisons of DV92 and G3116. We then used Edge R-package (v. 2.0.3) [62] to conduct differential gene expression analysis. We identified differentially expressed transcripts with a significance of P-value cutoff/FDR corrected P-value of 0.05. We also further filtered the differentially expressed genes by 2-fold cutoffs and those identified to be differentially expressed by the EdgeR. Principal components analysis (PCA) multidimensional scaling (MDS), and correlation matrix algorithms were used to assess and visualize a cross-sample comparisons. Both analyses show clustering based upon RPKM values for all genes among all replicates. The results, as expected, show four separate visualized clusters (DV92 light and dark replicates and G3116 light and dark replicates; Figure S7–9).

Data Access

Sequence files, assemblies, annotation files, SNP, SSR, transcript alignments, gene expression, network data files and results are available from the project's data site [63] ScholarsArchive at Oregon State University (http://hdl.handle.net/1957/47475). The transcriptome data are being integrated in the Barley Genome Browser available from the Ensembl Plants database (http://plants.ensembl.org). The data are also being provided to the small grains database GrainGenes (http://www.graingenes.org). The raw sequence files were submitted to the National Center for Biotechnology Information (NCBI) Sequence Read Archive under the accessions SRX283514/SRR924098 (DV92) and SRX257915/SRR922411 (G3116).

Supporting Information

Figure S1 The frequency distribution of transcripts of varying size (bp: base pair) in the *de novo* transcriptome assemblies of DV92, G3116 and the annotated transcriptomes of barley and wheat *T. urartu.*
(TIFF)

Figure S2 A view of the Ensembl Plants barley genome browser showing the comparison between the models of barley gene MLC 509496 and the homologous *T. monococcum* gene models derived from DV92 and G3116 transcriptomes. This alignment was generated using the Exonerate software package by allowing for gapped alignments (introns). The red arrows depict intron retention events and the blue arrow depicts intron-3 in the annotated barley gene model. Our data support barley MLOC_59496 gene model, including its 3' and 5' untranslated regions shown by open blocks.
(TIFF)

Figure S3 A view of the Ensembl Plants *T. urartu* genome browser showing the comparison between the *T. urartu* gene TUIUR30_2586-T1 model and the homologous *T. monococcum* gene models derived from DV92 and G3116 transcriptomes. This alignment was generated using the Exonerate software package by allowing for gapped alignments (introns). Our models show retention of introns (red arrows) in a couple of *T. monococcum* gene models, and the presence of an exon -4 (same as exon-3 in the barley model shown in figure-S2) missed in the *T. urartu* genome annotation (the dotted-line box). Our data do not support the presence of exon-3 in (blue arrow) in the annotated *T. urartu* gene TUIUR3_02586-T1.
(TIFF)

Figure S4 Seedling samples used for generating the transcriptomes of wheat accessions DV92 (left panel) and G3116 (right panel).
(TIFF)

Figure S5   TargetP analysis of the DV92 and G3116 peptides bearing the   Mitochondrial transcription termination factor-related domain. The proteins were predicted to be targeted to chlorplast (cTP) with a high confidence score of ~0.9. Both peptides were predicted to have a transit peptide length (Tplen) of 78aa.
(TIFF)

Figure S6   The line plot display of expression level in RPKM log2 values of transcripts that were grouped into light down regulated and light up-regulated co-expressed clusters (Figure 3 and 4) and have overlapping SNPs from the 9,808 SNP set. The table on the right shows homologous barley gene, functional annotation and the SNP variant effect on the transcript structure and/or function.
(TIFF)

Figure S7   Principal component analysis (PCA) analysis of RNA-Seq reads.
(TIFF)

Figure S8   Multidimensional scaling (MDS) analysis of RNA-Seq reads.
(TIFF)

Figure S9   Correlation matrix analysis of RNA-Seq reads.
(TIFF)

Table S1   Expression profiles of assembled transcripts from DV92 and G3116. It is a zip file with two tab-delimited files called Table S1 a.txt with DV92 gene expression data and Table S1 b.txt with G3116 gene expression data.
(ZIP)

Table S2   List of barley homologs clustered in a four-way Venn diagram (Figure 2C)
(XLSX)

Table S3   Enrichment of the Gene Otology-based fun-ctional annotation of the barley homologs clustered in a four-way Venn diagram (Figure 2C).
(XLSX)

Table S4   A short list of transcripts mapped to known and novel genes along with their expression datasets from the DV92 and G3116 accession.
(XLSX)

Table S5   Counts of Simple Sequence Repeats identified in the Triticum monococcum transcriptome data.
(XLSX)

Table S6   Enrichment of the Gene Otology based funct-ional annotation of the barley homologs overlapping the 9,808 SNP sites that had a different allele for DV92 and G3116 with reference to barley allele.
(XLSX)

Table S7   Number of transitions and   transversions resulting from SNP analysis with reference to the allele from barley the genome.
(XLSX)

Table S8   A list of DV92 and G3116 transcripts homo-logous to the barley genes overlapping the 9,800 SNP set. Table includes DV92 and G3116 transcript IDs, homologous barley gene ID, RPKM values, respective p-value scores, putative gene function annotation and the resultant SNP variant effect with reference to the barley gene models.
(XLSX)

Acknowledgments

We would like to thank Center for Genome Research and Biocomputing (CGRB) core facility staff, Anne-Marie Girard and Caprice Rosato for qualitative assessment of RNA, Mark Dasenko for Illumina cluster generation and sequencing and Matthew Peterson for computational support.

Author Contributions

Conceived and designed the experiments: PJ SEF MH VKT JL. Performed the experiments: MH SEF AS VKT MG GL JP JE CS AK DB PK CG FM SN PJ. Analyzed the data: SEF MG SN JE PJ. Contributed reagents/materials/analysis tools: SEF VKT MG GL JP JE CS AK DB PK CG FM PJ. Wrote the paper: SEF MG SN VKT JL PJ.

References

1. Harlan JR, Zohary D (1966) Distribution of Wild Wheats and Barley. Science 153: 1074–1080. doi:10.1126/science.153.3740.1074
2. Heun M, Schäfer-Pregl R, Klawan D, Castagna R, Accerbi M, et al. (1997) Site of Einkorn Wheat Domestication Identified by DNA Fingerprinting. Science 278: 1312–1314. doi:10.1126/science.278.5341.1312
3. Salamini F, Özkan H, Brandolini A, Schäfer-Pregl R, Martin W (2002) Genetics and geography of wild cereal domestication in the near east. Nat Rev Genet 3: 429–441. doi:10.1038/nrg817
4. Zoccatelli G, Sega M, Bolla M, Cecconi D, Vaccino P, et al. (2012) Expression of α-amylase inhibitors in diploid Triticum species. Food Chem 135: 2643–2649. doi:10.1016/j.foodchem.2012.06.123
5. Bennett MD, Leitch IJ (1995) Nuclear DNA Amounts in Angiosperms. Ann Bot 76: 113–176. doi:10.1006/anbo.1995.1085
6. Brenchley R, Spannagl M, Pfeifer M, Barker GLA, D'Amore R, et al. (2012) Analysis of the bread wheat genome using whole-genome shotgun sequencing. Nature 491: 705–710. doi:10.1038/nature11650
7. Jing H-C, Kornyukhin D, Kanyuka K, Orford S, Zlatska A, et al. (2007) Identification of variation in adaptively important traits and genome-wide analysis of trait–marker associations in Triticum monococcum. J Exp Bot 58: 3749–3764. doi:10.1093/jxb/erm225
8. Munns R, James RA, Xu B, Athman A, Conn SJ, et al. (2012) Wheat grain yield on saline soils is improved by an ancestral Na+ transporter gene. Nat Biotechnol 30: 360–364. doi:10.1038/nbt.2120
9. Shi F, Endo TR (1997) Production of wheat-barley disomic addition lines possessing an Aegilops cylindrica gametocidal chromosome. Genes Genet Syst 72: 243–248.
10. Saintenac C, Zhang W, Salcedo A, Rouse MN, Trick HN, et al. (2013) Identification of Wheat Gene Sr35 That Confers Resistance to Ug99 Stem Rust Race Group. Science. doi:10.1126/science.1239022
11. Dubcovsky J, Luo M-C, Zhong G-Y, Bransteitter R, Desai A, et al. (1996) Genetic Map of Diploid Wheat, Triticum monococcum L., and Its Comparison With Maps of Hordeum vulgare L. Genetics. 143: 983–999.
12. Feuillet C, Travella S, Stein N, Albar L, Nublat A, et al. (2003) Map-based isolation of the leaf rust disease resistance gene Lr10 from the hexaploid wheat (Triticum aestivum L.) genome. Proc Natl Acad Sci 100: 15253–15258. doi:10.1073/pnas.2435133100
13. Yan L, Loukoianov A, Tranquilli G, Helguera M, Fahima T, et al. (2003) Positional cloning of the wheat vernalization gene VRN1. Proc Natl Acad Sci 100: 6263–6268. doi:10.1073/pnas.0937399100
14. International Barley Genome Sequencing Consortium, Mayer KFX, Waugh R, Brown JWS, Schulman A, et al. (2012) A physical, genetic and functional sequence assembly of the barley genome. Nature 491: 711–716. doi:10.1038/nature11543
15. Ling H-Q, Zhao S, Liu D, Wang J, Sun H, et al. (2013) Draft genome of the wheat A-genome progenitor Triticum urartu. Nature 496: 87–90. doi:10.1038/nature11997
16. Jia J, Zhao S, Kong X, Li Y, Zhao G, et al. (2013) Aegilops tauschii draft genome sequence reveals a gene repertoire for wheat adaptation. Nature 496: 91–95. doi:10.1038/nature12028
17. Luo M-C, Gu YQ, You FM, Deal KR, Ma Y, et al. (2013) A 4-gigabase physical map unlocks the structure and evolution of the complex genome of Aegilops tauschii, the wheat D-genome progenitor. Proc Natl Acad Sci 110: 7940–7945. doi:10.1073/pnas.12190821_0

18. Arsovski AA, Galstyan A, Guseman JM, Nemhauser JL (2012) Photomorphogenesis. Arab Book Am Soc Plant Biol 10. Available: http://www.ncbi.nlm.nih.gov/pmc/articles/PMC3350170/. Accessed 2013 May 13.

19. Filichkin SA, Breton G, Priest HD, Dharmawardhana P, Jaiswal P, et al. (2011) Global Profiling of Rice and Poplar Transcriptomes Highlights Key Conserved Circadian-Controlled Pathways and cis-Regulatory Modules. PLoS ONE 6: e16907. doi:10.1371/journal.pone.0016907

20. Li J, Terzaghi W, Deng XW (2012) Genomic basis for light control of plant development. Protein Cell 3: 106–116. doi:10.1007/s13238-012-2016-7

21. Hanumappa M, Preece J, Elser J, Nemeth D, Bono G, et al. (2013) WikiPathways for plants: a community pathway curation portal and a case study in rice and arabidopsis seed development networks. Rice 6: 14. doi:10.1186/1939-8433-6-14

22. Dharmawardhana P, Ren L, Amarasinghe V, Monaco M, Thomason J, et al. (2013) A genome scale metabolic network for rice and accompanying analysis of tryptophan, auxin and serotonin biosynthesis regulation under biotic stress. Rice 6: 15. doi:10.1186/1939-8433-6-15

23. Li P, Ponnala L, Gandotra N, Wang L, Si Y, et al. (2010) The developmental dynamics of the maize leaf transcriptome. Nat Genet 42: 1060–1067. doi:10.1038/ng.703

24. Seo HS, Yang J-Y, Ishikawa M, Bolle C, Ballesteros ML, et al. (2003) LAF1 ubiquitination by COP1 controls photomorphogenesis and is stimulated by SPA1. Nature 423: 995–999. doi:10.1038/nature01696

25. Szekeres M, Németh K, Koncz-Kálmán Z, Mathur J, Kauschmann A, et al. (1996) Brassinosteroids Rescue the Deficiency of CYP90, a Cytochrome P450, Controlling Cell Elongation and De-etiolation in Arabidopsis. Cell 85: 171–182. doi:10.1016/S0092-8674(00)81094-6

26. Schulz MH, Zerbino DR, Vingron M, Birney E (2012) Oases: robust de novo RNA-seq assembly across the dynamic range of expression levels. Bioinformatics 28: 1086–1092. doi:10.1093/bioinformatics/bts094

27. Fox SE, Preece J, Kimbrel JA, Marchini GL, Sage A, et al. (2013) Sequencing and De Novo Transcriptome Assembly of Brachypodium sylvaticum (Poaceae). Appl Plant Sci 1: 1200011. doi:10.3732/apps.1200011

28. Fu N, Wang Q, Shen H-L (2013) De novo assembly, gene annotation and marker development using Illumina paired-end transcriptome sequences in celery (Apium graveolens L.). PLoS One 8: e57686. doi:10.1371/journal.pone.0057686

29. Wang Y, Zeng X, Iyer NJ, Bryant DW, Mockler TC, et al. (2012) Exploring the Switchgrass Transcriptome Using Second-Generation Sequencing Technology. PLoS ONE 7: e34225. doi:10.1371/journal.pone.0034225

30. Mount DW (2007) Using the Basic Local Alignment Search Tool (BLAST). Cold Spring Harb Protoc 2007: pdb.top17. doi:10.1101/pdb.top17

31. Conesa A, Götz S, García-Gómez JM, Terol J, Talón M, et al. (2005) Blast2GO: a universal tool for annotation, visualization and analysis in functional genomics research. Bioinformatics 21: 3674–3676. doi:10.1093/bioinformatics/bti610

32. Zhang F, Luo X, Hu B, Wan Y, Xie J (2013) YGL138(t), encoding a putative signal recognition particle 54 kDa protein, is involved in chloroplast development of rice. Rice 6: 7. doi:10.1186/1939-8433-6-7

33. Singla B, Chugh A, Khurana JP, Khurana P (2006) An early auxin-responsive Aux/IAA gene from wheat (Triticum aestivum) is induced by epibrassinolide and differentially regulated by light and calcium. J Exp Bot 57: 4059–4070. doi:10.1093/jxb/erl182

34. Li R, Li Y, Fang X, Yang H, Wang J, et al. (2009) SNP detection for massively parallel whole-genome resequencing. Genome Res 19: 1124–1132. doi:10.1101/gr.088013.108

35. Oyama T, Shimura Y, Okada K (1997) The Arabidopsis HY5 gene encodes a bZIP protein that regulates stimulus-induced development of root and hypocotyl. Genes Dev 11: 2983–2995. doi:10.1101/gad.11.22.2983

36. Fernandez-Silva P, Martinez-Azorin F, Micol V, Attardi G (1997) The human mitochondrial transcription termination factor (mTERF) is a multizipper protein but binds to DNA as a monomer, with evidence pointing to intramolecular leucine zipper interactions. EMBO J 16: 1066–1079. doi:10.1093/emboj/16.5.1066

37. Hyvärinen AK, Pohjoismäki JLO, Reyes A, Wanrooij S, Yasukawa T, et al. (2007) The mitochondrial transcription termination factor mTERF modulates replication pausing in human mitochondrial DNA. Nucleic Acids Res 35: 6458–6474. doi:10.1093/nar/gkm676

38. Robles P, Micol JL, Quesada V (2012) Unveiling Plant mTERF Functions. Mol Plant 5: 294–296. doi:10.1093/mp/sss016

39. Kim M, Lee U, Small I, Francs-Small CC des, Vierling E (2012) Mutations in an Arabidopsis Mitochondrial Transcription Termination Factor–Related Protein Enhance Thermotolerance in the Absence of the Major Molecular Chaperone HSP101. Plant Cell Online 24: 3349–3365. doi:10.1105/tpc.112.101006

40. Babiychuk E, Vandepoele K, Wissing J, Garcia-Diaz M, Rycke RD, et al. (2011) Plastid gene expression and plant development require a plastidic protein of the mitochondrial transcription termination factor family. Proc Natl Acad Sci 108: 6674–6679. doi:10.1073/pnas.1103442108

41. Peng J, Richards DE, Hartley NM, Murphy GP, Devos KM, et al. (1999) "Green revolution" genes encode mutant gibberellin response modulators. Nature 400: 256–261. doi:10.1038/22307

42. Harberd NP (2003) Relieving DELLA Restraint. Science 299: 1853–1854. doi:10.1126/science.1083217

43. Xiong L, Zhu J-K (2003) Regulation of Abscisic Acid Biosynthesis. Plant Physiol 133: 29–36. doi:10.1104/pp.103.025395

44. Liu A, Gao F, Kanno Y, Jordan MC, Kamiya Y, et al. (2013) Regulation of Wheat Seed Dormancy Is Mediated by Specific Transcriptional Switches That Induce Changes in Seed Hormone Metabolism and Signaling. PLoS ONE 8: e56570. doi:10.1371/journal.pone.0056570

45. Banerjee J, Maiti MK (2010) Functional role of rice germin-like protein1 in regulation of plant height and disease resistance. Biochem Biophys Res Commun 394: 178–183. doi:10.1016/j.bbrc.2010.02.142

46. Poland JA, Brown PJ, Sorrells ME, Jannink J-L (2012) Development of high-density genetic maps for barley and wheat using a novel two-enzyme genotyping-by-sequencing approach. PLoS One 7: e32253. doi:10.1371/journal.pone.0032253

47. Cavanagh CR, Chao S, Wang S, Huang BE, Stephen S, et al. (2013) Genome-wide comparative diversity uncovers multiple targets of selection for improvement in hexaploid wheat landraces and cultivars. Proc Natl Acad Sci 110: 8057–8062. doi:10.1073/pnas.1217133110

48. Poolman MG, Kundu S, Shaw R, Fell DA (2013) Responses to light intensity in a genome-scale model of rice metabolism. Plant Physiol 162: 1060–1072. doi:10.1104/pp.113.216762

49. Guerriero ML, Pokhilko A, Fernández AP, Halliday KJ, Millar AJ, et al. (2012) Stochastic properties of the plant circadian clock. J R Soc Interface R Soc 9: 744–756. doi:10.1098/rsif.2011.0378

50. Fox S, Filichkin S, Mockler TC (2009) Applications of Ultra-high-Throughput Sequencing. In: Belostotsky DA, editor. Plant Systems Biology. Methods in Molecular Biology. Humana Press. pp. 79–108. Available: http://link.springer.com/protocol/10.1007/978-1-60327-563-7_5. Accessed 2013 June 18.

51. Zerbino DR, Birney E (2008) Velvet: Algorithms for de novo short read assembly using de Bruijn graphs. Genome Res 18: 821–829. doi:10.1101/gr.074492.107

52. Min XJ, Butler G, Storms R, Tsang A (2005) OrfPredictor: predicting protein-coding regions in EST-derived sequences. Nucleic Acids Res 33: W677–W680. doi:10.1093/nar/gki394

53. Quevillon E, Silventoinen V, Pillai S, Harte N, Mulder N, et al. (2005) InterProScan: protein domains identifier. Nucleic Acids Res 33: W116–W120. doi:10.1093/nar/gki442

54. Hunter S, Jones P, Mitchell A, Apweiler R, Attwood TK, et al. (2012) InterPro in 2011: new developments in the family and domain prediction database. Nucleic Acids Res 40: 4725–4725. doi:10.1093/nar/gks456

55. Götz S, García-Gómez JM, Terol J, Williams TD, Nagaraj SH, et al. (2008) High-throughput functional annotation and data mining with the Blast2GO suite. Nucleic Acids Res 36: 3420–3435. doi:10.1093/nar/gkn176

56. Barrell D, Dimmer E, Huntley RP, Binns D, O'Donovan C, et al. (2009) The GOA database in 2009—an integrated Gene Ontology Annotation resource. Nucleic Acids Res 37: D396–D403. doi:10.1093/nar/gkn803

57. Du Z, Zhou X, Ling Y, Zhang Z, Su Z (2010) agriGO: a GO analysis toolkit for the agricultural community. Nucleic Acids Res 38: W64–W70. doi:10.1093/nar/gkq310

58. Temnykh S, DeClerck G, Lukashova A, Lipovich L, Cartinhour S, et al. (2001) Computational and Experimental Analysis of Microsatellites in Rice (Oryza sativa L.): Frequency, Length Variation, Transposon Associations, and Genetic Marker Potential. Genome Res 11: 1441–1452. doi:10.1101/gr.184001

59. Li R, Li Y, Kristiansen K, Wang J (2008) SOAP: short oligonucleotide alignment program. Bioinformatics 24: 713–714. doi:10.1093/bioinformatics/btn025

60. McLaren W, Pritchard B, Rios D, Chen Y, Flicek P, et al. (2010) Deriving the consequences of genomic variants with the Ensembl API and SNP Effect Predictor. Bioinformatics 26: 2069–2070. doi:10.1093/bioinformatics/btq330

61. Cumbie JS, Kimbrel JA, Di Y, Schafer DW, Wilhelm LJ, et al. (2011) GENE-Counter: A Computational Pipeline for the Analysis of RNA-Seq Data for Gene Expression Differences. PLoS ONE 6: e25279. doi:10.1371/journal.pone.0025279

62. Robinson MD, McCarthy DJ, Smyth GK (2010) edgeR: a Bioconductor package for differential expression analysis of digital gene expression data. Bioinforma Oxf Engl 26: 139–140. doi:10.1093/bioinformatics/btp616

63. Fox SE, Geniza MJ, Hanumappa M, Naithani S, Sullivan CM, et al. (2014) De novo transcriptome assembly and analyses of gene expression in diploid wheat Triticum monococcum. Oregon State University Libraries. Dataset. doi:10.7267/N92Z13FV.

19

Development of a Multi-Species Biotic Ligand Model Predicting the Toxicity of Trivalent Chromium to Barley Root Elongation in Solution Culture

Ningning Song[1], Xu Zhong[1], Bo Li[1,2], Jumei Li[1], Dongpu Wei[1], Yibing Ma[1]*

1 National Soil Fertility and Fertilizer Effects Long-term Monitoring Network/Institute of Agricultural Resources and Regional Planning, Chinese Academy of Agricultural Sciences, Beijing, P.R. China, **2** Institute of Plant Nutrition and Environmental Resources, Liaoning Academy of Agricultural Sciences, Shenyang, P.R. China

Abstract

Little knowledge is available about the influence of cation competition and metal speciation on trivalent chromium (Cr(III)) toxicity. In the present study, the effects of pH and selected cations on the toxicity of trivalent chromium (Cr(III)) to barley (*Hordeum vulgare*) root elongation were investigated to develop an appropriate biotic ligand model (BLM). Results showed that the toxicity of Cr(III) decreased with increasing activity of Ca^{2+} and Mg^{2+} but not with K^+ and Na^+. The effect of pH on Cr(III) toxicity to barley root elongation could be explained by H^+ competition with Cr^{3+} bound to a biotic ligand (BL) as well as by the concomitant toxicity of $CrOH^{2+}$ in solution culture. Stability constants were obtained for the binding of Cr^{3+}, $CrOH^{2+}$, Ca^{2+}, Mg^{2+} and H^+ with binding ligand: log K_{CrBL} 7.34, log K_{CrOHBL} 5.35, log K_{CaBL} 2.64, log K_{MgBL} 2.98, and log K_{HBL} 4.74. On the basis of those estimated parameters, a BLM was successfully developed to predict Cr(III) toxicity to barley root elongation as a function of solution characteristics.

Editor: Malcolm Bennett, University of Nottingham, United Kingdom

Funding: The work was financially supported by National Natural Science Foundation of China (grant number 40971262) and China Postdoctoral Science Foundation (grant number 2013M530783). The funders had no role in study design, data collection and analysis, decision to publish, or preparation of the manuscript.

Competing Interests: The authors have declared that no competing interests exist.

* Email: mayibing@caas.cn

Introduction

Chromium is one of the most widely used metals in modern industry [1], and it could be transferred into the environment through the waste products during various industrial processes [2], [3]. It has, therefore, become a common contaminant in waters and soils. Chromium occurs in the environment primarily in two common oxidation states: trivalent chromium (Cr(III)) and hexavalent chromium (Cr(VI)) [4]. Although Cr (III) is slightly less toxic than Cr (VI), exposure to excess Cr(III) could inhibit plant growth, and in humans it may decrease immune system activity [5]. Over recent years, several researches have been performed to determine the Cr(III) toxicity to plants [5], [6] and terrestrial invertebrates [7], [8]. However, most reported toxicity data were obtained based on total chromium concentration. Little focus on research into the influence of competition and speciation on Cr toxicity. Studies in aquatic toxicology have shown that the competition of other cations and the speciation of metals pose great influence on their toxicity [9], [10]. Therefore, a risk assessment considering the effects of cation competition and metal speciation is needed to properly assess the risk of Cr(III).

A biotic ligand model (BLM) has been developed to predict metal toxicity in aquatic systems [11], which incorporates metal complexation and speciation in the solution surrounding the organisms as well as interactions between metal ions and competing cations at the binding sites on the organism-water interface. The main assumption of the BLM is that metal toxicity is caused by free metal ions reacting with biological binding sites. The cations of H^+, Ca^{2+}, Mg^{2+}, Na^+ and K^+ might compete with metal ions for these binding sites and decrease the toxicity of the free metal ions [12], [13]. The complexation capacity of the BL and stability constants for the metal-BL and the cation-BL complexes have to be incorporated in a speciation model such as Visual MINTEQ [14] or WHAM (Windermere Humic Aqueous Model) [15] which allows to determine calculation of the free metal ion activity and speciation on basis of the water characteristics. So far the BLMs have been successfully applied in predicting the bioavailability and toxicity of several metals in aquatic systems and partially in terrestrial systems [9], [16], [17]. Thakali et al. [18], [19] have developed a terrestrial BLM and successfully applied it to Cu and Ni toxicity to several biological endpoints (such as barley root elongation) only in noncalcareous soils with pH≤7, because in calcareous soils, the free activity of metals predicted by speciation models has not been solved. Li et al. [9] refined a BLM to predict acute Ni toxicity to barley (*Hordeum vulgare*) root elongation in solution culture, and suggested that Ni^{2+} plus $NiHCO_3^+$ as toxic species and the competition of H^+, Mg^{2+} and Ca^{2+} with the binding sites of BL should be incorporated in the BLM. The general goal in the study was to develop a Cr(III)-BLM in solution culture, in order to further apply it to predict the toxicity of Cr(III) in aquatic and terrestrial systems.

To our knowledge, no data are available on the effect of pH, competing cations and Cr speciation on Cr(III) toxicity to plants in solution culture and there is no BLM applied to predict acute Cr(III) toxicity to plants. The objectives of the present study were: (1) to investigate the effect of H^+ competition on the toxicity of Cr(III) to barley root elongation across a wide range of pH values and to determine whether other Cr species are involved in toxicity responses; (2) to determine the effects of Ca^{2+}, Mg^{2+}, Na^+ and K^+ on Cr(III) toxicity to barley root elongation across a wide range of ion concentrations in order to obtain conditional binding constants for Cr^{3+} as well as other cations with BLs; and (3) to establish a multi-species BLM that can be used to predict Cr(III) toxicity to barely for a wide range of solution characteristics.

Materials and Methods

Experimental design

To assess the independent effect of different cations on Cr(III) toxicity, the target cation concentrations varied during one-set experiments, while all other cation concentrations were kept low and constant [8], [20]. Five sets of Cr(III) bioassays were performed: Ca-set, Mg-set, Na-set, K-set and pH-set (Table 1). Each set consisted of a series of tested solutions, in which only the concentration of target cation varied, while $CaCl_2$ was kept at 0.2 mM as background electrolyte. There were seven concentrations of Cr(III) (as $CrCl_3 \cdot 6H_2O$) plus one treatment without added Cr^{3+} as a control for all series, and the concentrations of Cr^{3+} in solution ranged from 0 to 25 μM. The concentrations of selected cations were based on the ranges that occur in natural soil pore waters [21].

Solution composition

The chemicals used were all analytical reagent, and deionized water was used during experiments. Tested solution cultures were prepared by adding different volumes of stock solutions of $CaCl_2$, $MgSO_4$, NaCl and KCl into deionized water. Except for pH-set, these media were adjusted to pH 5.50. For pH-set, the pH values were adjusted to a series of pH from 4.50 to 6.25. The value of pH was controlled using 1 mM MES-buffering (2-[N-morpholino] ethane sulfonic acid) and adding NaOH. MES was chosen because it does not form complexes with Cr (III) [22]. The values of pH and Eh in the nutrient solutions were tested before and after the bioassay using a pH meter (Delta 320, Mettler, Zurich, Switzerland) and a Eh meter (9678BNWP, Thermo, Chelmsford, America). To reach near-equilibrium conditions, media were prepared one day before the start of the bioassay. For all treatments, the pH values decreased by 0.05–0.20 (pH unit) when compared with the initial pH. The Eh values of different treatments were various, ranging from 328 to 452 mV. The chemical characteristics of the different tested solution cultures are summarized in Table 1.

Toxicity assays

The barely root elongation test was performed according to ISO guideline 11269-1 (ISO, 1993). Barley (*H. vulgare*) seeds were surface-sterilized in 2% NaClO for 30 min, after which they were thoroughly rinsed with deionized water and germinated on filter paper moistened with demonized water for 36 h at 20°C in the dark. When the radical emerged (approximately 2 mm in length), six seedlings were transferred to nylon net fixed on the surface of polypropylene pots containing 250 mL exposure solutions. There were three replicate pots for each exposure concentration. The culture containers were placed randomly in a growth chamber. The air temperature was maintained at 20°C during the 16 h

(22 k lux)/8 h dark cycles. Root length was measured after 5 d and elongation (RE, %) was calculated as percentage of the control using the equation as follows:

$$RE = \frac{REt}{REc} \times 100 \qquad (1)$$

where REt is the root length in the tested solution culture and REc is the root length in the control.

Chemical measurements

Atomic absorption spectrophotometry (Varian AA240FS/GTA120; Melbourne, Australia) was used to determine the concentration of Cu, Ca, Mg, Na and K.

The selective exchange resin Dowex-M4195 was used to evaluate whether Cr(III) was oxidized to Cr(VI) in the tested solution during the experiment period according to [23]. The Dowex-M4195 resin (particle size = 40 mesh) was immersed in deionized water for 2 days and washed with 1 M HCl. The resin was saturated with 500 mg L^{-1} $CuCl_2$ to reach Cu-saturated state in a glass column. The Cu-saturated resin was washed using deionized water until the effluent Cu concentration could no longer be detected. The Dowex-M4195 resin was transferred into separate flasks with the tested medium and shaken at 25°C for 24 h, and then washed with deionized water and placed into 100 mL of 10% NaCl to desorb the Cr adsorbed on resins until the effluent Cr concentration could no longer be detected. The Cr concentration in the desorbed solution was determined by inductively coupled plasma mass spectrometry (ICP-MS: 7500a, Agilent, Arcade, NY, USA).

Speciation of Cr in solutions

Speciation was calculated by Visual MINTEQ 3.0 (available at http://hem.bredband.net/b108693/). Input data for Visual MINTEQ were pH and the concentrations of Cr, Ca, Mg, K, Na, Cl and SO4. As the experiments were carried out in an open system, a CO_2 partial pressure of 3.5×10^{-4} atm (1 atm = 101.3 kPa) was assumed in the calculation of Visual MINTEQ.

Mathematical description of the BLM

Based on the BLM assumption, when the competing cations H^+, Ca^{2+}, Mg^{2+}, K^+ and Na^+ are considered, the fraction (f) of the total biotic ligand sites bound by Cr^{3+} is given by the following equation [7]:

$$f_{CrBL} = \frac{K_{CrBL}\{Cr^{3+}\}}{1 + K_{CrBL}\{Cr^{3+}\} + \sum K_{XBL}\{X^{n+}\}} \qquad (2)$$

where K_{CrBL} and K_{XBL} are conditional binding constants for the binding of Cr and cation X (e.g., Ca^{2+}, Mg^{2+}, K^+ or H^+) to the BL sites (M), respectively, and curly brackets {} indicate ion activity, such as $\{X^{n+}\}$, which is the activity of X^{n+} (M). {XBL} is the concentration of the specific cation-BL complex (M).

According to the methodology described in detail by De Schamphelaere and Janssen [13], when inhibition of barley root elongation is up to 50% of the control, Eq. (2) becomes:

$$EC50\{Cr^{3+}\} = \frac{f_{CrBL}^{50\%}}{\{1 - f_{CrBL}^{50\%}\}K_{CrBL}}\left(1 + \sum K_{XBL}\{X^{n+}\}\right) \qquad (3)$$

where $EC50\{Cr^{3+}\}$ is the free Cr^{3+} that results in 50% RE (50% of barley root elongation with respect to the control) and $f_{CrBL}^{50\%}$ is the

Table 1. Chemical composition of the solution cultures used in the different sets and the measured Cr toxicity threshold at 50% inhibition expressed by total concentration of Cr(III) (EC50[Cr_T]), free Cr activity (EC50{Cr^{3+}}) and $CrOH^{2+}$ activity (EC50{$CrOH^{2+}$}) for barely root elongation with 95% confidence intervals.

Sets	Treatments	The values of EC50 and 95% confidence intervals (nM)		
		EC50[Cr_T]	EC50{Cr^{3+}}	EC50{$CrOH^{2+}$}
Ca^{2+} (mM)	0.2	1333(1097–1620)	17.3(14.3–21.1)	1033(850–1256)
	1	2162(1710–2734)	25.0(19.8–31.5)	1489(1181–1877)
	2	3082(2579–3682)	32.7(27.4–39.1)	1951(1631–2335)
	5	4137(3659–4677)	37.7(33.3–42.6)	2246(1986–2540)
	10	5021(4295–5870)	41.6(34.1–50.8)	2480(2027–3033)
	15	6383(5299–7690)	44.9(37.3–53.9)	2674(2203–3219)
Mg^{2+} (mM)	0.05	1249(953–1636)	16.0(12.2–20.9)	952(726–1248)
	0.20	1521(1105–2095)	18.6(13.5–25.7)	1110(805–1530)
	0.50	2020(1491–2737)	23.3(15.9–34.1)	1387(947–2032)
	1.00	2785(2265–3423)	29.6(23.8–36.7)	1761(1420–2187)
	2.00	3367(2614–4337)	32.2(24.0–43.1)	1919(1433–2570)
	4.00	5259(4139–6682)	42.8(35.4–51.7)	2552(2112–3084)
K^+ (mM)	0.1	1386(1149–1602)	15.8(13.2–18.8)	939(789–1120)
	1.0	1570(1392–1770)	16.4(14.8–18.7)	957(728–1257)
	2.5	1678(1204–2338)	16.2 (11.6–22.5)	993(882–1171)
	5.0	1851(1347–2544)	16.6(12.1–22.8)	991(722–1362)
	7.5	2039(1816–2289)	17.3(15.4–19.4)	1032(919–1159)
	10	2097(1865–2358)	16.9(15.1–19.0)	1011(890–1136)
Na^+ (mM)	2.5	1194(961–1484)	15.4(12.4–19.2)	921(740–1146)
	5.5	1280(1061–1544)	15.7(13.0–18.9)	932(773–1124)
	10.5	1326(1185–1484)	15.3(13.7–17.1)	911(814–1019)
	15.5	1395(1156–1684)	15.0(12.4–18.1)	893(740–1077)
	20.5	1611(1365–1902)	16.4(13.9–19.3)	977(827–1153)
	25.5	1644(1371–1971)	16.0(13.3–19.2)	953(753–1143)
pH	4.5	537(464–622)	66.8(57.7–77.3)	398(343–462)
	5.0	764(662–881)	32.7(28.2–37.8)	621(538–717)
	5.5	1301(1189–1423)	16.9(15.5–18.5)	1011(922–1109)
	6.0	2457(2096–2880)	7.59(6.47–8.90)	1431(1218–1679)
	6.25	4037(3602–4525)	4.95(4.43–5.95)	1663(1488–1859)

fraction of the BLs that results in 50% RE when occupied by Cr. The barley root elongation is correlated to the fraction of the BLs (f_{CrBL}) and follows the log-logistic dose-response relationship according to Thakali et al. [18], [19].

$$RE = \frac{100}{1 + \left(\dfrac{f_{CrBL}}{f_{CrBL}^{50\%}}\right)^{\beta}} \quad (4)$$

where β is the shape parameter. Substituting f from Eq. (2) in Eq. (4) yields:

$$RE = \frac{100}{1 + \left(\dfrac{f_{CrBL}\{Cr^{3+}\}}{f_{CrBL}^{50\%}(1 - K_{CrBL}\{Cr^{3+}\} + K_{XBL}\{X^{n+}\})}\right)^{\beta}} \quad (5)$$

Eq. (5) provides the mathematical basis for the BLM that explicitly relates the biological response to the chemistry of the solution. Meanwhile, the free ion activity model (FIAM) is also fitted to the same dataset as the following equation for comparison with the BLM:

$$RE = \frac{100}{1 + \left(\frac{\{Cr^{3+}\}}{EC50\{Cr^{3+}\}}\right)^{\beta}} \quad (6)$$

The dose-response curves are plotted in terms of free Cr^{3+} activity (FIAM) and the fraction (f) of the barley root sites bound by toxic Cr species (BLM) by fitting a logistic model. The fitting parameters are conditional binding constants of all cations to BL (K_{MBL}), $f_{CrBL}^{50\%}$ and β for BLM and $EC50\{Cr^{3+}\}$ and β for FIAM. When comparing different models, the lower value of the root-mean-square error (RMSE) is used as an indicator of the better model:

$$RMSE = \sqrt{\frac{1}{N}\sum_{1}^{N}\left(R_{observed} - R_{predicted}\right)^2} \quad (7)$$

where N is the number of data, $R_{observed}$ the measured RE (as % of control) and $R_{predicted}$ the predicted RE (as % of control). The parameters of models were acquired by the mathematic model program in the DPS 9.5 statistical software [24].

Results

Distribution of chromium species in different pHs

Resin-extractable Cr by Dowex M4195 was not detected in the test medium, which implied that there was no Cr(III) oxidized to Cr(VI) during the experiment period. The distribution of Cr species in the solutions with pH from 4.5 to 6.25 is shown in Fig. 1. Free Cr^{3+} and $CrOH^{2+}$ were major species at pH 4.5, which were 12.4% and 74.1% of the total Cr, respectively. With increasing pH, the proportion of $CrOH^{2+}$ and Cr^{3+} in solution decreased sharply continuously concomitant with the increasing proportion of $Cr(OH)_2^+$ and $Cr(OH)_3$ (aq). At solution pH 6.25, the proportions of $Cr(OH)_2^+$ and $Cr(OH)_3$ (aq) reached 29.5% and 23.4% of total Cr, respectively. Other Cr species, such as $CrCl^{2+}$ were always quite low (<0.2% of total Cr) and were not considered for BLM development. Hence, the four main Cr species, Cr^{3+}, $CrOH^{2+}$, $Cr(OH)_2^+$ and $Cr(OH)_3$ (aq) were considered to test their effects on the toxicity to barley root elongation.

Effects of cations on Cr toxicity

The EC50 for barley root elongation expressed as free Cr^{3+} activity, ranged from 4.95 to 66.8 nM (Table 1). The values of $EC50\{Cr^{3+}\}$ increased linearly up to 2.59-fold with an increase of Ca^{2+} activity from 0.18 to 6.87 mM ($p<0.05$, $R^2=0.83$, Fig. 2C and Table 1). The increase of Mg^{2+} activity from 0.04 to 2.05 mM resulted in the increase of $EC50\{Cr^{3+}\}$ by a factor of 2.68. A linear relationship ($p<0.01$, $R^2=0.96$) was found between Mg^{2+} activity and $EC50\{Cr^{3+}\}$ (Fig. 2D and Table 1). However, no significant change in the $EC50\{Cr^{3+}\}$ was found when the activity varied from 0.10 to 8.97 for K^+, and from 2.35 to 21.7 mM for Na^+ (Table 1). Therefore, competition between K^+ and Na^+ with Cr^{3+} for binding sites on barley roots could be neglected when BLM was developed, and the values of $\log K_{KBL}$ and $\log K_{NaBL}$ could be approximately set to zero.

According to Eq. (3), if H^+ can compete with Cr^{3+} binding sites of barley root, then a linear relationship between $EC50\{Cr^{3+}\}$ and H^+ activity should exist in the pH-set. In the present study, the values of $EC50\{Cr^{3+}\}$ increased significantly with an increase of

H^+ activity in culture solution at $p<0.01$ level with $R^2=0.97$ (Fig. 2C), which could be explained by H^+ competition with Cr^{3+} for binding sites of barley root. In addition, from Cr species distribution, it was known that increasing pH from 4.50 to 6.25 resulted in an obvious decrease in the percentage of $CrOH^{2+}$ and an increase in the percentages of $Cr(OH)_2^+$ and $Cr(OH)_3$ (aq) to total Cr in solution. To determine whether $CrOH^{2+}$, $Cr(OH)_2^+$ and $Cr(OH)_3$ (aq) were toxic to barley root elongation, Eq. (3) was transformed to Eq. (8) when $CrOH^{2+}$, $Cr(OH)_2^+$ and $Cr(OH)_3$ (aq) were considered as toxic species as well as Cr^{3+} in the pH set:

$$\frac{1}{EC50\{Cr^{3+}\}} = \frac{1 - f_{CrBL}^{50\%}}{f_{CrBL}^{50\%}(1 + K_{HBL}\{H^+\} + K_{CaBL}\{Ca^{2+}\})}$$
$$\times (K_{CrBL} + K_{CrOHBL}K_{CrOH2+}\{OH^-\}$$
$$+ K_{Cr(OH)_2BL}K_{Cr(OH)_2^+}\{OH^-\}^2$$
$$+ K_{Cr(OH)_3BL}K_{Cr(OH)_3}\{OH^-\}^3) \quad (8)$$

where K_{CrOH2+}, $K_{Cr(OH)_2^+}$ and $K_{Cr(OH)_3}$ are stability constants for the formation of the $CrOH^{2+}$, $Cr(OH)_2^+$ and $Cr(OH)_3$ complexes, respectively. Based on equilibrium equations of $Cr^{3+} + OH^- = CrOH^{2+}$, $Cr^{3+} + 2OH^- = Cr(OH)_2^+$ and $Cr^{3+} + 3OH^- = Cr(OH)_3$, Eq. (8) could be transformed to Eq. (9):

$$\frac{1}{EC50\{Cr^{3+}\}} = \frac{1 - f_{CrBL}^{50\%}}{f_{CrBL}^{50\%}(1 + K_{HBL}\{H^+\} + K_{CaBL}\{Ca^{2+}\})}$$
$$\times (K_{CrBL} + K_{CrOHBL}\frac{\{CrOH^{2+}\}}{\{Cr^{3+}\}}$$
$$+ K_{Cr(OH)_2BL}\frac{\{Cr(OH)_2^+\}}{\{Cr^{3+}\}}$$
$$+ K_{Cr(OH)_3BL}\frac{\{Cr(OH)_3\}}{\{Cr^{3+}\}}) \quad (9)$$

The values of K_{HBL} and K_{CaBL} were set to $10^{4.74}$ and $10^{2.64}$ in the present study, respectively (see Table 2), and Eq. (9) can be transformed to a equation with $1/EC50\{Cr^{3+}\}$ as a dependent variable, and $\frac{\{CrOH^{2+}\}}{\{Cr^{3+}\}}$ as independent variables. The multiple regression between $1/EC50\{Cr^{3+}\}$ and $K_{HBL}\{H^+\}$, $CrOH^{2+}/Cr^{3+}$, $Cr(OH)_2^+/Cr^{3+}$ as well as $Cr(OH)_3/Cr^{3+}$ was calculated as:

$$\frac{1}{EC50\{Cr^{3+}\}} = \frac{1}{(1 + K_{HBL}\{H^+\})} \times ((38.6(\pm 2.30)^{***} +$$
$$0.511(\pm 0.011)^{***}\frac{\{CrOH^{2+}\}}{\{Cr^{3+}\}}) +$$
$$0.001(\pm 1.18))$$

$$(R^2 = 0.99) \quad (10)$$

According to Eq. (10), it was indicated that both the intercept and the coefficient of $CrOH^{2+}/Cr^{3+}$ were significant at $p<0.001$

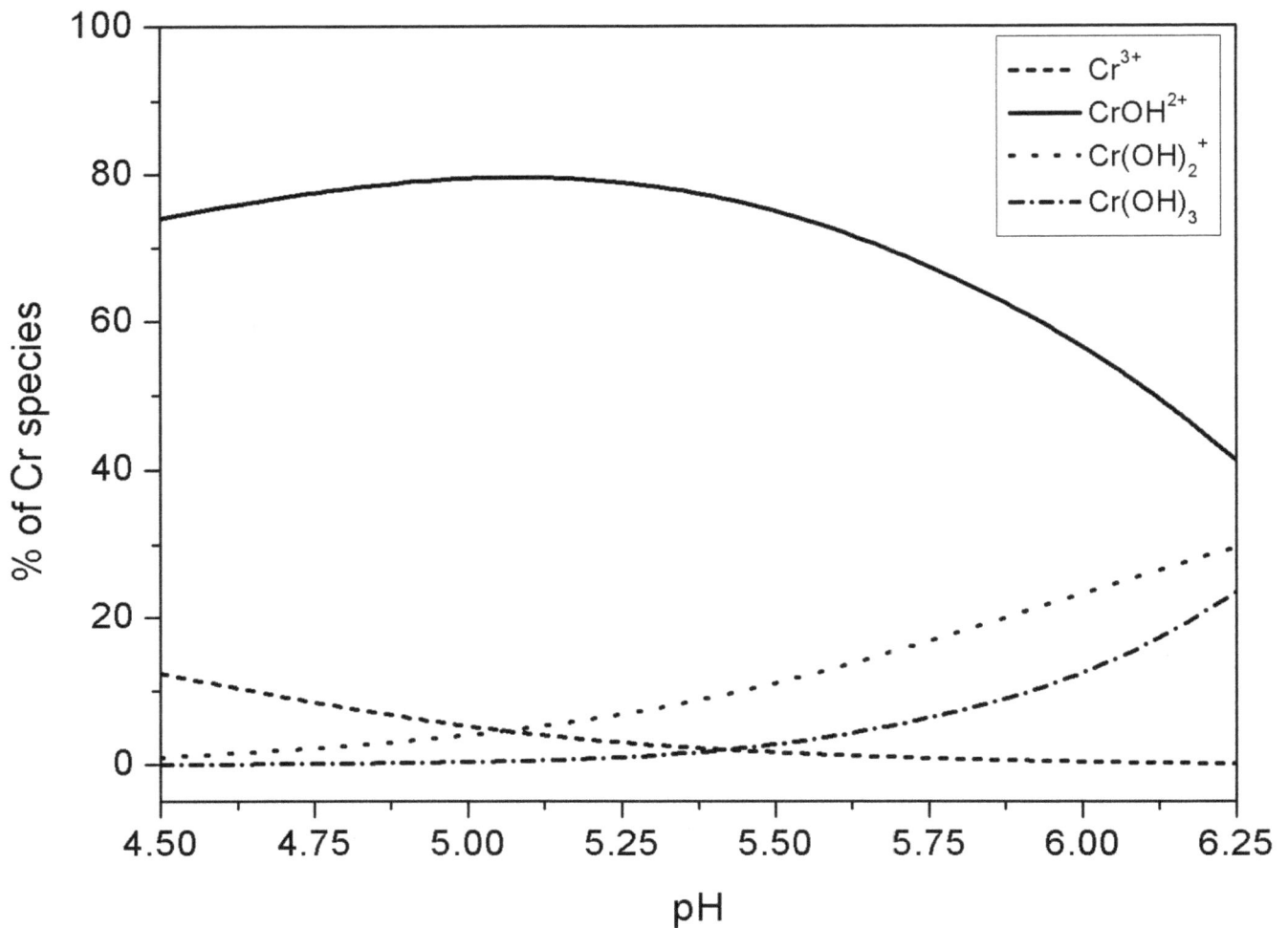

Figure 1. The distribution of different Cr(III) species (%) as a function of pH calculated by Visual MINTEQ.

level, which demonstrated that Cr toxicity to barely root elongation could be caused by Cr^{3+} plus $CrOH^{2+}$ when they exist in solution at certain pH values. These data suggested that the effects of $Cr(OH)_2^+$ and $Cr(OH)_3$ on the total Cr toxicity at pH 4.50–6.25 could be ignored and the toxicity of Cr^{3+} plus $CrOH^{2+}$ should be considered in the Cr(III)-BLM development.

Estimation of BLM parameters

When the toxicity of $CrOH^{2+}$ was considered, Eq. (4) can be transformed to Eq. (11) [13]:

$$f = \frac{K_{CrBL}\{Cr^{3+}\} + K_{CrOHBL}\{CrOH^{2+}\}}{1 + K_{CrBL}\{Cr^{3+}\} + K_{CrOHBL}\{CrOH^{2+}\} + \sum K_{XBL}\{X^{n+}\}} \quad (11)$$

Then barley root elongation could be written as follows:

$$RE = \frac{100}{1 + \left(\frac{K_{CrBL}\{Cr^{3+}\} + K_{CrOHBL}\{CrOH^{2+}\}}{f_{CrBL}^{50\%}(1 + K_{CrBL}\{Cr^{3+}\} + K_{CrOHBL}\{CrOH^{2+}\} + \sum K_{XBL}\{H^{n+}\})}\right)^{\beta}} \quad (12)$$

From Eq. (12), barley root elongation was affected by $\{Cr^{3+}\}$, $\{CrOH^{2+}\}$, $\{H^+\}$, $\{Mg^{2+}\}$ and $\{Ca^{2+}\}$, where $\{Na^+\}$ and $\{K^+\}$ were expelled from Eq. (12) because their effects on Cr toxicity were insignificant in the present study. So, Eq. (12) can be written as:

$$RE = \frac{100}{1 + \left(\frac{K_{CrBL}\{Cr^{3+}\} + K_{CrOHBL}\{CrOH^{2+}\}}{f_{CrBL}^{50\%}(1 + K_{CrBL}\{Cr^{3+}\} + K_{CrOHBL}\{CrOH^{2+}\} + K_{HBL}\{H^+\} + K_{CaBL}\{Ca^{2+}\} + K_{MgBL}\{Mg^{2+}\})}\right)^{\beta}} \quad (13)$$

The parameters, K_{CrBL}, K_{CrOHBL}, K_{HBL}, K_{CaBL}, K_{MgBL}, $f_{CrBL}^{50\%}$ and β can be obtained by data fitting the predicted RE (% of control) with minimal $RMSE$ and maximal R^2 for all sets using the DPS 9.5 statistic software. The conditional binding constants were obtained as follows: $\log K_{CrBL} = 7.43$, $\log K_{CrOHBL} = 5.61$, $\log K_{HBL} = 4.74$, $\log K_{CaBL} = 2.64$ and $\log K_{MgBL} = 2.98$ (Table 2). Those results indicate that toxicity across the wide range of pH and concentration of cations is closely related to activities of Cr^{3+} and $CrOH^{2+}$ as well as competition with H^+, Mg^{2+} and Ca^{2+} to barley root binding sites, which should be incorporated in the Cr(III)-BLM. Therefore the dose–response curves were plotted in terms of free Cr^{3+} activity based on FIAM, and in terms of f

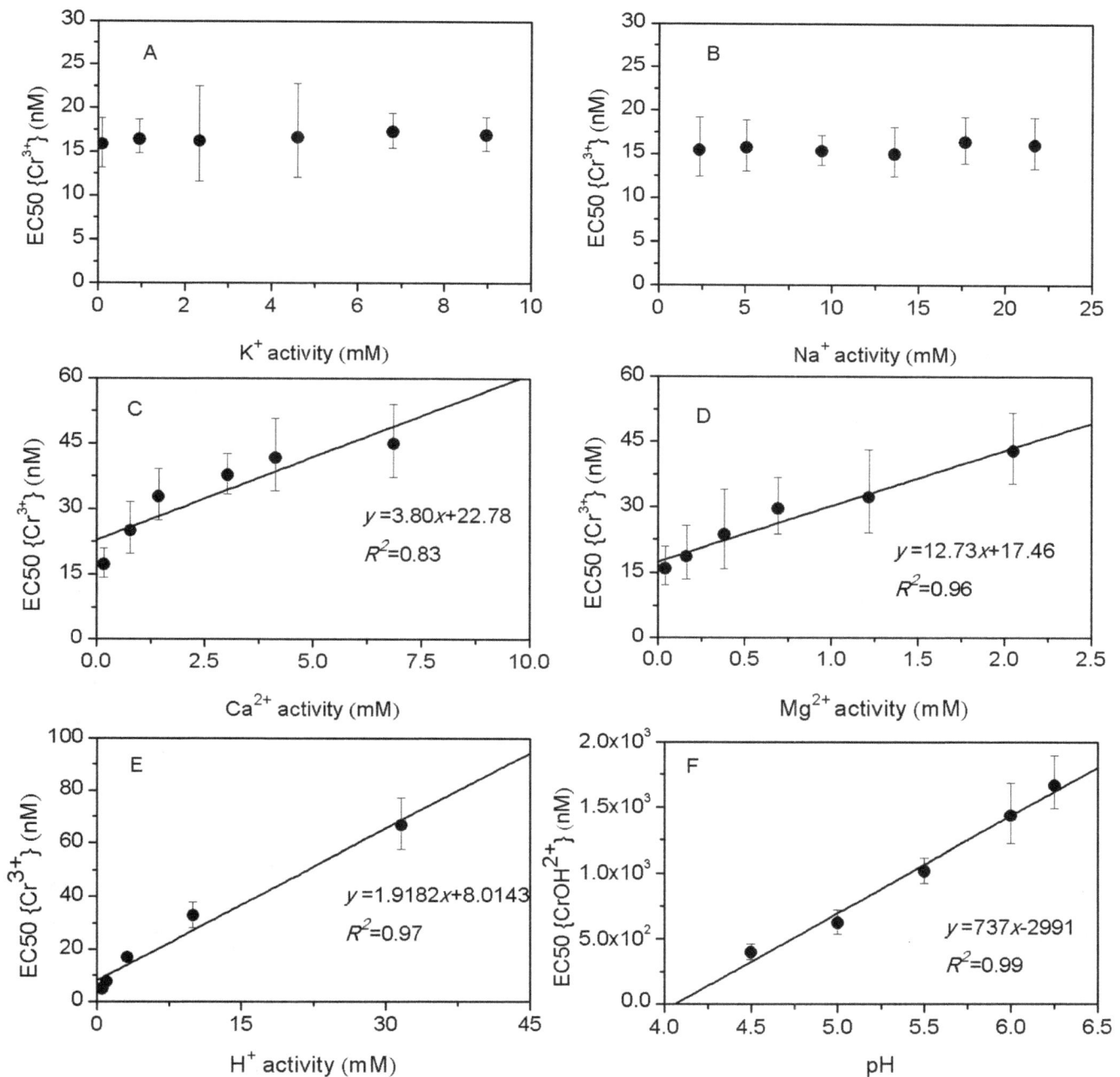

Figure 2. The EC50 values expressed as free Cr^{3+} activity (EC50{Cr^{3+}}) for (A, B, C, D, E) or free $CrOH^{2+}$ activity (EC50{$CrOH^{2+}$}) (F) for barley root elongation as a function of the free activity of K^+ (A), Na^+ (B), Ca^{2+} (C), Mg^{2+} (D), H^+ (E), and pH (F). Error bars indicate 95% confidence intervals. Solid lines represent significant correlations.

(fraction of the total barley root sites occupied by toxic Cr^{3+} and $CrOH^{2+}$ species) with considering the competitive effect of H^+, Ca^{2+} and Mg^{2+} based on the BLM (Fig. 3). Based on *RMSE* and R^2 values, the BLM considering the metal speciation and competing cations was able to predict Cr toxicity better than the FIAM. The *RMSE* decreased from 8.82 for FIAM to 5.15 for BLM, and the R^2 value increased from 0.91 for the FIAM to 0.97 for the BLM. Also, considering the influence of Cr^{3+}, $CrOH^{2+}$, H^+, Ca^{2+} and Mg^{2+}, the BLM clearly showed the best fit with the measured versus predicted values with intercept nearest 0 and the slope nearest 1 (Fig. 3D). The results indicate that BLM can

predict barley root elongation much better than FIAM when Cr^{3+} plus $CrOH^{2+}$ as toxic species and the competition of H^+, Mg^{2+} and Ca^{2+} with the binding sites of barley root are incorporated in the Cr(III)-BLM.

Validation of BLM

In attempt to examine the prediction ability of the developed Cr(III)-BLM for barley root elongation, an auto-validation was performed based on measured and predicted EC50{Cr^{3+}}. The predicted equation of EC50{Cr^{3+}} can be expressed as follows based on Eq. (3):

Table 2. Model fit summary and the optimum parameters (±standard errors) associated with the FIAM considering free Cr^{3+} activity only and BLM considering Cr^{3+} plus $CrOH^{2+}$ as toxic species and the competition of H^+, Mg^{2+} and Ca^{2+} for Cr(III) toxicity to barley root elongation.

Model	LogK					R^2	RMSE	$EC50\{Cr^{3+}\}$(nM) or $f_{CrBL}^{50\%}$	β
	Cr^{3+}	$CrOH^{2+}$	Ca^{2+}	Mg^{2+}	H^+				
FIAM						0.91	8.82	20.15±0.84	0.77±0.03
BLM	7.43±0.11	5.61±0.06	2.64±0.07	2.98±0.08	4.74±0.19	0.97	5.15	0.38±0.06	1.50±0.04

The parameters were estimated using the DPS 9.5 statistic software. RMSE represents the root-mean-squared error of the predicted % root elongation. R^2 was the determination coefficient of the models between the measured and the predicted % root elongation.

$$EC50\{Cr^{3+}\} = \frac{f_{CrBL}^{50\%}\left(1 + K_{HBL}\{H^+\} + K_{MgBL}\{Mg^{2+}\} + K_{CaBL}\{Ca^{2+}\}\right)}{\left(1 - f_{CrBL}^{50\%}\right)\left(K_{CrBL} + K_{CrOHBL}\frac{\{CrOH^{2+}\}}{\{Cr^{3+}\}}\right)} \quad (14)$$

Na^+ and K^+ were excluded from the $EC50\{Cr^{3+}\}$ prediction due to their insignificant effects on $EC50\{Cr^{3+}\}$ values of barley root elongation. The corresponding parameters (K_{HBL}, K_{MgBL}, K_{CaBL}, K_{CrBL}, K_{CrOHBL} and $f_{CrBL}^{50\%}$) are listed in Table 2. $EC50\{Cr^{3+}\}$ can be predicted when the activities of $\{H^+\}$, $\{Mg^{2+}\}$, $\{Ca^{2+}\}$, $\{CrOH^{2+}\}$ and $\{Cr^{3+}\}$ are obtained from Visual MINTEQ. Results from Fig. 4 showed that the predicted EC50s differed from the measured EC50s by less than a factor of 1.5 in the present study, indicating that the BLM can be used to predict Cr(III) toxicity to barley root elongation.

Discussion

In the present study, the Cr(III) toxicity threshold at 50% inhibition expressed by total concentration of Cr(III), i.e. $EC50[Cr_T]$, seemed to increase with increasing of K^+ or Na^+ activity. However, when the Cr toxicity threshold at 50% inhibition expressed by the activity of free Cr^{3+}, i.e. $EC50\{Cr^{3+}\}$, there was no significant effects of the activity of K^+ or Na^+ on Cr(III) toxicity (Table 1 and Fig. 2). The results suggested that the effects of K^+ and Na^+ activity on Cr(III) toxicity was attributed to the electrolyte-induced decreases of Cr(III) activity and not competition with Cr(III) for binding sites in *H. vulgare*. Protective effects of Ca^{2+}, Mg^{2+} and H^+ on Cr(III) toxicity to barley were found and the stability constants were derived (Table 2). Many researchers have reported protective effects of major cations and proton (i.e., Ca^{2+}, Mg^{2+}, K^+, and H^+) on the toxicity of several heavy metals [18], [19], [25]. For Cu toxicity, Kinraide et al. [26] reported that Ca^{2+} and Mg^{2+} had a protective effect against Cu toxicity to wheat (*Triticum aestivum*), while Le et al. [27] found that only H^+ could decrease Cu^{2+} toxicity to lettuce (*Lactuca sativa*) root elongation bioassay significantly. For Ni toxicity, Li et al. [9] found that $EC50\{Ni^{2+}\}$ was correlated significantly with the activity of Mg^{2+}, Ca^{2+} and H^+, not with the activity of Na^+ and K^+. In the case of Zn toxicity, it appeared that the increase of Mg^{2+} and K^+ activity could alleviate Zn toxicity to wheat (*T. aestivum*) and radish (*Raphanus sativus*) [28]. The protective effects of Mg^{2+}, Ca^{2+}, K^+ and H^+ on Zn^{2+} toxicity to barley were also found by Wang et al. [29]. The alleviating effects of cations such as Ca^{2+}, Mg^{2+}, K^+ and H^+ on metal toxicity can also be interpreted in terms of membrane-surface electrical potentials [30], [31]. Cell surfaces are negatively charged and these charges create negative potentials at the cell membrane surfaces. Changes in this surface electrical potential may influence the surface activities of free ions and the electrical driving force for ions and hence affect ion transport. Cations such as Ca^{2+} and Mg^{2+} depolarized the plasma membrane and reduced the negativity of the electrical potential at the outer surface of the plasma membrane and thereby alleviate uptake and effects of toxic metals [30].

The relative affinity of the BL sites for the cations, $H^+ > Mg^{2+} > Ca^{2+}$, was the same order as the results of an acute Ni-BLM for root elongation of *H. vulgare* developed by Li et al. [9] and an acute Zn-BLM for root elongation of *H. vulgare* developed by Wang et al. [28]. The binding constants log K_{HBL} (4.74) in the present study was found to be lower than that (log $K_{HBL} = 6.48$) reported by Thakali et al. [18], [19] in a terrestrial BLM for Cu

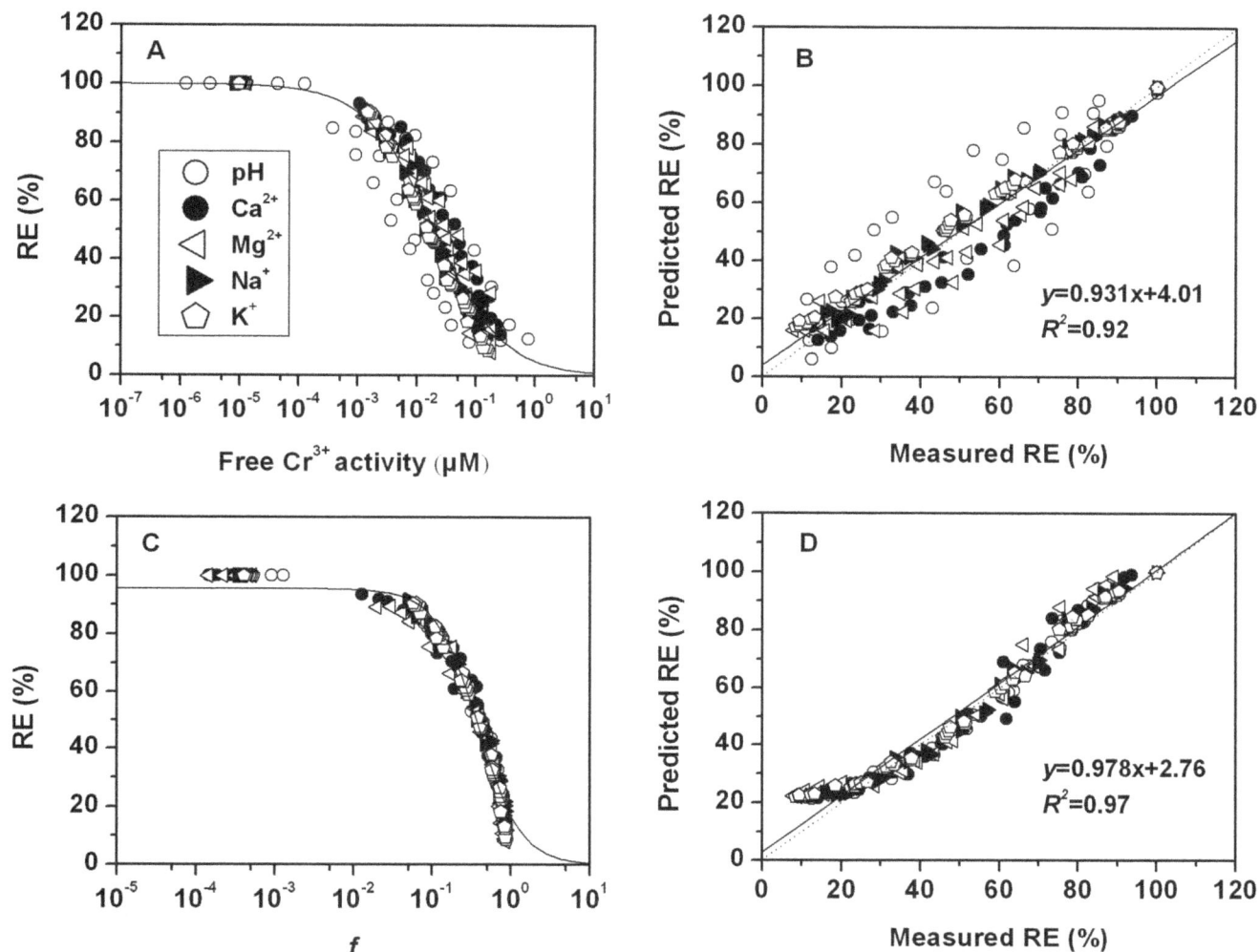

Figure 3. Toxicity of Cr to barley root elongation expressed as different dose–response curves: the dose as free Cr^{3+} activity only (A), as the fraction (f) of the total biotic ligand sites occupied by toxic Cr^{3+} and CrOH^{2+} considering the competitive effect of Ca^{2+} and Mg^{2+} activity (C). The measured versus predicted root elongation based on FIAM(B) and BLM(D).The dotted lines are 1:1 lines and the solid lines represent the linear regression relationships between the measured and predicted barley root elongation. The lines are the fitted logistic curves based on all sets.

toxicity to barley root elongation bioassay in soil solutions, whereas it was higher than (log K_{HBL} = 4.29) reported by Li et al. [9] in the BLM for acute Ni toxicity to barley root elongation and that (log K_{HBL} = 4.27) reported by Wang et al. [28] in a BLM for acute Zn toxicity to barley root elongation in culture solutions. The value of log K_{CaBL} (2.64) in the present study was similar with the result of acute Cu-BLM for root growth of *T. aestivum* developed by Luo et al. [29] (log K_{CaBL} = 2.43), but higher than that (log K_{CaBL} = 1.60) reported by Wang et al. [28]. The value of log K_{MgBL} (2.98) in the present study was similar with the result of acute Cu-BLM for root growth of *T. aestivum* developed by Wang et al. [20] (log K_{MgBL} = 2.92), whereas it was lower than that (log K_{MgBL} = 4.01) reported by Li et al. [9] and that (log K_{MgBL} = 3.72) reported by Wang et al. [28]. It was noted that the derived stability constants should be regarded as parameters that reflect the observed relations between the activity of Ca^{2+}, Mg^{2+} and H^{+} and the toxicity of metals. Differences in binding constants may, for example, result from different exposure duration, endpoint, target tissue or BL, or mechanisms of the toxicity of metals [8], [20]. More researches with chromium need to be done to investigate the

differences and similarities across organisms, endpoints and exposure duration.

The effect of solution pH on the metal activity can be explained, in part, by the competition of H^{+} and other heavy metal ions for the common binding sites, since the pH affects either the solubility and/or the speciation of many metal ions [32]. It has been indicated that besides free metal ions, the inorganic species of metals such as CuOH^{+}, ZnOH^{+} and NiHCO$_3^{+}$ were found also to be toxic to biota in the developed BLMs [9], [12], [13]. Heijerick et al. [33] observed an increase of the acute Zn toxicity to water flea *Daphnia magna* when effective concentrations were expressed as dissolved Zn but not as free Zn^{2+} activity and suggested that the effect of pH on acute Zn toxicity was a speciation effect. Li et al. [9] found that higher H^{+} activity decreased the Ni toxicity to barely through H^{+} competition with Ni^{2+} bound to biotic ligands at pH<7.0 or through the change of Ni species in solution at pH≥ 7.0, and also Ni^{2+} plus NiHCO$_3^{+}$ were toxic to barley root elongation in solution at pH≥7.0. Wang et al. [20] studied the acute Cu toxicity to barley root elongation in the pH range 5.98– 7.92 and found that the relation between H^{+} and EC50{Cu^{2+}}

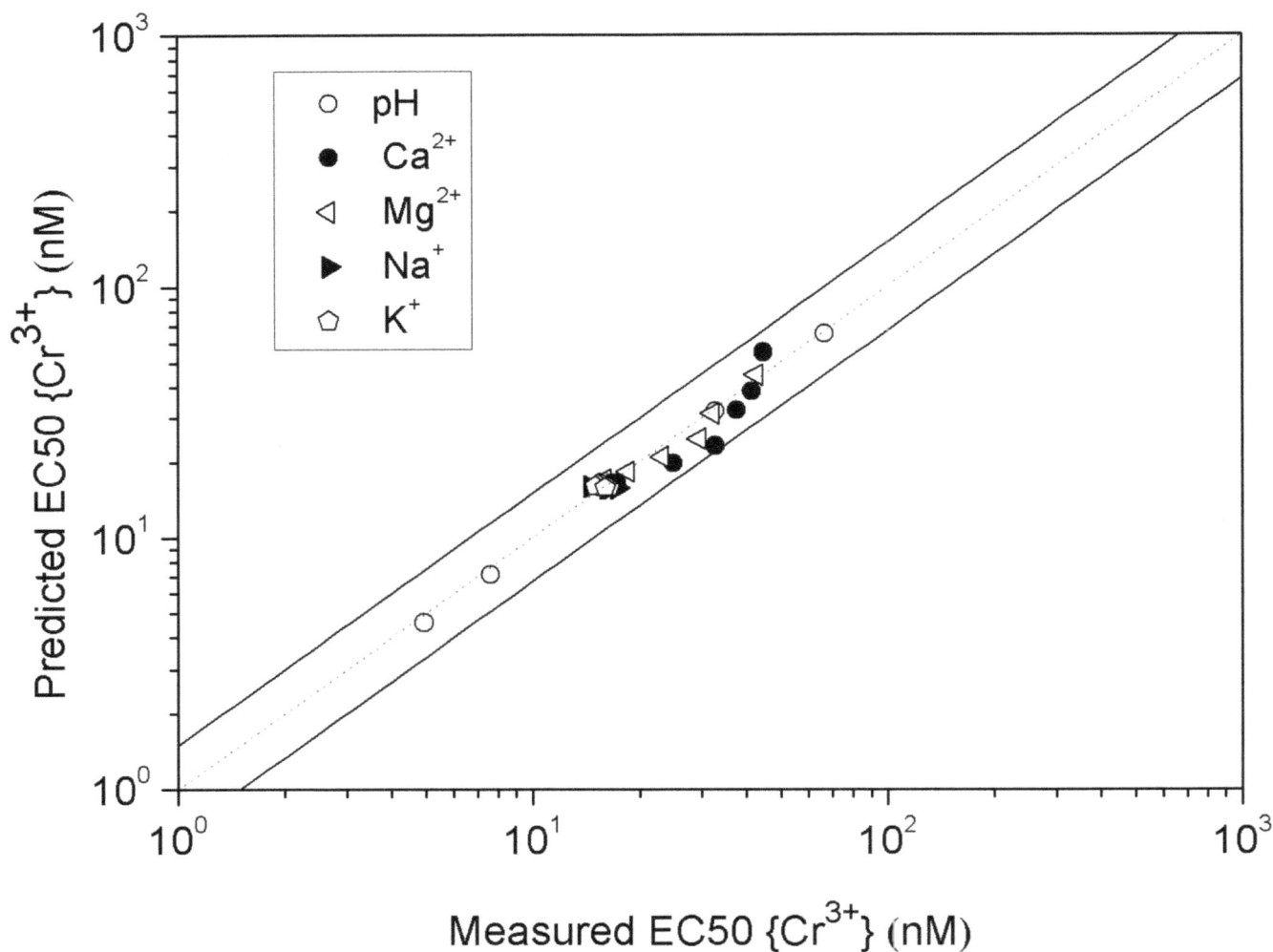

Figure 4. Relationship between the measured and predicted EC50{Cr³⁺} based on the BLM developed in the present study. The solid line indicates a perfect match between measured and predicted EC50{Cr³⁺} values, and the dashed lines indicate the range of a factor of 1.5 between observed and predicted EC50{Cr³⁺} values.

should rather be explained in terms of toxicity of Cu^{2+}, plus $CuHCO_3^+$, $CuCO_3$ (aq) and $CuOH^+$ than in terms of proton competition. In the present study, there was a linear relationship between H^+ activity and Cr^{3+} toxicity over the whole pH range, and the values of EC50{$CrOH^{2+}$} ranged from 398 to 1663 nM with the increasing pH from 4.50 to 6.25, indicating that the effect of pH on Cr metal toxicity was a significant competition effect as well as speciation effect between protons and metal ions. This finding was consistent with that reported by Cremazy et al. [34], who studied the uptake of a trivalent ion scandium (Sc) by *Chlamydomonas reinhardtii*, and found H^+ competitive for binding with Sc^{3+} transport sites within the pH range of 4.50 to 6.00, and also suggested that reasonable fit for BLM could also be obtained as a function of the first hydroxo-species ([$ScOH^{2+}$]) along with proton competition. The results from Table 2 showed that Cr^{3+} had a higher affinity to the biotic ligand than $CrOH^{2+}$, which can be correlated to the charge of the ion. It was in agreement with that of Yun et al. [35], who investigated biosorption of Cr(III) using protonated brown algae, *Ecklonia* biomass, and found chromium ions (Cr^{3+} and $CrOH^{2+}$) binding was attributed to

carboxylic groups, with values of $K_{Cr^{3+}} > K_{CrOH^{2+}}$ for the biosorption of Cr(III). Based on chemical complexation theory, the affinity of Cr^{3+} for ligands was much higher than $CrOH^{2+}$ which may result in Cr^{3+} being easier to bind to ligands with higher binding constant. In a study of Cr(III) biosorption onto protonated brown algae *Pelvetia canaliculata*, Vilar et al. [36] reported the modeling information on equilibrium and kinetics using the Cr(III) speciation in solution, and found that $CrOH^{2+}$ binding always remained lower than Cr^{3+} and diffused slower than Cr^{3+} even for pH values higher than 3.55, where the concentrations of ions was higher than Cr^{3+} ions. Although Cr^{3+} ions is not the dominated specie of the total Cr(III) at pH 4.50–6.25 (Fig. 1) in the present study, it was expected as one of the dominant toxic forms as well as $CrOH^{2+}$, since it has a higher affinity than $CrOH^{2+}$ to the binding sites.

The 5 d EC50{Cr^{3+}} for barely root elongation ranged from 4.95 to 66.8 nM for all treatments and varied about 13-fold, which clearly demonstrates the limitations of using free ion activity for predicting the toxicity of Cr(III). The BLM developed in this study could predict EC50s accurately (difference of factor of 1.5), indicating that it can be used to predict toxicity of Cr(III) to terrestrial plants. However, the application of this Cr(III)-BLM is hampered by the problematic of measuring or predicting metal speciation for the complex mixtures of organic matter in natural soil solutions. Also, when the constants derived in the present study are used to predict Cr(III) toxicity in soil by this Cr(III)-BLM, they still need be validated or further study by the experiments with dissolved organic matter (DOM) additions and with natural soils. The direct links between chemistry of metals in soils and their ecotoxicity might be a good approach in the future [18], [19].

Conclusions

In the present study, a BLM was developed for predicting the toxicity of Cr(III) to barley (*H. vulgare*) in nutrient solutions. It was found that Cr^{3+} plus $CrOH^{2+}$ as toxic species and competition

with H^+, Mg^{2+} and Ca^{2+} for the binding sites of BL should be incorporated into the BLM. The BLM parameters were derived and validated, and the developed BLM demonstrated good performance in predicting acute Cr(III) toxicity to barley root elongation. The BLM, therefore, may initiate a promising tool for improving the ecological relevancy of risk assessment procedures for trivalent metals such as Cr(III) as well as divalent metals in water and soils.

Author Contributions

Conceived and designed the experiments: YM NS. Performed the experiments: NS XZ. Analyzed the data: NS JL. Contributed reagents/materials/analysis tools: YM BL DW. Contributed to the writing of the manuscript: NS YM.

References

1. Stewart MA, Jardine PM, Barnett MO, Mehlhorn TL, Hyder LK, et al. (2003) Influence of soil geochemical and physical properties on the sorption and bioaccessibility of chromium (III). J Environ Qual 32: 129–137.
2. Bolan NS, Adriano DC, Natesan R, Koo BJ (2003) Effects of organic amendments on the reduction and phytoavailability of chromate in mineral soil. J Environ Qual 32: 120–128.
3. Yu PF, Juang KW, Lee DY (2004) Assessment of the phytotoxicity of chromium in soils using the selective ion exchange resin extraction method. Plant Soil 258: 333–340.
4. Ma YB, Hooda P (2010) Chromium Cobalt and Nickel. In: Hooda P, ed, Trace elements in soils. Wiley–Blackwell, Chichester, UK, pp 461–480.
5. Shanker AK, Cervantes C, Loza-Tavera H, Avudainayagam S (2005) Chromium toxicity in plants. Environ Int 31: 739–753.
6. López-Luna J, González-Chávez MC, Esparza-García FJ, Rodríguez-Vázquez R (2009) Toxicity assessment of soil amended with tannery sludge trivalent chromium and hexavalent chromium using wheat oat and sorghum plants. J Hazard Mater 163: 829–834.
7. Sivakumar S, Subbhuraam CV (2005) Toxicity of chromium(III) and chromium(VI) to the earthworm Eisenia fetida. Ecotoxicol Environ Saf 62: 93–98.
8. Lock K, De Schamphelaere KAC, Because S, Criel P, Van Eeckhout H, et al. (2006) Development and validation of an acute biotic ligand model (BLM) predicting cobalt toxicity in soil to the potworm Enchytraeus albidus. Soil Biol Biochem 38: 1924–1932.
9. Li B, Zhang X, Wang XD, Ma YB (2009) Refining a biotic ligand model for nickel toxicity to barley root elongation in solution culture. Ecotoxicol Environ Saf 72: 1760–1766.
10. Guo XY, Ma YB, Wang XD, Chen SB (2010) Re-evaluating the effects of organic ligands on copper toxicity to barley root elongation in culture solution. Chem Spec Bioavailab 22: 51–59.
11. Di Toro DM, Allen HE, Bergman HL, Meyer JS, Paquin PR, et al. (2001) Biotic ligand model of the acute toxicity of metals. 1. Technical basis. Environ Toxicol Chem 20: 2383–2396.
12. Santore RC, Di Toro DM, Paquin PR, Allen HE, Meyer JS (2001) Biotic ligand model of the acute toxicity of metals. 2. Application to acute copper toxicity in freshwater fish and Daphnia. Environ Toxicol Chem 20: 2397–2402.
13. De Schamphelaere KAC, Janssen CR (2002) A biotic ligand model predicting copper toxicity for Daphnia magna: the effects of calcium magnesium sodium potassium and pH. Environ Sci Technol 36: 48–54.
14. Jo HJ and Jung J (2009) Surface response model for prediction of the acute toxicity of Cu(II) and Cr Cr(VI) toward Daphnia magna. Toxicol Environ Health Sci 2: 141–147.
15. Wang X, Ma Y, Hua L, McLaughlin MJ (2009) Identification of hydroxyl copper toxicity to barley (Hordeum vulgare) root elongation in solution culture. Environ Toxicol Chem 28: 662–667.
16. An J, Jeong S, Moon HS, Jho EH, Nam K (2012) Prediction of Cd and Pb toxicity to Vibrio fischeri using biotic ligand-based models in soil. J Hazard Mater 203–204: 69–76.
17. Lock K, De Schamphelaere KAC, Because S, Criel P, Van Eeckhout H, et al. (2007) Development and validation of a terrestrial biotic ligand model predicting the effect of cobalt on root growth of barley (Hordeum vulgare). Environ Pollut 147: 626–633.
18. Thakali S, Allen HE, Di Toro DM, Ponizovsky AA, Rooney CP, et al. (2006) A terrestrial biotic ligand model. 1. Development and application to Cu and Ni toxicity to barley root elongation in soils. Environ Sci Technol 40: 7085–7093.
19. Thakali S, Allen HE, Di Toro DM, Ponizovsky AA, Rooney CP, et al. (2006) A terrestrial biotic ligand model terrestrial biotic ligand model. 2. Application to Ni
20. and Cu toxicities to plants invertebrates and microbes in soil. Environ Sci Technol 40: 7094–7100.
21. Wang XD, Ma YB, Hua L (2012) A biotic ligand model predicting acute copper toxicity for barley (Hordeum vulgare): Influence of calcium magnesium sodium potassium and pH. Chemosphere 89: 89–95.
22. Oorts K, Ghesquiere U, Swinnen K, Smolders E (2006) Soil properties affecting the toxicity of CuCl2 and NiCl2 for soil microbial processes in freshly spiked soils. Environ Toxicol Chem 25: 836–844.
23. Carbonaro RF, Stone AT (2005) Speciation of chromium(III) and cobalt(III) (amino)carboxylate complexes using capillary electrophoresis. Anal Chem 77: 155–164.
24. Chen CP, Juang KW, Lin TH, Lee DY (2010) Assessing the phytotoxicity of chromium in Cr(VI)–spiked soils by Cr speciation using XANES and resin extractable Cr(III) and Cr(VI). Plant Soil 334: 299–309.
25. Tang QY, Zhang CX (2013) Data Processing System (DPS) software with experimental design statistical analysis and data mining developed for use in entomological research. J Insect Sci 20: 254–260.
26. Jo HJ, Son J, Cho K, Jung J (2010) Combined effects of water quality parameters on mixture toxicity of copper and chromium toward Daphnia magna. Chemosphere 81: 1301–1307.
27. Kinraide TB, Pedler JF, Parker DR (2004) Relative effectiveness of calcium and magnesium in the alleviation of rhizotoxicity in wheat induced by copper zinc aluminum sodium and low pH. Plant Soil 259: 201–208.
28. Le TTY, Peijnenburg WJGM, Hendriks AJ, Vijver MG (2012) Predicting effects of cations on copper toxicity to lettuce (Lactuca sativa) by the biotic ligand model. Environ Toxicol Chem 31: 355–359.
29. Pedler JF, Kinraide TB, Parker DR (2004) Zinc rhizotoxicity in wheat and radish is alleviated by micromolar levels of magnesium and potassium in solution culture. Plant Soil 259: 191–199.
30. Wang XD, Li B, Ma YB, Hua L (2010) Development of a biotic ligand model for acute zinc toxicity to barley root elongation. Ecotoxicol Environ Saf 73: 1272–1278.
31. Kopittke PM, Kinraide TB, Wang P, Blamey FP, Reichman SM, et al. (2011) Alleviation of Cu and Pb Rhizotoxicities in Cowpea (Vigna unguiculata) as Related to Ion Activities at Root-Cell Plasma Membrane Surface. Environ Sci Technol 45: 4966–73.
32. Wang P, De Schamphelaere KA, Kopittke PM, Zhou DM, Peijnenburg WJ, et al. (2012) Development of an electrostatic model predicting copper toxicity to plants. J Exp Bot 63: 659–668.
33. Laurén DJ, McDonald DG (1985) Effects of copper on branchial ionoregulation in the rainbow trout, Salmo gairdneri Richardson. J Comp Physiol B 155: 635–644.
34. Heijerick DG, De Schamphelaere KAC, Janssen CR (2002) Predicting acute zinc toxicity for Daphnia magna as a function of key water chemistry characteristics: development and validation of a biotic ligand model. Environ Toxicol Chem 21: 1309–1315.
35. Cremazy A, Campbell PGC, Fortin C (2013) The biotic ligand model can successfully predict the uptake of a trivalent ion by a unicellular alga below pH 650 but not above: possible role of hydroxo-species. Environ Sci Technol 47: 2408–2415.
36. Yun YS, Park D, Park JM, Volesky B (2001) Biosorption of trivalent chromium on the brown seaweed biomass. Environ Sci Technol 35: 4353–4358.
37. Vilar VJP, Valle JAB, Bhatnagar A, Santos JC, De Souza SMAGU, et al. (2012) Insights into trivalent chromium biosorption onto protonated brown algae Pelvetia canaliculata: Distribution of chromium ionic species on the binding sites. Chem Eng J 200–202: 140–148.

20

High-Throughput Phenotyping to Detect Drought Tolerance QTL in Wild Barley Introgression Lines

Nora Honsdorf[1,2], Timothy John March[3], Bettina Berger[4], Mark Tester[5], Klaus Pillen[1]*

1 Chair of Plant Breeding, Institute of Agricultural and Nutritional Sciences, Martin-Luther University Halle-Wittenberg, Halle (Saale), Germany, **2** Interdisciplinary Center for Crop Plant Research (IZN), Halle (Saale), Germany, **3** School of Agriculture, Food and Wine, University of Adelaide, Waite Campus, Adelaide, Australia, **4** The Plant Accelerator, University of Adelaide, Waite Campus, Adelaide, Australia, **5** Center for Desert Agriculture, King Abdullah University of Science and Technology, Thuwal, Saudi Arabia

Abstract

Drought is one of the most severe stresses, endangering crop yields worldwide. In order to select drought tolerant genotypes, access to exotic germplasm and efficient phenotyping protocols are needed. In this study the high-throughput phenotyping platform "The Plant Accelerator", Adelaide, Australia, was used to screen a set of 47 juvenile (six week old) wild barley introgression lines (S42ILs) for drought stress responses. The kinetics of growth development was evaluated under early drought stress and well watered treatments. High correlation ($r = 0.98$) between image based biomass estimates and actual biomass was demonstrated, and the suitability of the system to accurately and non-destructively estimate biomass was validated. Subsequently, quantitative trait loci (QTL) were located, which contributed to the genetic control of growth under drought stress. In total, 44 QTL for eleven out of 14 investigated traits were mapped, which for example controlled growth rate and water use efficiency. The correspondence of those QTL with QTL previously identified in field trials is shown. For instance, six out of eight QTL controlling plant height were also found in previous field and glasshouse studies with the same introgression lines. This indicates that phenotyping juvenile plants may assist in predicting adult plant performance. In addition, favorable wild barley alleles for growth and biomass parameters were detected, for instance, a QTL that increased biomass by approximately 36%. In particular, introgression line S42IL-121 revealed improved growth under drought stress compared to the control Scarlett. The introgression line showed a similar behavior in previous field experiments, indicating that S42IL-121 may be an attractive donor for breeding of drought tolerant barley cultivars.

Editor: Tianzhen Zhang, Nanjing Agricultural University, China

Funding: This work was supported by the Interdisciplinary Centre for Crop Plant Research (IZN), Halle (Saale) (http://www.uni-halle.de/izn/), the German Plant Genome Research Initiative (GABI) of the Federal Ministry of Education and Research (BMBF, project 0313125B) (www.gabi.de/), and Group of Eight Australia – Germany Joint Research Co-operation Scheme, funded by the German Academic Exchange Service (www.daad.de) and the Group of Eight, Australia (http://www.go8.edu.au/). The funders had no role in study design, data collection and analysis, decision to publish, or preparation of the manuscript.

Competing Interests: The authors have declared that no competing interests exist.

* E-mail: klaus.pillen@landw.uni-halle.de

Introduction

Barley (*Hordeum vulgare* ssp. *vulgare*, hereafter abbreviated with *Hv*) is ranked fourth among the worldwide production of cereals. Due to its multipurpose use as animal feed, human food and substrate for malting it is one of the most important cereals worldwide [1]. Barley is known to be relatively tolerant to abiotic stresses among the major cereal crops and, thus, is often grown in more marginal sites [2]. However, the process of genetic erosion has been under way in barley since its domestication some 10,000 years ago and, in particular, since the advent of intensive modern elite breeding during the last century [3]. As a result of this process, diverse landraces have been replaced by modern elite cultivars with a much narrower gene pool. Therefore there is limited genetic diversity remaining in the elite barley gene pool for abiotic and biotic stress tolerance. The current loss of genetic variation in the elite gene pool tends to limit the breeding success of improved cultivars [4]. To overcome this problem several authors, e.g. Zamir [5], proposed to use wild relatives of crop species as donors of exotic germplasm to enhance elite varieties. Tanksley and Nelson [6] proposed the method of "advanced

backcross quantitative trait loci analysis" (AB-QTL) to introduce exotic genes into modern crop varieties. The method combines QTL detection and the introduction of favorable exotic alleles from a wild donor parent. Lines produced by advanced backcrosses ideally contain only one single introgression from the wild parent and are then referred to as introgression lines (ILs). This is achieved by several rounds of backcrossing to the recurrent parent and marker assisted selection (MAS). A set of ILs ideally represents the whole genome of a wild donor plant in the genetic background of a single elite variety [5].

Pillen et al. [7] published the first AB-QTL study in barley. Von Korff et al. [8] developed a BC_2DH population from a cross between the German spring barley cultivar Scarlett (*Hv*) and the Israeli wild barley accession ISR42-8 (*Hordeum vulgare* ssp. *spontaneum*, hereafter abbreviated with *Hsp*). The lines of this S42 population were used in several AB-QTL studies to identify QTL for yield, pathogen resistance and malting quality traits [9–13].

Schmalenbach et al. [14] used the S42 population to develop 59 ILs (S42ILs) by a further round of backcrossing with the recurrent parent Scarlett and subsequent selfing and MAS. Each of the S42ILs contains a single or a small number of *Hsp* introgressions.

Several QTL studies were conducted to verify QTL from AB-QTL studies and to identify new QTL for pathogen resistance, yield and quality parameters [14–16]. Naz et al. [17] studied root architecture of S42ILs and detected QTL for root dry weight and root volume. Later on, the S42IL population was extended to 73 lines and the lines were genotyped with a 1,536-SNP Illumina BOPA1 set [18]. Six hundred thirty-six informative SNPs and their known map order [19] allowed the precise localization of the *Hsp* introgressions. The S42IL set represents 87.3% of the wild barley donor genome. Moreover, Schmalenbach et al. [18] developed segregating high-resolution mapping populations (S42IL-HRs) for 70 S42ILs. Those lines are readily available to facilitate fine mapping and, ultimately, cloning of QTL.

Drought is one of the main factors limiting yield worldwide [20]. Due to climate change extreme weather events are predicted to occur more frequently and an altered pattern of drought occurrence is expected [21]. Therefore maintaining plant growth and yield under drought remains a major objective for plant breeding [22]. Many studies have been conducted on the impacts of terminal drought stress in cereals, while impacts of drought stress at early developmental stages are less well investigated [23]. However, several authors comment that yield may be enhanced by improved early vigor and a rapid development of maximum leaf area [24,25]. López-Castañeda and Richards [26] reported that on average barley has a higher yield in water limited environments compared to wheat, triticale, and oat. As part of a possible explanation, they pinpointed the faster and more vigorous growth of barley during vegetative development. Variation in this trait is, therefore, likely to be in direct relation to drought stress tolerance and yield.

Conventional methods to determine biomass and measure growth are time-consuming and labor intensive. Often they involve destructive harvest of plants and therefore make repeated measurements on the same plant impossible. New developments in plant imaging technologies allow the estimation of biomass and growth parameters as a non-destructive and rapid alternative to more traditional methods [27,28]. New phenotyping facilities enable automated imaging of plants. Several types of plant images can be taken, e.g. with infrared, near infrared, fluorescent and visible light. Scanning with infrared light gives information on plant or leaf temperature, while near infrared imaging sheds light on the plant water content and fluorescent pictures enable conclusions on plant health status. High resolution color pictures (RGB pictures), taken from the top and two side views are used to determine the projected shoot area of the plant. The projected shoot area serves as a measure for biomass. Hence, from RGB images taken at several time points, growth curves as well as growth rates can be calculated.

In fully automated greenhouses plants can be delivered via conveyor belts to watering, weighing and imaging stations. In these high-throughput phenotyping facilities several hundred individual plants can be imaged per day in a fully automated manner. High-throughput phenotyping facilities of this type are currently in use in various research institutes (e.g. The Plant Accelerator, Adelaide, Australia; CropDesign, Gent, Belgium; IPK Gatersleben, Germany, PhenoArch, Montpellier, France).

Such phenotyping facilities are ideal to combine controlled irrigation and phenotyping protocols [29]. A first application was given by Rajendran et al. [30] who used a manual imaging system (LemnaTecScanalyzer3D, Wuerselen, Germany) to screen *Triticum monococcum* accessions for salinity tolerance. They developed high-throughput quantification assays to distinguish sodium exclusion, sodium tissue tolerance and osmotic tolerance as the strategies plants use to establish salinity tolerance.

In this report, "The Plant Accelerator" was used to screen growth of wild barley ILs under well watered and drought treatments during vegetative growth. The aims were (1) to identify wild barley derived QTL within the set of S42ILs that control drought stress responses and (2) to test the use of non-destructive high-throughput imaging to measure vegetative stage drought response in barley.

We could show that high-throughput imaging provides accurate estimates for biomass development over time. Moreover several drought related QTLs were identified and genotypes detected that may be beneficial in future breeding programs.

Materials and Methods

Plant Material

Forty-seven wild barley ILs of the S42IL library and the recipient parent Scarlett were selected for the experiment. The S42ILs are derived from a cross between the German malting barley variety Scarlett and the Israeli wild barley accession ISR42-8. The 47 ILs possess few *Hsp* chromosome segments and were selected based on SSR and SNP genotyping to represent a large portion, 87.3%, of the ISR42–8 genome [18]. Repeated backcrossing and MAS are described in Schmalenbach et al. [14].

Glass House Cultivation

Two drought stress experiments, with duration of six weeks each, were conducted between end of March and mid of July 2011 in The Plant Accelerator greenhouse facilities in Adelaide, Australia (34°58′16.18″S; 138°38′23.88″E). Forty-eight barley genotypes were grown under a well watered and stress treatments with three replicates per genotype and treatment. Each experiment was designed in three randomized blocks. Control and stress treatments of each genotype were placed next to each other (Fig. 1, Table S1).

Single plants were grown in 2.5 L plastic containers with 2.1 kg of soil (50% UC Davis soil mix, 35% Coco-peat, 15% clay-loam). Three seeds per pot were directly sown into the soil and after germination thinned out, leaving one plant per pot. Plants were pre-grown for two weeks in a regular greenhouse and watering was performed manually to allow optimal germination and seedling establishment. Subsequently, the pots were transferred to the "smart house" where each pot was placed onto a cart on a conveyor belt and the two treatments were applied. Every second day, pots were weighed and watered automatically to 22% gravimetric water content for the well watered treatment and 15% for the stress treatment (Fig. 2). Based on the experience of the first

Figure 1. View of experiment 1 with five-week old barley S42IL plants growing in The Plant Accelerator.

Figure 2. Barley plants at the weighing and watering unit after leaving the imaging station.

experiment we adjusted the drought stress in the second experiment to 12% gravimetric water content to slightly increase drought effects. The experiments were carried out under natural lighting with the temperature in the greenhouse kept at a range between 15°C (night) and 22°C (day).

Phenotyping

With the onset of the stress treatment imaging of the plants started. Plant images were captured using a LemnaTec 3D Scanalyzer (LemnaTec, GmbH, Wuerselen, Germany). Every day, three RGB pictures (2056×2454 pixels) were taken of each barley plant, one top view image and two side view images with a $90°$ horizontal rotation. After background-foreground separation was applied to separate the plant tissue area from the background, pixel numbers per plant were counted and the pixel sum of the three pictures per plant was taken to define the projected shoot area. The shoot area measured over time was used to draw growth curves. For each growth curve, curve fitting with a 6th order polynomial was conducted to adjust for possible missing data points and absolute growth rate [dA/dt] and relative growth rate [(dA/dt)/A] were calculated. For each of the three curves the integral was determined and used as a trait in the statistical analysis. Moreover, six further traits were extracted from the images; caliper length, height, color (as hue angle in the HSI color scheme) and the two parameters shoot area top view and convex hull area to calculate compactness of each plant. At the end of the experiment, barley plants were harvested and above ground biomass, tiller number (TIL), and plant height (HEI) were

determined. Fresh biomass was weighed and, subsequently, oven dried to constant weight to determine dry biomass. Water use efficiency (WUE) was calculated by dividing dry biomass at the end of the experiment by the total amount of water added during the four weeks in the "smart house" [mg/g water]. Specific plant weight (SPW) was calculated from the dry weight and the maximum projected shoot area at the end of the experiment. In addition, simple stress indices (SSI) were calculated as follows: SSI = Ts/Tc, where Ts and Tc are the average trait performances of an IL under stress and control conditions, respectively. An overview of trait definitions is given in Table 1.

Genotyping

The S42ILs were genotyped with the 1,536-SNP barley BOPA1 set [19] of the Illumina GoldenGate assay [18]. Six hundred and thirty-six out of the tested 1,536 SNPs were polymorphic and used to characterize the extent of exotic *Hsp* introgressions in each S42IL (see Fig. S1).

Statistical Analysis

Statistical analyses were performed with SAS Enterprise Guide 4.2. [31]. Descriptive statistical parameters (Table S2) were calculated with procedure MEANS. Heritabilities across treatments were calculated as $h^2 = V_G/[V_G + V_{GT}/t + V_{GE}/e + V_{GET}/et + V_R/etr]$, and within treatments as: $h^2 = V_G/[V_G + V_{GE}/e + V_R/er]$. The terms V_G, V_{GT}, V_{GE}, V_{GET} and V_R represent the genotypic, genotype× treatment, genotype×environment, genotype×environment× treatment, and error variance components, respectively, calculated with procedure VARCOMP [31]. The terms t, e, and r indicate the number of treatments, experiments and replicates, respectively. Pearson correlation coefficients between traits were calculated with means across treatments, blocks and experiments and within drought stressed and control treatments, respectively, using the procedure CORR.

Analysis of variance was carried out with the procedure MIXED using model I to test for genotype main effects across treatments and experiments.

Model I:

$$Y_{ijkl} = \mu + L_i + T_j + E_k + L \times T_{ij} + L \times E_{ik} + B(E \times T_{kj})_l + \varepsilon_{ijkl}$$

and model II for genotype effects across experiments but within a single treatment.

Model II:

$$Y_{ikl} = \mu + L_i + E_k + L \times E_{ik} + B(E_k)_l + \varepsilon_{ikl}$$

Where μ is the general mean, L_i is the fixed effect of the ith line, T_j is the fixed effect of the jth treatment, E_k is the fixed effect of the kth experiment, $L \times T_{ij}$ is the fixed interaction between the ith line and the jth treatment, $L \times E_{ik}$ is the fixed interaction between the ith line and the kth experiment, $B(E \times T_{kj})_l$ is the random effect of the lth block nested in the interaction between kth experiment and jth treatment, $B(E_k)_l$ is the random effect of the lth block nested in the kth experiment and ε_{ijkl} and ε_{ikl} are the error of Y_{ijkl} and Y_{ikl}, respectively.

Following the mixed model analysis a Dunnett test was conducted where least square means (LSMEANS) of each IL were compared to the control Scarlett. In case an IL revealed a significant (P<0.05) deviation in trait performance from Scarlett, as main effect and/or as line×treatment interaction, a line×trait

Table 1. List of evaluated traits.

Trait	Abbreviation	Unit	Method of measurement
Imaging parameters			
Shoot area integral[a]	SAI	kPix[b]	Calculated from pixel sum of three images per plant per day; A
Absolute growth rate integral	AGRI	kPix/d	Calculated from pixel sum of three images per plant per day; dA/dt
Relative growth rate integral	RGRI	d^{-1}	Calculated from pixel sum of three images per plant per day; (dA/dt)/A
Height integral	HEII	kPix	Max. distance from bottom to top of plant
Caliper length integral	CALI	kPix	Max. distance between two points on the object boundary, top view image
Hull area integral	HULI	kPix	Smallest geometrical object without concave parts that covers whole plant, top view image
Shoot area top view integral	SATVI	kPix	Pixel number
Plant hue integral	HUEI	-	Average hue value calculated from all pixels per plant and day
Harvest parameters			
Tiller number	TIL	-	Number of tillers per pot
Height	HEI	cm	Plant height measured from bottom to leaf tip
Biomass dry	BMD	g	Weight of oven dried biomass per pot
Indices			
Water use efficiency	WUE	mg/g water	Harvested biomass per plant/total amount of irrigation water
Specific plant weight	SPW	mg/kPix	Harvested biomass per plant/pixel number per plant at end of experiment
Compactness integral	COMI	-	SATV/HUL per plant per day
Simple stress index	SSI	-	Trait performance under stress/trait performance under control treatment

[a]Integral: calculated for length of entire experiment, respectively.
[b]kilo Pixel.

association was assumed and the presence of a QTL was accepted. If several ILs with overlapping introgressions showed a similar effect, it was assumed that the ILs contained the same QTL. We consider this QTL as the most likely location of the effect and, thus, define a minimum number of QTL needed to explain all identified trait effects. The relative performance (RP) of an IL was calculated as RP (IL) = [LSMEANS (IL) – LSMEANS (Scarlett)] ×100/LSMEANS (Scarlett), where LSMEANS were calculated with model I across treatments, experiments and blocks or with model II across experiments and blocks, separately for each treatment. The detection of significant line by trait associations was conducted for every trait revealing a heritability with $h^2 > 0\%$ across treatments or within the two watering treatments, respectively.

Results

Trait Performance of S42ILs

For most traits, means were higher under well watered treatment than under drought stress (Fig. 3 and Fig. 4, Table S2), e.g. 2.2 g of biomass dry (BMD) vs. 0.9 g. There were, however, four exceptions. Compactness integral (COMI) was higher under drought treatment than under well watered treatment. The same was true for plant hue integral (HUEI), SPW, and WUE but differences were marginal.

Coefficients of variation (CV) differed strongly between traits (1.5 to 72.1%). Highest CV was calculated for BMD across treatments (72.1%). The lowest CVs were determined for HUEI, varying from 1.5 to 1.6% for the different treatments and the SSI. CV was generally higher under drought than under well watered treatment. The four exceptions were WUE, SPW and height integral (HEII) where CV under well watered treatment was higher and HUEI where CV was the same under both treatments.

Heritability was generally higher under well watered treatment. HEI, HEII and WUE were exceptions and showed higher heritabilities under drought treatment. Highest heritabilities were found for HEI, HEII and caliper integral (CALI) (between 46.2 and 76.8%). Low heritabilities were determined for WUE and HUEI under drought treatment and across treatments, as well as for BMD under drought treatment (15%). Most of the SSI showed heritabilities equal to 0. SSI (HEII), SSI (HUEI), SSI (SPW), SSI (WUE) revealed heritabilities between 4.7 and 14.5%.

Trait Correlations

Highest correlations were found among measured traits and among stress indices (Table S3). However, correlations between measured traits and stress indices were low. The measured traits showed the highest correlations between shoot area integral (SAI) and absolute growth rate integral (AGRI) (r = 0.99), BMD and SAI (Fig. 4), BMD and AGRI, shoot area top view integral (SATVI) and SAI and SATVI and AGRI (r = 0.98). Most correlations were positive and statistically significant. HEI showed negative correlation with relative growth rate integral (RGRI) and TIL, but values were not statistically significant. COMI showed negative correlations with all traits but HEII. Among stress indices highest correlations were found for SSI (SAI), SSI (AGRI) and SSI (SAI), as well as between SSI (WUE) and SSI (SPW) (with r>0.93). Interestingly, WUE, SPW and BMD showed only low correlations (r<0.43). Looking at the simple stress index, however, correlations between SSI (WUE), SSI (SPW) and SSI (BMD) were very high (r = 0.85 to 0.96). Autocorrelations between drought and well watered treatments of a single trait were high (r>0.61) for most traits. HEI with r = 0.85 had the highest correlation between treatments. RGRI showed a low but still significant correlation between the treatments with r = 0.33. Autocorrelations for SPW and WUE were not significant.

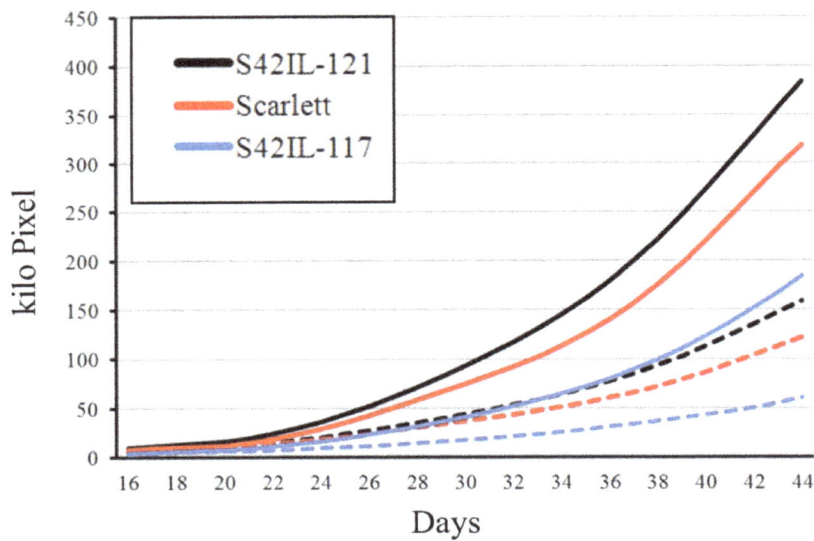

Figure 3. Development of shoot area of S42IL-121, Scarlett and S42IL-117 under well watered (solid line) and drought (dashed line) treatment, respectively.

Mixed Model Analysis of Variance

The mixed model analysis, including fixed line, treatment, and experiment effects (i.e. model I) revealed significant (P<0.05) line effects for all investigated traits (Table S4). Treatment had a clear impact on trait performance. For all traits, except WUE and SPW the effect was significant. Line by treatment interaction effects were not significant for any of the measured traits. The experiments had a significant effect on trait performance of all traits except leaf color measured as plant hue integral (HUEI). And line by experiment interaction was significant for all traits but RGRI, COMI, and HEII. In the mixed model analyses for single treatments including fixed line, and experiment effects (i.e. model II), line had a significant effect on trait performance for all traits but HUEI, RGRI, and WUE under well watered and HUEI and SPW under drought conditions (Tables S5 and S6). Also simple

stress indices were analyzed with model II (Table S7). The line effect was not significant for any of the simple stress indices.

QTL Detection

QTL were only determined for traits with heritability greater than 0. The Dunnett tests revealed, in total, 63 line effects for eleven out of 14 traits. These effects were detected either across treatments (39), within the drought treatment (15) or within the well watered treatment (9). Several of the measured effects were consistent between the different treatments. Thus, these line effects were summarized to a minimum of 44 QTL (Table 2 and Fig. 5). No QTL were identified for RGRI, SPW, and HUEI. Between two and nine QTL were identified for the traits AGRI, BMD, CALI, COMI, HEI, HEII, hull area integral (HULI), SAI,

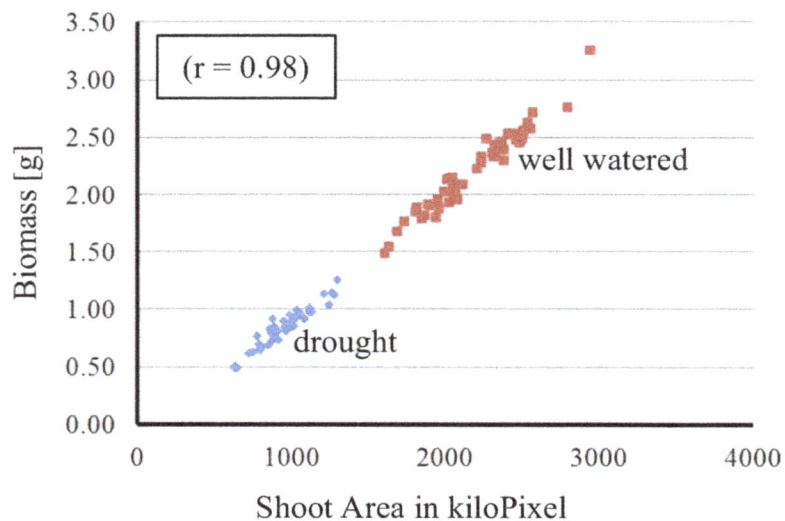

Figure 4. Correlation between biomass and shoot area integral under drought (blue dots) and well watered (red dots) treatment, respectively.

Table 2. List of 44QTL detected for 11 traits in the S42IL-population.

Trait[a]	QTL Name	Position of main introgression[b]	Line	Treatment[c]	LSMEANS Scarlett[d]	LSMEANS IL[e]	Dev. f. Sca	Dev. f. Sca %[g]	Candidate genes[h]	Studies with corresponding QTL[i]
AGRI	QAgri.S42IL-3H	3H; 204.48–255.13	S42IL-115	a	170.84	119.10	−51.74	−30.28		
	QAgri.S42IL-4H	4H; 027.52–064.77	S42IL-117	a, d	170.84	114.69	−56.15	−32.87		
	QAgri.S42IL-6H	6H; 073.90–133.47	S42IL-129	a	170.84	120.89	−49.95	−29.24		
BMD	QBmd.S42IL-3H	3H; 204.48–255.13	S42IL-115	a	1.66	1.10	−0.56	−33.64		
	QBmd.S42IL-4H	4H; 027.52–064.77	S42IL-117	a	1.66	0.99	−0.67	−40.30		
	QBmd.S42IL-4Hb	4H; 074.11–119.06	S42IL-121	a	1.66	2.26	0.60	35.97		
	QBmd.S42IL-6H	6H; 073.90–133.47	S42IL-129	a	1.66	1.10	−0.57	−33.94		IV
CALI	QCali.S42IL-1H	1H; 040.51–089.01	S42IL-103	a, w	19.39	15.83	−3.56	−18.35		
	QCali.S42IL-2H	2H; 102.66–104.81	S42IL-110	a, d	19.39	15.47	−3.92	−20.23		
	QCali.S42IL-3H	3H; 067.01–098.41	S42IL-111	a	19.39	16.52	−2.87	−14.80		
	QCali.S42IL-4H	4H; 027.52–064.77	S42IL-117	a, d	19.39	15.98	−3.42	−17.62		
	QCali.S42IL-4Hb	4H; 074.11–119.06	S42IL-121	a	19.39	22.34	2.94	15.18		
	QCali.S42IL-6H	6H; 073.90–133.47	S42IL-129	a, d, w	19.39	15.02	−4.37	−22.55		
COMI	QComi.S42IL-4H	4H; 027.52–064.77	S42IL-117	a, w	4.10	5.32	1.22	29.77		
	QComi.S42IL-6H	6H; 071.39–132.23	S42IL-128	w	3.82	5.23	1.42	37.13		
HEI	QHei.S42IL-1H	1H; 040.51–089.01	S42IL-103	a, w	49.00	42.75	−6.25	−12.76		
	QHei.S42IL-1Hb	1H; 130.68–173.49	S42IL-143	a	49.00	43.50	−5.50	−11.22	HvFT-3[3]	I
	QHei.S42IL-2H	2H; 063.96–110.84	S42IL-109	a	49.00	44.33	−4.67	−9.52	sdw3[2], HvFT4[3]	I, II, III, IV
	QHei.S42IL-3H	3H; 067.01–098.41	S42IL-111	a	49.00	44.50	−4.50	−9.18		
	QHei.S42IL-3Hb	3H; 154.99–253.73	S42IL-140	a, d, w	49.00	58.17	9.17	18.71	denso[1]	I, IV, V
	QHei.S42IL-4H	4H; 074.11–119.06	S42IL-121	a, d, w	49.00	57.75	8.75	17.86		II, IV
	QHei.S42IL-6H	6H; 073.90–133.47	S42IL-129	a	49.00	44.42	−4.58	−9.35		V
	QHei.S42IL-7H	7H; 134.43–193.89	S42IL-137	a, d	49.00	54.67	5.67	11.56		I, II, IV, V
HEII	QHeii.S42IL-1H	1H; 040.51–089.01	S42IL-103	a	11.67	9.85	−1.83	−15.64		
	QHeii.S42IL-1Hb	1H; 130.68–173.49	S42IL-143	a	11.67	10.06	−1.62	−13.85	HvFT-3[3]	I
	QHeii.S42IL-4H	4H; 061.15–119.06	S42IL-162	a	11.67	9.99	−1.68	−14.39		
	QHeii.S42IL-4Hb	4H; 171.25–183.54	S42IL-124	a	11.67	10.04	−1.64	−14.02		
	QHeii.S42IL-7H	7H; 176.37–229.66	S42IL-138	a	11.67	10.05	−1.63	−13.93		
HULI	QHuli.S42IL-1H	1H; 040.51–089.01	S42IL-103	a, w	6679	4443	−2236	−33.47		
	QHuli.S42IL-2H	2H; 102.66–104.81	S42IL-110	a, d	6679	4517	−2161	−32.36		
	QHuli.S42IL-3H	3H; 067.01–098.41	S42IL-111	a	6679	4617	−2062	−30.88		
	QHuli.S42IL-3Hb	3H; 204.48–255.13	S42IL-115	a	6679	4616	−2063	−30.89		
	QHuli.S42IL-4H	4H; 027.52–064.77	S42IL-117	a	6679	4327	−2352	−35.21		
	QHuli.S42IL-6H	6H; 073.90–133.47	S42IL-129	a	6679	4179	−2500	−37.42		
SAI	QSai.S42IL-2H	2H; 102.66–104.81	S42IL-110	d	995	625	−369	−37.14		I, (as biomass)

Table 2. Cont.

Trait[a]	QTL Name	Position of main introgression[b]	Line	Treatment[c]	LSMEANS Scarlett[d]	LSMEANS IL[e]	Dev. f. Sca	Dev. f. Sca %[g]	Candidate genes[h]	Studies with corresponding QTL[i]
	QSai.S42IL-4H	4H; 027.52-064.77	S42IL-117	a, d	1654	1116	-538	-32.53		
	QSai.S42IL-6H	6H; 073.90-133.47	S42IL-129	a, d	1654	1151	-504	-30.44		
SATVI	QSatvi.S42IL-2H	2H; 102.66-104.81	S42IL-110	d	398	214	-184	-46.18		
	QSatvi.S42IL-4H	4H; 027.52-064.77	S42IL-117	a	677	437	-240	-35.44		
	QSatvi.S42IL-6H	6H; 073.90-133.47	S42IL-129	a	677	450	-228	-33.64		
TIL	QTil.S42IL-3H	3H; 204.48-255.13	S42IL-115	a	8.17	5.92	-2.25	-27.55		
	QTil.S42IL-4H	4H; 171.25-183.54	S42IL-124	a, d, w	8.17	11.67	3.50	42.86	VRN-H2[3]	III, V
WUE	QWue.S42IL-4H	4H; 027.52-064.77	S42IL-117	d	1.97	1.25	-0.72	-36.53		
	QWue.S42IL-6H	6H; 073.90-133.47	S42IL-129	d	1.97	1.17	-0.80	-40.59		

[a] Trait abbreviations are given in Table 1.

[b] By chromosome and cM position.

[c] Treatment under which effect occured, a: main effect, d: under drought treatment, w: under well watered treatment.

[d] LSMEANS Scarlett for the indicated trait and treatment.

[e] LSMEANS for indicated IL for indicated trait and treatment, if under more than one treatment then a.

[f] Deviation from Scarlett = LSMEANS[IL] - LSMEANS [Scarlett].

[g] Deviation from Scarlett in % = (LSMEANS[IL] – LSMEANS [Scarlett])/LSMEANS[Scarlett] *100.

[h] References: [1] Laurie et al. (1995), [2] Gottwald et al. (2004), [3] Wang et al. (2010)

[i] Von Korff et al. (2006), II: Schmalenbach et al. (2009), III: Wang et al. (2010), IV: March et al. (in prep.), V: Honsdorf et al. (in prep.).

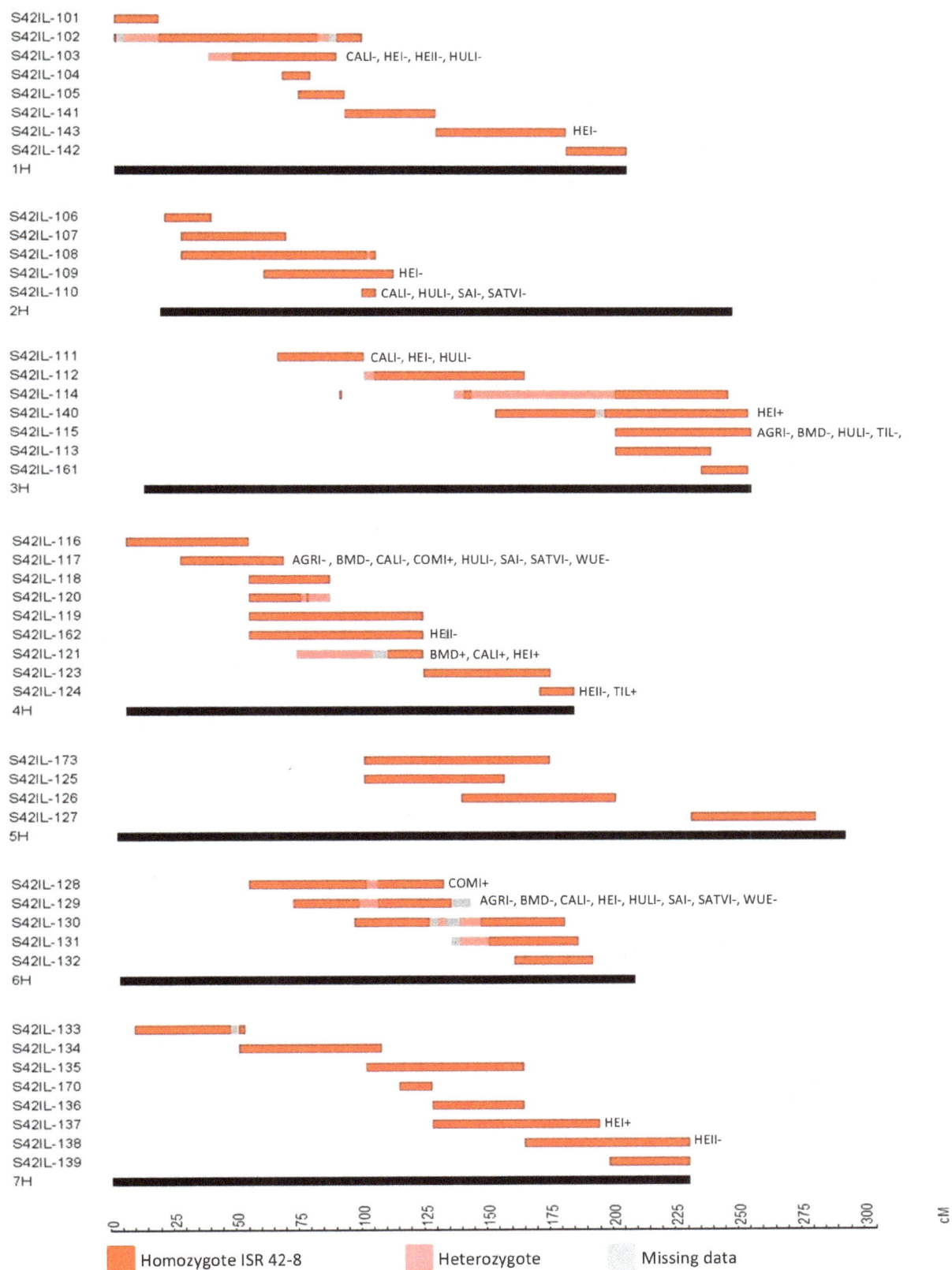

Figure 5. QTL map with indication of S42IL introgressions (Schmalenbach et al. 2011). SNP positions (in cM) are based on Close et al. (2009). QTL are placed right to the S42IL, indicated by trait abbreviations (see Table 1). The sign indicates an increasing (+) or a decreasing (−) *Hsp* effect.

SATVI, TIL, and WUE. In the following, the QTL are presented for each trait separately.

Absolute Growth Rate Integral (AGRI)

Three QTL were identified for AGRI. The QTL are located on chromosomes 3H, 4H and 6H and the *Hsp* allele in all three QTL reduced the trait performance. Across treatments the *Hsp* alleles reduced the integral of the absolute growth rate by approximately 30%. Under drought conditions the *Hsp* allele of QTL QAgri.S42IL-4H reduced the trait performance by almost 40%.

Biomass Dry (BMD)

For biomass four QTL were identified across treatments. The *Hsp* alleles at QTL QBmd.S42IL-3H and QBmd.S42IL-6H on chromosomes 3H and 6H reduced biomass by approximately 33%. The two QTL QBmd.S42IL-4H and QBmd.S42IL-4Hb on chromosome 4H showed contrary effects. While at the first QTL the *Hsp* allele reduced biomass by 40.3%, at the second QTL it increased it by 36.0%.

Caliper Length Integral (CALI)

Six QTL were detected for caliper length on chromosome 1H, 2H, 3H, 4H and 6H. All QTL were detected across treatments, three and two of them also showed effects for drought and well watered treatments, respectively. In five cases the *Hsp* allele had decreasing effects between 14.8 and 22.6% across treatments. In one case the *Hsp* allele at QTL QCali.S42IL-4Hb increased caliper length by 15.2% compared to Scarlett.

Compactness Integral (COMI)

For COMI two QTL were detected on chromosomes 4H and 6H. The effect of the *Hsp* allele at QComi.S42IL-4H was observed across treatments and under well watered treatment. It increased compactness across treatments by 29.8%. In case of QComi.S42IL 6H the effect was solely observed under well watered treatment and lead to an increase of 27.1% by the *Hsp* allele.

Height (HEI)

The highest number of QTL was detected for plant height. Eight QTL were detected across treatments. Of those, four also showed effects under either one or both of well watered and drought treatment. The QTL are located on all chromosomes except 5H. In five cases the *Hsp* alleles reduced plant height (9.2 to 12.8%). In three cases the *Hsp* allele increased plant height by 11.6 to 18.7%.

Height Integral (HEII)

Five QTL across treatments were detected for HEII on 1H, 4H and 7H. Two of those, QHeii.S42IL-1H and QHeii.S42IL-1Hb, coincided with QTL for manual measurement of HEI. At all detected QTL the *Hsp* alleles reduced HEII by between 13.9 to 15.6%.

Hull Area Integral (HULI)

For HULI six QTL were detected on chromosomes 1H, 2H, 3H, 4H, and 6H. In all cases the *Hsp* alleles reduced the hull area by between 30.9 to 37.4% across treatments. In addition, QHuli.S42IL-1H and QHuli.S42IL-2H showed effects under well watered and drought treatments, respectively.

Shoot Area Integral (SAI)

Three QTL were found for the integral of the projected shoot area. The QTL are located on chromosomes 2H, 4H, and 6H. All three QTL were detected under drought treatment, while, in addition, QSai.S42IL-4H and QSai.S42IL-6H showed effects across treatments. The presence of the *Hsp* allele reduced the projected shoot area between 30.4 and 37.1% across treatments.

Shoot Area Top View Integral (SATVI)

Three QTL were detected for shoot area top view that corresponded to the same QTL detected for SAI on chromosomes 2H, 4H, and 6H. However, only QSatvi.S42IL-2H was detected under drought treatment, while the two other QTL were detected across treatments. In all cases the *Hsp* alleles reduced SATVI between 33.6 and 46.2% compared to Scarlett.

Tiller Number (TIL)

For tiller number two QTL were identified on chromosomes 3H and 4H. The *Hsp* allele at QTil.S42IL-3H reduced tiller number by 27.6% across treatments. QTil.S42IL-4H was detected across treatments and separately within the two treatments. Across both treatments the *Hsp* allele increased the tiller number by 42.9%.

Water Use Efficiency (WUE)

Two QTL were detected for WUE. The *Hsp* allele of QTL QWue.S42IL-4H and QWue.S42IL-6H reduced WUE by 36.5 and 40.6% under drought treatment, respectively.

Relative Growth Rate Integral (RGRI) and Plant Hue Integral (HUEI), Specific Plant Weight (SPW), and Simple Stress Index (SSI)

For RGRI, HUEI, and SPW and the SSIs no QTL were detected in this study.

Discussion

The aim of the study was to validate the use of non-destructive high-throughput phenotyping to measure vegetative drought response in barley and to identify QTL derived from wild barley that control physiological traits related to drought stress. To the authors knowledge this is the first QTL report on drought stress that used a high-throughput phenotyping facility.

Plant growth and the biomass parameters tiller number, plant height, and shoot dry weight of 48 barley genotypes were investigated under drought and well watered treatments. Two week old barley plants were transferred into a high-throughput phenotyping greenhouse, where stress and control treatments were applied automatically. During the following four weeks of cultivation, plants were imaged daily in an automated manner. Images were processed and used as a measure for plant height, caliper length, biomass and, consequently, plant growth, and plant color. After a total of six weeks green plants were harvested. Tiller number, plant height, and shoot dry weight were determined for each plant. Moreover, plant compactness, water use efficiency and specific plant weight, as well as stress indices were calculated.

The mixed model ANOVA revealed a clear effect of the treatment on trait expression. Drought stressed plants had a lower growth rate and subsequently produced less biomass (see Fig. 4). However, there was no significant line×treatment interaction, indicating that the S42ILs reacted similar under drought stress and well watered conditions for the investigated traits. This finding is also supported by high autocorrelations for investigation of traits under drought and control treatments, e.g. 0.77 for biomass (see Table S3).

QTL Detection

In this study 44 QTL were identified in 15 ILs for eleven traits. In eight cases the *Hsp* alleles increased the performance of the trait, while in 36 cases there was a decreasing effect. This is to be expected since wild barley is known to carry many unfavorable alleles as well [5]. In six ILs only one QTL was determined, predominantly for HEI and HEII. Multiple QTL effects were found in nine ILs. For S42IL-115,-117, -121, and -129 QTL for BMD were detected in combination with one or more of the traits AGRI, HEI, SAI, and TIL which is in agreement with the high correlations found between those traits. SAI is a measure for biomass, and AGRI is the first derivative of SAI. Therefore, it was expected that ILs show effects for all three traits, simultaneously. However, this was not always the case. Thus, increasing the number of experiments and replications might be useful to increase the power of QTL detection. In the following, the traits or trait complexes are discussed separately.

Absolute Growth Rate Integral, Shoot Area Integral and Biomass Dry (AGRI, SAI, BMD)

The three traits were highly correlated with each other (r = 0.98 and r = 0.99). Since their relation may be functional, it appears likely that a single pleiotropic QTL may control the three traits AGRI, SAI and BMD, simultaneously.

For line S42IL-129 a biomass reduction of 33.9% was observed. March et al. (in prep.) found a similar decrease in biomass in that line measured under terminal drought stress. This suggests that biomass production may be partly controlled by similar genes during early and late drought stress occurrence.

Three QTL were detected for SAI. Two of those, namely QSai.S42IL-4H and QSai.S42IL-6H, in line S42IL-117 und S42IL-129, respectively, were in accordance with BMD QTL. The *Hsp* allele, in both lines caused a decrease in the projected shoot area. Due to the high correlation of the traits it can be assumed that QTL for biomass correspond to QTL for shoot area. The *Hsp* allele of the QTL QSai.S42IL-2H on chromosome 2H caused a decrease of 37% in projected shoot area, compared to Scarlett. Von Korff et al. [10] described a QTL related to biomass reduction (QMas.S42-2H.a) in the same region in an AB-QTL field trial. Since this AB-QTL population is the parent population of the S42ILs used in this study, it is likely that the same QTL was detected in both the greenhouse and field trials.

AGRI is directly related to SAI. This might be seen as a reason for the detection of similar QTL for AGRI and SAI and, consequently, BMD. Many QTL studies on growth focus on relative growth rate instead of absolute growth rate. In the present study RGRI showed only a weak correlation to AGRI (r = 0.35) and other biomass parameters. Poorter et al. [33] pointed out that in their study QTL for RGR rather co-located with QTL for seed mass than with QTL for biomass. This fits well to the weak correlations found between biomass parameters and RGRI. However, two of the QTL detected for AGRI coincided with locations where previous studies mapped QTL for RGR. Yin et al. [34] reported a minor effect for relative growth rate associated with the *denso* locus on chromosome 3H in a spring barley recombinant inbred line (RIL) population of the cross Prisma×Apex, which may co-localize with QAgri.S42IL-3H. Poorter et al. [33] conducted QTL studies in a F_2 population derived from a cross between two *Hsp* accessions. They mapped QTL for relative growth rate on chromosomes 1H, 2H, 5H and a minor QTL on 6H. The latter one might be in accordance with QAgri.S42IL-6H.

Tiller Number (TIL)

Two QTL were detected for the trait tiller number on chromosomes 3H and 4H. Wang et al. [35] identified the *VRN-H2* gene on chromosome 4H in introgression line S42IL-124. Whereas S42IL-124 carries a dominant winter-type allele, Scarlett carries the recessive and deleted spring type allele at *Vrn-H2*. S42IL-124 showed an increased tiller number compared to Scarlett. Since *Vrn-H2* is known to have a pleiotropic effect on tiller number [36], we assume that this gene explains the underlying effect of the QTL. Studies on other populations revealed QTL for tiller number on chromosomes 4H as well. In a cross between two wild barley accessions Elberse et al. [37] found a QTL for tiller number on that chromosome. Baum et al. [38] identified a QTL on chromosome 4H where the *Hsp* allele increased the number of tillers and a QTL on chromosome 3H where the *Hsp* allele had a decreasing impact in an Arta×*Hsp* 41-1 RIL population. Those effects might correspond to the QTL detected in this study. Both QTL occurred irrespective of the treatment. Especially QTil.S42IL-4H appears to be a very stable QTL. It was detected across and within treatments and was detected in several studies under varying conditions, in field studies as well as under greenhouse conditions. Moreover, von Korff et al. [10] detected QTL for number of ears, which is directly related to tiller number, in the same genomic regions. On 4H the *Hsp* allele increased the number of ears, while on 3H it has a decreasing effect. This supports the observation of a stable QTL.

Height (HEI) and Height Integral (HEII)

Plant height was determined in two ways. First, height (HEII) was modeled from the images taken during four weeks and the integral of the height growth curve was calculated. Second, height (HEI) was measured manually when plants were harvested after six weeks at the end of the experiment. The correlation between HEI and HEII was relatively low with r = 0.72, compared to the correlation between SAI and BMD with r = 0.98. While SAI shows a constant increase over time, HEII shows an overall increase, but fluctuation between days may be strong. When a new leaf is unfolded the plant grows higher, however, when the leaf becomes too heavy and bends down, the height of the plant appears to be shorter. At the end of the experiment the length of the stretched plant was measured, which is longer than the upright standing plant. Nevertheless two coinciding QTL were found between HEI and HEII on chromosome 1H.

Six out of eight QTL were already identified in previous field studies with the S42 population and/or the S42ILs, exhibiting similar effects of the same direction in all three studies. QTL QHei.S42IL-3Hb was already detected in von Korff et al. [10] and March et al. (in prep.). In this region also the *denso* dwarfing gene was mapped [39], which may be identical with the semi-dwarf gene *sdw1* [40]. The second largest effect, after QHei.S42IL 3Hb, was associated with QHei.S42IL-4H in S42IL-121. This QTL corresponds to QHei.S42IL-4H.a in Schmalenbach et al. [15]. In both studies the *Hsp* allele increased plant height by 18%. March et al. (in prep.) mapped a QTL for height for S42IL-121 as well. A third QTL (QHei.S42IL-7H) with an increasing effect of the *Hsp* allele was detected on chromosome 7H. Here an effect that was already found in the studies of von Korff et al. [10], Schmalenbach et al. [15] and March et al. (in prep.) could be verified. Moreover two QTL where the *Hsp* allele had a decreasing effect on plant height [10] were verified. In S42IL-143 HEI was reduced by 11% (QHei.S42IL-1Hb) and HEII (QHeii.S42IL-1Hb) by 14%. Von Korff et al. [10] detected a QTL in the same region on chromosome 1H. The flowering time gene *HvFT3* is mapped in the same region and known to have a pleiotropic effect on plant height [35]. QHei.S42IL-2H in S42IL-109 had reduced height by 9.5%. March et al. (in prep.) and Schmalenbach et al. [15] found the same effect in their studies. Von Korff et al. [10] found a similar effect in the region where the *Hsp* introgression of S42IL-109 was mapped. Moreover two candidate genes are mapped to the chromosomal region. These are the dwarfing gene *sdw3* [41] and the flowering gene *HvFT4*, which is known to have an effect on plant height [35].

All of the HEI QTL in the present experiments were detected across treatments. Six out of eight QTL were also found in previous field and glasshouse studies. The QTL therefore seem to be very stable across locations as well as across treatments. Moreover they seem to be independent of the developmental stage. The present experiments, thus, allowed the verification of effects after six weeks that were previously screened in field experiments after flowering, indicating that phenotyping juvenile plants may be predictive for adult plant performance, at least in regard to growth parameters. The high heritability of 76.8% supports this finding.

For HEII two QTL coincided with previous studies. Besides QHeii.S42IL-1Hb mentioned above, this was QHeii.S42IL-4Hb where the *Hsp* allele reduced height by 14%. This QTL was also detected by von Korff et al. [10] and Wang et al. [35]. Heritability for digitally determined height was lower (61.4%) than for the manually measured one. Determining height by multiple measurements apparently was not an advantage here. However, this may change at a later stage of plant development. After shooting, the plant height is less subjected to bending of leaves and therefore can be measured more precisely by the imaging technique.

Water Use Efficiency (WUE)

Water use efficiency indicates how much biomass a plant can produce per unit water supplied. Thus, increased WUE has the potential to improve yield under drought stress conditions. Measuring WUE in regular greenhouse experiments is time-consuming. Therefore the high-throughput phenomics facility greatly assisted in scoring of water use efficiency through automated watering of pots to specific weights.

In this study the two S42ILs -117 and -129, with wild barley introgressions on chromosomes 4H and 6H, respectively, showed significant differences in WUE compared to Scarlett. Both ILs showed reduced water use efficiency compared to Scarlett. These ILs also produced less biomass. S42IL-117 and S42IL-129, thus, clearly carry unfavorable alleles for this trait. Chen et al. [42] pointed out that WUE itself is difficult to measure under field conditions and that a suitable tool to measure WUE efficiency is missing. Carbon isotope discrimination is a commonly used technique to measure WUE. Teulat et al. [43] used this method and identified QTL for WUE on chromosome 6H in a set of 167 RILs from a cross between Tadmor and Er/Apm and likewise Diab et al. [44] identified a QTL for the same trait on chromosome 4H. The QTL detected in this study may correspond to the ones found in the studies mentioned before and suggest the results from both techniques are correlated.

Compactness Integral, Shoot Area Top View Integral, Hull Area Integral (COMI, SATVI and HULI)

The compactness of a plant describes how much of the hull area is covered by leaves. It was calculated as the ratio of SATVI to HULI. The more compact a plant is, the more ground cover it has with regard to the hull area. Two QTL were detected for this trait.

SATVI and HULI showed a high correlation of r = 0.9, however, correlations between COMI and HULI and between COMI and SATVI were only moderate and negative. This indicates that in general, bigger plants take more space and have a lower compactness compared to smaller plants. In the present experiments this was observed by comparing drought stressed and well watered plants. Drought stressed plants showed on average a higher compactness than well watered plants. Jansen et al. [45] report the same effect on a study in *Arabidopsis thaliana*. Compactness shows negative correlation with all other traits evaluated in this experiment, with the exception of HEII (r = 0.16). An example for this is S42IL-117. This introgression line has a higher compactness, but reduced biomass, and other growth parameters compared to Scarlett.

SATVI is one of the three parameters that control SAI and, thus, is highly correlated with this trait as well as with BMD and AGRI. As one may expect, for SATVI the same QTL were detected as for SAI. For HULI a total of six QTL were detected. Three of those may be due to high correlations in accordance with SATVI, AGRI and BMD.

Caliper Length Integral (CALI)

Caliper length describes the maximum diameter of the plant. For this trait six QTL were detected. Those were in accordance with QTL for HULI. This can be explained by the close connection of both traits. Hull area is taken as the basis to calculate caliper length and both traits are highly correlated (r = 0.94). CALI also shows positive and high correlations with AGRI, SAI, SATVI, HEI, and BMD. Plants with a large diameter cover a larger area, tend to be bigger, have a higher growth rate and a higher biomass than plants with a smaller diameter. Therefore, a lot of information on plant structure can be deduced from the plant diameter.

Stress Indices

Simple stress indices were calculated for each trait as the ratio of the mean plant performance under drought stress versus well watered treatments. In this study no QTL for a SSI was detected. Additionally the authors used two more complex stress indices (modified after Fischer and Maurer [46]), but were not able to detect QTL with those either. A stress index states how well a genotype performs under stress conditions relative to its performance under control conditions. Therefore, to see differences between genotypes for a stress index, a line by treatment interaction is necessary. If all genotypes show a similar growth reaction under stress and control conditions, the initially existing differences between the genotypes may be drastically reduced. In the present experiments line by treatment interactions were not significant and autocorrelations were high between the treatments. This may be the reason why no significant effect for the stress indices was found. This notion is supported by Wang et al. [47]. In their study on "mathematically-derived traits in QTL mapping" the authors pointed out that an increased complexity of the genetic architecture of derived traits (e.g. stress indices) may reduce the power of QTL detection.

High-throughput Phenotyping using The Plant Accelerator

Determination of biomass by manual harvest is tedious and time-consuming. In addition, destructive harvest makes repeated measurements on the same plant impossible. Visual light imaging technologies applied in this study can solve these problems by utilizing the strong correlation between the projected shoot area and the actual biomass [27]. Imaging technologies have been successfully used in several studies in *Arabidopsis thaliana*, e.g. Granier et al. [48] and Leister et al. [49]. The first study investigated nine accessions under different levels of water deficit in the phenotyping facility "PHENOPSIS". Reaction to water deficits was, amongst other traits, characterized by leaf area growth determined through images. The authors pointed out the importance of the automated watering in their experiment, which enables equal conditions for all plants. A characteristic that was also found very important in the present experiments. Leister et al. [49] described a first approach of using an image based technology for high-throughput growth analysis. They calculated plant area from top view images and found high correlations to plant fresh weight.

In this study the sum of three two-dimensional pictures was used as a measure of plant biomass. In these experiments correlation between SAI and BMD and between AGRI and BMD were very high (r = 0.98). The results with six-week old barley plants proved that the sum of three pictures accounts sufficiently for overlapping leaves during early development. Rajendran et al. [30] found the same for *T. monococcum*. However, as Munns et al. [27] pointed out, accuracy may decline when plants become larger and produce multiple shoots. The results of the present study approved that The Plant Accelerator is suited to enable detailed growth analysis of barley plants. The prediction of biomass by the image-based leaf sum gave accurate results when comparing to actual biomass. Growth curves can give detailed information on differences in development of genotypes. For instance, the maximum of the absolute growth rate gives insight into the change from vegetative to generative phase of plant development. The present experiments ran only for six weeks. Therefore not all plants have reached this point. In future experiments this factor should be accounted for by adjusting the duration of the experiment. Automated imaging and the appropriate analysis pipeline make the detection of different developmental stages of plants feasible in high-throughput. With this technique it is possible to detect differences in stress responses between genotypes not only at different time points, but also to account for differences in development at those time points.

Rajendran et al. [30] used non-destructive imaging to screen for different response mechanisms of *T. monococcum* to salt stress. In contrast to conventional salt stress experiments, where tolerance is measured as total biomass production of stressed plants compared to unstressed plants, the growth curves provided through daily imaging gave detailed insight into the tolerance mechanisms of the plants. While osmotically tolerant plants showed a constant growth rate, the growth rate of sodium excluders first dropped than increased after a couple of days. Moreover plant color was analyzed. No stress symptoms occurred on leaves in the present experiments. Due to little variation between the genotypes no QTL was detected for leave color. However, in the experiments by Rajendran et al. [30], color analysis was successfully used to screen leaf damages due to high salt concentrations.

Fluorescence imaging gives information on the health statues of a plant. It allows for detection of leaf senescence and necrosis. However, such symptoms were not observed in the present experiments and, thus, this parameter was not applied. Nonetheless, the technique is readily available. In addition, near infrared (NIR) and infrared (IR) imaging may be useful for future plant growth evaluations and QTL studies. NIR enables the observation of the water status of a plant, while IR is used to determine shoot temperature.

Conclusion

In this study the use of a non-destructive high-throughput phenotyping platform was implemented to map QTL controlling vegetative drought stress responses in barley. Several QTL where the exotic *Hsp* allele had a positive effect on trait performance were detected. In particular, introgression line S42IL-121 showed improved growth under drought stress compared to the recurrent parent Scarlett. The line showed the same behavior in previous field experiments. Thus, this introgression line might be interesting for further breeding.

Moreover, several QTL were detected where the *Hsp* allele had a decreasing effect on trait performance. Especially two QTL for water use efficiency might be interesting for further investigation. In future, interesting effects of S42IL-121 and other S42ILs will be fine mapped with already available high-resolution progeny [18] to further narrow down the QTL region and, ultimately, clone the underlying genes, which caused the observed QTL effects.

Supporting Information

Figure S1 Map of 47 S42ILs, the map contains 636 BOPA1 SNPs.
(PPTX)

Table S1 Experimental layout with genotypes referred to positions (1–100) per replication (1–3).
(XLSX)

Table S2 Descriptive statistics for S42Ils.
(XLSX)

Table S3 Correlations between traits and stress indices.
(XLSX)

Table S4 ANOVA (Model I) results of studied traits.
(XLSX)

Table S5 ANOVA (Model II drought) results of studied traits.
(XLSX)

Table S6 ANOVA (Model II well watered) results of studied traits.
(XLSX)

Table S7 ANOVA (Model II SSI) results of studied traits.
(XLSX)

Acknowledgments

We are grateful to the team of "The Plant Accelerator, Australian Plant Phenomics Facility" for carrying out the experiments.

Author Contributions

Conceived and designed the experiments: TJM BB MT KP. Performed the experiments: NH TJM BB. Analyzed the data: NH BB. Contributed reagents/materials/analysis tools: BB KP. Wrote the paper: NH KP. Provided comments and corrected the manuscript: TJM BB MT KP.

References

1. Druka A, Sato K, Muehlbauer GJ (2011) Genome Analysis: The State of Knowledge of Barley Genes. In: Ullrich SE, editor. Barley: Production, Improvement, and Uses. 1 ed: Blackwell Publishing Ltd.

2. Jana S, Wilen RW (2005) Breeding for Abiotic Stress Tolerance in Barley. In: Ashraf M, Harris PJC, editors. Abiotic stresses: plant resistance through breeding and molecular approaches: Food Products Press.

3. Tanksley SD, McCouch SR (1997) Seed banks and molecular maps: Unlocking genetic potential from the wild. Science 277: 1063–1066.

4. Zhao J, Sun HY, Dai HX, Zhang GP, Wu FB (2010) Difference in response to drought stress among Tibet wild barley genotypes. Euphytica 172: 395–403.

5. Zamir D (2001) Improving plant breeding with exotic genetic libraries. Nature Reviews Genetics 2: 983–989.

6. Tanksley SD, Nelson JC (1996) Advanced backcross QTL analysis: A method for the simultaneous discovery and transfer of valuable QTLs from unadapted germplasm into elite breeding lines. Theoretical and Applied Genetics 92: 191–203.

7. Pillen K, Zacharias A, Leon J (2003) Advanced backcross QTL analysis in barley (*Hordeum vulgare* L.). Theoretical and Applied Genetics 107: 340–352.

8. von Korff M, Wang H, Leon J, Pillen K (2004) Development of candidate introgression lines using an exotic barley accession (*Hordeum vulgare* ssp. *spontaneum*) as donor. Theoretical and Applied Genetics 109: 1736–1745.

9. von Korff M, Wang H, Leon J, Pillen K (2005) AB-QTL analysis in spring barley. I. Detection of resistance genes against powdery mildew, leaf rust and scald introgressed from wild barley. Theoretical and Applied Genetics 111: 583–590.

10. von Korff M, Wang H, Leon J, Pillen K (2006) AB-QTL analysis in spring barley: II. Detection of favourable exotic alleles for agronomic traits introgressed from wild barley (*H. vulgare* ssp. *spontaneum*). Theoretical and Applied Genetics 112: 1221–1231.

11. von Korff M, Wang H, Leon J, Pillen K (2008) AB-QTL analysis in spring barley: III. Identification of exotic alleles for the improvement of malting quality in spring barley (*H. vulgare* ssp. *spontaneum*). Molecular Breeding 21: 81–93.

12. von Korff M, Leon J, Pillen K (2010) Detection of epistatic interactions between exotic alleles introgressed from wild barley (*H. vulgare* ssp. *spontaneum*). Theoretical and Applied Genetics 121: 1455–1464.

13. Saal B, von Korff M, Leon J, Pillen K (2011) Advanced-backcross QTL analysis in spring barley: IV. Localization of QTL x nitrogen interaction effects for yield-related traits. Euphytica 177: 223–239.

14. Schmalenbach I, Koerber N, Pillen K (2008) Selecting a set of wild barley introgression lines and verification of QTL effects for resistance to powdery mildew and leaf rust. Theoretical and Applied Genetics 117: 1093–1106.

15. Schmalenbach I, Leon J, Pillen K (2009) Identification and verification of QTLs for agronomic traits using wild barley introgression lines. Theoretical and Applied Genetics 118: 483–497.

16. Schmalenbach I, Pillen K (2009) Detection and verification of malting quality QTLs using wild barley introgression lines. Theoretical and Applied Genetics 118: 1411–1427.

17. Naz AA, Ehl A, Pillen K, Leon J (2012) Validation for root-related quantitative trait locus effects of wild origin in the cultivated background of barley (Hordeum vulgare L.). Plant Breeding 131: 392–398.

18. Schmalenbach I, March TJ, Bringezu T, Waugh R, Pillen K (2011) High-Resolution Genotyping of Wild Barley Introgression Lines and Fine-Mapping of the Threshability Locus *thresh-1* Using the Illumina GoldenGate Assay. G3: Genes, Genomes, Genetics 1: 187–196.

19. Close TJ, Bhat PR, Lonardi S, Wu YH, Rostoks N, et al. (2009) Development and implementation of high-throughput SNP genotyping in barley. BMC Genomics 10: 13.

20. Pennisi E (2008) Plant genetics: The blue revolution, drop by drop, gene by gene. Science 320: 171–173.

21. Tester M, Langridge P (2010) Breeding Technologies to Increase Crop Production in a Changing World. Science 327: 818–822.

22. Cattivelli L, Rizza F, Badeck FW, Mazzucotelli E, Mastrangelo AM, et al. (2008) Drought tolerance improvement in crop plants: An integrated view from breeding to genomics. Field Crops Research 105: 1–14.

23. Tyagi K, Park MR, Lee HJ, Lee CA, Rehman S, et al. (2011) Fertile crescent region as source of drought tolerance at early stage of plant growth of wild barley (*Hordeum vulgare* L. ssp. *spontaneum*). Pakistan Journal of Botany 43: 475–486.

24. El Hafid R, Smith DH, Karrou M, Samir K (1998) Root and shoot growth, water use and water use efficiency of spring durum wheat under early-season drought. Agronomie 18: 181–195.

25. Lu ZJ, Neumann PM (1998) Water-stressed maize, barley and rice seedlings show species diversity in mechanisms of leaf growth inhibition. Journal of Experimental Botany 49: 1945–1952.

26. López-Castañeda C, Richards RA (1994) Variation in temperate cereals in rainfed environments II. Phasic development and growth. Field Crops Research 37: 63–75.

27. Munns R, James RA, Sirault XRR, Furbank RT, Jones HG (2010) New phenotyping methods for screening wheat and barley for beneficial responses to water deficit. Journal of Experimental Botany 61: 3499–3507.

28. Golzarian MR, Frick RA, Rajendran K, Berger B, Roy S, et al. (2011) Accurate inference of shoot biomass from high-throughput images of cereal plants. Plant Methods 7: 11.

29. Berger B, Parent B, Tester M (2010) High-throughput shoot imaging to study drought responses. Journal of Experimental Botany 61: 3519–3528.

30. Rajendran K, Tester M, Roy SJ (2009) Quantifying the three main components of salinity tolerance in cereals. Plant Cell and Environment 32: 237–249.

31. SAS Institute (2008) The SAS Enterprise guide for Windows, release4.2. SAS Institute, Cary, NC, USA.

32. Becker H (2011) Pflanzenzüchtung. Stuttgart: Verlag Eugen Ulmer. 368 p.

33. Poorter H, van Rijn CPE, Vanhala TK, Verhoeven KJF, de Jong YEM, et al. (2005) A genetic analysis of relative growth rate and underlying components in *Hordeum spontaneum*. Oecologia 142: 360–377.

34. Yin X, Stam P, Dourleijn CJ, Kropff MJ (1999) AFLP mapping of quantitative trait loci for yield-determining physiological characters in spring barley. Theoretical and Applied Genetics 99: 244–253.

35. Wang G, Schmalenbach I, von Korff M, Leon J, Kilian B, et al. (2010) Association of barley photoperiod and vernalization genes with QTLs for flowering time and agronomic traits in a BC(2)DH population and a set of wild barley introgression lines. Theoretical and Applied Genetics 120: 1559–1574.

36. Karsai I, Meszaros K, Szucs P, Hayes PM, Lang L, et al. (2006) The influence of photoperiod on the Vrn-H2 locus (4H) which is a major determinant of plant development and reproductive fitness traits in a facultative X winter barley (Hordeum vulgare L.) mapping population. Plant Breeding 125: 468–472.

37. Elberse IAM, Vanhala TK, Turin JHB, Stam P, van Damme JMM, et al. (2004) Quantitative trait loci affecting growth-related traits in wild barley (Hordeum spontaneum) grown under different levels of nutrient supply. Heredity 93: 22–33.

38. Baum M, Grando S, Backes G, Jahoor A, Sabbagh A, et al. (2003) QTLs for agronomic traits in the Mediterranean environment identified in recombinant inbred lines of the cross 'Arta' x *H. spontaneum* 41–1. Theoretical and Applied Genetics 107: 1215–1225.

39. Laurie DA, Pratchett N, Bezant JH, Snape JW (1995) RFLP mapping of 5 major genes and 8 quantitative trait loci controlling flowering time in a winter x spring barley (*Hordeum vulgare* L.) cross. Genome 38: 575–585.

40. Jia Q, Zhang J, Westcott S, Zhang X-Q, Bellgard M, et al. (2009) GA-20 oxidase as a candidate for the semidwarf gene sdw1/denso in barley. Functional & Integrative Genomics 9: 255–262.

41. Gottwald S, Stein N, Borner A, Sasaki T, Graner A (2004) The gibberellic-acid insensitive dwarfing gene *sdw3* of barley is located on chromosome 2HS in a region that shows high colinearity with rice chromosome 7L. Molecular Genetics and Genomics 271: 426–436.

42. Chen J, Chang SX, Anyia AO (2011) Gene discovery in cereals through quantitative trait loci and expression analysis in water-use efficiency measured by carbon isotope discrimination. Plant Cell and Environment 34: 2009–2023.

43. Teulat B, Merah O, Sirault X, Borries C, Waugh R, et al. (2002) QTLs for grain carbon isotope discrimination in field-grown barley. Theoretical and Applied Genetics 106: 118–126.

44. Diab AA, Teulat-Merah B, This D, Ozturk NZ, Benscher D, et al. (2004) Identification of drought-inducible genes and differentially expressed sequence tags in barley. Theoretical and Applied Genetics 109: 1417–1425.

45. Jansen M, Gilmer F, Biskup B, Nagel KA, Rascher U, et al. (2009) Simultaneous phenotyping of leaf growth and chlorophyll fluorescence via GROWSCREEN FLUORO allows detection of stress tolerance in Arabidopsis thaliana and other rosette plants. Functional Plant Biology 36: 902–914.

46. Fischer RA, Maurer R (1978) Drought Resistance in Spring Wheat Cultivars. 1. Grain Yield Responses. Australian Journal of Agricultural Research 29: 897–912.

47. Wang Y, Li HH, Zhang LY, Lu WY, Wang JK (2012) On the use of mathematically-derived traits in QTL mapping. Molecular Breeding 29: 661–673.

48. Granier C, Aguirrezabal L, Chenu K, Cookson SJ, Dauzat M, et al. (2006) PHENOPSIS, an automated platform for reproducible phenotyping of plant responses to soil water deficit in Arabidopsis thaliana permitted the identification of an accession with low sensitivity to soil water deficit. New Phytologist 169: 623–635.

49. Leister D, Varotto C, Pesaresi P, Niwergall A, Salamini F (1999) Large-scale evaluation of plant growth in Arabidopsis thaliana by non-invasive image analysis. Plant Physiology and Biochemistry 37: 671–678.

Permissions

The contributors of this book come from diverse backgrounds, making this book a truly international effort. This book will bring forth new frontiers with its revolutionizing research information and detailed analysis of the nascent developments around the world.

We would like to thank all the contributing authors for lending their expertise to make the book truly unique. They have played a crucial role in the development of this book. Without their invaluable contributions this book wouldn't have been possible. They have made vital efforts to compile up to date information on the varied aspects of this subject to make this book a valuable addition to the collection of many professionals and students.

This book was conceptualized with the vision of imparting up-to-date information and advanced data in this field. To ensure the same, a matchless editorial board was set up. Every individual on the board went through rigorous rounds of assessment to prove their worth. After which they invested a large part of their time researching and compiling the most relevant data for our readers.

The editorial board has been involved in producing this book since its inception. They have spent rigorous hours researching and exploring the diverse topics which have resulted in the successful publishing of this book. They have passed on their knowledge of decades through this book. To expedite this challenging task, the publisher supported the team at every step. A small team of assistant editors was also appointed to further simplify the editing procedure and attain best results for the readers.

Apart from the editorial board, the designing team has also invested a significant amount of their time in understanding the subject and creating the most relevant covers. They scrutinized every image to scout for the most suitable representation of the subject and create an appropriate cover for the book.

The publishing team has been an ardent support to the editorial, designing and production team. Their endless efforts to recruit the best for this project, has resulted in the accomplishment of this book. They are a veteran in the field of academics and their pool of knowledge is as vast as their experience in printing. Their expertise and guidance has proved useful at every step. Their uncompromising quality standards have made this book an exceptional effort. Their encouragement from time to time has been an inspiration for everyone.

The publisher and the editorial board hope that this book will prove to be a valuable piece of knowledge for researchers, students, practitioners and scholars across the globe.

List of Contributors

Xin Chen
Chengdu Institute of Biology, Chinese Academy of Sciences, Chengdu, Sichuan, China
College of Life Sciences, Sichuan University, Chengdu, Sichuan, China
University of Chinese Academy of Sciences, Beijing, China

Hai Long, Guangbing Deng, Zhifen Pan, Junjun Liang and Maoqun Yu
Chengdu Institute of Biology, Chinese Academy of Sciences, Chengdu, Sichuan, China

Ping Gao
College of Life Sciences, Sichuan University, Chengdu, Sichuan, China

Yawei Tang and Nyima Tashi
Tibet Academy of Agricultural and Animal Husbandry Sciences, Lhasa, Tibet, China

Sachin Rustgi, Janet Matanguihan, Richa Gemini, Rhoda A. T. Brew-Appiah, Nuan Wen, Claudia Osorio, Nii Ankrah and Kevin M. Murphy
Department of Crop & Soil Sciences, Washington State University, Pullman, Washington, United States of America

Jaime H. Mejías
Department of Crop & Soil Sciences, Washington State University, Pullman, Washington, United States of America
Instituto de Investigaciones Agropecuarias INIA, Vilcún, Chile

Diter von Wettstein
Department of Crop & Soil Sciences, Washington State University, Pullman, Washington, United States of America
School of Molecular Biosciences, Washington State University, Pullman, Washington, United States of America
Centre for Reproductive Biology, Washington State University, Pullman, Washington, United States of America

Xiao-Qing Zeng, Xiao-Hong Liu and Huan-Bin Shi
State Key Laboratory for Rice Biology, Biotechnology Institute, Zhejiang University, Hangzhou, China

Guo-Qing Chen
State Key Laboratory of Rice Biology, China National Rice Research Institute, Hangzhou, China

Bo Dong
Institute of Virology and Biotechnology, Zhejiang Academy of Agricultural Science, Hangzhou, China

Jian-Ping Lu
College of Life Sciences, Zhejiang University, Hangzhou, China

Fucheng Lin
State Key Laboratory for Rice Biology, Biotechnology Institute, Zhejiang University, Hangzhou, China
China Tobacco Gene Research Center, Zhengzhou Tobacco Institute of CNTC, Zhengzhou, China

Jianbin Zeng, Xiaoyan He, Dezhi Wu, Bo Zhu, Shengguan Cai, Umme Aktari Nadira, Zahra Jabeen and Guoping Zhang
Department of Agronomy, Key Laboratory of Crop Germplasm Resource of Zhejiang Province, Zhejiang University, Hangzhou, China

Mohammad Nasir Uddin, Agnieszka Kaczmarczyk and Eva Vincze
Department of Molecular Biology & Genetics, Faculty of Science & Technology, Aarhus University, Slagelse, Denmark

Xinjia Dai, Suxia Gao and Deguang Liu
State Key Laboratory of Crop Stress Biology for Arid Areas (Northwest A&F University), Yangling, Shaanxi Province, China
Key Laboratory of Integrated Pest Management on Crops in Northwestern Loess Plateau, Ministry of Agriculture, Yangling, Shaanxi Province, China
College of Plant Protection, Northwest A &F University, Yangling, Shaanxi Province, China

Renee M. Petri and Tim A. McAllister
Lethbridge Research Centre, Agriculture and Agri-Food Canada, Lethbridge, Alberta, Canada

Cletos Mapiye
Lacombe Research Centre, Agriculture and Agri-Food Canada, Lacombe, Alberta, Canada
Department of Animal Sciences, Faculty of AgriSciences, Stellenbosch University, Matieland, Western Cape, South Africa

Mike E. R. Dugan
Lacombe Research Centre, Agriculture and Agri-Food Canada, Lacombe, Alberta, Canada

Nataília Corniani, Edivaldo D. Velini and Ferdinando M. L. Silva
São Paulo State University, Faculty of Agronomic Sciences, Botucatu, SP, Brazil

Matthias Witschel
BASF SE, GVA/HC-B009, Ludwigshafen, Germany

N. P. Dhammika Nanayakkara
National Center for Natural Products Research, School of Pharmacy, University of Mississippi, University, MS, United States of America

Franck E. Dayan
USDA-ARS Natural Products Utilization Research Unit, University, MS, United States of America

Van Tran, Diana Weier, Ruslana Radchuk, Johannes Thiel and Volodymyr Radchuk
Institute of Plant Genetics and Crop Plant Research (IPK), Gatersleben, Germany

Francesca Sparla, Paolo Trost and Claudia Pirone
Department of Pharmacy and Biotechnology FABIT, University of Bologna, Bologna, Italy

Giuseppe Falini
Department of Chemistry "G. Ciamician", University of Bologna, Bologna, Italy

Ermelinda Botticella and Francesco Sestili
Department of Agriculture, Forestry, Nature & Energy, University of Tuscia, Viterbo, Italy

Valentina Talamé, Riccardo Bovina, Silvio Salvi and Roberto Tuberosa
Department of Agricultural Sciences, University of Bologna, Bologna, Italy

Yung-Fen Huang, Charlene P. Wight and Nicholas A. Tinker
Eastern Cereal and Oilseed Research Centre, Agriculture and Agri-Food Canada, Ottawa, Ontario, Canada

Jesse A. Poland
Department of Plant Pathology, Kansas State University, Manhattan, Kansas, United States of America

Eric W. Jackson
General Mills Crop Biosciences, Manhattan, Kansas, United States of America

Hungyen Chen and Hirohisa Kishino
Graduate School of Agricultural and Life Sciences, The University of Tokyo, Tokyo, Japan

Junko Yamagishi
Institute for Sustainable Agro-ecosystem Services, The University of Tokyo, Tokyo, Japan

Inge E. Matthies and Marion S. Röder
Department of Gene and genome mapping, Leibniz Institute of Plant Genetics and Crop Plant Research (IPK), Gatersleben, Sachsen-Anhalt, Germany

Marcos Malosetti and Fred van Eeuwijk
Biometris, Wageningen University and Research Centre, Wageningen, Gelderland, The Netherlands

María Alonso-Ayuso, José Luis Gabriel and Miguel Quemada
School of Agriculture Engineering, Technical University of Madrid, Madrid, Spain

Erik Lysøe and Even S. Riiser
Department of Plant Health and Plant Protection, Bioforsk - Norwegian Institute of Agricultural and Environmental Research, Ås, Norway

Linda J. Harris
Eastern Cereal and Oilseed Research Centre, Agriculture and Agri-Food Canada, Ottawa, Canada

Sean Walkowiak and Rajagopal Subramaniam
Eastern Cereal and Oilseed Research Centre, Agriculture and Agri-Food Canada, Ottawa, Canada Department of Biology, Carleton University, Ottawa, Canada

Hege H. Divon
Section of Mycology, Norwegian Veterinary Institute, Oslo, Norway

Carlos Llorens
Biotechvana, Valéncia, Spain

Toni Gabaldón
Bioinformatics and Genomics Programme, Centre for Genomic Regulation, Barcelona, Spain
Universitat Pompeu Fabra, Barcelona, Spain
Institució Catalana de Recerca i Estudis Avançats, Barcelona, Spain

H. Corby Kistler and Wilfried Jonkers
ARS-USDA, Cereal Disease Laboratory, St. Paul, Minnesota, United States of America

Anna-Karin Kolseth
Department of Crop Production Ecology, Swedish University of Agricultural Sciences, Uppsala, Sweden

Kristian F. Nielsen, Ulf Thrane and Rasmus J. N. Frandsen
Department of Systems Biology, Technical University of Denmark, Lyngby, Denmark

Atta Soliman
Department of Plant Science, Faculty of Agricultural and Food Sciences, University of Manitoba, Winnipeg, Manitoba, Canada
Department of Genetics, Faculty of Agriculture, University of Tanta, Tanta, El-Gharbia, Egypt

Belay T. Ayele and Fouad Daayf
Department of Plant Science, Faculty of Agricultural and Food Sciences, University of Manitoba, Winnipeg, Manitoba, Canada

Barbara U. Metzler-Zebeli
Institute of Animal Nutrition and Functional Plant Compounds, Department for Farm Animals and Veterinary Public Health, Vetmeduni Vienna, Vienna, Austria
University Clinic for Swine, Department for Farm Animals and Veterinary Public Health, Vetmeduni Vienna, Vienna, Austria

Kathrin Deckardt and Qendrim Zebeli
Institute of Animal Nutrition and Functional Plant Compounds, Department for Farm Animals and Veterinary Public Health, Vetmeduni Vienna, Vienna, Austria

Margit Schollenberger and Markus Rodehutscord
Institute of Animal Nutrition, University of Hohenheim, Stuttgart, Germany

Ningning Song, Xu Zhong, Jumei Li, Dongpu Wei and Yibing Ma
National Soil Fertility and Fertilizer Effects Long-term Monitoring Network/Institute of Agricultural Resources and Regional Planning, Chinese Academy of Agricultural Sciences, Beijing, P.R. China

Bo Li
National Soil Fertility and Fertilizer Effects Long-term Monitoring Network/Institute of Agricultural Resources and Regional Planning, Chinese Academy of Agricultural Sciences, Beijing, P.R. China
Institute of Plant Nutrition and Environmental Resources, Liaoning Academy of Agricultural Sciences, Shenyang, P.R. China

Nora Honsdorf
Chair of Plant Breeding, Institute of Agricultural and Nutritional Sciences, Martin-Luther University Halle-Wittenberg, Halle (Saale), Germany
Interdisciplinary Center for Crop Plant Research (IZN), Halle (Saale), Germany

Timothy John March
School of Agriculture, Food and Wine, University of Adelaide, Waite Campus, Adelaide, Australia

Bettina Berger
The Plant Accelerator, University of Adelaide, Waite Campus, Adelaide, Australia

Mark Tester
Center for Desert Agriculture, King Abdullah University of Science and Technology, Thuwal, Saudi Arabia

Klaus Pillen
Chair of Plant Breeding, Institute of Agricultural and Nutritional Sciences, Martin-Luther University Halle-Wittenberg, Halle (Saale), Germany

Index